Solar Prominences

Astrophysics and Space Science Library

More information about this series at
http://www.springer.com/series/5664

Jean-Claude Vial • Oddbjørn Engvold
Editors

Solar Prominences

Springer

Editors
Jean-Claude Vial
Institute of Astrophysics Space
Université Paris-Sud 11
Orsay, France

Oddbjørn Engvold
Institute of Theoretical Astrophysics
University of Oslo, Blindern
Oslo, Norway

ISSN 0067-0057 ISSN 2214-7985 (electronic)
ISBN 978-3-319-10415-7 ISBN 978-3-319-10416-4 (eBook)
DOI 10.1007/978-3-319-10416-4
Springer Cham Heidelberg New York Dordrecht London

Library of Congress Control Number: 2014955674

Cover illustration: A solar prominence began rising and expanding in a graceful, floating style during a little less than four hours (Mar. 16, 2013). The sequence was captured in extreme ultraviolet light. A large cloud of the electrically charged gas appeared to hover further out above the surface before it faded away. Credit: Solar Dynamics Observatory/NASA.

Printed on acid-free paper

Springer is part of Springer Science+Business Media (www.springer.com)

This book is dedicated to Einar Tandberg-Hanssen who was a leader in the field

Preface

Since the publication in 1995 of the latest book devoted to solar prominences ("The nature of solar prominences" by Einar Tandberg-Hanssen, Kluwer) very important new results have been obtained from space missions and ground-based observatories. A tremendous progress in instrumentation has led to new results and visions which theoreticians and modelers have difficulty to catch up with. From its Lagrangian L1 viewpoint, SOHO (Solar and Heliospheric Observatory) has provided 24 h a day continuous spectacular images and spectra of the Sun in general and prominences in particular. It has allowed registering of Eruptions of Prominences and associated Coronal Mass Ejections whose role in the Earth environment has been assessed through about 20 years of observations, complemented by the high spatial resolution TRACE (Transition Region and Coronal Explorer) mission. The overarching question of the actual geometry of these "photogenic" structures could be addressed with the STEREO (Solar TErrestrial RElations Observatory) mission with its two viewpoints allowing for a stereoscopic study of prominences. With its Solar Optical Telescope (SOT—the largest solar telescope ever flown), the Hinode mission addressed the dynamics and the fine structure of prominences. Since 2010, the SDO (Solar Dynamics Observatory) mission has provided continuous high-cadence sets of full-Sun spectacular images in a dozen of wavelengths spanning the solar atmosphere from the photosphere to the high temperature corona and flares. Since 2013, the UV spectro-imager IRIS (Interface Region Imaging Spectrograph) has been answering the question of the thermodynamic and flows diagnostics of prominences. On the ground, the SST (Swedish Solar Telescope) has been upgraded and now fully benefits from its 1m entrance aperture and unique spatial resolution with the implementation of adaptive optics and software tools to correct for the distortion by the Earth atmosphere. Along with the DOT (Dutch Open Telescope) it has recently provided unique images of prominences and filaments.

In the meantime, theoreticians and modelers have made tremendous progress in many fields of plasma physics of which prominences are a very special case: MHD (magnetohydrodynamics), radiative transfer, radiation polarization, interaction matter-radiation, spectroscopic diagnostics, etc. Although these topics are at the forefront of modern solar physics and are consequently sometimes difficult to explain, an exceptional effort has been made in this book to present a pedagogical view for readers at many levels. The dynamics of prominences, whether at their formation days, or during their fine structure variations within less than one minute, or during their final erupting fate, is now fully treated and modelers now address the question raised in 1978 by E. Jensen "With all these observations so full of beautiful dynamics and happenings in velocity space – how come that so much effort is devoted to the study of static configurations?".

The 18 chapters address the main issues related to the various types of prominences: quiescent, intermediate, active region prominences, coronal clouds, eruptive prominences. They describe the tools used to image and diagnose the prominence plasma, with some emphasis on the major and most difficult-to-measure component: the magnetic field. They show the relations between this 1 A.U.-away bubbles of gas and our Earth. And the reader will see many light-years away how exoplanets are affected by the erupting prominences of their host stars.

The book is intended for advanced students in astrophysics, post-graduates, solar physicists and more generally astrophysicists. Being based on many spectacular images, it can also be useful for amateur astronomers interested in this enigmatic feature of our fascinating Sun.

This book is dedicated to Einar Tandberg-Hanssen who was a leader in the field.

Orsay, France Jean-Claude Vial
Oslo, Norway Oddbjørn Engvold

Contents

Chapter 1
Historical Background and Introduction

Jean-Claude Vial

> *Protubérances: "Ces amas de matière lumineuse ayant une grande vivacité et possédant une activité photogénique très remarquable" (A. Secchi "Le Soleil", 1877, I. 385)*
> *Prominences: "These masses of luminous matter with a high brilliance and with a very remarkable photogenic activity"*

Abstract Forty and twenty years after the two books published by Einar Tandberg-Hanssen (Solar prominences (Geophysics and astrophysics monographs), Vol. 12. Dordrecht: D. Reidel Publishing Co., 1974; The nature of solar prominences, astrophysics and space science library, Vol. 199. Dordrecht: Kluwer Academic Publishers, 1995) on solar prominences, it is time to update our knowledge and understanding of these fascinating solar structures. After a brief history which overviews first eclipse observations (drawings and then photography), spectrographic, coronagraphic and later on polarimetric measurements, the chapter presents samples of the most spectacular results of the last two decades, obtained whether from space or on the ground. It discusses the contents of the book in order to encourage the reader to dip into the following 17 chapters which provide comprehensive and detailed observations, information about the methods used, and interpretation of the results on the basis of the latest theoretical and modelling works.

1.1 A Brief History

1.1.1 The Era of Eclipses

In his book (Solar Prominences 1974), Tandberg-Hanssen (1974) traced back the first observations of prominences to the eclipse of 1239 when an observer, Muratori, reported a "burning hole" in the corona. Since then, strange structures were seen

J.-C. Vial (✉)
Université Paris-Sud & CNRS, Institut d'Astrophysique Spatiale, Orsay, France
e-mail: jean-claude.vial@ias.u-psud.fr

J.-C. Vial, O. Engvold (eds.), *Solar Prominences*, Astrophysics and Space Science Library 415, DOI 10.1007/978-3-319-10416-4_1

Fig. 1.1 (**a**) Drawing by the English astronomer Baily during the solar eclipse of 1842 (reproduced by A. Secchi, Le Soleil, I, p. 310). (**b**) Engraving made by Eugene Bouvard during the 1842 eclipse at Digne (France) (courtesy S. Koutchmy)

during total solar eclipses, labelled "burning holes", "red flames", ... sometimes thought to be clouds or mountains on the Moon (Vassenius 1733).

Clever observers had time to make drawings which are valuable records of the state of the solar corona at the time of the eclipse (see Fig. 1.1a where prominences are the red off-the-limb structures). But during the eclipse of July 8, 1842, as mentioned by A. Secchi, observers were so surprised by these pink colored structures that some confusion followed, where observers, such as Airy, Arago or Baily saw different structures at different locations! (see Fig. 1.1b). Some observers even considered them to be the result of optical illusions! After the 1851 eclipse in Sweden, it was admitted that these bright pieces were clouds (i.e. gases), but clouds truly belonging to the Sun ... or to the Moon? However, most observers (A. Secchi being at the forefront) noticed the change of altitude of the structures during the motion of the Moon, which they interpreted as proof that they were not attached to the Moon.

Moreover, we have evidence that when the 1851 eclipse passed over southern Scandinavia on July 28, the observers working for Airy in Oslo, Norway and in Gothenburg, Sweden, actually discovered the eruption of a prominence (Fig. 1.2, top). Compare this one and a half centuries-old observation with a modern observation showing in great details the coronal environment of prominences (Fig. 1.2, bottom).

A major step forward occurred in 1860 with the use of photography, which allowed for a permanent and "objective" record of these structures. Photography

Fig. 1.2 *Top*: view of the corona and a prominence whose shape is known to be typical for erupting during the July 28, 1851 eclipse at Gothenburg, Sweden. The observer was working for Airy (courtesy O. Engvold). *Bottom*: Image taken during the total solar eclipse of 11 July 2010. The dark hook-shaped feature on the east limb is part of an erupting prominence. One prominence at the west limb is shown in *red*. Note the helmet coronal structuring above prominences. Eclipse image, courtesy of Prof. M. Druckmüller and S. Habbal

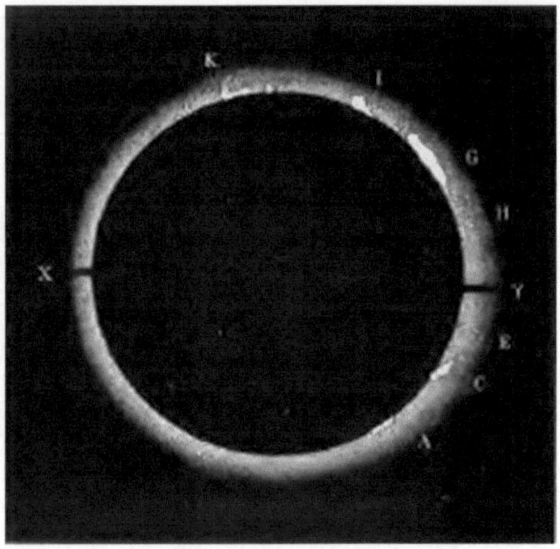

Fig. 1.3 Photograph of the eclipse of 1860 at the Desertios de las Palmas by A. Secchi (Le Soleil, I, p. 378). At least seven structures were detected

even allowed observations of prominences that were invisible to the naked eye (Fig. 1.3). The observers derived important parameters such as the shape, altitude, emissivity and also noticed that these structures were mostly connected to an underlying layer (to be identified later as the chromosphere).

At that time, photography did not provide colours, so it was important to visually observe prominences. Their commonly adopted colour was red with some mixture of violet that A. Secchi compared to the colour of the peach tree flower. Some yellow colour was occasionally noticed at the prominence tops.

1.1.2 The Use of the Spectroscope

Another important advance was achieved with spectral analysis (the foundations had been built by Bunsen and Kirchhoff) of the emitting gases.

Spectroscopes were used during the 1868 eclipse, with their slits being placed on the observed prominences, and the presence of emission lines (a hydrogen line in particular) confirmed the gaseous nature of prominences (Fig. 1.4).

But, since the observers had no wavelength scale, there still were some uncertainties concerning the identification of these lines. Should the observers wait for the next eclipse? No, the day after, J. Janssen carefully positioned the slit of his spectroscope on an off-limb suspected structure and confirmed the presence of the two lines of hydrogen (C and F) and later the D lines (two from Sodium and a mysterious third one corresponding to an unknown element). This element corresponding to the third D3 line observed for the first time in prominences was coined Helium (Helios is the Sun God in Greek) because of its solar origin. It was detected in the Earth's atmosphere 27 years later. It was identified as atomic number 2, the most abundant element in the Universe after Hydrogen.

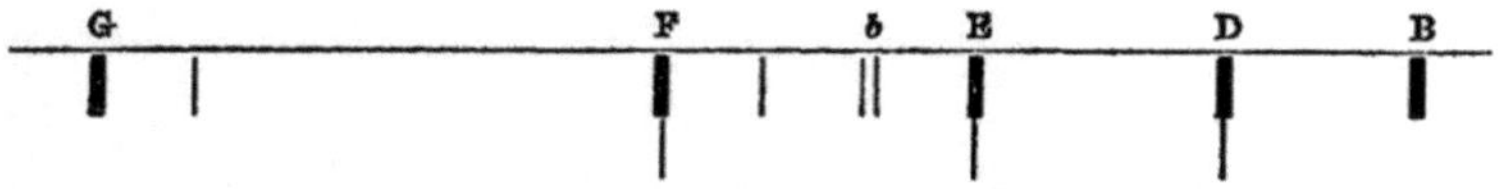

Fig. 1.4 Prominence spectrum obtained by G. Rayet (the Astronomer well-known for the Wolf-Rayet stars) during the eclipse of 1868 in Malaisia (Rayet 1869). The spectrum is marked by emission lines. The letters correspond to the Fraunhofer classification where the F line is the Hβ line of hydrogen

In "Le Soleil" (1877, II.1), Secchi wrote: "L'éclipse de 1868 sera une date mémorable dans l'histoire de l'Astronomie car c'est alors que M. Janssen apprit aux savants à étudier en tous temps les protubérances" or "The 1868 eclipse will remain a never-to-be-forgotten date in the history of Astronomy for this is when M. Janssen taught scientists how to study prominences for all times". Since then the spectroscope was used daily for the study of prominences in two different modes: a narrow slit allowing for spectroscopy and a wide slit allowing for monochromatic imaging. This means that actually the issue of spectro-imaging was already discussed and solved by observers of that time!

Moreover, the observations being more systematic over a longer period of time, it was noticed that the number and the position of prominences varied with solar activity. Classifications appeared (see Engvold 2014), mostly based upon their shapes and the proximity of active regions.

In spite of the progress brought by photography and spectrography, some observers continued visual observations. One of them, Prof. C. Fearnley in Oslo, recorded the images on black paper (Fig. 1.5). Note the delicate thread structuring, today called "fine structure".

Early observers (e.g. Vassenius 1733) had not only noted the (red) emission but also its temporal variation in prominence, which explains the use of terms such as "red flames" to characterize them. A. Secchi (Le Soleil, II. p. 42) mentions "the velocity of their motions when they are submitted to an eruptive force which launches them from the interior above the surface of the Sun". Observers were amazed by the altitude that these structures could reach (up to 250 arcsec above the limb) and already wondered about the nature of the force driving the material. Actually, what they were observing was the eruption of prominences, often above active regions, for which velocities of 100 km/s (or more) were measured with the use of a chronometer. The magnitude of the velocities even led A. Secchi (Le Soleil, II, p. 108) to raise questions that we think to be still valid today: "The nebulous masses are so quickly illuminated and they disappear within such a short time that one wonders whether it is a temporary transformation instead of an actual transport of mass".

The observers also measured the width of the line profiles from which they coarsely derived some temperatures.

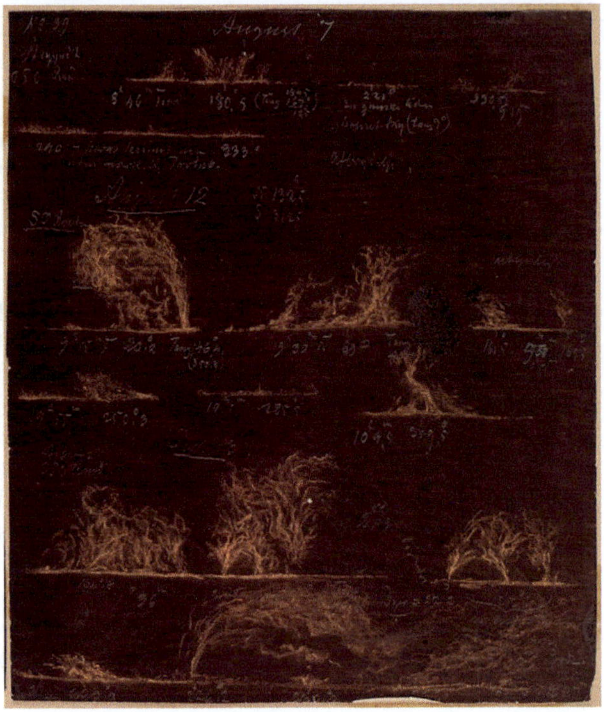

Fig. 1.5 An example of the chalk drawings made by Prof. C. Fearnley between 1871 and 1873 in Oslo, with a wide slit spectrohelioscope. Note the large variety of morphological structures (Courtesy O. Engvold)

1.1.3 Identification of Prominences Observed at the Limb with Structures Observed on the Disk

The observers moved the (wide) slit of the spectroscope around the limb which allowed them to build an image within an hour (Fig. 1.6). Systematic observations between 1871 and 1876 allowed the detection of a correlation between some prominences and the proximity of sunspots on one hand, and the declining number of prominences when the minimum of solar activity was reached (in 1875) on the other hand. Moreover, observers such as A. Secchi already classified prominences as steady (to be named "quiescent", later on) and active (to be named "active region prominences" later) (see Engvold 2014).

However, a precise identification of off-limb prominences with on-disk structures was not possible until the use of spectroheliographs allowing the isolation of (cool) chromospheric lines, called "black lines", such as the Hβ of hydrogen and the UV H and K lines of ionized calcium (Hale and Ellerman 1903 at Yerkes Observatory with a 1.05 m refractor, and Deslandres 1910 at Meudon Observatory initially with a 0.3 m mirror). It was then realized that the dark elongated features (called "elongated dark flocculi") were the very same ones than the off-limb prominences seen bright on the dark sky background. They were called *filaments* (Fig. 1.7). With the simple use of the third law of Kirchhoff, one could already derive that the temperature of

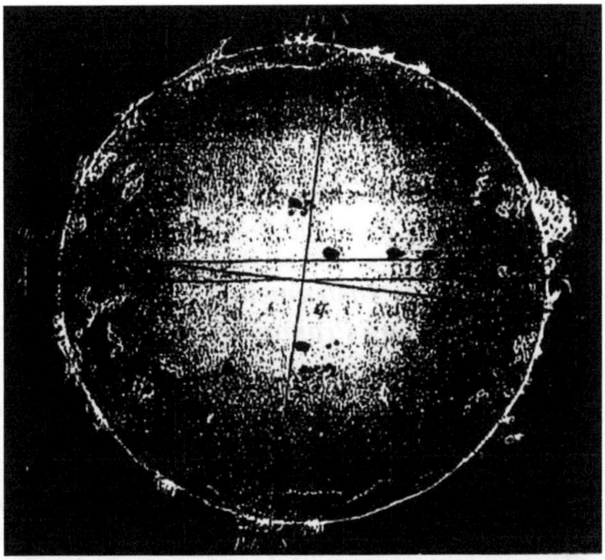

Fig. 1.6 The composite of limb drawings obtained with the wide slit spectroscope combined with disk drawings depicting sunpots and faculae (Secchi, Le Soleil, II, p. 164)

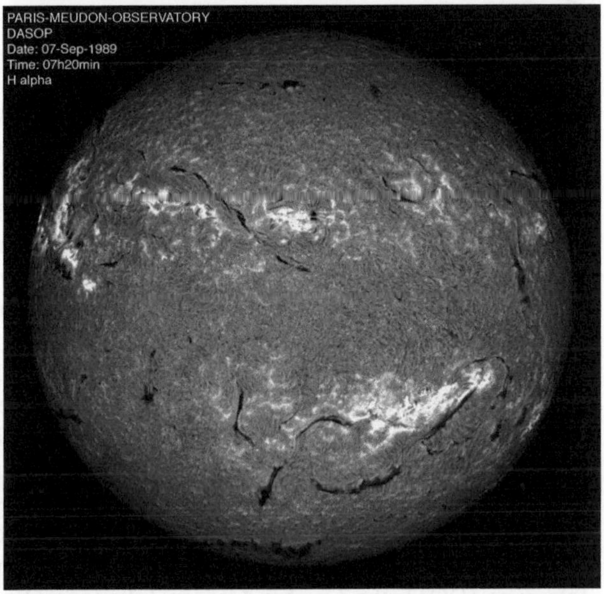

Fig. 1.7 Spectroheliogram in the Hα line obtained at Paris-Meudon Observatory. The dark elongated structures seen on the disk, called filaments, will appear as prominences when, rotating with the sun, they reach the limb (courtesy LESIA/Observatoire de Paris/CNRS/UPMC/Univ. Paris Diderot/)

the filament which acts as a screen between the chromosphere and the observer is lower than the temperature of the emitting chromosphere (10^4 K).

The systematic recording of K3 and Hα spectroheliograms at Meudon Observatory from 1919 to 1930–1937 has led to an impressive body of synoptic data from which many spatial and statistical characteristics have been derived (D'Azambuja and D'Azambuja 1948). An excellent example is provided by the study of the law of rotation of filaments as a function of latitude (Fig. 25 of D'Azambuja and D'Azambuja 1948) which is an accurate proof of the solar differential rotation. Similar observations and identifications were made in the mm radio domain by Khangil'din (1964) who stated that "the presence of local regions of reduced radio brightness above dark filaments is established". According to Schmahl et al. (1981) two-thirds of the radio depressions are associated with Hα filaments.

1.1.4 A Major Instrumental Progress for Continuously Observing Prominences: The Coronagraph

The slit of the spectroscope allowed the observer to "ignore" the considerably stronger emission from the solar disk. But if one wanted to have an instantaneous picture of the whole corona (including prominences), as is possible during eclipses, it was necessary to block the light from the disk. This was the idea of Bernard Lyot who invented the coronagraph in 1930. The instrument was primarily designed for a permanent observation of the solar corona. But, in 1938, with the proper (Hα) filter and a camera, B. Lyot, could perform the first cinematography of prominences shown at the International Astronomical Union General Assembly in Stockholm in 1938. After his death in 1952, his colleagues built a movie called "Les flammes du Soleil" (or "The flames of the Sun") which demonstrates the variety of structures and their extreme variability (see, e.g., http://www.canal-u.tv/video/cerimes/flammes_du_soleil.9171). In the 1960s, R. Dunn (1960, 1965) used the fine Hα structure of a prominence for measuring the instrumental profile of his instrument (Dunn 1965, Fig. 36, p. 80). He also performed cinematography of prominences which showed that the thin threads or "ropes" (about 300 km across) were the location of downward and also upward vertical motions in the range 5–10 kms^{-1} (Engvold 1976). It is worth mentioning that R. Dunn built the first evacuated solar telescope (the famous solar tower at Sacramento Peak), which solved the issue of the internal telescope seeing and allowed for a spatial resolution below 1 arcsec. In doing so, it contributed much to the study of prominences, along with the solar telescope at Big Bear Solar Observatory and now the New Vacuum Solar Telescope at the Yunnan Observatory (Xu et al. 2014). The vacuum solution was used for the Solar Swedish Telescope (see below) and others but the entrance window was limited to about 1 m. The difficulty is now being overcome with open telescopes such as the Dutch Open Telescope (DOT), the German Gregor Telescope (1.5 m) and the planned ATST (4 m, see below).

For the fine structure (imaging) studies of prominences, the coronagraph remains the main instrumental tool, with the proper filter tuned in a "cool" line appropriate for observing prominences. For instance, since 1994, in support to SOHO, the Pic-du-Midi coronagraph has continuously provided daily images in the Hα line and now in the 1083 nm line of Helium. This program (http://www.climso.fr/index.php/fr/) is run every day by amateurs organized within the "Observateurs Associés" Society. Among the many coronagraphs installed all over the world let us note the 40 cm coronagraph at the John W. Evans Solar Facility in Sacramento Peak Observatory in the U.S.A., the 52 cm coronagraph at Kislovodsk (then Soviet Union) and now the Large and Small coronagraphs installed in Wroclaw (Poland) and Rhozen (Bulgaria) by Bogdan Rompolt.

1.1.5 A (Relatively) New Technique: Polarimetry for Accessing the Magnetic Field

The connection between prominences and magnetic field was quickly recognized when Babcock and Babcock (1955) noted that filaments mostly coincided with the neutral line between regions with opposite polarities, i.e. the line where the vertical component of the magnetic field is null. This led to the natural question: what is the magnitude and orientation of the magnetic field? The endeavour of its measurement started with observations at the Crimean Solar Tower where H. Zirin (then in sabbatical year) and A. Severny (1961) derived from Zeeman splitting in the Hβ line, magnetic field values of the order of 200 G in an active prominence (and lower values in a quiescent one).

By 1964, an HAO-designed magnetograph was implemented at the HAO Climax Observatory. Systematic measurements were then performed and a more versatile magnetograph was installed by Einar Tandberg-Hanssen (Harvey and Tandberg-Hanssen 1968) which led to two Ph.D. thesis (Rust 1966; Harvey 1969).

In the 1970s, the Hanlé effect, well adapted for the measurement of weak fields, was used through polarimetric observations made at the Pic-du-Midi Observatory (Leroy et al. 1977; Sahal-Bréchot et al. 1977). Details on the techniques and the results are given in Chap. 8 (Lopez Ariste 2014) devoted to the magnetometry of prominences. A new era in the study of prominences opened up thanks to gradually larger instruments collecting more photons with higher polarimetric sensitivities. A next step in this direction will be the 4-m Advanced Technology Solar Telescope (ATST) and its spectropolarimeters (see Sect. 1.2).

1.1.6 Prominences from Space: The Beginnings

Since the very beginning of the solar space era, prominences were systematically observed with all of the advantages linked to space: access to new wavelength windows (especially the UV and EUV), continuity of observations (up to 24 h a day), coupled with gradually better spatial resolution due to progress in the pointing stability. Since the spacecraft were not recovered, the detectors could only be photoelectric devices (photomultipliers, channeltrons, …) which had the serious drawback of being monopixel, a major difficulty for building an image! The whole series (one to eight) of Orbiting Solar Observatories and the Solar Maximum

Mission of NASA had to rely upon a method of rastering, i.e. moving the whole instrument and consequently the solar image on the entrance slit of the spectrometer, a motion which could take typically about an hour. In the spectrometer itself, the spectroscopy was performed by rotating a diffraction grating. The spatial resolution was limited by the width and the height of the integrating slit: with a too high slit, the spatial resolution was too poor while with a too short slit, the collected number of photons was too small and the time for scanning an area was too long. The difficulty was (temporarily) overcome with the Skylab mission, when the US astronauts used a dedicated telescope, the Apollo Telescope Mount (ATM), associated with films as detectors, films which were brought back to Earth for processing. Beautiful results were obtained until 1974 through the three visiting periods by the astronauts. Then the Skylab station was put in a sleeping mode and finally desintegrated in the Earth atmosphere, probably because of the high level of solar activity which resulted in an increase in atmospheric drag that affected the orbit of this spacecraft.

Another difficulty was related to data processing, and in particular, the production of representative images by (big) computers on films which were then visualized on dedicated machines comparable to a slide or a film projector in a movie theater.

The latest space images recorded on film in the Lα line by Bonnet et al. (1980) on a rocket flight already gave evidence of a sub-arcsecond fine structure (Fig. 1.8).

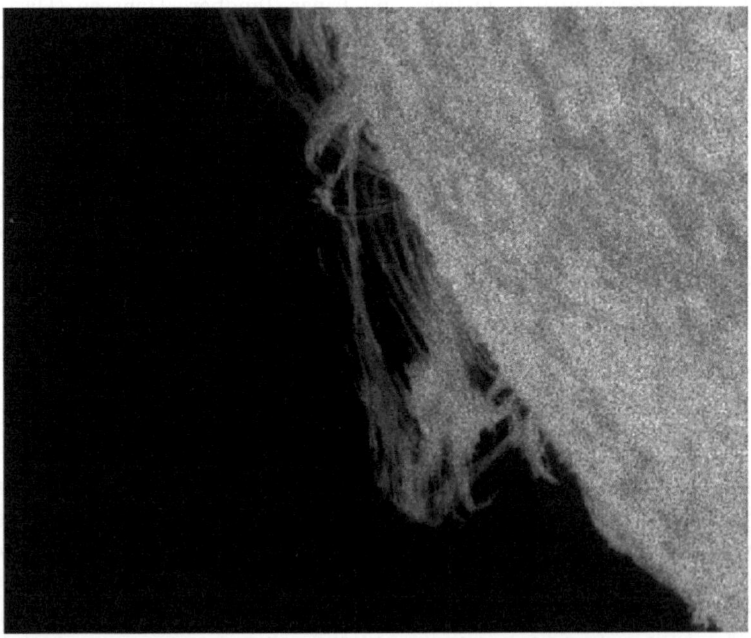

Fig. 1.8 Subimage of a prominence recorded by the Transition Region Camera in the Lα line of hydrogen at 121.6 nm (courtesy R.M. Bonnet)

Fig. 1.9 Collage of four Eruptive Prominences caught in action by EIT on SOHO. The observed radiation is the resonance line of He II (singly ionized helium) at 30.4 nm, a line that is formed at about 60,000 K. The images have been obtained in 2000 and 2001 (courtesy, EIT/SOHO consortium)

1.2 The Modern Era

A revolution started in the 1980s with the use of two-dimensional detectors such as CCDs (Charge-Coupled Devices) in astronomy. As with photographic film, it is possible to obtain instantaneous images (at the focus of the telescope or the coronagraph) or spectra at each pixel of the slit (in the focal plane of the spectrograph). This major advance occurred first in ground-based instrumentation and then extended to space where detectors working in the EUV-UV were developed. A beautiful example is provided by the Extreme ultra-violet Imaging Telescope (EIT) on SOHO, which has been observing the Sun (including prominences) since 1996. Since SOHO was positioned around the Lagrangian point L1 between the Sun and Earth, continuous observations became possible; this allowed to catch all dynamic phenomena during their evolution such as eruptive prominences, samples of which are given in Fig. 1.9.

This unique position of SOHO benefited the whole package of instruments (imagers, coronagraphs, spectrographs, magnetograph and in situ experiments) which were extensively used by scientists working on solar prominences.

SOHO detected thousands of Coronal Mass Ejections (CME) (mostly associated with eruptive prominences) which were followed by the EIT imager and then by the LASCO coronagraphs.

But scientists were feeling some frustration because they always viewed these structures only as projected in the plane of sky. The idea of having two viewpoints led to the STEREO mission when two spacecraft at 1 AU separated on October 2006, one being Ahead (A) and the other Behind (B) the Earth. Actually, one has access to a third viewpoint at the Earth orbit, provided by EIT on SOHO and since

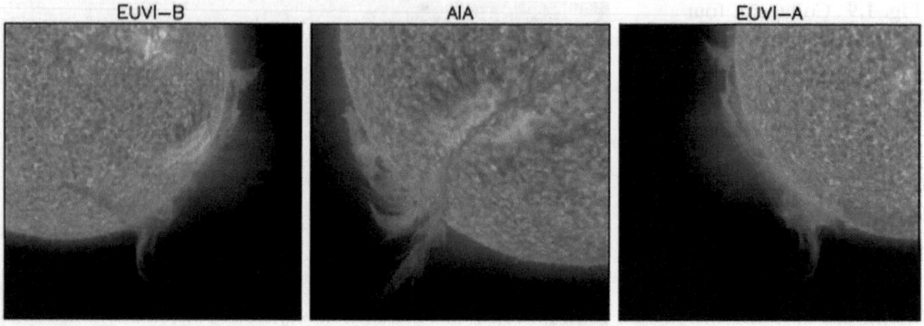

Fig. 1.10 An eruptive prominence observed on December 6, 2010, during its ascent from three different viewpoints by the STEREO A (Ahead) and B (Behind) spacecraft (separated by 172°) and the AIA telescope on SDO in Earth orbit. Note that movies (movie_with_aia.mov and prom_20101206b.mov) are available in the Electronic Supplementary Materials of the paper (from Thompson 2013)

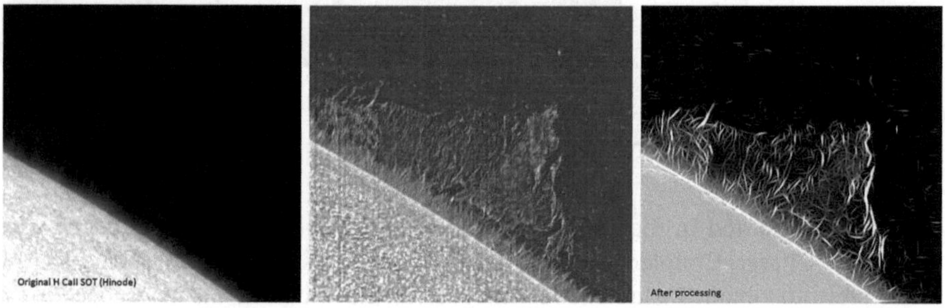

Fig. 1.11 Prominence seen in H Ca II by the Hinode SOT on 22 June 2010 (*left*) and processed by OMC (Octodirectional Maxima of Convexities) (*middle*) and additionally denoised using the 2-D Morlet wavelet (*right*) in order to suppress artefacts and enhance the contrast (from Tavabi et al. 2013)

2010, the AIA telescope of Solar Dynamics Observatory in the same EUV 30.4 nm line (Fig. 1.10).

In the same year, on 23 September 2006, Japan launched the Hinode mission with a set of instruments (Kosugi et al. 2007), one of them being the Solar Optical Telescope with its 50 cm aperture and spectropolarimeters. High resolution images were obtained in the Hα line of hydrogen and the H line of ionized calcium (Fig. 1.11, left).

Major progress has also been made in the field of image processing, as shown in Fig. 1.11 middle and tentatively in Fig. 1.11 right. Actually, image processing is now part of instrumentation as recently demonstrated by a dedicated volume of Solar Physics (2013, Vol. 283/1). For instance, automated detection and tracking of filaments are now routinely performed, an information which ends up in complete databases easy to access through their catalogs (see e.g. Bonnin et al. 2013).

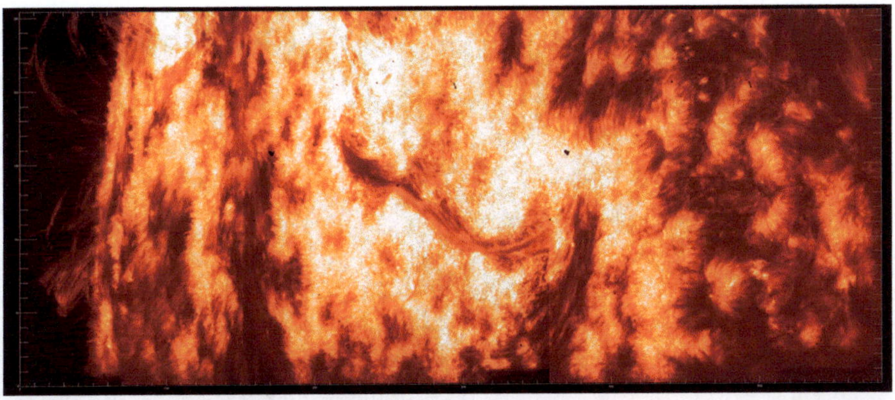

Fig. 1.12 Composite of images recorded by the VAULT rocket experiment in the Lα line of Hydrogen at 121.6 nm. The total field of view is 583 × 234 arcsec (courtesy A. Vourlidas and Vourlidas et al. 2010)

In order to fully appreciate the dynamical behavior of prominences (even the quiescent ones) movies are built and accessible on the www. For instance, one can find a spectacular movie at: http://www.youtube.com/watch?v=l5T1s7Wf5P0 and also the mpeg animation of Fig. 1 of Berger et al. (2008).

As one can see in Fig. 1.11, the field-of-view of the SOT on Hinode is somewhat limited (which allows for its high spatial resolution). The TRACE (Transition Region and Coronal Explorer) satellite also provided a wealth of 8 by 8 arcmin UV and EUV images between 1998 and 2010 from which a huge database was built including filament/prominence movies (http://trace.lmsal.com/POD/bigmovies/filaments/).

The VAULT rocket (Korendyke et al. 2001) also recorded the very fine structure of an active region filament (and prominence) in the Lα line with a 0.4 arcsec resolution but through a set of different exposures with a wider field of view (Vourlidas et al. 2010; Vial et al. 2012). See Fig. 1.12 and compare the fine structure in the prominence with Fig. 1.8.

In order to get a global view of the solar atmosphere (including its prominences), one needs a full-Sun imager providing information about different levels in the atmosphere, corresponding to different lines and consequently temperatures. In order to have a continuous view at high temporal resolution, one needs the proper orbit and a very high telemetry rate. These are the main features of the NASA Solar Dynamics Observatory, launched on February 11, 2010. Being in a geosynchronous orbit, it can look at the Sun permanently and it can transmit science data to the station below at the amazing rate of 130 Mbps. One of the imagers (AIA for Atmospheric Imaging Assembly) working in the He II line at 30.4 nm (see Fig. 1.13), can provide an image of prominences and filaments on the disk every 12 s, 24 h a day! With such a cadence, many movies show spectacular prominence eruptions on the basis of images such as the one of Fig. 1.13 (Movies can be found at: http://sdo.gsfc.nasa.gov/gallery/main). In the framework of SDO observing plans

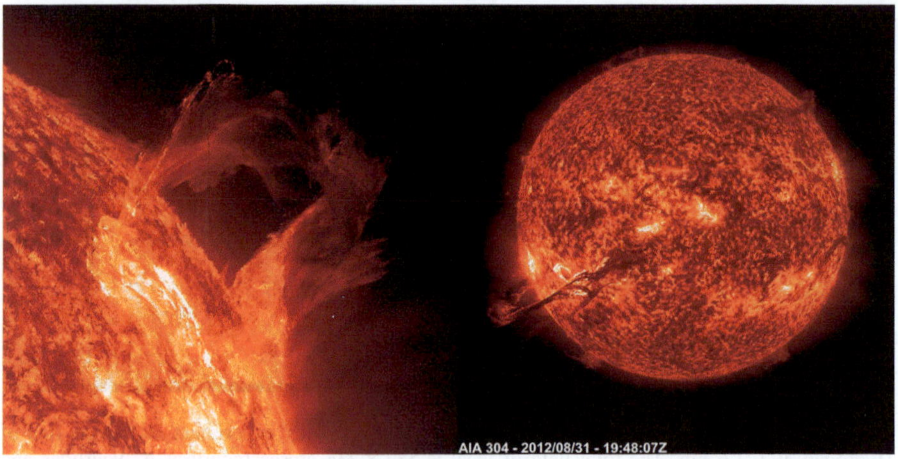

Fig. 1.13 *Left*: Prominence eruption on March 16, 2013, in the He II 30.4 nm line, as viewed by AIA on SDO. *Right*: Prominence eruption following a C-class flare as seen by AIA/SDO in the He II line at 30.4 nm. Note bright and dark regions of the prominence depending on their orientations, the bright ones being illuminated by the solar chromosphere and the dark ones being shielded by the prominence material. Also note that filaments are clearly visible as dark features on the disk in this EUV line (Courtesy AIA/SDO)

and database, catalogs of filaments and prominences have been built for a systematic detection of eruptions (see e.g. http://www.helioviewer.org/ where one can find catalogs of filaments, filament activations and filament eruptions).

Space instrumentation is limited to (relatively) small apertures, the SOT and its 50 cm being an exception. On the ground, most telescopes are, as we shall see, in the 1 m-class. Such large apertures allow for a better spatial resolution (if the distortions by the Earth atmosphere are properly corrected) or a higher photons collecting power (very useful for spectroscopy and polarimetry which are photon-starved).

We give a few examples of ground-based instrumentation below.

Recent improvements in ground-based observatories include speckle processing which is currently performed at the Dutch Open Telescope (DOT) in Canary Islands (Fig. 1.14). The peculiarity of this telescope of 45 cm aperture is that it is open to the wind which homogenises the air temperature and provides better seeing. Through despeckling (performed post-observation) it can reach a spatial resolution of 0.3 arcsec (Fig. 1.15).

The major revolution was the introduction of adaptive optics (AO) which corrects in real-time the image distortion from the Earth atmosphere. This solution has been used for decades by night-time astronomers but it was a challenge to make it work in solar astronomy because of the large number of low-contrast features. An innovative solution has been implemented at the end of the 1990s at the Dunn Solar Telescope at Sacramento Peak (USA) (see e.g. Rimmele 2000).

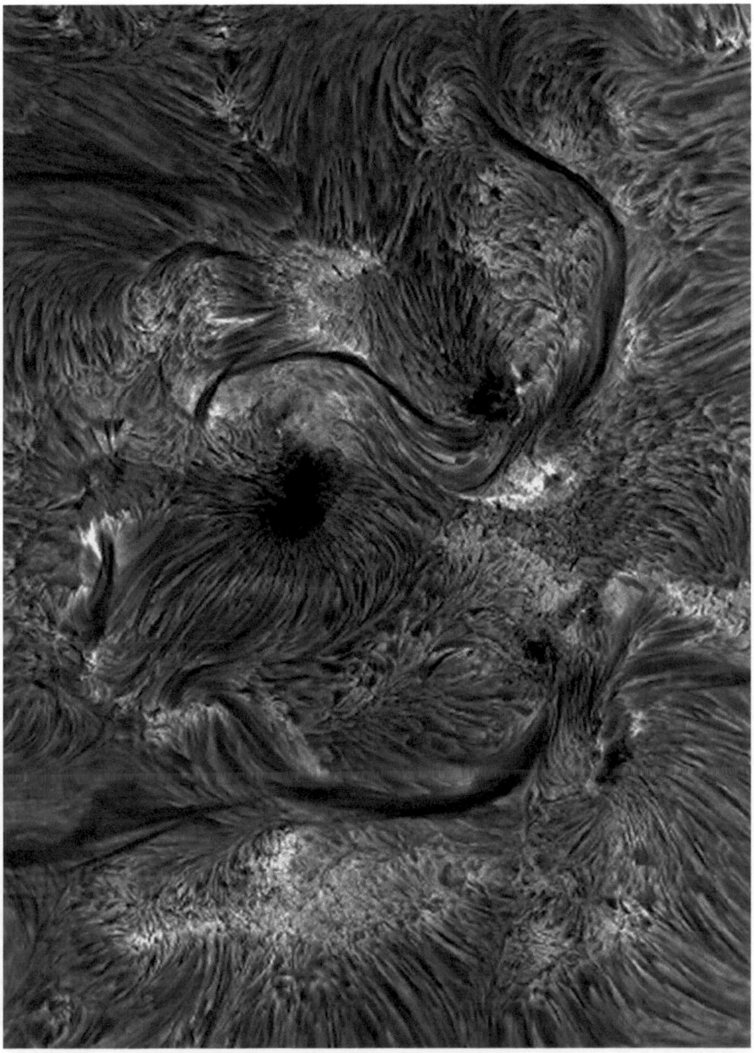

Fig. 1.14 Solar active region AR10786 in an image mosaic obtained by P. Sütterlin with the Dutch Open Telescope on July 8, 2005. The field of view measures 133 × 182 arcsec. The sunspot umbrae remain dark in Hα. The long slender dark structures are active region filaments (courtesy R. Rutten)

This technique is now currently used by the 1-m Swedish Solar Telescope (SST) also operating in the Canary Islands (Fig. 1.16) and the 1.6 m New Solar Telescope at Big Bear. The SST is in addition set up for the application of a post-processing technique called Multi-Frame Blind Deconvolution, enabling diffraction-limited observations (about 70 km) (Scharmer et al. 2003; van Noort et al. 2005).

Fig. 1.15 The Dutch Open Telescope (DOT) with its canopy closed

An example is given in Fig. 1.17 where one can see quasi linear threads probably tracing the magnetic field in a special part called a "barb" found sticking out from the main body of filaments (Lin et al. 2005).

Imagery, even performed simultaneously in different wavelengths, allows us to trace the apparent motions of the prominence but it does not provide an as precise plasma diagnostics as spectroscopy which also adds information on plasma motions along the line-of-sight. Major advances in spectroscopic instrumentation have been obtained with the Extreme Ultraviolet Imaging Spectrometer (EIS) on Hinode (Culhane et al. 2007), a spectrograph which focused on EUV coronal lines but could record the Lβ line of He II at 25.6 nm in prominences.

Fig. 1.16 Swedish 1-m Solar Telescope (SST) in Canary Islands

Another major step was recently obtained with the Interface Region Imaging Spectrograph (IRIS) which was launched on June 27, 2013 by a Pegasus rocket from an Orbital L-1011 aircraft. It is a spectrograph working in the UV (140 nm) and Near UV (280 nm) fed by a 20 cm telescope (Fig. 1.18).

This aperture and the pointing stability allow for a spatial resolution close to 300 km on the Sun. The orbit being heliosynchronous, it is possible to observe the Sun continuously all the year round (except for some eclipses from November to February). Actually, imaging is possible on IRIS in two different ways:

1. through motion of the entrance slit (called raster mode) which provides the full spectrum at each spatial position,
2. through a permanent solar image at the (reflecting) "jaws" of the slit (called slit-jaw imaging). This latest device is well-known by ground-based observers who usually rely upon an Hα image. Its implementation on IRIS has the peculiarity of recording the image in UV lines, which provide the right context for UV spectra.

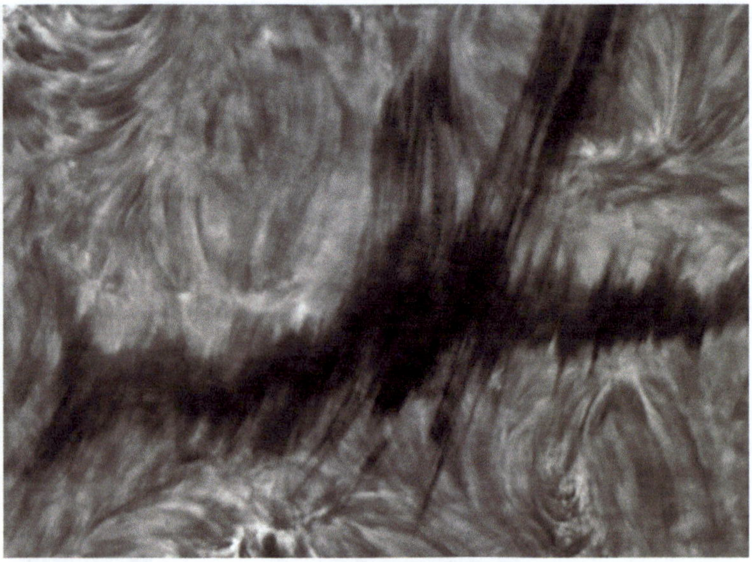

Fig. 1.17 Hα picture of the "barb" of a filament taken with the SST (courtesy O. Engvold)

Fig. 1.18 A view of the IRIS instrument (telescope and spectrograph) mounted on the S/C. Note the guide telescope on top of the telescope (courtesy A. Title)

The combination of slit-jaw images and spectra at high cadence allows the construction of spectacular movies available at: http://www.lmsal.com/hek/hcr?cmd=view-recent-events&instrument=iris.

The UV channel records spectral lines such as the resonance lines of Si IV and C IV formed in the Prominence Corona Transition Region (PCTR), at temperatures of a few 10,000 K (see Chap. 3, Parenti 2014). The NUV channel focuses upon an interesting doublet of Mg II: the "h" (280.271 nm) and "k" (279.553 nm) lines named by analogy with the H and K lines of Ca II. Similarly, they are formed at temperatures lower than 15,000 K in the cool core of prominences. Figure 1.19 provides an example of Mg II h and k spectra obtained at the South Pole of the Sun.

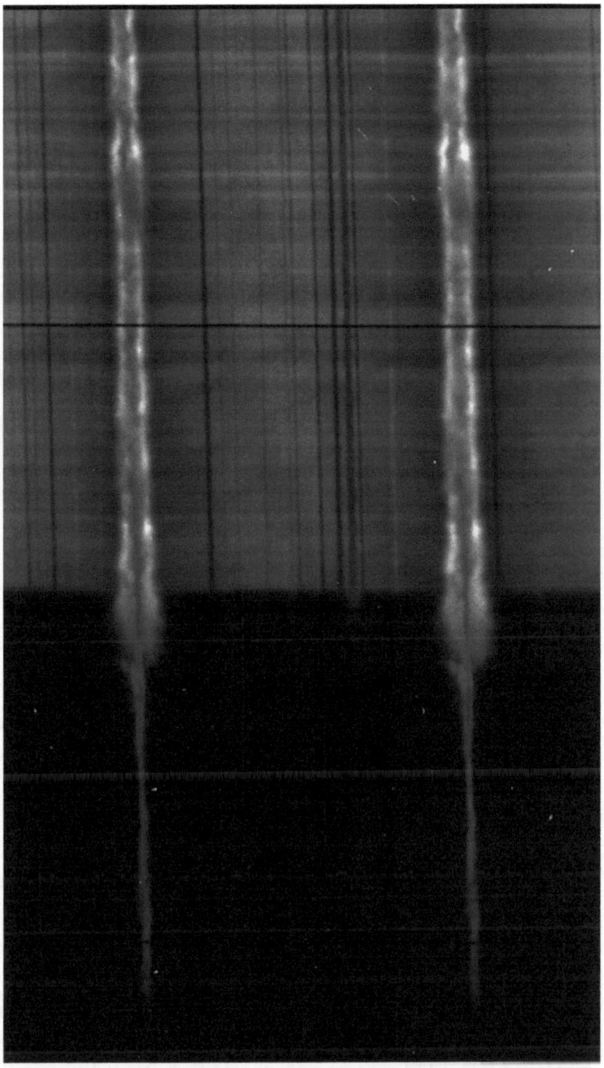

Fig. 1.19 Uncalibrated stigmatic spectra of the Mg II k and h lines observed by IRIS. The lower part of the figure where the profiles are narrow corresponds to a prominence of the polar crown seen above the South Pole limb. The upper part of the figure where profiles are wide and reversed corresponds to the chromosphere close to (and just above) the limb (courtesy A. Title—LMSAL) (see also Fig. 1 of Heinzel et al. 2014)

Note the very narrow unreversed profiles in the prominence, which correspond to very thin threads as shown in Chaps. 2 and 6 (Engvold 2014; Labrosse 2014).

Actually, the slit-jaw observing mode has the advantage of providing a context but has the drawback of ignoring the spectral profiles outside the slit. In the case of IRIS, one can also move the image on the slit, a mode which provides instantaneous spectra. Other modes are possible, on the ground and in space, where one records simultaneous spectra and simultaneous monochromatic images.

This is currently done with a dual Fabry-Perot in the Interferometric BIdimensional Spectrometer (IBIS), or the Universal Birefringent Filter at the Dunn Solar Telescope of Sacramento Peak. These devices have allowed a tremendous progress

Fig. 1.20 THEMIS (Heliographic Telescope for the Study of the Magnetism and Instabilities on the Sun) in Canary Islands

not only in the recording of broad chromospheric lines but also prominence/filament line profiles because they allow for a rather easy derivation of the line-of-sight velocities through the comparison of "blue" and "red" parts of the lines. Another original scheme consists in a multichannel substractive double-pass spectro-imaging (MSDP) which is a grating spectrometer providing simultaneous images (with a typical field-of-view of 30 by 250 arcsec) in nine bands in the selected line (e.g. Hα). It has been implemented in various observatories including the 90-cm aperture THEMIS (Heliographic Telescope for the Study of the Magnetism and Instabilities on the Sun) in Canary Islands (Fig. 1.20).

Let us also mention the powerful CRisp Imaging SpectroPolarimeter (CRISP) of the SST (first described by Scharmer 2006). It provides 2-D monochromatic images in more than ten wavelength positions in a line such as H-alpha in the course of a very few seconds.

A major project, now under construction in Hawaï, is the Advanced Technology Solar Telescope (ATST now named Daniel K. Inouye Solar Telescope) built by the National Solar Observatory in the U.S.A. Its primary mirror is an off-axis paraboloid with a 4.24 m diameter (limited to a 4-m aperture) which will be the first monolithic solar mirror of this size ever built. The heat stop assembly will define a field of view of 5 by 5 arcmin. The secondary is a 65 cm off-axis aspheric concave mirror, in SiC, a material with excellent thermal and physical properties. A very useful device for coronagraphic observations (including prominences, Rimmele et al. 2014) is a Lyot stop located at the first pupil image. ATST benefits from adaptive and active optics system at its Coudé focus (the Multi Conjugate Adaptive Optics, a tool used to extend the field of view, see e.g. Langlois et al. 2013) and will provide a spatial resolution eight times better than the SOT on Hinode. The following instrumentation is a complex set of imagers and spectropolarimeters (see Fig. 1.21). ATST is expected to see its first light in 2019. (Note that Europe also has a 4-m

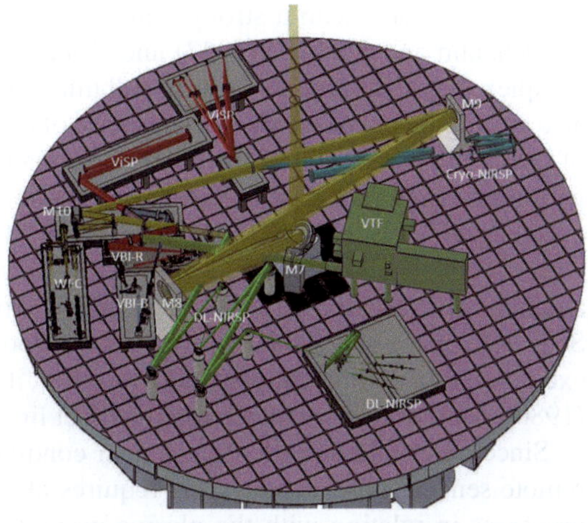

Fig. 1.21 ATST instrumentation at the Coudé level. It includes a Visible Broad-band Imager, a Visible Spectro-Polarimeter, two Near-IR Spectro-Polarimeters, and a Visible Tunable Filter (see http://atst.nso.edu) (courtesy of NSF/AURA/NSO/DKIST)

telescope project EST: see http://www.est-east.eu/, India proposes a 2-m National Large Solar Telescope and China has plans for a 6–8 m Giant Solar Telescope).

At that time, the Solar Orbiter mission of ESA (and NASA) and the Solar Probe Plus of NASA, will have started their journeys close to the Sun and other S/C from Russia, Japan (and other Agencies) will be monitoring the Sun. It will then be essential to coordinate the observations from these space and ground-based instruments as has been done in the past. With the SOHO mission, periodic campaigns based on Joint Observing Programs included SOHO, TRACE and many ground-based observatories. This allowed coverage of a large range of wavelengths and to perform a multi-temperature analysis. The joint analysis was facilitated by the use of common FITS formats and software such as SolarSoft (http://www.lmsal.com/solarsoft/). The same procedures were used for Hinode which was operated from Japan.

1.3 The Rich Physics Associated with Prominences

As the reader will see throughout the book, prominences provide an amazingly large playground for different fields of physics. The very fact that they exist as cool and dense structures in the hot and diffuse coronal "furnace" is a challenge for plasma physicists. Although downward motions are observed, why does the material not drop in free fall, as already noticed by Rothschild et al. (1955)? The challenge is increased since their gas (or plasma) is far from being totally ionized, which means that many plasma processes can occur which can lead to instabilities (plasma gradient, Rayleigh-Taylor, MHD wave damping, ambipolar diffusion). Since, as will be seen later on, the density and more importantly the magnetic field are still not precisely determined, it is not clear how much the magnetic pressure dominates the kinetic pressure. The ionization also depends on the proximity of an active region (or a flare) with a strong ionizing radiation field. Since the early works of Kippenhahn and Schlüter (1957) and Kuperus and Raadu (1974) more and more complex magnetic models have been built. The structuring in very thin threads is also an enigma increased by our ignorance of the actual size of these threads. In view of the large diversity of these structures (already noticed by early observers), along with their temporal variability, it is not possible to build detailed realistic models, even empirical. But progress has been made towards a description of prominences with a set of slender cylinders and loops, as we shall see in the book. Apparent flows and intensity variations are still discussed in about the same terms as by Secchi (see Sect. 1.1.2): are they actual (fast) mass transports or simply local variations of the excitation of a permanent plasma? Moreover, it has been demonstrated by Rompolt (1980) that radial flows may strongly affect line emissivity.

Since determination of the physical conditions in these structures relies upon remote sensing observations, this requires also the understanding of the radiative processes in relation with the plasma properties. And it just happens that prominences are strongly out of (local) thermodynamic equilibrium from the standpoints of populations of atoms and ions and consequently of the emitted radiation. In some lines, the plasma is optically thick, which complicates the diagnostics. Another basic fact which still is not understood is the heating of prominences. This can be seen as a paradox since the prominence plasma is much cooler than the surrounding corona. But in terms of radiative equilibrium (e.g. in the Lyman continuum) prominences should be cooler than observed (Heasley and Mihalas 1976; Fontenla 1979), which implies a heating process still unidentified in spite of many candidates (Alfvèn waves energy deposit, ambipolar diffusion, . . .).

The very existence of prominences is still not well understood. The understanding of their formation, their stability and their eruption requires a great deal of information about the plasma and the magnetic field, inside, beneath and around the structures. This requires a comprehensive set of multi-wavelength observations requiring the best spectral, spatial and temporal resolutions that are often impossible to perform. MHD modelers and theoreticians have to live with this incomplete information that they circumvent by clever schemes such as field extrapolation. The very fact that prominences are dynamic also increases the difficulty of their task.

In summary, the study of prominences calls for an interdisciplinary approach involving various techniques (imaging, spectroscopy, polarimetry), matter-radiation interaction physics, plasma physics, etc. This approach is the essence of the book.

1.4 The Aim of the Book and to Whom It Is Addressed

Since the publication in 1995 of the latest book devoted to solar prominences ("The nature of solar prominences" by Einar Tandberg-Hanssen (1995), Kluwer) very important new results have been obtained from space missions and ground-based observatories. During more than 18 years, SOHO has provided 24 h a day continuous observations of the Sun in general and prominences in particular. SOHO was complemented by the high spatial resolution TRACE. The STEREO mission has provided two viewpoints allowing for a stereoscopic view of the Sun . . . and prominences. The Hinode mission offered a unique opportunity of studying the fine structure of prominences. More recently, the Solar Dynamics Observatory (SDO) has provided continuous high-cadence sets of spectacular images from which movies have been built, now available on the internet. In 2013, the UV spectro-imager IRIS (Interface Region Imaging Spectrograph) was launched and since then has observed prominences. On the ground, the new Swedish Solar Telescope (SST) and the Dutch Open Telescope (DOT) have recently provided unique images of prominences and filaments. Consequently, the ambition of the book is to present a comprehensive and pedagogical approach of prominences based upon the latest results obtained by the most recent instrumentation along with the up-to-date theories and modelling.

The book is intended for advanced students in astrophysics, post-graduates, solar physicists and more generally astrophysicists. Being based on many spectacular images, it can also be useful for amateur astronomers interested in this enigmatic feature of our fascinating Sun.

1.5 Table of Contents of the Book

After the present general introduction, Chap. 2 describes prominences in detail, their structures and their environment and discusses the various classifications, a difficult task when one considers the variety and the temporal variation of structures that we have seen in this Introduction. In this chapter, we alternately zoom in on the fine structure and zoom out on the large-scale features. Among these, the large-scale spine (which follows the Polarity Inversion Line) and the small-scale barbs (which connect to the chromosphere) are presented.

Once prominences are well identified, Chaps. 3–7 present their thermodynamic properties along with the ingenuous methods used to derive them.

Chapter 3 is devoted to the spectral diagnostic of optically thin plasma, in the cool core and in the hotter Prominence-Corona Transition Region (PCTR). Although the methods used seem to be "simple", the results (e.g. the "Hvar model" see Engvold et al. 1990 and also Patsourakos and Vial 2002) largely depend on the observed structures and in particular their sizes limited by the spatial resolution of the observation. However, the results derived from the latest missions put more constraints on the temperature, the neutral and electron densities, ionization degree and other thermodynamic quantities.

Chapter 4 covers the issues of the derivation of mass and velocity flows allowing determination of mass flows. These physical quantities are the key to understanding the mass and energy equilibria and consequently the stability of prominences. The presentation of the methods used leads to the uncertainties in the mass and flows which are derived. Moreover, the orientation and the magnitude of the flows largely depend on the observed structures (e.g. spine and barbs). Among some outstanding issues, the chapter discusses the origin of mass flows, the possible role of barbs, the relation with cavities (see Chap. 13) and with the magnetic structuring.

Chapter 5 is an introductory tutorial on radiative transfer and Non-Local Thermodynamic Equilibrium (NLTE) forward modelling, whether in one or two dimensions. This tutorial is focused on the specific case of prominences i.e. structures illuminated from below (photospheric and chromospheric radiation) and from all around (EUV coronal radiation) which moreover can be submitted to a changing Doppler-shifted illumination depending on their motion. It addresses the specific case of radiative transfer in radio wavelengths. It presents new techniques for coupling radiation and magneto-hydrostatics. It allows to better understand the following Chap. 6 which presents the results of the NLTE modelling and compares them to observations. Note that this chapter also addresses the prominence radio emission.

Chapter 6 presents the (direct) method used for deriving the major properties of prominences through non-LTE modeling. It shows that the direct method is both necessary and difficult because of the complex coupling between radiation and plasma. No thermodynamic quantity can be easily derived from the spectroscopic observation of optically thick lines (and continua). But some laws between observed radiation and emitting plasma can be derived as a result of systematic modeling. The chapter shows how major progresses have been made when moving from one-dimensional (1D) "monolithic" models to a set of 1D slabs simulating a bunch of threads and when comparing with 2D models. The modeling concerns basically the hydrogen and helium elements (which determine the ionization ratio) but also the "trace" elements (e.g. Ca II, Mg II) which provide still more constraints.

Chapter 7 addresses the fundamental issue of stability of prominences with a thorough description of all the terms entering the energy equation. These terms have been derived and discussed in the case of the chromosphere and the chromosphere–corona transition region. The basic approach consists in establishing the list of all possible losses and to derive the required energy input. Essentially two regions are considered: a cool core (less than 10^4 K) and a prominence–corona transition region (PCTR) (from 10^4 to 10^6 K). The conductive flux is negative in the PCTR and positive in the cool core. The enthalpy flux is important in the PCTR as a heating process, along with the radiative losses which peak at a few 10^4 K at constant pressure. Other heating mechanisms (waves, . . .) are also considered. Finally, Chapter 7 discusses the energy balance of quiescent prominences (cool core and PCTR) and the various models which satisfy the observational constraints.

Chapters 8–13 introduce a very important and recently studied ingredient, the magnetic field coupled with the dynamics of prominences.

With Chap. 8, we are introduced to the techniques and the results of prominence magnetometry and its forward and inverse solutions. Actually it is a tutorial on the two main modifications of the radiation polarization induced by the presence of magnetic fields. Zeeman and Hanle effects are described and compared in the context of prominences. It is shown that the Hanle effect is the main tool and that old Zeeman measurements provided values of the magnetic field which were, by chance, not far from the actual ones. In the case of the forward problem, the scattering is treated rigorously for the He lines for both linear and circular polarization. The techniques related to the inverse (mathematically ill-posed) problem are also presented. This chapter ends with a critical discussion about the ambiguities, sources of error and uncertainties in prominences magnetometry. It also proposes paths for improving the field derivation such as inclusion of radiative transfer, 3D geometry, etc.

Chapter 9 covers the issue of the magnetic field structure, as derived from direct and indirect observations. It provides the detailed specific characteristics of the two main types of prominences (channel and coronal clouds) prominences. As far as channel filaments are concerned, they are essentially defined by the presence of a polarity reversal boundary as revealed by magnetograms. The usual three classes (active region filaments, intermediate and quiescent filaments) are precisely defined and their magnetic properties derived, one would dare to say surprisingly, from patient temporal sequences of spectroheliograms. Amazing pictures at high spatial resolution are complemented by labels which provide support to the text. The chapter also puts the emphasis on a new physical quantity: the chirality. Overall the reader will have access to a complete and consistent picture on the essential rôle of the magnetic field in structuring, stabilizing and destabilizing prominences at all scales.

Chapter 10 covers the dynamics (models and observations) related to the formation of prominences. It provides some answers to the important question: where does the material come from? from the "sky" above or from the "ground" below? It recalls the many observational constraints imposed by observations such as the observed flows (and their directions). Then it reviews the four main models of mass formation: injection (from the chromosphere), levitation (by the magnetic field), evaporation-condensation (by which the plasma is heated and then cooled by radiative losses) and magneto-thermal (based on Kelvin–Helmholtz instability) models which are discussed in terms of observational signatures. The review comes with dedicated didactic cartoons and, where necessary, the presentation of basic physical processes such as magnetic reconnection.

In Chap. 11, some basics of MHD waves are introduced and the properties of MHD waves fully described. The different modes of oscillations are discussed in two different geometries: slabs and threads. The comparison of observed oscillation periods and damping times with the predicted values offers a new diagnostic tool: prominence seismology. This analysis has been recently improved (and complexified) with the inclusion of flows in the geometrical models and the coupling of MHD waves and radiative transfer (also see Chap. 5). Obviously, the determination of plasma parameters and the precise geometry of the prominence would be benefitting for the seismology (e.g. derivation of the magnetic field) and reciprocally.

Chapter 12 focuses on the MHD simulations of the emergence of twisted magnetic flux tubes which can ultimately lead to prominence eruption through build-up of free energy and helicity in the corona. It distinguishes between ideal MHD instabilities (essentially torus and kink instabilities which are treated both analytically and with numerical simulations) and fast magnetic reconnection in current sheets. Many modeling results are convincingly compared with observations. The issue of the hypothesis of force-free coronal structures is also discussed.

Chapter 13 focuses on coronal cavities whose importance has been stressed with the new Hinode (SOT) observations along with MHD models. The difficulty of observing cavities should not conceal the fact that they are probably ubiquitous. Their location, morphology and thermal properties are thoroughly presented, along with plane-of-sky and line-of-sight flows. For each observation, a MHD model is proposed. Polarimetry puts forward an essential ingredient: the magnetic field parallel to the underlying neutral line. Cavities are shown to be dependable predictors of prominence eruption. The whole chapter leads to the conclusion that cavities actually are magnetic flux ropes.

Chapter 14 describes large patterns and filament channels taking into account the large number of filaments/prominences and their evolution over time spans of the order of a solar rotation (and more). The chapter is based upon the three categories defined in Chap. 2: quiescent filaments (QF), intermediate filaments (IF) and active region filaments (ARF). It first focuses on the formation locations which most often involve multiple bipole interactions. The chirality follows an hemispheric pattern with some exceptions. From patient (but not so numerous) observations, it is possible to distinguish two mechanisms at work: reconfiguration of pre-existing coronal fields vs. emergence of horizontal flux tubes. The chapter provides an extensive list of models which can explain the observations and concludes on the statement that different mechanisms can be at work for the three classes of prominences.

Chapters 15–17 pay attention to the final phase of prominence eruptions (PE) and their impacts on Earth.

The dynamics of eruptive prominences (EP), up to 1 A.U. are detailed in Chap. 15. It is shown that EPs cannot be separated from flares and Coronal Mass Ejections (CMEs), as evidenced by statistical associations, correlated cycle variations and joint kinematics. The (cool) ejected material can be followed far in the heliosphere with such in situ signatures as low-charge Fe ions or increased He^{+}/He^{++} ratio. Attention is paid to high-latitude prominences whose "rush to the pole" and further eruption play role in the polarity sign reversal at the poles. The chapter also shows that polar CMEs are similar to low-latitude CMEs, an indication of a common bipolar origin. Finally the non-radial motions of EPs and CMEs are compared, an important issue in the frame of Space Weather.

Chapter 16 is also devoted to the relation between Eruptive Prominences and Coronal Mass Ejections from the point of view of CMEs, with focus on detailed observations (and properties) of EPs rather far in the heliosphere through remote-sensing techniques. The three-component classical model of CMEs is presented with the (new) evidence of the presence of a flux rope. Respective mass and energy values are discussed. Farther in the interplanetary space, it is more and more difficult to identify the EP cool material with remote-sensing. However, apart from in situ detection of magnetic clouds, it has been recently made possible to identify EP with the help of Heliospheric Imagers (HIs) on board STEREO and SMEI missions.

Chapter 17 further details the propagations of EPs and CMEs in the interplanetary medium with the help of the unique HI observations on STEREO. The simulations which take into account the (magnetic) interaction with the solar wind confirm the increase of mass, by a 2–3 factor, up to 1 AU. The chapter then focuses on the different impacts of EPs and CMEs on our Earth as measured in situ. It discusses the three mains sources of Space Weather: radiation (EUV, X-ray), high-energy particles accelerated in the interplanetary medium and high-energy particles accelerated inside Earth's magnetosphere. As a conclusion, the Chapter raises the issue of the difficulty of detecting EP material at 1 AU on one hand, and the rôle this cool material could play in CME geo-effectiveness on the other hand.

Chapter 18 opens a new window into non-solar astronomy: stellar prominences and stellar CMEs along with their influence on planetary evolution. Firstly, the chapter reviews some major features of stellar activity. Then it presents the major properties of the more than one thousand exoplanets detected to this day. It introduces the new concept of habitability summarized in a mass/orbit diagram. Because of the impact of stellar radiation and plasma flows on planetary atmospheres, the stellar activity is thoroughly discussed for late-type stars in terms of luminosity, wind and CMEs. The importance of exoplanets magnetic fields is not ignored, even if they cannot be measured! Scaling laws between the exoplanet magnetic dipole, the rotation rate and the radius of the dynamo region are proposed. The chapter derives a value of the minimum magnetospheric radius which provides an effective shielding of the planet. Hot Jupiters are also detailed. As the reader will discover, the issue of prominences on our near Sun naturally leads to the issue of the possibility of life on planets located many light-years away from us.

Acknowledgments I thank O. Engvold, S. Koutchmy and J.W. Leibacher for kindly providing useful material and helpful suggestions and corrections.

References

Babcock, H. D., & Babcock, H. W. (1955). The sun's magnetic field, 1952–1954. *The Astrophysical Journal, 121*, 3.

Berger, T. E., Shine, R. A., Slater, G. L., Tarbell, T. D., Title, A. M., Okamoto, T. J., Ichimoto, K., Katsukawa, Y., Suematsu, Y., Tsuneta, S., Lites, B. W., & Shimizu, T. (2008). Hinode SOT observations of solar quiescent prominence dynamics. *The Astrophysical Journal, 676*(1), L89–L92.

Bonnet, R. M., Decaudin, M., Bruner, E. C., Jr., Acton, L. W., & Brown, W. A. (1980). High-resolution Lyman-alpha filtergrams of the sun. *Astrophysical Journal, Part 2 – Letters to the Editor, 237*, L47–L50.

Bonnin, X., Aboudarham, J., Fuller, N., Csillaghy, A., & Bentley, R. (2013). Automation of the filament tracking in the framework of the HELIO project. *Solar Physics, 283*, 49.

Culhane, J. L., et al. (2007). The EUV imaging spectrometer for Hinode. *Solar Physics, 243*, 19.

D'Azambuja, L., & D'Azambuja, M. (1948). Etude d'ensemble des protubérances solaires et de leur évolution, Annales de l'Observatoire de Paris, Tome VI, Gauthiers-Villars.

Deslandres, H. (1910). Ann. Obs. Paris-Meudon 4 (III), pp. 54–55.

Dunn, R. (1960). Photometry of the solar chromosphere, Ph.D. Thesis, Harvard University.

Dunn, R. (1965). Sacramento Peak Obs. Contr. Report #87, Fig. 36, p. 80.
Engvold, O. (1976). The fine structure of prominences. I – Observations – H-alpha filtergrams. *Solar Physics, 49*, 283.
Engvold, O. (2014). Description and classification of prominences. In J.-C. Vial, & O. Engvold (Eds.), *Solar prominences, ASSL* (Vol. 415, pp. 31–60). Springer.
Engvold, O., Hirayama, T., Leroy, J.-L., Priest, E. R., & Tandberg-Hanssen, E. (1990). Hvar reference atmosphere of quiescent prominences. In V. Ruzdjak & E. Tandberg-Hanssen (Eds.), *Dynamics of quiescent prominences. Proceedings of I.A.U. Colloquium #117, held in Hvar, SR Croatia, Yugoslavia, September 25–29, 1989*. New York: Springer-Verlag.
Fontenla, J. M. (1979). A prominence model based on spectral observations. *Solar Physics, 64*, 177.
Hale, G.E., & Ellerman, F. (1903). Publ. Yerkes Obs. 3(I), 3.
Harvey, J.W. (1969). *Magnetic fields associated with solar active-region prominence*. Ph.D. Thesis, University of Colorado.
Harvey, J., & Tandberg-Hanssen, E. (1968). The magnetic field in some prominences measured with the He I, 5876 Å line. *Solar Physics, 3*, 316.
Heasley, J. N., & Mihalas, D. (1976). Structure and spectrum of quiescent prominences – energy balance and hydrogen spectrum. *Astrophysical Journal, 205*, 273.
Heinzel, P., Vial, J.-C., & Anzer, U. (2014). On the formation of Mg II h and K lines in solar prominences. *Astronomy and Astrophysics, 564*, 132.
Khangil'Din, U. V. (1964). Characteristics of solar active regions obtained from observations on millimeter wavelengths. *Soviet Astronomy, 8*, 234.
Kippenhahn, R., & Schlüter, A. (1957). Eine Theorie der solaren Filamente. Mit 7 Textabbildungen. *Zeitschrift fur Astrophysik, 43*, 36.
Korendyke, C. M., Vourlidas, A., Cook, J. W., Dere, K. P., Howard, R. A., Morrill, J. S., Moses, J. D., Moulton, N. E., & Socker, D. G. (2001). High-resolution imaging of the upper solar chromosphere: First light performance of the very-high-resolution advanced ultraviolet telescope. *Solar Physics, 200*, 63–73.
Kosugi, T., et al. (2007). The Hinode (solar-B) mission: An overview. *Solar Physics, 243*, 3.
Kuperus, M., & Raadu, M. A. (1974). The support of prominences formed in neutral sheets. *Astronomy and Astrophysics, 31*, 189.
Labrosse, N. (2014). Derivation of the major properties of prominences using non-LTE modeling. In J.-C. Vial & O. Engvold (Eds.), *Solar prominences, ASSL* (Vol. 415, pp. 129–153). Springer.
Langlois, M., Moretto, G., Béchet, C., Montilla, I., Tallon, M., Goode, P., Gorceix, N., & Shumko, S. (2013). Concept for solar multi-conjugate adaptive optics at big bear observatory. In S. Esposito & L. Fini (Eds.), *Proceedings of the third AO4ELT conference*. Firenze, Italy, May 26–31, 2013. Online at http://ao4elt3.sciencesconf.org/, id. #62.
Leroy, J.-L., Ratier, G., & Bommier, V. (1977). The polarization of the D3 emission line in prominences. *Astronomy and Astrophysics, 54*, 811.
Lin, Y., Engvold, O., Rouppe van der Voort, L. H. M., & Wiik, J. E. (2005). Thin threads of solar filaments. *Solar Physics, 226*, 239–254.
Lopez Ariste, A. (2014). Magnetometry of prominences. In J.-C. Vial & O. Engvold (Eds.), *Solar prominences, ASSL* (Vol. 415, pp. 177–202). Springer.
Parenti, S. (2014). Spectral diagnostics of cool prominence and PCTR optically thin plasmas. In J.-C. Vial & O. Engvold (Eds.), *Solar prominences, ASSL* (Vol. 415, pp. 61–76). Springer.
Patsourakos, S., & Vial, J.-C. (2002). SOHO contribution to prominence science. *Solar Physics, 208*, 253.
Rayet, G. (1869). Report of M. Rayet. *Astronomical Register, 7*, 133–134.
Rimmele, T. R. (2000). Solar adaptive optics. In P. L. Wizinowich (Ed.), *Adaptive Optical Systems Technology, Proceedings of SPIE* (pp. 218–231).
Rimmele, T., Berger, T., Casini, R., Elmore, D., Kuhn, J., Lin, H., Schmidt, W., & Wöger, F. (2014). Prominence science with ATST instrumentation. In *Nature of prominences and their role in space weather, proceedings of the International Astronomical Union (IAU) symposium* (Vol. 300, pp. 362–369).

Rompolt, B. (1980). Doppler brightening effect in H-alpha line for optically thin moving prominences. *Hvar Observatory Bulletin, 4*(1), 39.

Rothschild, K., Pecker, J.-C., & Roberts, W. O. (1955). The prominence of July 25, 1951. *The Astrophysical Journal, 121*, 224.

Rust, D.M. (1966). *Measurements of the magnetic fields in quiescent solar prominences.* Ph.D. Thesis, University of Colorado.

Sahal-Bréchot, S., Bommier, V., & Leroy, J.-L. (1977). The Hanle effect and the determination of magnetic fields in solar prominences. *Astronomy and Astrophysics, 59*, 223.

Scharmer, G. B. (2006). Comments on the optimization of high resolution Fabry-Pérot filtergraphs. *Astronomy and Astrophysics, 447*, 1111.

Scharmer, G. B., Bjelksjo, K., Korhonen, T. K., Lindberg, B., & Petterson, B. (2003). The 1-meter Swedish solar telescope. In S. L. Keil & S. V. Avakyan (Eds.), Innovative Telescopes and Instrumentation for Solar Astrophysics, *Proceedings of SPIE* (Vol. 4853, p. 341).

Schmahl, E. J., Bobrowsky, M., & Kundu, M. R. (1981). Observations of solar filaments at 8, 15, 22 and 43 GHz. *Solar Physics, 71*, 311–328.

Tandberg-Hanssen, E. (1974). (Geophysics and astrophysics monographs), *Solar prominences* (Vol. 12). Dordrecht: D. Reidel Publishing Co.

Tandberg-Hanssen, E. (1995). Astrophysics and Space Science Library, *The nature of solar prominences,* (Vol. 199). Dordrecht: Kluwer Academic Publishers.

Tavabi, E., Koutchmy, S., & Ajabshirzadeh, A. (2013). Increasing the fine structure visibility of the Hinode SOT Ca II H filtergrams. *Solar Physics, 283*, 187–194.

Thompson, W. T. (2013). Alternating twist along an erupting prominence. *Solar Physics, 283*, 489–504.

van Noort, J. M., Rouppe van der Voort, L. H. M., & L̈ofdahl, M. G. (2005). Solar image restoration by use of multi-frame blind deconvolution with multiple objects and phase diversity. *Solar Physics, 228*, 191–215.

Vassenius, B. (1733). Observation de l'éclipse totale de Soleil avec demeure, faite à Gothebourg en Suède à 57° 40' 54 le 2 Mai 1733, *Philosophical Transactions of the Royal Society of London, 38*, 134.

Vial, J.-C., Olivier, K., Philippon, A. A., et al. (2012). High spatial resolution VAULT H-Lyα observations and multiwavelength analysis of an active region filament. *Astronomy and Astrophysics, 541*, 108.

Vourlidas, A., Sanchez Andrade-Nu˜no, B., Landi, E., et al. (2010). The structure and dynamics of the upper chromosphere and lower transition region as revealed by the subarcsecond VAULT observations. *Solar Physics, 261*, 53.

Xu, Z., Jin, Z. Y., Xu, F. Y., Liu, Z. (2014). Primary observations of solar filaments using the multi-channel imaging system of the New Vacuum Solar Telescope. In B. Schmieder, J.-M. Malherbe & S. T Wu (Eds.), *Nature of prominences and their role in space weather. Proceedings of the International Astronomical Union, IAUSymposium* (Vol. 300, pp. 117–120).

Zirin, H., & Severny, A. B. (1961). Measurement of magnetic fields in solar prominences. *Observatory, 81*, 155.

Chapter 2
Description and Classification of Prominences

Oddbjørn Engvold

Abstract Solar prominences are bright cloud-like structures when observed beyond the solar limb and they appear as dark filamentary objects which are termed filaments when seen against the solar disk. The aims of prominence classifications were from the start to establish references and frameworks for understanding the physical conditions for their formation and development through interplay with the solar magnetic environment. The multi-thermal nature of solar prominences became fully apparent once observations from space in UV, VUV, EUV and X-rays could be made. The cool prominence plasma is thermally shielded from the much hotter corona and supported in the field of gravity by small- and large-scale magnetic fields of the filament channels. High cadence, subarcsecond observing facilities on ground and in space have firmly proven the highly dynamic nature of solar prominences down to the smallest observed structural sizes of 100 km. The origin of the ubiquitous oscillations and flowing of the plasma over a variety of spatial and temporal scales, whether the cool dense plasma originates from below via levitation, injections by reconnection or results from condensation processes, are central issues in prominence research today. The unveiling of instabilities leading to prominences eruptions and Coronal Mass Ejections is another important challenge. The objective of this chapter is to review the main characteristics of various types of prominences and their associated magnetic environments, which will all be addressed in details in the following chapters of this book.

2.1 Introduction

Early solar astronomers could observe prominences only at the rare occasions of total solar eclipses until J. Janssen and Sir Norman Lockyer independently discovered with the use of spectroscopes that the luminous prominences radiated in a very few spectral lines. Their reddish color was due to the dominant Hα line of hydrogen at λ6562.8 Å. Hale (1903) and Deslandres (1910) realized both that dark

O. Engvold (✉)
University of Oslo, Blindern, Institute of Theoretical Astrophysics, Oslo, Norway
e-mail: oengvold@astro.uio.no

J.-C. Vial, O. Engvold (eds.), *Solar Prominences*, Astrophysics and Space Science Library 415, DOI 10.1007/978-3-319-10416-4_2

filaments seen in absorption on the disk were prominences seen against a brighter background. In this book the term filament is generally synonymous to prominence seen on the disk.

The EUV, far-EUV and X-ray spectral regions contain numerous emission lines formed at temperatures ranging from the photospheric to the coronal ones. Vial (2014) provides an overview of various space instruments used for prominence observations, from the early OSO satellites to the recent SDO/AIA. As an example, the main UV and EUV lines recorded with instruments on-board SOHO cover a range in ionization stages from Si II λ1259 Å and C II λ1037 Å, formed at T = 13–25 $\times$ 10^3 K, to O IV λ554 Å and O VI λ1037 Å representing, respectively, 2 $\times$ 10^5 K and 4 $\times$ 10^5 K (Labrosse et al. 2010). The dominant bright Hα, Ca II H and K lines, the He I at λ5787 Å and the two Na I lines at λ5892 and λ5896 Å are all emitted from the cool core prominence plasma at electron temperatures from 7.5 $\times$ 10^3 to 10^4 K (Poland and Tandberg-Hanssen 1983; Hirayama 1985; Tandberg-Hanssen 1995). The fact that the cool prominence structures may also be recognized in lines emitted from higher temperature plasma demonstrates that all prominences are covered by a fairly thin temperature layer which is commonly referred to as the *Prominence Corona Transition Region* (*PCTR*) (Vial 1990).

The thermodynamic parameters of the prominence plasma are derived from observed line intensities and polarization, often in combination with radiative transfer calculation and modeling (see Labrosse et al. 2010). The commonly cited cool plasma densities are 10^{10}–10^{11} cm^{-3} and gas pressure 0.1–1 dyn cm^{-2} (Hirayama 1985; Parenti and Vial 2007; Labrosse et al. 2010; Parenti 2014a). This implies that the cool prominence material is roughly 100-fold cooler than the corona gas and about 100-fold more dense. The corresponding parameters of the high temperature regions are derived from multi-wavelength observations of UV and EUV line emission by taking into account the volumes already occupied by lower temperature plasma (Anzer et al. 2007; Heinzel et al. 2008).

The first direct measurements of magnetic fields of solar prominences which were based on the Zeeman Effect are summarized in Tandberg-Hanssen (1974). Later measurements using the Hanlé effect were pioneered by the French group at Pic-du-Midi and Meudon Observatory (Leroy 1981; Bommier et al. 1994). The generally accepted typical magnetic field strength in the cool plasma is 3–30 G (Leroy 1989) implies that the ratio of the plasma pressure ($p = 2nkT$) cited above to the magnetic pressure ($p_{mag} = B^2/2\,\mu_0$), usually referred to as *plasma-β*, then becomes 0.01–1. This implies that the magnetic fields represent the dominant force in the prominence plasma and thereby largely control their structure and dynamics. A fundamental condition for formation and development of solar prominences is clearly the local magnetic field topology rooted in the photosphere below. It is firmly established that filaments are located above the border between negative and positive magnetic fields on the Sun's surface (Martin 1998a). Comprehensive discussions of magnetic fields of solar prominences are given by Mackay et al. (2010) and López Ariste (2014).

Advancement in spatial resolution in observations from ground- and space-based telescopes provides a timely reminder that on a small-scale, and over a large range of temperatures, prominences are highly dynamic and rapidly changing in contrast to their apparent stability on the more global scale, where quiescent prominences appear relatively stable with lifetimes ranging from days up to approximately a month. Longer reported lifetimes have not been verified due to the absence of observations from the back side of the Sun and the tendency for prominences to repeatedly develop at nearly the same sites. Active prominences occurring in the vicinity of active regions are more dynamic and usually more short-lived. Active intervals or activations also occur in quiescent prominences but less frequently. The majority of prominences eventually undergo instabilities that lead to eruption and disappearance (*disparitions brusques*), most often associated with *Coronal Mass Ejections* (*CMEs*). A few prominences end their life simply by the draining of all their mass back to the chromosphere. The formation and development of solar prominences, including their fascinating variations of shape and dynamics, will be summarized in the following sections of this chapter, while the full and detailed discussions are handled in the following chapters of this book.

2.2 Classifications

A number of schemes of solar prominence classification have been proposed and are still in general use. Classifications are based on combinations of their morphology, dynamic properties and relative locations. It was early realized that prominences located close to active regions were changing quite rapidly compared with more slowly changing ones that appear in regions well removed from active regions, and at higher solar latitudes. The Italian astronomer Secchi and contemporary scientists like Respighi and Fearnley made a number of prominence drawings using wide slit spectroscopes (cf. Vial 2014). They were all struck by the variety in shapes and activity. Secchi concluded that prominences fell into two categories, one for short-lived prominences which he called *eruptives* and long-lived ones which were denoted *quiescents*.

George E. Hale's realization of the spectrohelioscope (Hale 1929) permitted more systematic observations and analysis of shapes and motions of prominences at the limb and on the disk. His instrument provided also radial velocities of the observed features. Regular spectrohelioscopic observations were subsequently initiated both at Greenwich and at Mt Wilson Observatory. Newton (1935) concluded from his studies that solar filaments and prominences were of two types; (1) those that are not associated with sunspot, and (2) those that are associated with sunspots and active regions. His measurements of radial velocities up to 100 km s^{-1} and higher were clearly associated with erupting cases. Edwin Pettit at Mt Wilson Observatory suggested a much more detailed classification scheme consisting of six main classes and several sub-classes. Pettit's classification (Pettit 1932) is illustrated below. All prominences were first divided into those connected with spots and

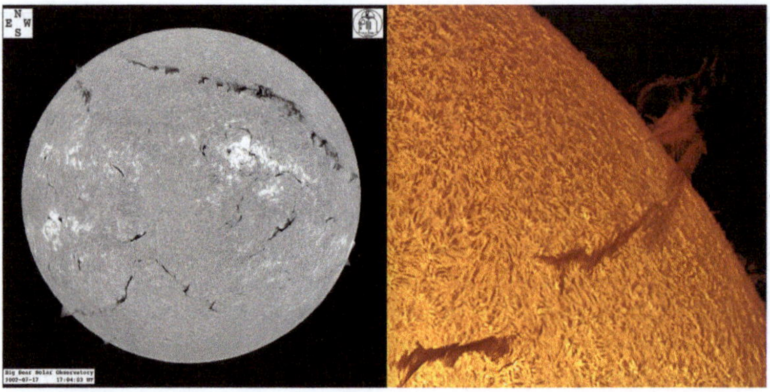

Fig. 2.1 *Left*: The full disk image observed in Balmer Hα 2002 July 17 shows a variety of filaments some located in active regions and others at high solar latitudes and away from active regions (Credit: BBSO). The *right* image (also Hα) illustrates the change from the appearance of absorbing to emitting as a filament observed on November 25, 2011 crosses the solar limb (Credit: Tom Wolfe)

those that were not. Figure 2.2 is a self-explaining comprehensive illustration of the categories and sub-categories of Pettit's classification system.

Further progress in instrumentation and photographic techniques, like Bernard Lyot's coronagraph and monochromatic filter based on birefringence of quartz and the selective transmission of polaroids (see Vial 2014), enabled Donald H. Menzel to establish routine observations of the chromosphere and prominences at the High Altitude Observatory in Colorado. A notable collection of systematic observations of prominences which also included motion pictures, led Menzel and Evans (1953) to suggest a classification scheme, which differed somewhat from Pettit's scheme. They assigned the letter *A* to prominences in which matter flows downwards from above and the letter *B* to prominences with matter flowing into the corona from below. The letter *S* was assigned to prominences connected with sunspots and *N* to the others (non-spot). There were subcategories to each of the four types as shown in the following resulting scheme:

A. Prominences originating from above in coronal space

S. Spot prominences:

l. Loops
f. Funnels

N. Non-spot prominences:

a. Coronal rain
b. Tree trunks
c. Trees
d. Hedgerows
e. Suspended clouds
m. Mounds

B. Prominences originating from below in the chromosphere

S. Spot prominences

s. Surges
p. Puffs

N. Non-spot prominences

s. Spicules

In a subsequent classification scheme proposed by de Jager (1959) prominences were classified as either (I) *Quiescent* or (II) *Moving Prominences*. The quiescent prominences were grouped further into *Normal* (low to medium latitudes) and *Polar* (high latitudes). The class of *Moving Prominences* included *Active*, *Eruptive* and *Spot (associated) Prominences* plus *Surges* and *Spicules* (Fig. 2.2)

Zirin's (1966, 1988) classification of class 2 *Long-lived, Quiescent Prominences* was identical to de Jager's Class I, while he grouped *Loops*, *Coronal Rain*, *Surges* and *Sprays* under his Class 1 *Flare-associated, Short-lived Prominences*.

Any prominent solar feature seen above the rim of the Sun was historically associated with the term prominence. Increased awareness of related phenomena led to a rich "zoo" of solar features that were subsequently classified under a prominence umbrella. This included features like flare loops, surges, coronal loops or arches, various types of mass ejections, large spicules, coronal rain and coronal cloud prominences. *Surges* and *Loops*, which occur in conjunction with flare activity, are now recognized as active region jet phenomena and post flare loops, respectively, and regarded quite different from solar prominences. Furthermore, spicules constitute rather the main structure of the chromosphere. In the following chapters *Coronal Rain* and *Coronal Cloud Prominences* remain as a significantly different types of prominence.

It is quite common today to divide prominences into *Quiescent*, *Intermediate* (combined) and *Active Region Prominences*, which also will be the adopted classification in the following subsections. This simple classification evolved from the early active and quiescent designations and has remained practical with the recognition of intermediates which fill-in a broad continuum of filaments ranging from low and narrow ones in the active regions to high and wide quiescent ones. All three groupings have varied lengths from very short to extremely long (Fig. 2.1).

Although many former classifications have mostly historical interest (cf. Tandberg-Hanssen 1995) they have all served to identify and understand the physical conditions for their formation and development over a large range of spatial and temporal scales. Only coronal cloud prominences and coronal rain are sufficiently different to be under a separate heading as discussed in Sect. 2.5.

In addition to spectacular differences in morphology and dynamics, solar prominences also show notable variations in spectroscopic characteristics.

The use of spectral classifications was initiated by Martin Waldmeier (cf. Waldmeier 1970) who compared observed intensities of Mg I lines (λ5184, λ5172 and λ5167 Å) and the Fe II line λ5186 Å and showed that these line ratios were

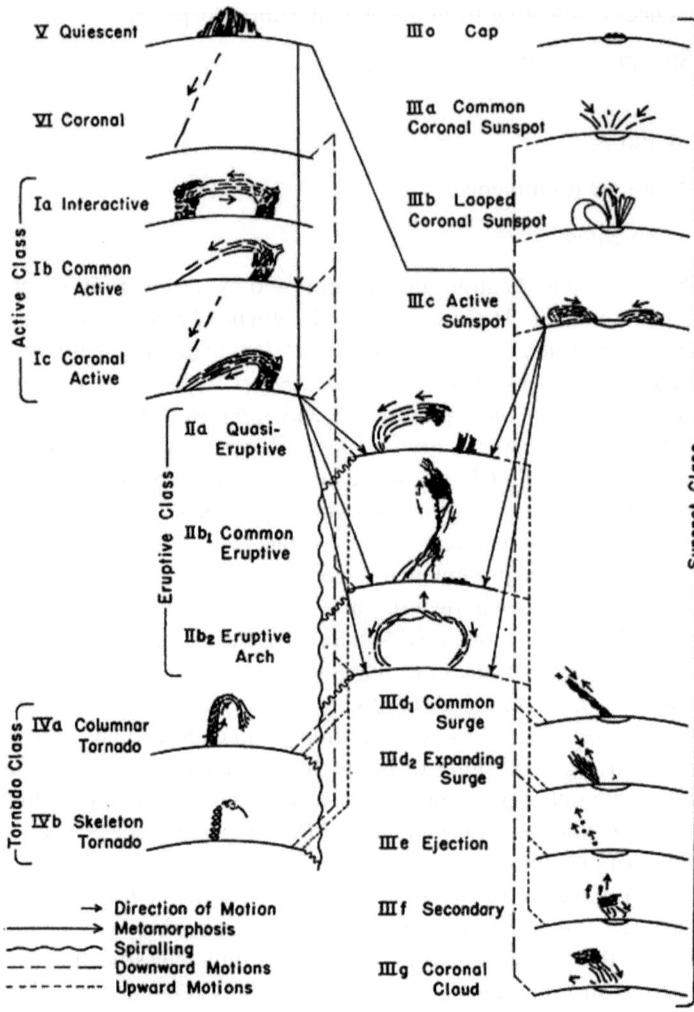

Fig. 2.2 Drawings of various types of solar prominences made from photographs. (*Credit*: Petit, 1932)

not the same in solar prominences and flares. A more detailed classification scheme was introduced by Zirin and Tandberg-Hanssen (1960) where they could distinguish quiescent prominences from active prominences and flares. Using a well-known variation with height in the chromosphere of relative intensity of neutral and ionized lines they could separate prominences into two major categories, i.e. quiescent prominences and active prominences (and flares). Their study showed that I (He II λ4686 Å) << I (He I λ4713 Å) both in the low chromosphere and in quiescent prominences, whereas these two lines are equally bright in the high chromosphere and in active prominences.

Fig. 2.3 Examples of various types of solar prominences. *Left image*: "hedgerow prominence". Middle image: "suspended cloud", which is also the same class of prominence that Pettit called coronal cloud (III g in Fig. 2.1). *Right image*: "tree prominence" (*Credit*: Richard B. Dunn)

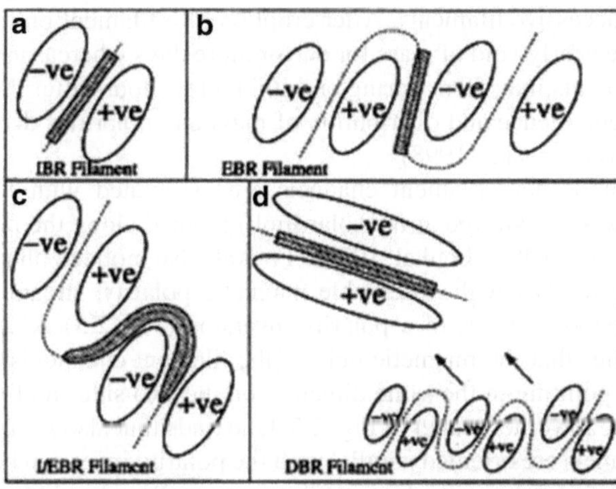

Fig. 2.4 Classification scheme for solar filaments developed by Tang (1987) and Mackay et al. (2008). (**a**) Filaments that form above the internal PIL (*Polarity Inversion Line*) of single bipoles are classified as IBR. (**b**) Those forming on the external PIL between bipoles or between bipoles and unipolar regions of flux are classified as EBR. (**c**) Filaments that lie both above the internal PIL within a bipole and the external PIL outside the bipoles are classified I/EBR. (**d**) Finally, those filaments that form in diffuse bipolar distributions resulting from flux emergence and the diffuse region can no longer be associated with any single bipole emergence are classified as DBR (*Credit*: Mackay et al. 2008)

Today's studies and analysis of the spectral emissions of solar prominences are subject to complex non-LTE radiative transfer modeling which will be covered in details in this book (Heinzel 2014; Labrosse 2014).

A new and different classification system was proposed initially by Tang (1987) and developed further by Mackay et al. (2008), with the aim to understand better where filaments form relative to the magnetic configuration in the photosphere below. The proposed four categories which are presented in Fig. 2.4 and valid for large, stable filaments, are discussed in detail by Mackay (2014).

2.3 Environments of Active Region, Intermediate and Quiescent Prominences

2.3.1 Filament Channels

Filament channels provide the magnetic environment in the low corona where filaments may form and be supported against gravity and thermally shielded from the surrounding hot corona. Channels follow along the division between opposite polarities in the line-of-sight magnetic fields measured in the photosphere, which has variously been referred to as the neutral line, *Polarity Inversion Line* (PIL) and *Polarity Reversal Boundary* (*PRB*). Filament channels tend to be long-lived and may spawn many successive filaments. After eruption of a filament on the quiet Sun, a channel may be nearly void of mass for one or more days whereas in active regions, successive filaments may form during or within a few hours after the eruption of a filament. The emergence and distribution of magnetic polarities determines where channels form (Gaizauskas 1998).

In the chromosphere filament channels are associated with fibril structures (called spicules when viewed at the solar limb) aligned along the polarity reversal boundaries (Smith 1968; Foukal 1971). Foukal also noticed that fibrils which are rooted in *plagettes* with observable magnetic polarity, stream in antiparallel directions on opposite sides of a polarity inversion (Fig. 2.5). The orientation of the fibrils implies that the magnetic field of the filament channel is predominantly horizontal and pointing in the same direction on the two sides a of the channel, as illustrated in the lower left panel of Fig. 2.5. One finds that also small coronal loops within the channels are oriented parallel with the polarity inversion boundary which implies that channel fields extend into the low corona (Wood and Martens 2003) and thereby that the most of filament axis is embedded in this horizontal field.

Similar systematic orientations of coronal cells are noticed in 1.2 MK data in the Fe XII λ193 Å line observed with the SDO/AIA instrument (Sheeley et al. 2013) (Fig. 2.5). These coronal cells have the approximate diameter of photospheric *supergranules* $\sim$30,000 km (Simon and Leighton 1964) but are centered over network fields at the vertices of *supergranules* rather directly over supergranules.

Martin et al. (1992) introduced the concept of chirality of filament channels. The channels were classified as either dextral or *sinistral* depending on the axial field direction observed from the positive polarity side of the channel as illustrated in the right panel of Fig. 2.6. The two columns in the left panel of Fig. 2.6 illustrate the one-to-one chirality relationships for fibril pattern (upper frames), filament spines and barbs (middle frames) and the overlying coronal loops (bottom frames).

A next major discovery was reported by Martin et al. (1994) who found a strong tendency of hemispheric dependence in location of the two chiral systems in the sense that a majority of dextral channels were observed in the northern hemisphere while the southern hemisphere harbored mainly sinistral channels. This systematic difference in the orientation of the magnetic fields of filament channels in the two hemispheres holds fundamental information on the channel formation and on the

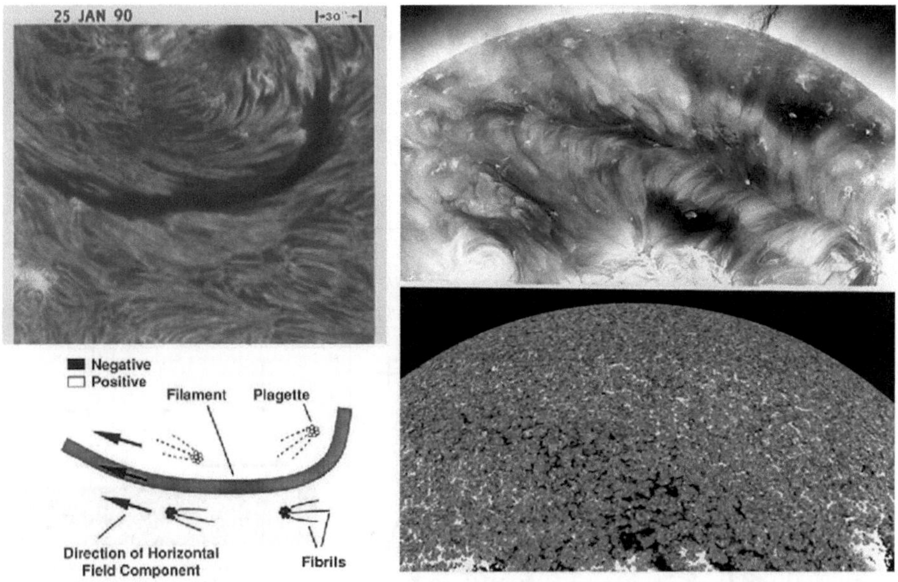

Fig. 2.5 *Left*: Illustration of opposite orientation of fibrils on the two sides of an AR Hα filament (*Credit*: Martin et al. 1992). *Right*: SDO images of a northern-hemisphere filament channel on April 23, 2012, showing cellular plumes leaning in opposite directions on the two sides a of the channel (*Credit*: Sheeley et al. 2013)

origin of solar filaments (Martin 1998a). These issues are discussed in detail by Martin (2014) and Mackay (2014).

In addition to the coronal arcades above filament channels are coronal fields seen in white-light that extend more or less radially outwards and form so-called *helmet streamers* which necessarily consist of oppositely directed magnetic fields. The magnetic arcades reach from 50,000 to 70,000 km into the solar corona while streamers extend to a solar radius or more as seen in images taken during solar eclipses.

2.3.2 Coronal Cavities

Early eclipse and coronagraphic observations showed that the filament channels contained regions of notably reduced emission between prominences and their surrounding coronal loop systems, which were subsequently referred to as coronal cavities. Some authors chose to call them prominence cavities. The total eclipse picture in Fig. 2.7 shows a bright coronal helmet streamer extending from the northeast solar limb and low-density cavity at the helmet base.

The global structure of coronal cavities is evidently shaped like tunnels, which implies that in order to measure their true brightness they must be oriented more

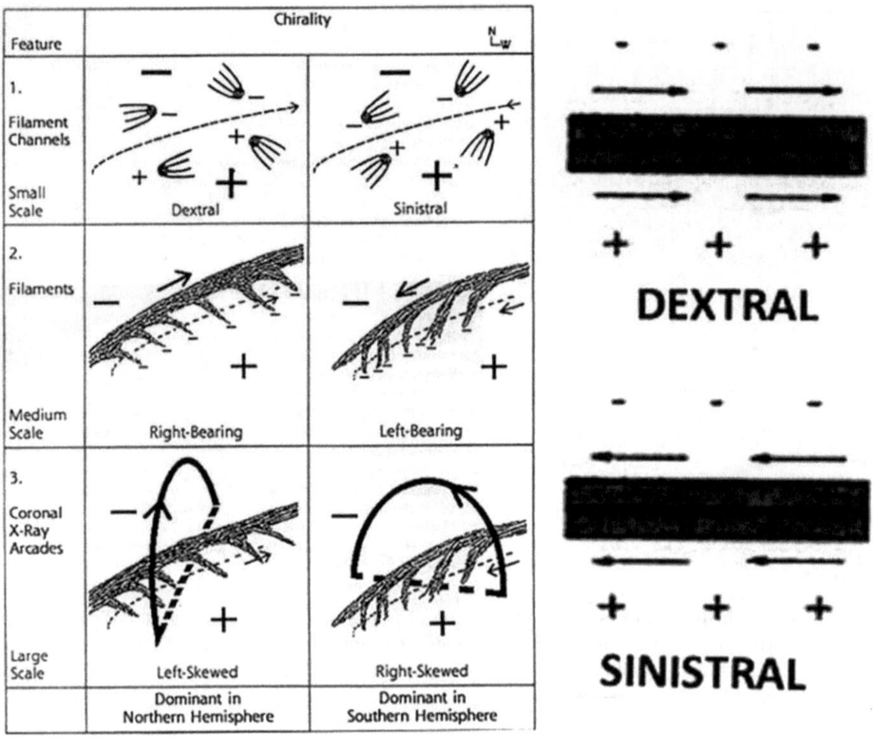

Fig. 2.6 The *left panel* shows schematic representations of the general magnetic fields of filament channels (*Credit*: Martin 1998b). The direction of the magnetic field in relation to the chirality of a dextral and a sinistral filament channel is emphasized in the *right panel* (*Credit*: Mackay et al. 2010)

or less along the line-of-sight. Therefore, largely East–west oriented high latitude channels provide most favorable conditions for brightness measurements. Emission of EUV lines formed at coronal temperatures proves that the cavities are not truly empty. Dudik et al. (2012) observed notable emission in the Fe XII λ193 Å line channel of SDO/AIA in a coronal cavity which indicates gas temperatures around 1.6 MK. Gibson (2014) shows evidence for a more multithermal situation. The reduced brightness of the cavities agrees with a plasma density that is about 30 % less than in the surrounding coronal regions (Fuller et al. 2008).

Spectral observations from the EIS instrument show the presence of large-scale flows with line-of-sight velocities $\sim$8 km s^{-1} in coronal cavities (Schmit et al. 2009). In addition, a noticeable swirling behavior of the flows is consistent with the view that cavities are filled and controlled by helically shaped magnetic flux ropes (Okamoto et al. 2010; Habbal et al. 2010; Liu et al. 2012a; Kucera et al. 2012). Three-dimensional models of coronal prominence cavity morphology are developed and discussed by Gibson et al. (2010, 2014).

Fig. 2.7 *White*-light total eclipse of 1988 March 18 showing a well-formed, bright coronal helmet streamer extending from the northeast limb, with a localized, low-density cavity at the helmet base. Within that cavity, a spatially unresolved, quiescent prominence appears as a bright blot (*Credit*: HAO/UCAR)

Recent studies by Berger et al. (2012) indicate that cavities may play a signification role in prominence formation and development. They show a pre-existing prominence disappearing slowly as a bright emission cloud forms in the regions immediately above. A subsequent prominence reformation follows a steady loss of mass by downward streaming from the cloud.

2.4 Structure and Dynamics of Active Region, Intermediate and Quiescent Solar Prominences

Typical solar prominences and filaments are composed of a *spine*, *barbs* and two extreme *ends*. The spine defines the upper main body that is oriented largely in the channel direction. The *barbs* diverge from the spines, much like exit roads of a highway, and bend down into the chromosphere and photosphere below. The ends of filaments also bend down towards the photosphere similar to regular barbs. Spines and barbs are common to both quiescent and active region prominences but the spines are much higher for quiescent prominences and the barbs are therefore also higher and can extend outward further from the spine. Barbs are not a ubiquitous feature of prominences as there are smaller short-lived active region features with no barbs while long-lasting quiescent prominences have very large barbs (Martin et al. 2008).

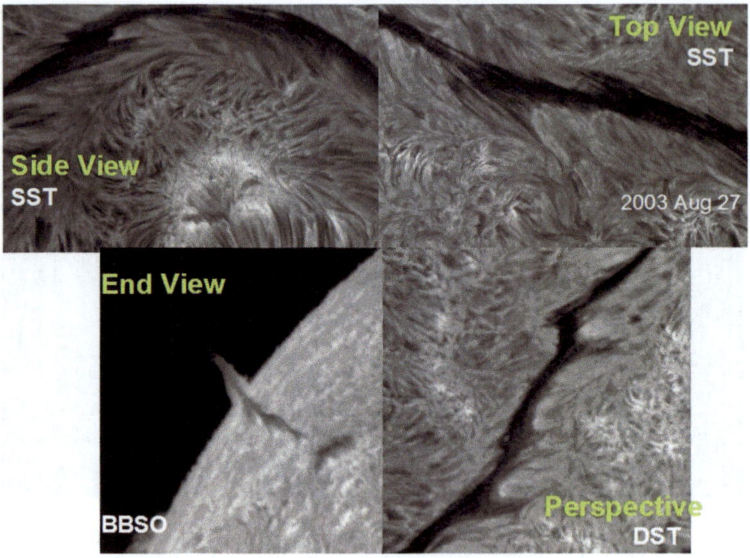

Fig. 2.8 Hα filtergrams of major sections of four intermediate filaments with a continuous spine and barbs viewed from various perspectives based on observations from the Swedish Solar Telescope (SST), the Big Bear Solar Observatory (BBSO) and the Dunn Solar Tower (DST). (*Credit*: Lin et al. 2008)

High-resolution Hα images demonstrate that spines and barbs are both composed of thin threads which constitute the fundamental structures of all solar filaments (Lin et al. 2008). Figure 2.8 contains examples of four similar prominences, intermediate between active region and quiescent filaments seen from four perspectives to provide a 3-D impression of the relative orientation of barbs to their associated spine. The upper left panel shows the side view of an active region filament. The upper right panel illustrates the top view of an intermediate filament with two independent barbs on either side of the spine. Many threads are stacked along the spine and the two barbs. The lower left panel shows an end view of a filament crossing the east limb. Here one clearly sees several barbs extending from both sides of the spine into the chromosphere. This view reveals the narrowness of the spine and shows it in absorption above the limb because it is optically thick due to many threads in the line-of-sight. In the lower right panel, a quiescent filament with several barbs is viewed partly from the side and partly from above.

Several studies have shown a notable correspondence between filament barbs and enhanced concentrations of magnetic flux located at *superganulation* cell boundaries in the photosphere below (Plocieniak and Rompolt 1973; Martin and Echols 1994; Lin et al. 2005b). The study of Martin and Echols (1994) suggested that the barbs tend to be rooted in or next to minority polarity magnetic fields on either side of the PIL.

Pevtsov and Neidig (2005) found that fragmented filaments represent the early evolution of quiescent filament development in Hα. These filaments began their formation with a few individual "clumps" which later grow and develop interconnecting *spines* which thereby form a continuous filament body. One may assume that these "clumps" represent the start of barb formation. The magnetic structure is evidently already developed; the higher prominence body is often rather faint in Hα at the early stage in formation. Several studies have shown that the higher regions of quiescent prominences are more pronounced in the hotter He II 304 Å line compared to Hα in absorption as well as in emission (Wang et al. 1998; Lin 2000; Xu et al. 2010). This difference is most probably due to an increase with height in ionization of Hydrogen. The highly resolved Hα image in Fig. 2.11 demonstrates also that the barb consists of a number of thin threads. One notes that threads connecting with the two neighboring threads within barbs appear to be rooted in separate but closely spaced locations in the chromosphere. At the assumed bottom part of this barb the volume density of the threads becomes so high that the individual threads cannot be resolved.

2.4.1 Active Region Prominences

Active region (AR) prominences are located adjacent to sunspots. The characteristics of AR prominences are their relatively thin and straight spines. Their barbs are in general very few and less pronounced. AR prominences are relatively short-lived and subject to eruptions or major "activation" events resulting in lifetimes from several minutes to a few hours (Berger 2013).

Being closely associated with sunspot groups AR prominences correlate well in numbers and activity with the solar cycle.

Chae et al. (2001) could follow the formation of an AR filament resulting from reorientations of the local magnetic configuration in the photosphere below due to converging and shearing flows. The typical smooth, blade-like structured AR filament is displayed in the right image of Fig. 2.9, and the left image shows an AR prominence with the typically horizontal thread-structures in the spine.

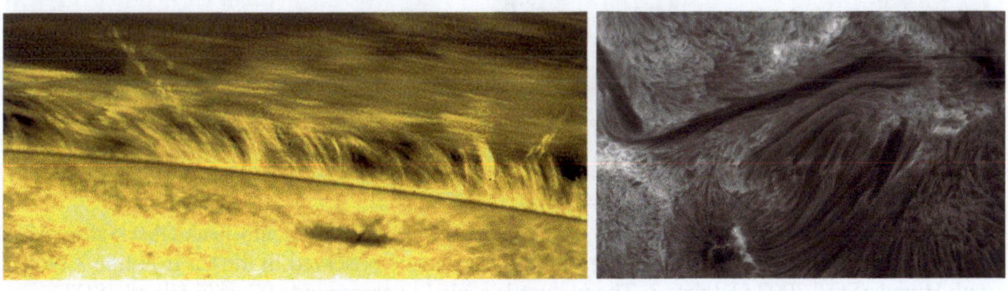

Fig. 2.9 *Right*: A slender AR filament seen in Hα obtained at the SST on 22 August 2003 is seen to have barbs extending a short distance to each side of the spine seen from above (*Credit*: The Swedish 1-m Solar Telescope). *Left*: Thin threads of an active region prominence in Ca II H line (λ3968 Å) bandpass observed with Hinode/SOT 2007 February 8 (*Credit*: Okamoto et al. 2007)

2.4.2 Quiescent Type Prominences

A commonly existing *quiescent prominence* is the hedgerow type consisting of long and tall blade-like palisades along the filament channels. The dimensions of a well-developed *quiescent prominence* are typically less than 5,000 km wide, 30,000 km high by 200,000 km long but both longer and shorter examples are readily found. Quiescent prominences are commonly located in high latitude regions ($\geq 50°$) in the polar crown filament channels which vary slightly in latitude and orientation throughout the solar sphere and over the solar cycle. The dominating structures of quiescent filaments are barbs with largely vertical threads. Some curtains of barb threads often end on arcs at the prominence base (cf. Martin et al. 2009), as shown in the left image of Fig. 2.10, which possibly also are related to so-called "*bright rims*" (Paletou 1997). The horizontally oriented spines in the higher regions of prominence body are generally rather faint in Hα but they appear a lot more pronounced in the hotter He II λ304 Å line (Wang et al. 1998; Lin 2000; Xu et al. 2010).

The persistent quiescent filaments occurring at high latitudes are commonly referred to as polar crown prominences. D'Azambuja and D'Azambuja (1948) concluded from comprehensive and careful investigations that quiescent prominences in their global appearance are exceedingly stable structures appearing at high latitudes and may last from weeks to several months. However, the continuous recent He II 304 Å observations from SOHO and SDO show that eruptions of segments of polar crown filaments are much more common than indicated from earlier and less frequent observations from ground-based observatories and quiescent prominences can develop at low latitudes as well as high latitudes. At low latitudes they are more likely to be destabilized as a result of being within 30 heliographic degrees of the site of a new active region (Feynman and Martin 1995).

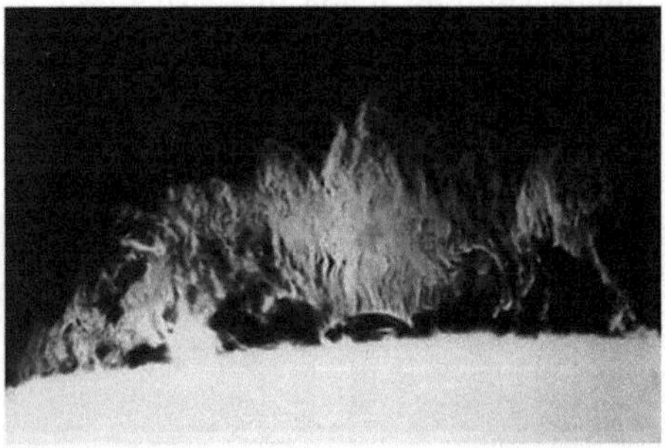

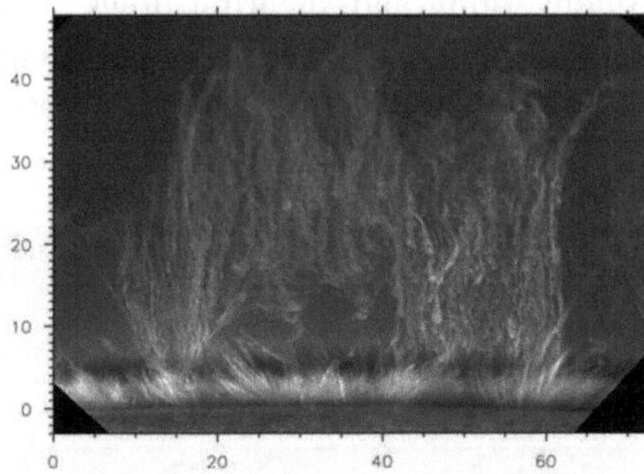

Fig. 2.10 *Left*: Quiescent prominences observed in Hα at the Sacramento Peak Observatory 1970 December 7 (*Credit*: NSO/NOAO). *Right*: Tall prominence observed in the Ca II H line λ3968 Å with Hinode/SOT 2007 October 3. The picture is scaled in arcseconds (*Credit*: Hinode/SOT)

2.4.3 *Intermediate (Combined) Type Prominences*

Intermediate filaments form between weak unipolar background fields regions and active regions or active region complexes and constitute a class in-between the other two. They also occur between and within decaying active regions. They have lengths of ~100,000 km along the extended filament channels and do not necessarily occupy the full length of a channel. One part of an intermediate filament may have the appearance of the quiescent type while another part may have the resemblance of an AR filament. The upper right image of Fig. 2.8 shows part of a long *Intermediate* type filament located at N22E18 and at some distance from an active region on 27 August 2003. This filament has the characteristic continuous, slender body of the AR type and well developed fine-structured barbs typical for quiescent filaments. From many such examples, representative of I/EBR class of Mackay et al. (2008), it is clear that the differences in AR, intermediate and quiescent filaments are ones of scale or degree of activity rather than fundamental difference in their nature and physics.

2.4.4 *Substructures*

2.4.4.1 Threads

The characteristic fine structures of solar prominences were clearly noticeable in the fine drawings by the early solar observers (cf. Vial 2014), but these became more fully appreciated after the famous high resolution observations of Dunn (1960) displayed in time-lapsed movies. Similar type of prominence movies were earlier made by Robert McMath at the McMath–Hulbert Solar observatory. Recent high spatial and temporal resolution observations from ground-based observatories and telescopes in space have confirmed that the entire bodies of solar prominences consist of complex, rapidly changing fine structures. Spines consist of bundles of largely horizontally oriented threads and blobs subjected to *counterstreaming* motions (Zirker et al. 1998; Lin et al. 2003; Ahn et al. 2010; Berger 2013). The same fine threads and blobs continue into the barbs, which diverge from the spine at intervals resembling the photospheric supergranular cell sizes, and bend down into the photosphere below. High-resolution data shows that the structural sizes, e.g. thickness, of the small-scale structures vary from truly several arc sec down to the resolution limit of the best instruments, e.g. $\geq$0.15 arc sec (~100 km), which implies that some structures may even be thinner.

The top view of two adjacent multi-thread filament barbs are displayed in Fig. 2.11. The Doppler image on the right demonstrates the presence of *counter-streaming*, both down- and up-flows in separate but adjacent, interleaved threads of plasma. The darkest regions in the intensity image correspond to the transition where the sharp and clear spine threads are curving steeply downward into the barbs

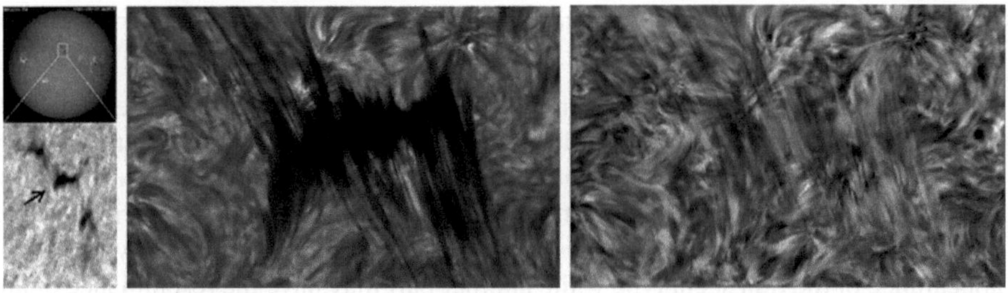

Fig. 2.11 The *middle panel* shows a high resolution Hα image of a barb of a fragmented hedgerow filament observed with the SST on August 22, 2004. The *arrow* in the *lower left image*, which was recorded with LESIA, Observatoire de Paris, indicates which one is the observed fragment. The high resolution image reveals a dark multi-thread, multi-footpoint barb on the *left* with sparse spine threads extending out of the image to the north while the dark barb threads on the *right* are associated with thin spine threads extending out of the image to the south. A few spine threads in the *middle* could be superposed against the dark barbs rather than being connected to them. The image to the right shows the corresponding Doppler image which is derived by subtracting the *red wing image* ($\Delta\lambda = +0.3$ Å) from the *blue wing* ($\Delta\lambda = -0.3$ Å) that make *blue-shifted* elements appear *bright* and *red-shifted dark* (*Credit*: Lin et al. 2007)

or vice-versa, the transition of the steep barb threads coming out in our line-of-sight into the horizontal spine. Therefore, the column density of the barb threads in the line-of-sight is higher than in the horizontal threads of the spine. At the left and right sides of the adjacent dextral barbs, one can see some locations where the barb threads connect to the chromosphere. However, in the bottom part of the barbs, many of the fine threads are too densely packed to be resolved even in these high quality SST images.

Berger (2013) points out that the threads appear somewhat thicker and more structured in prominences at the limb compared with the smooth threads seen against the disk (Lin et al. 2005a). Such differences between the on-disk filament threads and threads in off-limb prominences pose a challenge in interpretation and modeling of prominences. A possible solution to this problem could be that on-disk absorption in Hα is largely dependent on the population of the $n = 2$ energy level in the hydrogen atoms while the off-limb emission structures depend more on the $n = 3$ level population. The latter population is much more sensitive to variations in the thermodynamic parameters of the cool prominence plasma.

Besides the apparent internal flowing of plasma along the threads one observes sideways (swaying) motions of individual threads. Individual threads in barbs move sideways with speed 2–3 km s^{-1} (Lin et al. 2005a) which also compares well with the observed small-scale flow velocities of magnetic flux elements in the photosphere. Line-of-sight (LOS) Doppler motions at speeds of 5–10 km s^{-1} of individual prominence substructures were studied by Zirker and Koutchmy (1991).

The highly inclined oriented fine structure in barbs remains a key mystery in studies of prominence barbs. The appearance of smooth and elongated fine structures in combination with flow velocities up to 15 km s^{-1} and higher would suggest that the flows are field-aligned and the orientation of the threads reflects the orientation of the magnetic fields, which includes the highly inclined ones as well. One does not observe free fall speeds of cool prominence plasma in highly inclined barbs. The vertical extent of barbs is much longer than the gravitational scale height of prominence plasma ($\sim$200 km) which implies that the plasma must somehow be supported against gravity (Mackay et al. 2010). Steele and Priest (1992) and Aulanier and Démoulin (1998) modeled the thread structures as a series of sharply dipped magnetic field lines under magnetostatic conditions. However, static magnetic topologies seem incompatible with the morphological character of the thin threads as well as the observed flowing and counterstreaming of the plasma.

The ubiquitous presence of oscillations in solar prominences, off-limb as well as on-disk, led Pécseli and Engvold (2000) to study the possibility that damping of MHD waves might serve to accelerate the partly ionized cool plasma and thereby counteract and/or balance gravity. The presence of a necessary high frequency waves for this mechanism to work is still beyond the current limit of detection in solar observations.

2.4.4.2 Filling Factor

As discussed and shown above and elsewhere in this chapter (Figs. 2.9, 2.10 and 2.11) solar prominences are made up of numerous thin threads and small-scale droplets. The angular widths of the thinnest threads and other small-scale structures are comparable to the resolution limit of the best instruments today, i.e. $\sim$0.15 arcsec, which suggests that some threads may be even thinner. Thermodynamic modeling based on observed emission of prominences depends on the proper knowledge of the true volumes of the radiating plasma. Zirker and Koutchmy (1990) assumed a clustering of moving, unresolved, uniform, threads that reproduced the observed structures and concluded that the observed fine structures might consist of up to 20 single thinner threads along the line-of-sight. It is generally believed that the thinnest volumes of the cool plasma have thread-like shapes whereas the presumed thin *Prominence Corona Transition Region* (PCTR) must inevitably be more tube-like.

The effective radiating volume is referred to as the *filling factor* which thereby becomes a central parameter in interpretation and modeling of observed line emission from both the cool core and the PCTR of prominences. Mariska et al. (1979) and Widing et al. (1986) derived volume filling factors in the range 0.018–0.024. Cirigliano et al. (2004) concluded from observations of the PCTR with the SUMER instrument on SOHO that the *filling factor* may be as low as 10^{-3}.

The filling factor has remained an issue of concern in prominence modeling, which is discussed by Parenti (2014b) and Labrosse (2014).

2.4.4.3 Minifilaments

In the era of moderate spatial resolution (~1 arcsec) and temporal resolution (~1 min) observers took note of a small-scale analogue to large-scale filaments which are referred to as miniature filaments or commonly shortened to *minifilaments* (Hermans and Martin 1986). From a detailed study of time-lapse datasets in Hα obtained at Big Bear Solar Observatory Wang et al. (2000) concluded that a typical *minifilament* of projected length around 20,000 km has a lifetime of 50 min from first appearance through disappearance and eruption. Similar to large-scale filaments, also *minifilaments* reside above local PILs. *Minifilaments* have a variety of characteristics in common with AR filaments and quiescent filaments and may serve as a proxy in studies of more complex systems (Denker and Tritschler 2009).

2.4.4.4 Pillars and "Tornadoes"

Tornado-like prominences resembling terrestrial tornadoes in shape when seen on the solar limb were noticed by several observers (cf. Panasenco et al. 2014). Pettit (1932) described these structures as "Vertical spirals or tightly twisted ropes" and introduced tornado-like prominences as a separate class.

A group of tornado-like prominences structures at the solar limb shown in the left panel of Fig. 2.12 were observed by the space instrument TRACE (cf. Vial 2014) 1999 November 27 (Panasenco et al. 2014). These pillar-looking structures which appear to fan out in the tree-shaped structures are typical for barbs of large quiescent prominences (Pevtsov and Neidig 2005; Lin et al. 2008) that are also classed as hedgerow prominences by Menzel and Evans (1953).

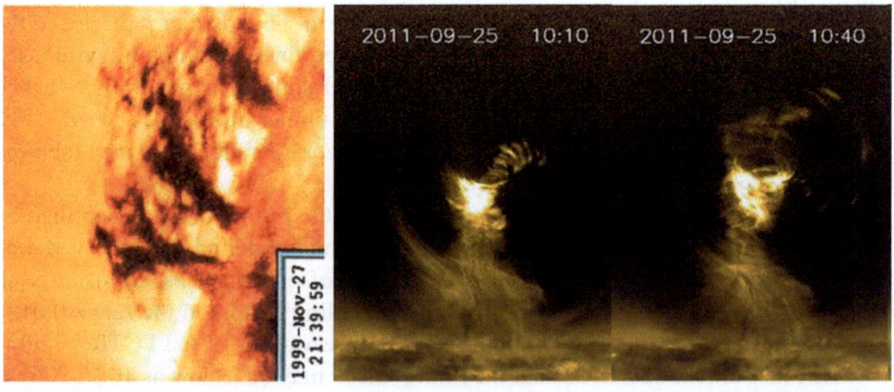

Fig. 2.12 *Left image*: A group of tornado-like prominences observed by TRACE in the λ171 Å line 1999 November 27 (*Credit*: Panasenco et al. 2014). The *right images* of a huge tornado-like feature were captured by the Solar Dynamic Observatory which show a spectacular formation of a dynamic event in the coronal cavity above a solar prominence (*Credit*: NASA/Li et al. 2012)

The development of a large tornado-like event was recorded with the SDO/AIA during 2011 September 24 through 26 (Li et al. 2012). Two examples of this feature recorded in the 171 Å line channel are also displayed in Fig. 2.12. The fascinating, long time coverage of this event showed its formation as a result of upward material flow from below which penetrated into the cavity above the prominence. The same event was studied by Panesar et al. (2013) who found that flare activity in a neighboring active region had an apparent causal relationship with this tornado-like event.

Su et al. (2012) and Wedemeyer-Böhm et al. (2012) concluded that tornado-like barbs are rooted in vortexes located at intersections of supergranulation cells where rotating magnetic structures could develop. In the following study Wedemeyer et al. (2013) concluded from combined 171 Å data of SDO/AIA and Hα observations with the SST that the legs (barbs) of prominences in pre-eruption phase appear associated with rotating tornados. The sideways oscillating appearance in 2-D of a such event does not necessarily prove the presence of spiraling motion which should be expected in the case of plasma motion in a tornado-like helical magnetic structure. Further clarification of this issue is foreseen. Panasenco et al. (2014) find that the apparent tornado-like structure and motion in hedgerow quiescent prominences may be fully explained as a combination of counterstreaming and oscillation.

Tornado-like features discussed by Li et al. (2012) and Panesar et al. (2013), and some of the tornados described by Pettit (1932), are transient and rapidly changing in overall structure compared with the apparently more stable pillars recorded with TRACE and presented above in Fig. 2.12 (Panasenco et al. 2014). Such differences in character, degree of activity and associated events may be indicators of different physical processes among the variety of features that have been called tornado prominences and point to the need for Doppler images from spectral data for more definitive interpretations.

2.4.5 Dynamics

2.4.5.1 Flows

High-resolution time series reveal ubiquitous flowing of the cool plasma along the thread directions. Zirker et al. (1998) detected a steady bidirectional streaming with typical speeds of 10–20 km s^{-1} everywhere along closely spaced threads in a large filament. The pattern was observed in both wings of Hα which confirmed that the flows are mass motions and not caused by some kind of excitation wave. This flow pattern, which is being referred to as *counterstreaming*, was confirmed in a later study by Lin et al. (2003). The same flow pattern is seen both in *spines* and in *barbs*. Engvold et al. (1985) detected systematic flows in the PCTR. Time series of Ca II H images from the filter pass band of Hinode/SOT confirm the presence of flows along spines and up and down in barbs (Ahn et al. 2010). Recent studies by

Alexander et al. (2013) using simultaneous observations of an active region filament in the λ193 Å line with the ultra-high spatial resolution (0.2 arcsec) and temporal resolution of the Hi–C (Cirtain et al. 2013) and the SDO/AIA instrument (He II λ304 Å and continuum $\lambda \sim 1{,}600$ Å) find anti-parallel flows in threads ($\sim$0.8 arcsec thick) at velocities as high as 70–80 km s^{-1}, which is notably higher than reported for the cool prominence plasma (see Fig. 4.6 in Kucera 2014).

It is generally accepted that mass flows at the speeds quoted above in assumed low-β plasma must inevitably be field-aligned and that the flow pattern thereby reflects the structure and orientation of the local magnetic fields.

A consequence of continuous streaming of the plasma through the entire prominence body is that in order to maintaining the mass through the observed lifetimes implies an approximate global balance between loss and inflow of plasma. Assuming typical lengths of quiescent prominences between 30,000 and 100,000 km and flow speed of $\sim$10 km s^{-1}, the entire mass of a prominence will be exchanged in the course of 1–3 h. Understanding the apparent ever-present flowing and its consequences remains a central issue in prominence studies.

Haerendel and Berger (2011) observed isolated knots or droplets of plasma from quiescent prominences to fall at near free fall speed at about 100 km s^{-1}. Similar features were studied and discussed by Hillier et al. (2012). Also most erupting prominences have similar rapidly streaming down flows of mass concurrent with the outward bodily transport of all or part of the prominence.

2.4.5.2 Oscillations

The oscillating nature of solar filaments was first noticed as bodily "winking filaments" with velocity amplitudes of 20 km s^{-1} and higher, shaken by flare generated waves (Ramsey and Smith 1966). Information on smaller-amplitude oscillations is usually derived from Doppler velocity data, in addition to high-resolution time series which also permit measurements of transverse (side-ways) swaying motion of filament threads (Lin et al. 2009). High resolution time series all show an ever-present oscillatory pattern in solar filaments.

Studies of small-amplitude periodic variations in line-of-sight motions in prominences, with the aim to understand the magnetic structures and interaction with the plasma, revealed the presence of a wide range of oscillatory periods (Molowny-Horas et al. 1999; Banerjee et al. 2007; Engvold 2008). Periods (P) $<$ 10 min are referred to as short, while intermediate and long periods are, respectively, 10 min $<$ P $<$ 40 min and P $>$ 40 min. Small-amplitude oscillations, Δv $=$ 0.1–3 km s^{-1}, are detected at all periods, whereas large-amplitudes (20–40 km s^{-1}) are commonly observed at long periods. Small-amplitude oscillations are generally associated with individual threads, but they appear in addition to partly involve the entire filament body (Lin et al. 2003).

Lin et al. (2003) found evidence for traveling waves in the thread structure to move in the same direction as the mass flows. The presence of continuously generated, propagating groups of waves perpendicular to prominence magnetic

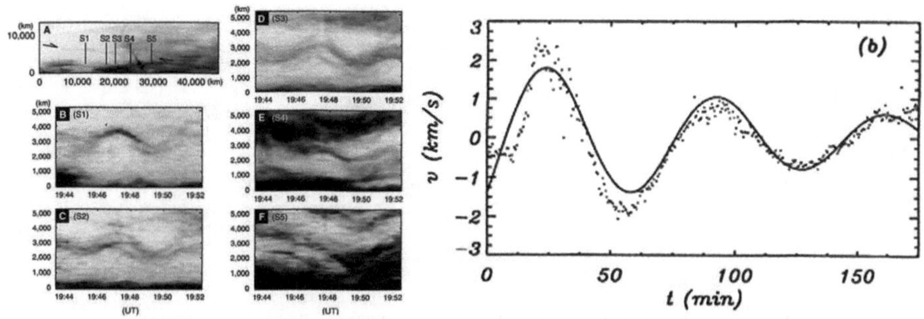

Fig. 2.13 *Left panel*: Examples of prominence threads undergoing synchronous oscillations along the spine of an AR prominence observed with Hinode/SOT 2007 February 8. Lines S1 to S5 indicate the locations of height versus time plots in the panels B to F (*Credit*: Okamoto et al. 2007). Right diagram: Damped long period oscillation in a quiescent prominence derived from Doppler velocity (*dots*) and fitted function (*continuous line*) versus time. The period is 70 min and the damping time is 101 min (*Credit*: Molowny-Horas et al. 1999)

field is confirmed also in a study by Schmieder et al. (2013). From Ca II H line (λ3968 Å) band pass movies of an active region prominence (Fig. 2.13) Okamoto et al. (2004) examined six threads and detected vertical oscillatory motions with amplitudes in the plane of the sky ranging from 400 to 1,800 km. The horizontally oriented threads appeared to contain continuous horizontal flows at speeds in the range 15–46 km s^{-1}. These authors propose that the observed oscillations might represent propagating Alfvén waves along the horizontally oriented magnetic fields of this prominence. Chen et al. (2009) found evidence from EUV data that up-flows connected with *counterstreaming* are associated with stronger magnetic fluxes, e.g. brighter *plage* areas, while down-flows seem connected to a weaker flux.

The oscillatory amplitudes in solar filaments and prominences decrease with time and die out in the course of a few periods (right panel Fig. 2.14). This phenomenon is referred to as wave damping. The damping times range usually from one to three times the corresponding period (Oliver and Ballester 2002). The loss in wave energy indicated by wave damping in filaments might possibly be involved in accelerating the ever-present flowing of the partly ionized plasma. Alternative damping mechanisms are discussed in the review by Soler et al. (2014).

The main aim of prominence seismology is to infer and understand the internal structure and physical properties of solar prominences. Further details are given in reviews by Ballester (2006, 2014) and Lin (2011).

2.4.5.3 Prominence Plumes

Visible-light spectral observations of quiescent prominences exhibit plume-like features rising through them from the chromosphere/photosphere with the shape of "mushroom caps" at velocities in the range 20–30 km s^{-1}. They may start as a single

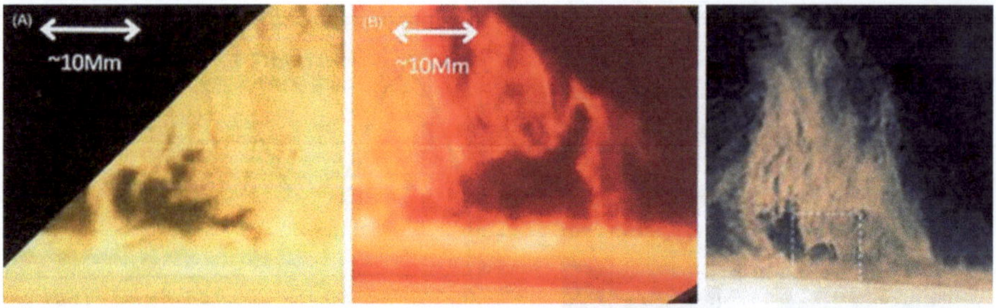

Fig. 2.14 *Left image*: Observations in Ca II H line of dark up-flows in a quiescent prominence on 2007 August 8. Middle image: A similar feature observed in Hα 2010 June 22. (*Credit*: Hiller et al. 2012). *Right image*: Quiescent prominence observed in 2007 in Hα at Mauna Loa Solar Observatory 2007 April 25, at 90W 36S heliographic coordinates. The white dashed box highlights an area of up-flow development in the prominence (*Credit*: T. Berger et al. 2010)

~10,000 km large plume, or bubble, which occasionally breaks up into smaller ones. In SDO/AIA λ193 Å images the plumes appear slightly brighter than in the prominence itself but notably less bright than the corona outside the prominences (Dudik et al. 2012).

Plumes in quiescent prominences were first reported by Stellmacher and Wiehr (1973) and later studied in detail by Berger et al. (2008, 2010, 2011) in observations from the Hinode satellite. It is generally thought that prominence plumes represent under-dense plasma relative to the ordinary prominence plasma and give rise to Rayleigh–Taylor buoyancy instability (Ryutova et al. 2010). The assumed magneto-convective (plasma-$\beta \approx 1$) plume features have been observed to rise into the overlying coronal cavities but their influence on cavity evolution is yet unclear (Berger et al. 2011).

Plumes are not yet identified in *Intermediate* and *AR* type prominences (Berger 2013).

2.4.5.4 The Eruptive Phase of Solar Prominences

The early phase of prominence eruptions is noticed as a *slow rise* at speeds of about 0.1–1 km s^{-1} several hours before its actual eruption (Sterling and Moore 2004; Isobe et al. 2007), after which it undergoes rapid upward acceleration to velocities ranging from 100 to 1,000 km s^{-1}. Erupting prominences leave behind concurrently formed flare loops and post-flare loops that straddle the vacated filament channel. They are accompanied by the expulsion of overlying and surrounding coronal loop systems that develop into *Coronal Mass Ejections* (*CMEs*). In the final stage the CME structure expands at nearly constant speed (Liu et al. 2009). The leading front continues its fast outward motion while fractions of the core material are occasionally seen to collapse back towards the Sun (Wang and Sheeley 2002). In two thirds of observed cases prominences reform in the same prominence or filament channel, with a similar shape in the course of 1–7 days.

Occasionally, and very likely depending on the magnetic environment, only parts of a prominences erupt as one part stays anchored in the channel while the rest undergoes eruption (Liu et al. 2009). In other eruptive events only the higher filament body may take part in the pre-eruptive *slow rise* and subsequent eruption (Liu et al. 2012c).

A number of structural and dynamic changes, in addition to the slow rise, signal the beginning of an erupting event. Some polar crown filaments exhibit large amplitude oscillations during the pre-eruption slow-rise phase (Isobe and Tripathi 2006). Spectral changes are often observed both in emission and absorption in the early phase of an eruption. Both active region and quiescent prominences exhibit enhanced non-thermal motions and become darker (when viewed on the disk as filaments) and brighter (when viewed at the limb). Observations of highly ionized EUV lines show increased emission in conjunction with prominence eruption (Engvold et al. 2001) (Fig. 2.15).

Much attention has been given to understanding the triggering of eruptive prominences (van Driel-Gesztelyi and Culhane 2009; Parenti 2014a). Some of the common mechanisms proposed as triggers for solar events are interactions with emerging magnetic flux (Bruzek 1952; Feynman and Martin 1995; Wang and Sheeley 1999), interactions of prominence fields with overlying coronal fields (Antiochos et al. 1999), interactions of the root fields of prominences with adjacent magnetic fields (Nagashima et al. 2007), and in relatively rare cases, being hit by flare waves (Okamoto et al. 2004; Isobe et al. 2007) and long-term effects associated with observed cancelling magnetic fields (Martin et al. 1985, 2012). The review by Aulanier (2014) provides a detailed evaluation of various proposed mechanisms for prominence eruptions.

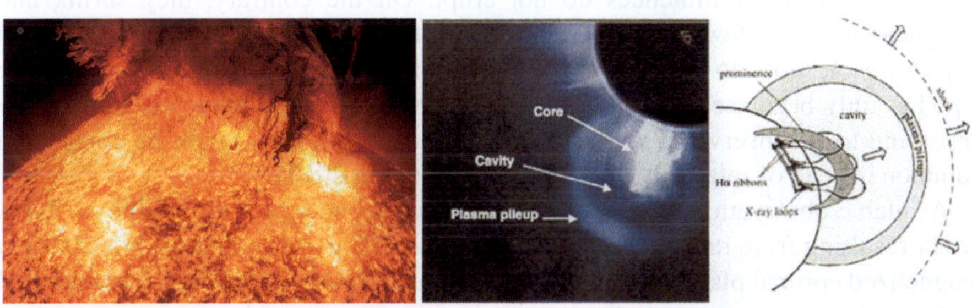

Fig. 2.15 *Left image*: A huge prominence eruption observed with SDO on June 7, 2011 (*Credit*: NASA/SDO/J. Major). *Middle image*: A Solar Maximum Mission archive image showing the principal features of an erupting prominence and a CME (Hundhausen 1999). *Right panel*: Schematic view of a CME (Credit: Forbes 2000)

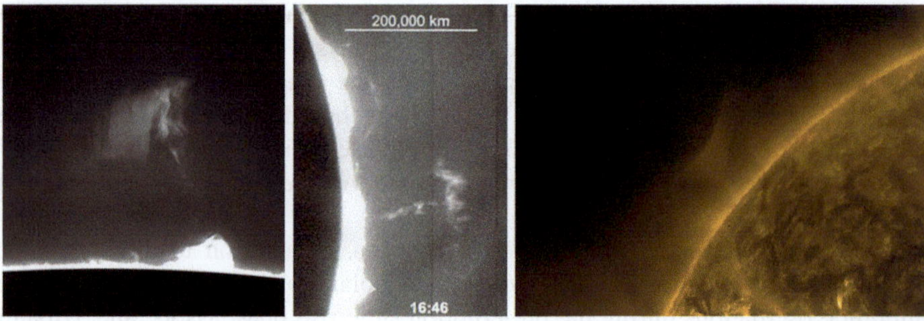

Fig. 2.16 Hα images Coronal Cloud Prominences; the *left image* of 26 August 2013 (*Credit*: James Ferreira) and the *middle image* was obtained at Helio Research on September 17, 2004 (*Credit*: Helio Research). The *right image* shows a funnel prominences of 28 April 2012 obtained in 171Å with SDO/AIA (*Credit*: NASA/SDO)

2.5 Coronal Cloud Prominences and Coronal Rain

Coronal cloud prominences are cool material suspended up to 200,000 km in the corona. This is rather high in comparison to stable quiescent prominences which rarely exceed 35–50,000 km during their non-erupting state. Allen et al. (1998) studied the structure and kinematics of a number of such prominences and referred to them as "*coronal spiders*" due to their characteristic shape. Coronal cloud prominences have also been termed "*funnel prominences*" because of a characteristic V-shaped structure (Liu et al. 2012b) which might be due to one particular view angle of an asymmetric structure. In some cases the V-shape is preceded by or followed by an expansion of the cloud feature into a spider-like shape.

Coronal cloud prominences do not erupt. On the contrary, they shrink and disappear within a few hours to a day from drainage along well-defined curved trajectories at close to free-fall speeds resembling coronal rain. This prominence type has only been seen above the limb to date and they are evidently too weakly absorbing to be observable against the disk. Otherwise, they might be considered common instead of relatively uncommon (see Martin 2014).

Available observations suggest that cloud prominences become visible in 304 Å or Hα resulting from radiative cooling instabilities (Karpen and Antiochos 2008) in magnetized coronal plasma when thermal conduction becomes effectively inhibited by local changes in magnetic field configuration. The formation process is not yet fully understood and is a subject for further investigation (Fig. 2.16).

The physical nature of Coronal Cloud Prominences appears to differ in several respects from more common, regular channel associated type prominences. Also, there is so far no evidence for them to be associated with PILs.

"*Coronal rain*" is observed to come from coronal cloud prominences and in addition seen to condense directly out of the thin, hot corona. This phenomenon was first observed in Hα as cascades of small, bright packets of matter streaming down along trajectories that closely outline the orientation of the pervasive magnetic fields. From recent observations of fine structured loops Antolin and Rouppe van der Voort (2012) conclude that *coronal rain* is a common phenomenon seen in the low temperature lines Hα and Ca II H, with an average falling speed around 70 km s^{-1} and an acceleration notably below free fall. Schrijver (2001) followed the various shapes of coronal rain formation at speed up to 100 km s^{-1} in coronal loops from pass bands of the TRACE instruments, from a few million degrees down to less than 100,000 K. Coronal rain appears closely associated to solar flares which suggest that the triggering mechanisms of these two phenomena are connected. The two seemingly different sources of coronal rain may result from variations in magnetic topology in the coronal regions at stake, which either may support the formation of a coronal cloud prominence that subsequently are drained via the rain, or the condensing matter is drained as quickly as the apparent condensation takes place.

Acknowledgments The author is grateful for helpful discussions with Jean-Claude Vial and suggestions from Sara F. Martin and Jack B. Zirker in preparation of this chapter.

References

Ahn, K., Chae, J., Cao, W., & Goode, P. R. (2010). Patterns of flows in an intermediate prominence observed by Hinode. *The Astrophysical Journal, 721*, 74–79.

Alexander, C. E., Walsh, R. W., Régnier, S., Cirtain, J., et al. (2013). Anti-parallel EUV flows observed along active region filament threads with Hi-C. *The Astrophysical Journal, 775*, L32–L38.

Allen, U. A., Bagenal, F., & Hundhausen, A. J. (1998). Analysis of Hα observations of high altitude coronal condensations, new perspectives on solar prominences. In D. F. Webb, B. Schmieder, & D. M. Rust (Eds.), *ASP conference series, IAU colloquium 167* (Vol. 150, p. 290).

Antiochos, S. K., DeVore, C. R., & Klimchuk, J. A. (1999). A model for solar coronal mass ejections. *The Astrophysical Journal, 510*, 485–493.

Antolin, P., & Rouppe van der Voort, L. (2012). Observing the fine structure of loops through high-resolution spectroscopic observations of coronal rain with the CRISP instrument at the Swedish solar telescope. *The Astrophysical Journal, 745*, 152–173.

Anzer, U., Heinzel, P., & Fárnik, F. (2007). Prominences on the limb: Diagnostics with UV EUV lines and the soft X-ray continuum. *Solar Physics, 242*, 43–52.

Aulanier, G. (2014). The physical mechanisms that initiate and drive solar eruptions. In *IAU symposium* (Vol. 300, pp. 184–196).

Aulanier, G., & Demoulin, P. (1998). 3-D magnetic configurations supporting prominences. I. The natural presence of lateral feet. *Astronomy and Astrophysics, 329*, 1125–1137.

Ballester, J. L. (2006). Seismology of prominence-fine structures: Observations and theory. *Space Science Reviews, 122*, 129–135.

Ballester, J. L. (2014). Magnetism and solar prominences: MHD waves. In J.-C. Vial, & O. Engvold (Eds.), *Solar Prominences, ASSL* (Vol. 415, pp. 257–294). Springer.

Banerjee, D., Erdélyi, R., Oliver, R., & O'Shea, E. (2007). Present and future observing trends in atmospheric magnetoseismology. *Solar Physics, 246*, 3–29.

Berger, T. (2013). Solar prominence fine structure and dynamics. Nature of prominences and their role in space weather. In *Proceedings of the IAU symposium* (Vol. 300, pp. 15–29).

Berger, T. E., Shine, R. A., Slater, G. L., et al. (2008). Hinode SOT observations of solar quiescent prominence dynamics. *The Astrophysical Journal, 676*, L89–L92.

Berger, T. E., Slater, G., Hurlburt, N., et al. (2010). Quiescent prominence dynamics observed with the Hinode solar optical telescope. I. Turbulent upflow plumes. *The Astrophysical Journal, 716*, 1288–1307.

Berger, T., Testa, P., Hillier, A., et al. (2011). Magneto-thermal convection in solar prominences. *Nature, 472*, 197–200.

Berger, T. E., Liu, W., & Low, B. C. (2012). SDO/AIA detection of solar prominence formation within a coronal cavity. *The Astrophysical Journal, 758*, L37.

Bommier, V., Landi Degl'Innocenti, E., Leroy, J.-L., & Sahal-Brechot, S. (1994). Complete determination of the magnetic field vector and of the electron density in 14 prominences from linear polarization measurements in the HeI D3 and Hα lines. *Solar Physics, 154*, 231–260.

Bruzek, A. (1952). Die Ausbreitung von "Euptionsstörungen". Mit 2 Textabbildungen. *Zeitschrift für Astrophysik, 31*, 111.

Chae, J., Wang, H., Qiu, J., Goode, P. R., Strous, L., & Yun, H. S. (2001). The formation of a prominence in active region NOAA 8668. I. SOHO/MDI observations of magnetic field evolution. *The Astrophysical Journal, 560*, 476–489.

Chen, H., Jiang, Y., & Ma, S. (2009). An EUV jet and Hα filament eruption associated with flux cancelation in a decaying active region. *Solar Physics, 255*, 79–90.

Cirigliano, D., Vial, J.-C., & Rovira, M. (2004). Prominence corona transition region plasma diagnostics from SOHO observations. *Solar Physics, 223*, 321–351.

Cirtain, J. W., Golub, L., Winebarger, A. R., et al. (2013). Energy release in the solar corona from spatially resolved magnetic braids. *Nature, 493*, 501–503.

D'Azambuja, L., & D'Azambuja, M. (1948). *A comprehensive study of solar prominences and their evolution from spectroheliograms obtained at the observatory and from synoptic maps of the chromosphere published at the institution* (Vol. 6, part 7). Ann. Obs. Paris-Meudon.

de Jager, C. (1959). Structure and dynamics of the solar atmosphere. *Handbuch der Physik, 52*, 80.

Denker, C., & Tritschler, A. (2009). Mini-filaments – small-scale analogues of solar eruptive events*?* In *IAU symposium* (Vol. 259, pp. 223–224).

Deslandres, H. (1910). *Recherches Sur l'Atmosphère Solaire; Photographies des Couches Gazeuses Supèriures* (IV(I), pp. 1–139) Ann. Obs. Paris-Meudon.

Dudík, J., Aulanier, G., Schmieder, B., Zapiór, M., & Heinzel, P. (2012). Magnetic topology of bubbles incbrr quiescent prominences. *The Astrophysical Journal, 761*, 9–22.

Dunn, R. (1960). *Photometry of the solar chromosphere*, Ph.D. Thesis, Harvard University.

Engvold, O. (2008). Observational aspects of prominence oscillations. In *IAU symposium* (Vol. 247, pp. 152–157).

Engvold, O., Tandberg-Hanssen, E., & Reichmann, E. (1985). Evidence for systematic flows in the transition region around prominences. *Solar Physics, 96*, 35–51.

Engvold, O., Jakobsson, H., Tandberg-Hanssen, E., Gurman, J. B., & Moses, D. (2001). On the nature of prominence absorption and emission in highly ionized iron and in neutral hydrogen. *Solar Physics, 202*, 293–308.

Feynman, J., & Martin, S. F. (1995). The initiation of coronal mass ejections by newly emerging magnetic flux. *Journal of Geophysical Research, 100*, 3355–3367.

Forbes, T. G. (2000). A review on the genesis of coronal mass ejections. *Journal of Geophysical Research, 105*, 23153–23166.

Foukal, P. (1971). Morphological relationships in the chromospheric Hα fine structure. *Solar Physics, 19*, 59–71.

Fuller, J., Gibson, S. E., de Toma, G., & Fan, Y. (2008). Observing the unobservable? Modeling coronal cavity densities. *The Astrophysical Journal, 678*, 515–530.

Gaizauskas, V. (1998). Filament channels: Essential ingredients for filament formation (review). In *ASP conference series* (Vol. 150, pp. 257–264).

Gibson, S. (2014). Coronal cavities: Observations and implications for the magnetic environment of prominences. In J.-C. Vial, & O. Engvold (Eds.), *Solar prominences*. Springer.

Gibson, S. E., Kucera, T. A., Rastawicki, D., et al. (2010). Three-dimensional morphology of a coronal prominence cavity. *The Astrophysical Journal, 724*, 1133–1146.

Habbal, S. R., Druckmüller, M., Morgan, H., et al. (2010). Total solar eclipse observations of hot prominence shrouds. *The Astrophysical Journal, 719*, 1362–1369.

Haerendel, G., & Berger, T. (2011). A droplet model of quiescent prominence downflows. *The Astrophysical Journal, 731*, 82.

Hale, G. E. (1903). The snow horizontal telescope. *The Astrophysical Journal, 17*, 314.

Hale, G. E. (1929). The spectrohelioscope and its work. *The Astrophysical Journal, 70*, 265.

Heinzel, P. (2014). Radiative transfer in solar prominences. In J.-C. Vial, & O. Engvold (Eds.), *Solar prominences, ASSL* (Vol. 415, pp. 101–128). Springer.

Heinzel, P., Schmieder, B., Fárník, F., et al. (2008). Hinode, TRACE, SOHO, and ground-based observations of a quiescent prominence. *The Astrophysical Journal, 686*, 1383–1396.

Hermans, L. M., & Martin, S. F. (1986). Small-scale eruptive filaments on the quiet sun. *BAAS, 18*, 991.

Hillier, A., Isobe, H., Shibata, K., & Berger, T. (2012). Numerical simulations of the magnetic Rayleigh–Taylor instability in the Kippenhahn–Schlüter prominence model. II. Reconnection-triggered downflows. *The Astrophysical Journal, 756*, 110–120.

Hirayama, T. (1985). Modern observations of solar prominences. *Solar Physics, 100*, 415–434.

Hundhausen, A. (1999). Coronal mass ejections. In K. T. Strong, J. L. R. Saba, B. M. Haisch, & J. T. Schmelz (Eds.), *The many faces of the sun: A summary of the results from NASA's solar maximum mission* (p. 143). New York: Springer.

Isobe, H., & Tripathi, D. (2006). Large amplitude oscillation of a polar crown filament in the pre-eruption phase. *Astronomy and Astrophysics, 449*, L17–L20.

Isobe, H., Tripathi, D., Asai, A., & Jain, R. (2007). Large-amplitude oscillation of an erupting filament as seen in EUV, Hα, and microwave observations. *Solar Physics, 246*, 89–99.

Karpen, J. T., & Antiochos, S. K. (2008). Condensation formation by impulsive heating in prominences. *The Astrophysical Journal, 676*, 688.

Kucera, T. A. (2014). Derivations and observations of prominence bulk motions and mass. In J.-C. Vial, & O. Engvold (Eds.), *Solar prominences, ASSL* (Vol. 415, pp. 77–99). Springer.

Kucera, T. A., Gibson, S. E., Schmit, D. J., Landi, E., & Tripathi, D. (2012). Temperature and extreme-ultraviolet intensity in a coronal prominence cavity and streamer. *The Astrophysical Journal, 757*, 73.

Labrosse, N. (2014). Derivation of major properties of prominences using non-LTE modeling. In J.-C. Vial, & O. Engvold (Eds.), *Solar prominences, ASSL* (Vol. 415, pp. 129–153). Springer.

Labrosse, N., Heinzel, P., Vial, J.-C., Kucera, T., Parenti, S., Gunár, S., Schmieder, B., & Kilper, G. (2010). Physics of solar prominences: I—Spectral diagnostics and non-LTE modelling. *Space Science Reviews, 151*, 243–332.

Leroy, J.-L. (1981). Simultaneous measurement of the polarization in Hα and D3 prominence emissions. *Solar Physics, 71*, 285–297.

Leroy, J. L. (1989). Observation of prominence magnetic fields. *Astrophysics and Space Science Library, 150*, 77.

Li, X., Morgan, H., Leonard, D., & Jeska, L. (2012). A solar tornado observed by AIA/SDO: Rotational flow and evolution of magnetic helicity in a prominence and cavity. *The Astrophysical Journal, 752*, L22–L27.

Lin, Y. (2000). *Comparison of Hα and He II 304 Å brightness variation in solar prominences*. MA Thesis, Institute of Theoretical Astrophysics, University of Oslo.

Lin, Y. (2011). Filament thread-like structures and their small-amplitude oscillations (invited review). *Space Science Reviews, 158*, 237.

Lin, Y., Engvold, O., & Wiik, J. E. (2003). Counterstreaming in a large polar crown filament. *Solar Physics, 216*, 109–120.

Lin, Y., Engvold, O., Rouppe van der Voort, L., Wiik, J. E., & Berger, T. E. (2005a). Thin threads of solar filaments. *Solar Physics, 226*, 431–451.

Lin, Y., Wiik, J. E., Engvold, O., Rouppe van der Voort, L., & Frank, Z. A. (2005b). Solar filaments and photospheric network. *Solar Physics, 227*, 283–297.

Lin, Y., Engvold, O., Rouppe van der Voort, L. H. M., & van Noort, M. (2007). Evidence of traveling waves in filament threads. *Solar Physics, 246*, 65–72.

Lin, Y., Martin, S. F., & Engvold, O. (2008). Filament substructures and their interrelation. In *ASP conference series* (Vol. 383, p. 235).

Lin, Y., Soler, R., Engvold, O., Ballester, J. L., Langangen, Ø., Oliver, R., & Rouppe van der Voort, L. H. M. (2009). Swaying threads of a solar filament. *The Astrophysical Journal, 704*, 870–876.

Liu, R., Alexander, D., & Gilbert, H. R. (2009). Asymmetric eruptive filaments. *The Astrophysical Journal, 691*, 1079–1091.

Liu, J., Zhou, Z., Wang, Y., Liu, R., et al. (2012a). Slow magnetoacoustic waves observed above a quiet-sun region in a dark cavity. *The Astrophysical Journal, 758*, L26–L32.

Liu, W., Berger, T. E., & Low, B. C. (2012b). First SDO/AIA observation of solar prominence formation following an eruption: Magnetic dips and sustained condensation and drainage. *The Astrophysical Journal, 745*, L21–L29.

Liu, R., Kliem, B., Török, T., et al. (2012c). Slow rise and partial eruption of a double-decker filament. I. Observations and interpretation. *The Astrophysical Journal, 756*, 59–73.

Lopez Ariste, A. (2014). Magnetometry of prominences. In J.-C. Vial, & O. Engvold (Eds.), *Solar prominences, ASSL* (Vol. 415, pp. 177–202). Springer.

Mackay, D. (2014). Formation of large-scale pattern of filament channels and filaments. In J.-C. Vial, & O. Engvold (Eds.), *Solar prominences, ASSL* (Vol. 415, pp. 353–378). Springer.

Mackay, D. H., Gaizauskas, V., & Yeates, A. R. (2008). Where do solar filaments form?: Consequences for theoretical models. *Solar Physics, 248*, 51–65.

Mackay, D. H., Karpen, J. T., Ballester, J. L., Schmieder, B., & Aulanier, G. (2010). Physics of solar prominences: II—Magnetic structure and dynamics. *Space Science Reviews, 151*, 333–399.

Mariska, J. T., Doschek, G. A., & Feldman, U. (1979). Extreme-ultraviolet limb spectra of a prominence observed from SKYLAB. *The Astrophysical Journal, 232*, 929–939.

Martin, S. F. (1998a). Conditions for the formation and maintenance of filaments (invited review). *Solar Physics, 182*, 107–137.

Martin, S. F. (1998b). Filament Chirality: A Link Between Fine-Scale and Global Patterns (Review). *ASP Conference Series, 150*, 419–429.

Martin, S. (2014). The magnetic field structure of prominences from direct and indirect observations. In J.-C. Vial, & O. Engvold (Eds.), *Solar prominences, ASSL* (Vol. 415, pp. 203–233). Springer.

Martin, S. F., & Echols, C. R. (1994). An observational and conceptual model of the magnetic field of a filament. In R. J. Rutten & C. J. Schrijver (Eds.), *Solar surface magnetism [NATO Advanced Research Workshop]* (p. 339). Dordrecht: Kluwer Academic Publishers.

Martin, S. F., Livi, S. H. B., & Wang, J. (1985). The cancellation of magnetic flux. II – In a decaying active region. *Australian Journal of Physics, 38*, 929–959.

Martin, S. F., Marquette, W. H., & Bilimoria, R. (1992). The solar cycle pattern in the direction of the magnetic field along the long axes of polar filaments, the solar cycle. In *ASP conference series* (Vol. 27, p. 53).

Martin, S. F., Bilimoria, R., & Tracadas, P. W. (1994). Magnetic field configurations basic to filament channels and filaments. In R. J. Rutten & C. J. Schrijver (Eds.), *Solar surface magnetism* (p. 303). Dordrecht: Kluwer Academic Publishers.

Martin, S. F., Lin, Y., & Engvold, O. (2008). A method of resolving the 180-degree ambiguity by employing the chirality of solar features. *Solar Physics, 250*, 31–51.

Martin, S. F., Panasenco, O., Agah, Y., Engvold, O., & Lin, Y. (2009). Relating a prominence observed from the solar optical telescope on the Hinode satellite to known 3-D structures of filaments. In *ASP conference series* (Vol. 415, p. 183).

Martin, S. F., Panasenco, O., Berger, M. A., et al. (2012). The build-up to eruptive solar events viewed as the development of chiral systems. In *ASP conference proceedings* (Vol. 463, p. 157).

Menzel, D. H., & Evans, J. W. (1953). *Acad. Naz. Lincei. Conv. Volta, 11*, 119.

Molowny-Horas, R., Wiehr, E., Balthasar, H., Oliver, R., & Ballester, J. L. (1999). *Prominence Doppler oscillations*. JOSO annual report, 1998, pp. 126–127.

Nagashima, K., Isobe, H., Yokoyama, T., Ishii, T. T., Okamoto, T. J., & Shibata, K. (2007). Triggering Mechanism for the Filament Eruption on 2005 September 13 in NOAA Active Region 10808. *The Astrophysical Journal, 668*, 533–545.

Newton, H. W. (1935). Note on two allied types of chromospheric eruption. *MNRAS, 95*, 650.

Okamoto, T. J., Nakai, H., Keiyama, A., Narukage, N., UeNo, S., Kitai, R., Kurokawa, H., & Shibata, K. (2004). Filament oscillations and Moreton waves associated with EIT waves. *The Astrophysical Journal, 608*, 1124–1132.

Okamoto, T. J., Tsuneta, S., Berger, T. E., et al. (2007). Coronal transverse magnetohydrodynamic waves in a solar prominence. *Science, 318*, 1577.

Okamoto, T. J., Tsuneta, S., & Berger, T. E. (2010). A rising cool column as a signature of helical flux emergence and formation of prominence and coronal cavity. *The Astrophysical Journal, 719*, 583–590.

Oliver, R., & Ballester, J. L. (2002). Oscillations in quiescent solar prominences observations and theory (invited review). *Solar Physics, 206*, 45–67.

Paletou, F. (1997). On Hα source function vertical variations in filaments and bright rims visibility. *Astronomy and Astrophysics, 317*, 244–247.

Panasenco, O., Martin, S. F., & Velli, M. (2014). Apparent solar tornado-like prominences. *Solar Physics, 289*, 603–622.

Panesar, N. K., Innes, D. E., Tiwari, S. K., & Low, B. C. (2013). A solar tornado triggered by flares? *Astronomy and Astrophysics, 549*, 105–110.

Parenti, S. (2014a). Solar prominences: Observations. Living Reviews in *Solar Physics, 11*, 1–88.

Parenti, S. (2014b). Spectral diagnostics of cool and PCTR optically thin plasma. In J.-C. Vial, & O. Engvold (Eds.), *Solar prominences, ASSL* (Vol. 415, pp. 61–76). Springer.

Parenti, S., & Vial, J.-C. (2007). Prominence and quiet-sun plasma parameters derived from FUV spectral emission. *Astronomy and Astrophysics, 469*, 1109–1115.

Pécseli, H., & Engvold, O. (2000). Modeling of prominence threads in magnetic fields: Levitation by incompressible MHD waves. *Solar Physics, 194*, 73–86.

Pettit, E. (1932). Characteristic features of solar prominences. *The Astrophysical Journal, 76*, 9.

Pevtsov, A. A., & Neidig, D. (2005). Accumulation of filament material at the boundaries of supergranular cells. In K. Sankarasubramanian, M. Penn, & A. Pevtsov (Eds.), *ASP conference series* (Vol. 346, p. 219).

Plocieniak, S., & Rompolt, B. (1973). Positions of filament feet in relation to the supergranular calcium network. *Solar Physics, 29*, 399–401.

Poland, A. I., & Tandberg-Hanssen, E. (1983). Physical conditions in a quiescent prominence derived from UV spectra obtained with the UVSP instrument on the SMM. *Solar Physics, 84*, 63–70.

Ramsey, H. E., & Smith, S. F. (1966). Flare-initiated filament oscillations. *Astronomical Journal, 71*, 197.

Ryutova, M., Berger, T., Frank, Z., Tarbell, T., & Title, A. (2010). Observation of plasma instabilities in quiescent prominences. *Solar Physics, 267*, 75–94.

Schmieder, B., Kucera, T. A., Knizhnik, K., Luna, M., Lopez-Ariste, A., & Toot, D. (2013). Propagating waves transverse to the magnetic field in a solar prominence. *The Astrophysical Journal, 777*, 108–119.

Schmit, D. J., Gibson, S. E., Tomczyk, S., Reeves, K. K., Sterling, A. C., Brooks, D. H., Williams, D. R., & Tripathi, D. (2009). Large-scale flows in prominence cavities. *The Astrophysical Journal, 700*, L96–L98.

Schrijver, C. J. (2001). Catastrophic cooling and high-speed downflow in quiescent solar coronal loops observed with TRACE. *Solar Physics, 198*, 325–345.

Sheeley, N. R., Jr., & Warren, H. P. (2012). Coronal cells. *The Astrophysical Journal, 749*, 40–54.

Sheeley, N. R., Jr., Martin, S. F., Panasenco, O. & Warren, H. P. (2013). Using coronal cells to infer the magnetic field structure and chirality of filament channels. *Astrophysical Journal, 772*, 88–99.

Simon, G. W., & Leighton, R. B. (1964). Velocity fields in the solar atmosphere. III. Large-scale motions, the chromospheric network, and magnetic fields. *The Astrophysical Journal, 140*, 120.

Smith, S. F. (1968). The formation, structure and changes in filaments in active regions. In *IAU symposium* (Vol. 35, p. 267).
Soler, R., Oliver, R., & Ballester, J. L. (2014). The damping of transverse oscillations of prominence threads: A comparative study. In *IAU symposium* (Vol. 300, pp. 48–51).
Steele, C. D. C., & Priest, E. R. (1992). A model for the fibril structure of normal-polarity solar prominences. *Solar Physics, 140*, 289–306.
Stellmacher, G., & Wiehr, E. (1973). Observation of an instability in a "Quiescent" prominence. *Astronomy and Astrophysics, 24*, 321.
Sterling, A. C., & Moore, R. L. (2004). Evidence for gradual external reconnection before explosive eruption of a solar filament. *The Astrophysical Journal, 602*, 1024–1036.
Su, Y., Wang, T., Veronig, A., Temmer, M., & Gan, W. (2012). Solar magnetized "Tornadoes:" Relation to filaments. *The Astrophysical Journal, 756*, L41–L48.
Tandberg-Hanssen, E. A. (1974). *Solar prominences*. Dordrecht: D. Reidel Publ. Co.
Tandberg-Hanssen, E. (1995). The nature of solar prominences. In *Astrophysics and Space Science Library* 199. Kluwer Academic Publishers.
Tang, F. (1987). Quiescent prominences – Where are they formed? *Solar Physics, 107*, 233–237.
van Driel-Gesztelyi, L., & Culhane, J. L. (2009). Magnetic flux emergence, activity, eruptions and magnetic clouds: Following magnetic field from the sun to the heliosphere. *Space Science Reviews, 144*, 351–381.
Vial, J. C. (1990). The prominence–corona interface. *Lecture Notes in Physics, 363*, 106–119.
Vial, J.-C. (2014). Historical background and introduction. In J.-C. Vial, & O. Engvold (Eds.), *Solar prominences, ASSL* (Vol. 415, pp. 1–29). Springer.
Waldmeier, M. (1970). The structure of the monochromatic corona in the surroundings of prominences. *Solar Physics, 15*, 167–175.
Wang, Y.-M., & Sheeley, N. R., Jr. (1999). Filament eruptions near emerging bipoles. *The Astrophysical Journal, 510*, L157–L160.
Wang, Y.-M., & Sheeley, N. R., Jr. (2002). Observations of core fallback during coronal mass ejections. *The Astrophysical Journal, 567*, 1211–1224.
Wang, H., Chae, J., Gurman, J. B., & Kucera, T. A. (1998). Comparison of prominences in Hα and He II 304 Å. *Solar Physics, 183*, 91–96.
Wang, J., Li, W., Denker, C., Lee, C., Wang, H., Goode, P. R., McAllister, A., & Martin, S. F. (2000). Minifilament eruption on the quiet sun. I. Observations at Hα central line. *The Astrophysical Journal, 530*, 1071–1084.
Wedemeyer, S., Scullion, E., Rouppe van der Voort, L., Bosnjak, A., & Antolin, P. (2013). Are giant tornadoes the legs of solar prominences? *The Astrophysical Journal, 774*, 123–138.
Wedemeyer-Böhm, S., Scullion, E., Steiner, O., et al. (2012). Magnetic tornadoes as energy channels into the solar corona. *Nature, 486*, 505–508.
Widing, K. G., Feldman, U., & Bhatia, A. K. (1986). The extreme-ultraviolet spectrum (300–630 A) of an erupting prominence observed from SKYLAB. *The Astrophysical Journal, 308*, 982–992.
Wood, P., & Martens, P. (2003). Measurements of flux cancellation during filament formation. *Solar Physics, 218*, 123–135.
Xu, Y., Jing, J., & Wang, H. (2010). Measurements of filament height in Hα and EUV 304 Å. *Solar Physics, 264*, 81–91.
Zirin, H. (1966). *The solar atmosphere*. Waltham, MA: Blaisdell-Ginn.
Zirin, H. (1988). *Astrophysics of the sun*. Cambridge: Cambridge University Press.
Zirin, H., & Tandberg-Hanssen, E. (1960). Physical conditions in limb flares and active prominences. IV. Comparison of active and quiescent prominences. *The Astrophysical Journal, 131*, 717.
Zirker, J. B., & Koutchmy, S. (1990). Prominence fine structure. *Solar Physics, 127*, 109–118.
Zirker, J. B., & Koutchmy, S. (1991). Prominence fine structure. II – Diagnostics. *Solar Physics, 131*, 107–118.
Zirker, J. B., Engvold, O., & Martin, S. F. (1998). Counter-streaming gas flows in solar prominences as evidence for vertical magnetic fields. *Nature, 396*, 440–441.

Chapter 3
Spectral Diagnostics of Cool Prominence and PCTR Optically Thin Plasmas

Susanna Parenti

Abstract This chapter is dedicated to introduce information we can derive from optically thin emission of prominences plasma. This emission comes mostly from the interface region with the corona, the prominence-corona transition region, and it is observable in the UV-EUV wavebands. After a general introduction to the formation of the optically thin emission, we present the diagnostics methods which are used to infer the thermal properties of the emitting plasma under isothermal and multi-thermal hypothesis. We then describe a diagnostics technique to infer the electron density. For each method presented we give advantages and limitations, together with the main results. We then discuss the diagnostics at small, unresolved scales introducing the filling factor and conclude with some final remarks.

3.1 Introduction to the Thin Emission of Prominences

As introduced by Vial (2014) and Engvold (2014), prominences are cool and dense structures of the solar corona. As we move out from their central body towards the surrounding corona, their density decreases while the temperature increases. Under these conditions the regime of the emitting plasma changes from optically thick to thin. If the plasma is optically thick it means that it partially or totally absorbs the self-produced and incident light. For this reason filaments on the disk appear darker than their surrounding. The diagnostic of optically thick plasma is detailed in Heinzel (2014) and Labrosse (2014). On the contrary, when the plasma is optically thin it is transparent to the self-produced or incident radiation, meaning that all the radiation going through the structure escapes from it and eventually reaches the observer. Thus for an observed structure, the total photons of a spectral line and its profile measured by the observer are due to the sum of the contributions to this line from the plasma along the line of sight. In this case the structure appears as bright as, or brighter than, the surrounding environment.

S. Parenti (✉)
Royal Observatory of Belgium, 3 Av. Circulaire, 1180 Brussels, Belgium
e-mail: s.parenti@oma.be

J.-C. Vial, O. Engvold (eds.), *Solar Prominences*, Astrophysics and Space Science Library 415, DOI 10.1007/978-3-319-10416-4_3

The external envelope of prominences, where the plasma becomes hotter than about 10^5 K, less dense and mostly optically thin, is called prominence-corona transition region (PCTR, Parenti and Vial 2014). This is the interface layer with the 1MK ambient corona. Observations of filaments suggest that their large scale magnetic field is mainly oriented horizontally (see López Ariste 2014; Martin 2014 in this book). This implies that the very low perpendicular thermal conduction makes the thermal gradient along such a direction very steep, producing a thin PCTR layer.

In view of the different properties of the plasma forming the prominence core and the PCTR, we need to apply different diagnostic techniques to derive information from the plasma emission. This emission is the result of various atomic processes which are more or less dominant depending on the plasma conditions. In the next sessions we will give an overview of the diagnostics methods applied to optically thin plasmas.

3.2 Spectral Lines Formation

Most of the optically thin plasma in prominences emits in the EUV (100–1000 Å) and UV (1000–4000 Å) bands, producing spectral lines and continuum. In this chapter we describe the plasma diagnostics for spectral lines (bound-bound transitions), while the continuum is treated in Kucera (2014) and Labrosse (2014).

To extract the plasma properties from remote sensing data, as in the case for spectral lines, we need to make some assumptions on the source. Such assumptions are more or less strongly depending on the inversion technique used and the quality of the data used. In case of the optically thin emission of prominences (mostly the emission of the PCTR), we can use the commonly adopted diagnostics for EUV emission.

To formulate the following diagnostics we assume for the emitting plasma:

- to be in ionization equilibrium. This means that the result of all the processes for ionization and recombination of the emitting ion is in balance;
- to be in a steady state. This means that the sum of the excitation and de-excitation processes acting on an energy level of the ion keeps its population constant in time. Such a population is found satisfying the statistical equilibrium equation for all the energy levels involved by the atomic processes;
- the electron number density, n_e, is high enough so that the time scales for excitation and de-excitation of the energy levels of the emitting ions are much faster than the ionization-recombination processes. In this case the two problems can be treated separately;
- the impacting electrons are assumed to have a thermal distribution at temperature T_e.

A first approximation, which is often valid, is to consider a two levels atom (e.g. the *ground* level g and level j, like in an *allowed transition*). To this we add that the spectral line formed at the wavelength λ_{jg} is the result of excitation by the dominant electron collision mechanism, while the spontaneous radiative decay mechanism dominates the de-excitation. The population of these two levels is obtained by solving the simplified statistical equilibrium equation

$$n_e n_g C_{gj} = n_j A_{jg} \tag{3.1}$$

where C_{gj} is the collisional excitation rate coefficient, which is a function of the temperature, n_j is the number density of the upper level and A_{jg} is the spontaneous radiative decay rate coefficient.

The emissivity of the line is given by

$$P(\lambda_{jg}) = \frac{hc}{\lambda_{jg}} n_j A_{jg} \ [erg\ cm^{-3}\ s^{-1}]. \tag{3.2}$$

The line intensity is

$$I(\lambda_{jg}) = \frac{hc}{4\pi\lambda_{jg}A} \int_V n_e n_g C_{gj} dV \ [erg\ cm^{-2} s^{-1} sr^{-1}] \tag{3.3}$$

with A being the cross-sectional area of the emitting volume V and C_{gj}:

$$C_{gj} = \frac{8.63 \times 10^{-6} \Upsilon_{gj}(T_e)}{\omega_g} T_e^{-1/2} \exp(\frac{-hc}{\lambda_{jg} k T_e}) \ [cm^3 s^{-1}] \tag{3.4}$$

where $\Upsilon_{\mathrm{gj}}(T_{\mathrm{e}})$ is the thermally averaged collision strength and ω_{g} is the statistical weight of the ground level.

The number density of the ground level, n_g, can be rewritten as function of known quantities:

$$n_g = \frac{n_g}{n_{ion}} \frac{n_{ion}}{n_{el}} \frac{n_{el}}{n_H} \frac{n_H}{n_e} n_e \tag{3.5}$$

where $n_{\mathrm{g}}/n_{\mathrm{ion}} \sim 1$ for resonant transitions, $n_{\mathrm{ion}}/n_{\mathrm{el}}$ is the ionization ratio of the ion relative to the total number density of the element, $n_{\mathrm{el}}/n_{\mathrm{H}}$ is the abundance of the element (here after called *Ab*, see for instance Asplund et al. 2009), with respect to hydrogen and $n_{\mathrm{H}}/n_{\mathrm{e}} = 0.83$ as we assume hydrogen and helium to be completely ionized.

Using Eq. (3.5) and collecting all the atomic physics parameters under a unique function called *contribution function*, $G(T_e)$, we can rewrite Eq. (3.3) as function of quantities easier to establish:

$$I(\lambda_{jg}) = \frac{1}{4\pi A} \int_V Ab\ G(T_e)\ n_e^2 dV \tag{3.6}$$

where $G(T_e)$ is (in c.g.s. units)

$$G(T_e) = \frac{8.63 \times 10^{-6} \Upsilon_{gj}}{\omega_g} \frac{n_{ion}}{n_{el}} 0.83 T_e^{-1/2} \exp(\frac{-hc}{\lambda_{jg} k T_e}) \frac{hc}{\lambda_{jg}} \tag{3.7}$$

The spectral techniques discussed here invert Eq. (3.6) or a set of these to derive the averaged value (in the emitting volume) of n_e, T_e, Ab. Depending on the approximation used, we could also infer $n_e^2(T_e)$, V and the fraction of the volume occupied by the unresolved structure, called *filling factor*.

For instance a simple diagnostics for the averaged n_e^2 can be to deduce the volume V from observations assuming a simple geometry. To do this, we need to impose a value for the element abundance (which is also assumed to be constant within the volume) and use tabulated values for $G(T_e)$. But these are quite important assumptions, and we need to be aware of the limitations of the results obtained. However, as it will shown in the following, a more reliable technique can be used to derive the plasma parameter from Eq. (3.6).

To invert Eq. (3.6) we also need to know the $G(T_e)$ function. This value, as all the atomic information needed for the spectroscopic diagnostics, is provided by theoretical calculations and/or laboratory measurements collected in atomic databases. Among the existing, the most used and freely available for the EUV emission in solar physics is the CHIANTI database (Dere et al. 1997; Landi et al. 2013) which also provides a free package of software for the calculation of the spectra and data inversion, all well documented within the CHIANTI home page.[1] The basic CHIANTI model includes the following multi-level processes: electron, proton excitation and de-excitation together with spontaneous radiative decay, photoexcitation and stimulated emission. Protons and electrons have a Maxwellian distribution at the same temperature. The plasma ionization is dominated by collisions and photo-ionization is not included. The calculation for the continuum includes the free-free, free-bound and the two photon processes. More complex models for the spectral line and continuum formation are also available in this database.

An example of the output of CHIANTI is shown in Fig. 3.1. This is the result of a calculation for the $G(T_e)$ function of three lines from the C II-III-IV ions. These lines are emitted by the PCTR and observed by the SOHO/SUMER and IRIS spectrometers. All the diagnostics techniques described in this chapter can be treated using this database. However, some caution should be taken particularly using intensities from the low ionization ions, as not all the physical processes for the line formation are included. For instance, optically thick or/and photo-ionization processes are not treated by CHIANTI.

[1] http://www.chiantidatabase.org/.

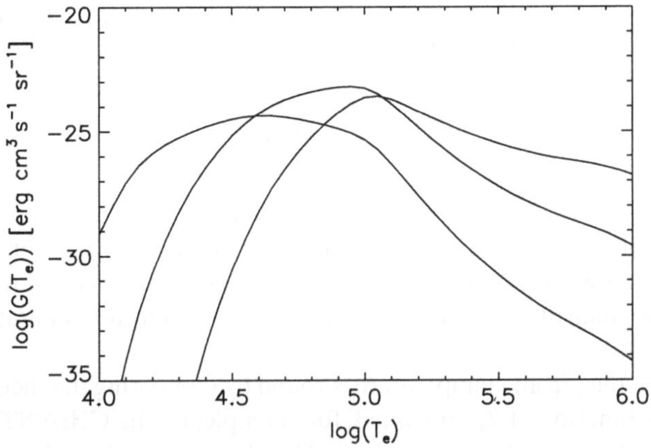

Fig. 3.1 Theoretical values of the logarithm of the contribution function as function of the logarithm of temperature for the C II (1335.7080 Å), C III (977.0200 Å) and C IV (1550.7750 Å) lines respectively which peak at increasing temperatures. These functions are calculated using the CHIANTI atomic database and software (v.7.1.3) at a plasma pressure of 0.014 dyn cm^{-2}

3.3 Diagnostics for UV-EUV Data

As the properties of the prominence plasma emitting in the UV-EUV are quite similar to those of the solar chromosphere and transition region, the diagnostic techniques that can be applied are often the same. Using the total measured radiation and profile of one or more spectral lines, we can infer the electron temperature, the electron density (and deduce the mass) and emission measure, the elements abundance, and study the different plasma motions. These quantities are the key elements to characterize prominences, as they provide the ranges within which certain physical processes may be at work. For instance, they play a role in the mechanical and thermal stability (see Kucera 2014; Gilbert 2014; Heinzel 2014 in this book).

3.3.1 Emission Measure and Differential Emission Measure

One of the simplest ways to deduce the electron temperature is to measure the total intensity of two spectral lines (I_1 and I_2) from the same element (to eliminate the dependence from the element abundance in Eq. (3.6)) at close stages of ionization (to sample the plasma in a relatively small temperature range) and formed by allowed transitions (to ensure the same electron density dependence of the line intensity). In this case we can write

$$\frac{I_1}{I_2} = \frac{EM_1}{EM_2} \times \frac{G_1(T_e)}{G_2(T_e)} \tag{3.8}$$

where we have introduced the *column emission measure* (EM, Ivanov-Kholodnyi and Nikol'Skii 1963; Pottasch 1963) along the line of sight l, as

$$EM = \int_l n_e^2 dl \quad [cm^{-5}] \tag{3.9}$$

For this technique we use the assumption that the two lines are formed in the same volume of plasma along the line of sight, which is a reasonable assumption if we use lines from adjacent stages of ionization. Under this condition, the ratio of the two line intensities is only a function of the electron temperature, given by $G_1(T_e)/G_2(T_e)$.

The value of the plasma temperature is found by comparing the theoretical values of this ratio as function of T_e (obtained, for example, by the CHIANTI database) to the measured ratio of the lines intensities. The closest value found to the theoretical curve is the averaged temperature along the line of sight.

This technique is generally used when we don't have enough measured spectral lines to apply a more sophisticated diagnostics or when we are dealing with large spectral bands (which generally collect information within a large temperature range). For instance, with the SDO/AIA multi-band and high spatial resolution instrument (Lemen et al. 2012) it was found a temperature $> 2.5 \times 10^5$ K within buoyant structures below prominences, suggesting the local presence of a source of plasma heating (Berger et al. 2010).

It is more realistic to think that the plasma along the line of sight is multi-thermal, as l is generally crossing multiple structures filled by optically thin and thick plasmas. If we are interested in sampling the temperatures where the plasma is optically thin ($> 10^4$ K) we can use the function called *differential emission measure* (DEM). The DEM is proportional to $n_e^2(T_e)$ in the temperature intervals dT_e and it is defined here as

$$DEM(T_e) = n_e^2 \frac{dl}{dT_e} \quad [cm^{-5}\ K^{-1}] \tag{3.10}$$

Be aware of that there exist different definitions of the DEM. For instance, Eq. (3.10) can be written considering the fraction of emitting volume (dV) along the line of sight.

The relation between the DEM and the total intensity of an optically thin line [Eq. (3.6)] is given by

$$I(\lambda_{jg}) = \frac{1}{4\pi} \int_{T_e} AbG(T_e)DEM(T_e)dT_e \tag{3.11}$$

Even if using the DEM technique is a step forward in inferring the multi-temperature information of the plasma, we still cannot derive its spatial distribution along the line of sight. This, however, is a recurrent problem when we are dealing with optically thin plasma.

The DEM of an observed structure is found by inverting Eq. (3.11) for a set of spectral lines which are formed in the interested temperature range. Several inversion methods to solve this problem are used by the solar community, and the best solution is found minimizing a criterion which involves comparing the measured intensities with their theoretical values for a candidate DEM solution. To calculate the theoretical intensities it is generally assumed that the element abundance is a constant, while the $G(T_e)$ function is calculated using an atomic database.

One of the key elements for a successful inversion is to have enough data points to cover a wide temperature range and properly constrain the inversion. For instance, to sample the PCTR of prominences we need to reach a temperature possibly up to 1MK, since there are observational indications of still a weak emission at this temperature (Parenti et al. 2012; Kucera et al. 2012; Parenti 2014).

Figure 3.2 shows the DEM of the top part of an off-limb quiescent prominence observed by Parenti and Vial (2007) obtained using the intensities of a prominence atlas by Parenti et al. (2004) and Parenti et al. (2005) using data from the SOHO/SUMER UV spectrometer (Wilhelm et al. 1995). This solution (solid line) was reached using a large dataset (symbols in the figure) in order to cover the temperature range from the lower PCTR to the corona.

The profile of this DEM resembles closely that of the chromosphere-corona transition region (CCTR) suggesting that the physical processes acting in their

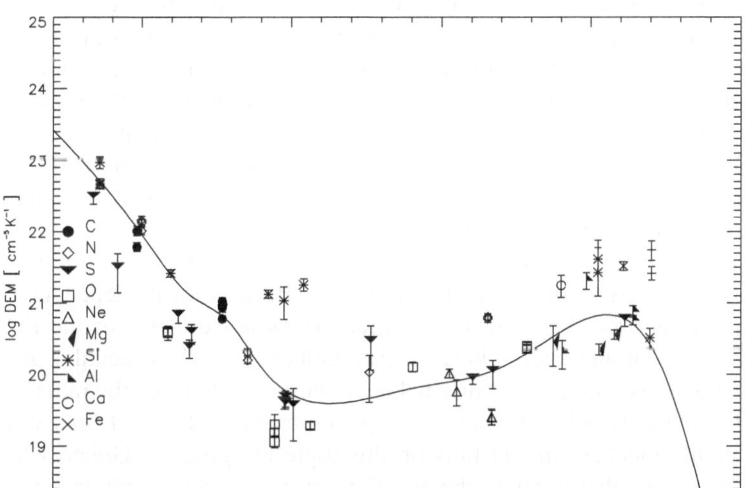

Fig. 3.2 Logarithm of the differential emission measure (*solid line*) as function of the logarithm of the temperature for a quiescent prominence observed by SOHO/SUMER. The different symbols represent the data points used to constrain the solution. Each symbol refer to a different chemical element. Credit: Parenti and Vial, A&A, 469, 1109, 2007, reproduced with permission © ESO

plasmas are similar. Generally, the DEM of the PCTRs is lower in amplitude than in the CCTR (2.5–10 times, depending on the temperature, for the case shown in Fig. 3.2). Such a difference may have several origins, primarily the view angle and the line of sight, as the prominence is observed at the limb while the CCRT is measured on the disk. It is indeed possible that the thicknesses of the large scale transition regions of the two structures are different. More difficult is to assess a difference in their density, as the measurements in prominences spread over a quite large interval of values. In a similar way, a difference in the properties of the small scale structures between the CCTR and PCTR can lead to infer different averaged densities. This later problem could, however, be solved with sufficient high spatial resolution measurements (see also the next two sections for details on the density and filling factor).

Unfortunately, we do not have a large enough set of derived DEMs to assess recurrent properties such as DEM gradient or temperature where the minimum occurs. From the few cases published (see Labrosse et al. (2010) for a list of references), we notice some variability of such properties, even though the uncertainties in the data and in the atomic physics calculations, as well as in the inversion technique used, can have a role in it (see also Sect. 3.3.3).

The use of UV-EUV observations for an off-limb prominence gives uncertainties in the DEM solution at the low and high extremes of the temperature interval. To constrain the low temperature we need to use lines from ions in a very low ionization stage. In this case most of the emission becomes optically thick and Eq. (3.11) is not valid anymore. Also, the statistical equilibrium equation becomes more complex than Eq. (3.1) increasing the uncertainties in the solution. However, this low temperature part is important as its plasma surrounds the prominence core and we need to link the properties of these two regions to correctly study the whole structure. At high temperatures the ambient corona emission dominates. In this case it is needed an extra step in the data analysis before deriving the DEM, which is the background emission subtraction from the prominence data. Unfortunately this ambient data is not always available, as in the case of Fig. 3.2.

Properly constraining the DEM solution means to better constrain prominence models both for its formation and equilibrium. Because of its dependence on n_e^2, this function is used, for instance, to infer the radiative losses of the PCTR and study the thermal and energy balance (see Gilbert (2014) for details). At present the investigations indicate that thermal conduction can balance the radiative losses above $T_e \approx 10^5$ K, while for lower temperatures we still do not fully understand the balancing mechanisms. Details on this topic are given in Gilbert (2014), but we can anticipate that to study the equilibrium of the whole prominence we need to link the properties of the core to that of the PCTR, including the losses. This is still not achieved, as the data from the two regions have to be treated differently and, in particular for the core, the data inversion requires the use of some modeling. The problem is difficult and the determination of the radiative losses of the whole prominence is still under study.

The multi-thermal information given by the DEM can also be used to infer the unresolved fine structure (see Sect. 3.3.2.1) and study the formation of prominences (see Karpen (2014) in this book).

3.3.2 Electron Density and Gas Pressure

Besides the allowed transitions which produce the brightest lines of the EU-EUV spectrum, under certain plasma conditions forbidden and intercombination transitions can also emit bright enough lines to be observed. Such transitions (involving the so called *metastable* levels) are interesting because they are only efficient in a particular density and temperature range, and the spectral lines produced can be used, for instance, for density diagnostics (their contribution function is then a function of density also: $G(n_e, T_e)$). Metastable levels have a lifetime larger than an allowed transition ($> 10^{-4}$ s, Phillips et al. 2012). In particular, they have a small value of the radiative decay coefficient (A_{ij}) which allows the collisional de-excitation process to compete in the line formation for high densities. Lines from allowed transitions having a metastable level as the lower level are also density sensitive. Figure 3.3 shows the theoretical variation of the fractional population (the ratio between the electrons populating a given energy level to the total number of electrons of the ion) of four metastable levels of the C III ion with the electron density. The maximum ionization fraction of this ion is reached at about 7×10^4 K, and it produces bright lines within the PCTR. These can be used for its density diagnostics, as the range of density-dependence of the lines intensity is within the typical PCTR values. Figure 3.4 shows the theoretical variation of the ratio of two C III lines commonly observed by, for instance, SOHO/SUMER.

The electron density can be inferred by using the ratio of two lines intensity from the same ion, where at least one involves a metastable level m. Using Eq. (3.6), and assuming the second line is due to a transition to the ground level, this ratio can be written as

$$\frac{I(\lambda_{jg})}{I(\lambda_{mk})} \propto \frac{n_e^2}{F(T_e, n_e)n_e} \tag{3.12}$$

Knowing the electron temperature value (or using, for instance, the temperature corresponding to the maximum of the ionization fraction), the line ratio is a function of the density only. The density is inferred by matching the ratio derived from observations with the theoretical value calculated at different densities. An example of this theoretical curve is shown in Fig. 3.4. The lower levels of the transitions at the origin of these lines are the no. 4 and 1 in Fig. 3.3.

Similarly to the other diagnostics methods already described, this requires the use of specific assumptions in the atomic calculations. When different assumptions are taken or different theoretical calculations are used, we can expect differences in the results. The experience has shown that such differences are also responsible for

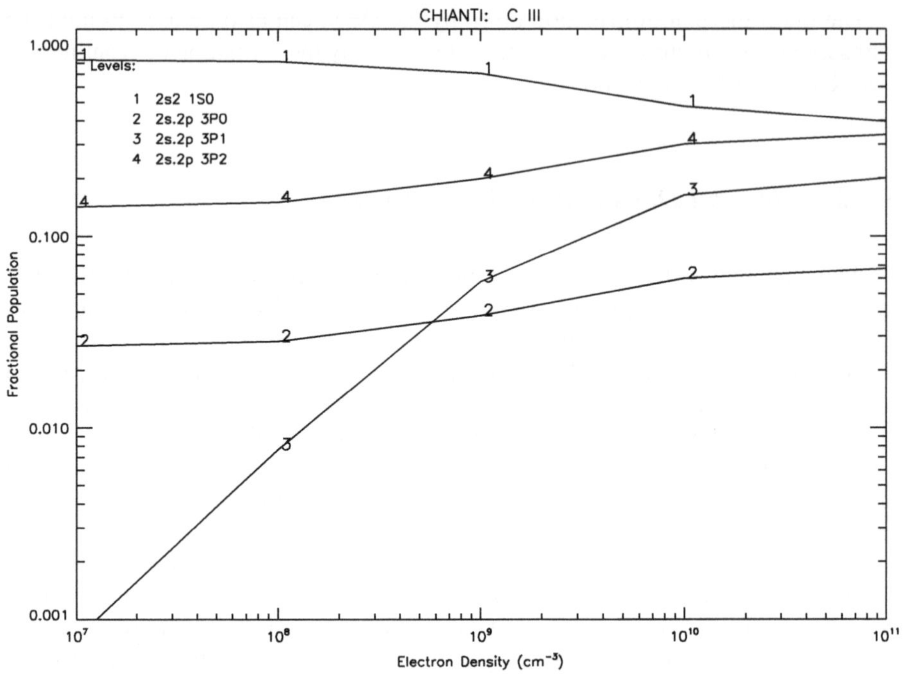

Fig. 3.3 Fractional population of four metastable levels for C III. These have been calculated using the CHIANTI database (v. 7.1.3) at a constant temperature of 10^5 K

the wide range of results found in the literature. Finally, we have to remember, that the density inferred with this method provides an averaged value, $< n_e >$, along the line of sight.

The literature reports (Labrosse et al. 2010) electron density values of $6 \times 10^8 < n_e < 5 \times 10^{11}$ cm^{-3} for temperatures of $7 \times 10^4 < T_e < 2 \times 10^5$ K derived using the line ratio technique of EUV data for quiescent and $3 \times 10^8 < n_e < 3 \times 10^{11}$ cm^{-3} for activated and erupting prominences. The gas pressure in the quiescent state is generally $0.01 < p < 0.3$ dyn cm^{-2}. The most used lines are from C III, Si III-IV, O III-IV corresponding to the temperature emission of the PCTR.

In the same range of densities fall the values found using the EM method. This can be applied when the averaged electron temperature is known ($< T_e >$) and an isothermal emitting plasma is assumed. In such a case we can invert the intensity of only one spectral line (or a large band data) making assumptions on the emitting volume or on the line of sight (depending on how the EM has been defined, see Eq. (3.9)):

$$< EM >= \frac{I}{G(< T_e >)} \tag{3.13}$$

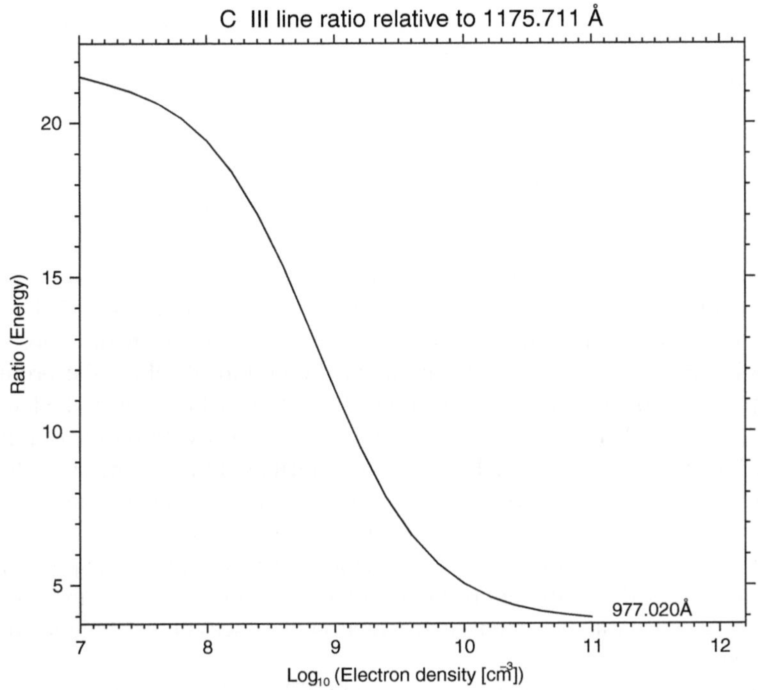

Fig. 3.4 Variation of the lines ratio C III 977.020/1175.711 Å with the electron density for a temperature of 10^5 K calculated using the CHIANTI atomic database (v. 7.1.3)

and from Eq. (3.9)

$$< n_e^2 >= \frac{< EM >}{l} \tag{3.14}$$

3.3.2.1 Fine Structure and Filling Factor

Our capability to infer the thermodynamic parameters from the emitting volume is limited by the spatial resolution of the UV-EUV instruments. Measurements of prominences core emission with ground-based high resolution telescopes have established that it is made of a thin structure of about 0.3" thick (Lin et al. 2008). Similar information for the PCTR has been obtained from very few measurements from rocket missions with about 0.3" spatial resolution, while systematic observations, until recently, had only 1"–2" resolution. These direct measurements at high resolution of PCTR threads have been done by VAULT (imaging the H-Lyα,

(Vourlidas et al. 2010; Vial et al. 2012)), and by Hi-C (imaging the 193 Å band, (Cirtain et al. 2013)) of on-disc active region filaments. They give results of the PCTR fine structure thickness of about, respectively, 0.4" and 0.8". This fine structure has the shape of elongated threads parallel to the filament spine (see Fig. 3.5).

At present there are no such measurements in off-limb prominences, which will certainly add information. Hopefully these different view angles will reveal new details of the fine structure. As the apparent fine structure of prominences core observed at the limb shows vertical or horizontal threads, there is a debate on the geometry of their fine magnetic structure (see also Labrosse 2014; Martin 2014). Possible interpretations suggest: the presence of magnetic "dips" where the cool material is found, or sheared aligned cool threads along the prominence spine partially filled by the cool material. If each cool thread is embedded in its own PCTR (panel (a) of Fig. 3.5), an equivalent high resolution measurement of the PCTR fine structure at the limb should contribute in resolving such dilemma. Probably more difficult would be to interpret the observations in one or the other of the two magnetic models in the case where a common PCTR exists for multiple cool threads (panel (b) of Fig. 3.5). Indeed, the discrepancy between optical and UV resolutions has led also to the open question on how much is structured the PCTR and on its geometry at the spatial scale (fraction of arcsec) of the cool threads of the prominence core.

The cited rocket measurements suggest a PCTR enveloping multiple cool threads (see Fig. 3.5 right). To confirm these few results using data at high resolution and completely answer all the above open questions, we need to expand such observations to a larger number of prominences over different lines of sight.

To account for the lack of information on the fine structure in the UV from the direct observation, we introduce the *filling factor* parameter, which is a measure

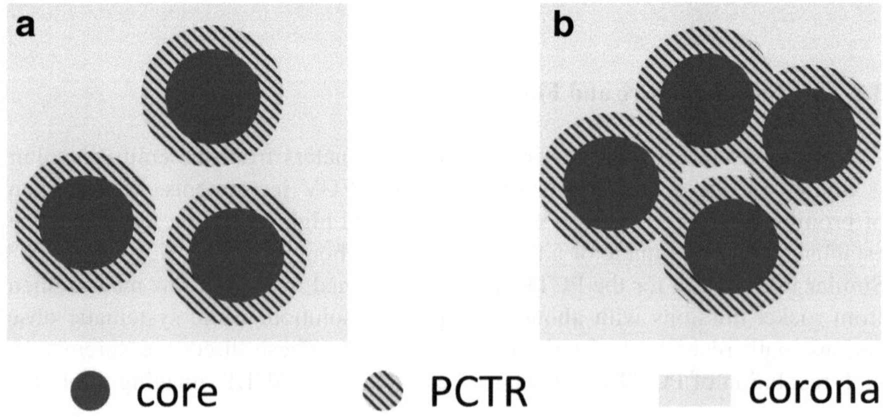

Fig. 3.5 Two possible configurations of the fine structure of prominences seen across their axes. (**a**): each cool thread has its own PCTR; (**b**): several cool threads have a common PCTR

of the fraction of the observed volume emitting the observed spectral lines. In the literature exist various definitions which slightly change depending on how the data are used, which information are accessible (i.e. spatial resolution or temperature range) or which hypothesis are used (i.e. it exists a non resolved structure perpendicular to the line of sight or along it). A discussion is made in Labrosse et al. (2010).

One of the more common techniques defines the filling factor as the ratio between the density obtained using Eq. (3.14) from the inferred emission measure (from Eq. (3.13)), and the density derived applying the line ratio technique (from which we have the averaged density, Eq. (3.12)):

$$f = \frac{< n_e^2 >}{< n_e >^2} \tag{3.15}$$

It is to be noticed that using this method requires estimating the thickness of the observed structure along the line of sight, as indicated in Eq. (3.14).

Results of this kind of analysis give values around 0.02 (Labrosse et al. 2010). Similar results were also obtained using the measure of spectral lines width coupled with a fine structure model (Labrosse et al. 2010; Parenti 2014). Knowing the filling factor and estimating the volume occupied by the PCTR, we can derive its mass (Kucera 2014). The filling factor in the PCTR has to be compared with that for the cool part of the prominence, as it will be an element for evaluating the global fine structuring of the prominence (Labrosse 2014). The results from the high resolution measurements of VAULT (Vial et al. 2012), give a PCTR thicker and less structured than the cool threads. These imply a filling factor at warm temperature ($> 2\times10^4$ K) higher than the one for the cool thread, probably closer to 1.

The following step in this analysis of the fine structure consists in inferring the number of unresolved threads, which will indicate the PCTR magnetic structuring. This requires the use of a model capable of reproducing the plasma parameters inferred from the observations (i.e. density or/and line width). The results indicate an apparent number of < 35 threads along the line of sight (Parenti 2014). Such results however, depend on the kind of data (type of prominence, line of sight, sampled temperature and spatial resolution) and the model assumed, so that it is reasonable to expect the real number of threads to be higher.

To summarize the results on this topic, all the investigations agree in finding the PCTR fine structured, but we have different values concerning its thickness. This limitation, mainly due to the spatial resolution of the UV-EUV instruments, can be overcome with more systematic high resolution observations of prominences seen over different lines of sight. The NASA/IRIS spectrometer (De Pontieu et al. 2014), launched in 2013 and which reaches a spatial resolution of about 0.3", has all the potential to allow a step forward on this issue. It will possibly indicate if the PCTR envelopes a single or multiple cool threads (as suggested by different models and schematically shown in Fig. 3.2). It would be also extremely interesting for modeling the small scale PCTR to derive the DEM and the density at such a resolution.

3.3.3 Final Remarks

To conclude this chapter, it is important to realize that the diagnostics discussed here are affected by uncertainties which should be taken into account in estimating the error bar in the inferred plasma parameters. The theoretical values for $G(T_e)$, for instance, have uncertainties due to how well the atomic parameters are known. In addition, this function can change depending on the hypothesis used for its calculation. As a result different databases may provide different $G(T_e)$ for the same spectral line.

In prominences various factors may break the ionization equilibrium condition (see also Heinzel 2014). For instance, the behavior of Hydrogen (neutrals and protons) has been investigated and found that ambipolar diffusion is one of these (Fontenla et al. 1996). Also, observations of the fine scale revel flows and apparent temporal changes that can be at the origin of local density and temperature changes. These variations have a time scale of the order of few minutes. When this time becomes shorter that the time needed for ionization and recombination of an ion, the equilibrium brakes. Engvold (1980), for instance, already noticed that at the densities found in prominences, the H emission may be formed in out of equilibrium conditions.

Similar investigations have to be done also for the PCTR plasma, where a break in the equilibrium could arise. For instance, let us take the C III ion. For this ion the collisional ionization time scale at 10^5 K and $n_e = 10^9\,\mathrm{cm}^{-3}$ is 51 s, and 100 times smaller for a density of $10^{11}\,\mathrm{cm}^{-3}$. The photoionization has been estimated to be between 10–50 s depending on the solar illumination (E. Landi, private communication). The recombination time scale is, for the two mentioned densities, 23.4 s and 0.2 s respectively. These tell us that density variations within the measured range do not affect ionization equilibrium in prominences. It would be useful if this example will be followed by more systematic investigations.

We have mentioned that element abundance is generally taken as a constant for the theoretical calculations of the line intensity but this has still to be confirmed by the observations. The abundances of H and He are discussed in Labrosse (2014), while there are very few studies concerning measurements of the abundance of less common elements. These suggest that prominences have a composition closer to that of the photosphere than the corona (e.g. Parenti and Vial 2007). This is a key factor for understanding prominences formation, depending on whether the prominence plasma has a coronal or photospheric origin.

The DEM and EM techniques until now have been used for off-limb prominences. The main reason for this is that most of the available EUV-UV data are affected by continuum absorption once observed on the disk. At the same time it would be very interesting to establish the thermal structure of the PCTR through simultaneous observations from different viewpoints. This will be achieved by the out-of-ecliptic observations, as those provided by the next solar mission Solar Orbiter.

The temporal information of the thermal structure of prominences, particularly in the erupting case, is another interesting parameter which is not much treated in the literature. For the quiescent case it requires the following of the structure for days/weeks. For the erupting case, often the temperature is obtained applying the line ratio technique with the assumption of ionization equilibrium, which can be far from being valid.

Finally, in order to understand the local and global equilibrium conditions of prominences, we highlight the importance not only of deriving all the thermodynamic parameters at small scales, but to identifying any spatial variation within the prominence and its PCTR.

Acknowledgements The author acknowledges the financial support from the Belgian Federal Science Policy Office through the international cooperation programs and the ESA-PRODEX program. The author is also grateful to Enrico Landi who provided the photoionization rate for the C III ion, and to J. M. Fontenla for providing his solar irradiance spectral model. CHIANTI is a collaborative project involving George Mason University, the University of Michigan (USA) and the University of Cambridge (UK).

References

Asplund, M., Grevesse, N., Sauval, A. J., & Scott, P. (2009). The chemical composition of the sun. *ARA&A*, *47*, 481–522. doi:10.1146/annurev.astro.46.060407.145222, 0909.0948.

Berger, T., Slater, G., Hurlburt, N., Shine, R., Tarbell, T., Title, A., et al. (2010). Quiescent prominence dynamics observed with the hinode solar optical telescope. i. turbulent upflow plumes. *Astrophysical Journal*, *716*, 1288–1307. doi:10.1088/0004-637X/716/2/1288.

Cirtain, J. W., Golub, L., Winebarger, A. R., de Pontieu, B., Kobayashi, K., Moore, R. L., et al. (2013). Energy release in the solar corona from spatially resolved magnetic braids. *Nature*, *493*, 501–503. doi:10.1038/nature11772.

De Pontieu, B., Title, A. M., Lemen, J. R., et al. (2014). The Interface Region Imaging Spectrograph (IRIS). *Solar Physics*, *289*, 2733–2779. doi:10.1007/s11207-014-0485-y, 1401.2491.

Dere, K. P., Landi, E., Mason, H. E., Monsignori Fossi, B. C., & Young, P. R. (1997). CHIANTI - an atomic database for emission lines. *Astronomy and Astrophysics Supplement*, *125*, 149–173. doi:10.1051/aas:1997368.

Engvold, O. (1980). Thermodynamic models and fine structure of prominences. *Solar Physics*, 67:351–355, doi:10.1007/BF00149812

Engvold, O. (2014). Description and classification of prominences. In: J.-C. Vial & O. Engvold (Eds.), *Solar prominences, ASSL* (Vol. 415, pp. 31–60). Springer.

Fontenla, J. M., Rovira, M., Vial, J. C., & Gouttebroze, P. (1996). Prominence thread models including ambipolar diffusion. *Astrophysical Journal*, *466*, 496. doi:10.1086/177527.

Gilbert, H. R. (2014). Energy balance. In J.-C. Vial & O. Engvold (Eds.), *Solar prominences* (pp. 157–178). Springer.

Heinzel, P. (2014). Radiative transfer in solar prominences. In J.-C. Vial & O. Engvold (Eds.), *Solar prominences, ASSL* (Vol. 415, pp. 101–128). Springer.

Ivanov-Kholodnyi, G. S., & Nikol'Skii, G. M. (1963). Extreme ultraviolet solar radiation and the structure of the solar atmosphere in active and undisturbed regions. *Soviet Astronomy 6*, 609.

Karpen, J. T. (2014). Plasma structure and dynamics. In J.-C. Vial & O. Engvold (Eds.), *Solar prominences, ASSL* (Vol. 415, pp. 235–255). Springer.

Kucera, T. A. (2014). Derivations and Observations of Prominence Bulk Motions and Mass. In J.-C. Vial & O. Engvold (Eds.), *Solar prominences, ASSL* (Vol. 415, pp. 77–99). Springer.

Kucera, T. A., Gibson, S. E., Schmit, D. J., Landi, E., & Tripathi, D. (2012). Temperature and extreme-ultraviolet intensity in a coronal prominence cavity and streamer. *Astrophysical Journal, 757*, 73. doi:10.1088/0004-637X/757/1/73.

Labrosse, N. (2014). Derivation of the major properties of prominences using NLTE modelling. In J.-C. Vial & O. Engvold (Eds.), *Solar prominences, ASSL* (Vol. 415, pp. 431–451). Springer.

Labrosse, N., Heinzel, P., Vial, J. C., Kucera, T., Parenti, S., Gunár, S., et al. (2010). Physics of solar prominences: I – spectral diagnostics and non-lte modelling. *Space Science Reviews, 151*, 243–332. doi:10.1007/s11214-010-9630-6, 1001.1620.

Landi, E., Young, P. R., Dere, K. P., Del Zanna, G., & Mason, H. E. (2013). CHIANTI—An atomic database for emission lines. xiii. soft x-ray improvements and other changes. *Astrophysical Journal, 763*, 86. doi:10.1088/0004-637X/763/2/86.

Lemen, J. R., Title, A. M., Akin, D. J., et al. (2012). The Atmospheric Imaging Assembly (AIA) on the Solar Dynamics Observatory (SDO). *Solar Physics, 275*, 17–40. doi:10.1007/s11207-011-9776-8.

Lin, Y., Martin, S., & Engvold, O. (2008). Filament substructures and their interrelation. In: R. Howe, R. Komm, K. Balasubramaniam, & G. Petrie (Eds.), Subsurface and atmospheric influences on solar activity, astronomical society of the pacific (Vol. 383, p. 235). San Francisco: ASP Conference Series.

López Ariste, A. (2014). Magnetometry of prominences. In J.-C. Vial & O. Engvold (Eds.), *Solar prominences, ASSL* (Vol. 415, pp. 177–202). Springer.

Martin, S. (2014). The magnetic field structure of prominences from direct and indirect observations. In J.-C. Vial & O. Engvold (Eds.), *Solar prominences, ASSL* (Vol. 415, pp. 203–233). Springer.

Parenti, S. (2014). Solar prominences: Observations. *Living Reviews in Solar Physics, 11*, 1. doi:10.12942/lrsp-2014-1.

Parenti, S., & Vial, J. C. (2007). Prominence and quiet-sun plasma parameters derived from fuv spectral emission. *Astronomy and Astrophysics, 469*, 1109–1115. doi:10.1051/0004-6361:20077196.

Parenti S, & Vial JC (2014) On the nature of the prominence - corona transition region. In: *IAU Symposium* (Vol. 300, pp. 69–78). doi:10.1017/S1743921313010764.

Parenti, S., Vial, J. C., & Lemaire, P. (2004). Prominence atlas in the sumer range 800–1250 å: I. observations, data reduction and preliminary results. *Solar Physics, 220*, 61–80. doi:10.1023/B:sola.0000023444.58697.e7.

Parenti, S., Vial, J.C., & Lemaire, P. (2005). Prominence atlas in the SUMER range 800–1250 Å. II. Line profile properties and ions identifications. *Astronomy and Astrophysics, 443*, 679–684. doi:10.1051/0004-6361:20053122.

Parenti, S., Heinzel, P., & Schmieder, B. (2012). On the nature of prominence emission observed by sdo/aia. *Astrophysical Journal, 754*, 66. doi:10.1088/0004-637X/754/1/66.

Phillips, K. J. H., Feldman, U., & Landi, E. (2012). Ultraviolet and X-ray spectroscopy of the solar atmosphere. Cambridge, UK: Cambridge University Press.

Pottasch, S. R. (1963). The lower solar corona: Interpretation of the ultraviolet spectrum. *Astrophysical Journal, 137*, 945+.

Vial, J. C. (2014). Historical background and introduction. In J.-C. Vial & O. Engvold (Eds.), *Solar prominences, ASSL* (Vol. 415, pp. 1–29). Springer.

Vial, J. C., Olivier, K., Philippon, A. A., Vourlidas, A., & Yurchyshyn, V. (2012). High spatial resolution VAULT H-Lyα observations and multiwavelength analysis of an active region filament. *Astronomy and Astrophysics, 541*, A108. doi:10.1051/0004-6361/201118275.

Vourlidas, A., Sanchez Andrade-Nuño, B., Landi, E., Patsourakos, S., Teriaca, L., Schühle, U., et al. (2010). The structure and dynamics of the upper chromosphere and lower transition region as revealed by the subarcsecond vault observations. *Solar Physics 261*, 53–75. doi:10.1007/s11207-009-9475-x, 0912.2272.

Wilhelm, K., Curdt, W., Marsch, E., Schuhle, U., Lemaire, P., Gabriel, A., et al. (1995) Sumer - solar ultraviolet measurements of emitted radiation. *Solar Physics*, *162*, 189–231. doi:10.1007/BF00733430.

Chapter 4
Derivations and Observations of Prominence Bulk Motions and Mass

Terry A. Kucera

Abstract In this chapter we review observations and techniques for measuring both bulk flows in prominences and prominence mass. Measuring these quantities is essential to development and testing of models discussed throughout this book. Prominence flows are complex and various, ranging from the relatively linear flows along prominence spines to the complex, turbulent patterns exhibited by hedgerow prominences. Techniques for measuring flows include time slice and optical flow techniques used for motions in the plane of the sky and the use of spectral line profiles to determine Doppler velocities along the line of sight. Prominence mass measurement is chiefly done via continuum absorption measurements, but mass has also been estimated using cloud modeling and white light measurements.

4.1 Bulk Motions

Prominences have long been known to be dynamic structures, displaying internal motions of various kinds even when globally at rest. Here we discuss bulk motions and methods used for measuring them, although the measurement methods are also relevant to the oscillatory motions discussed in Ballester (2014; Chapter 11).

A good understanding of flows is highly important for testing and constraining models of prominence formation and stability, determining the role of flows in the force and energy equilibria of prominences, and casting light on closely related questions concerning prominence magnetic field structure. For instance, evaporative-condensation models predict that flows of cool material should originate in the corona, while injection and levitation models involve cool material flowing upwards form the chromosphere. If we assume that flows are moving along magnetic field lines then the flow trajectories can tell us about the magnetic structure of the prominence, but it could also be the case that the material is moving with a changing field or even in some cases diffusing across the field. The various

T.A. Kucera (✉)
Laboratory for Solar Physics, Code 671, NASA GSFC, Greenbelt, MD 20771, USA
e-mail: therese.a.kucera@nasa.gov

J.-C. Vial, O. Engvold (eds.), *Solar Prominences*, Astrophysics and Space Science Library 415, DOI 10.1007/978-3-319-10416-4_4

theoretical models of prominence magnetic field and related dynamics and their predictions are discussed in more detail in Karpen (2014; Chapter 10).

To obtain the required information we need detailed trajectories of prominence plasma features, including the origin of the plasma, and any changes in temperature and velocity. How do observed motions compare to those expected from various models of the prominence magnetic field and actual magnetic field measurements? What connections, if any, are there between flows at different temperatures? What processes can explain these observed flows?

In the last few decades, new instrumentation has yielded more information concerning flows. In the visible regime, the new high temporal and spatial resolution data combined with spectral information have revealed counter-streaming flows in filaments on the disk in observations such as those from the Swedish Solar Telescope (SST) at La Palma. The similarly good spatial resolution of the *Hinode*/Solar Optical Telescope (SOT) combined with excellent long term image stability allowed by its space based platform have allowed detailed studies of the complex motions observed in prominences at the limb.

In the UV and EUV range spectrographs such as *Solar and Heliospheric Observatory (SOHO)'s* Coronal Diagnostic Spectrometer (CDS) and Solar Ultraviolet Measurements of Emitted Radiation (SUMER) and *Hinode's* EUV Imaging Spectrometer (EIS) have provided information in a range of lines formed at chromospheric, transition region, and coronal temperatures. The new *Interface Region Imaging Spectrograph (IRIS)* mission provides higher resolution spectroscopic as well as plane-of-sky information concerning the motions of plasma at chromospheric and transition region temperatures. High cadence UV and EUV imaging information from the *Solar Dynamics Observatory (SDO)'s* Atmospheric Imaging Assembly (AIA) and, earlier, from the *Transition Region and Coronal Explorer (TRACE)* and *SOHO*/Extreme ultraviolet Imaging Telescope (EIT) have been important as well.

4.1.1 Measurements of Motions

Motions in the plane of the sky are measured by tracking actual features, while line of sight (LOS) motions are detected using Doppler shifts. Sometimes the plane-of-sky (POS) and line position methods can be combined to good effect. With either LOS or POS measurements alone, we have no direct information on the 3D structure. However, in some cases we can make estimates based on the knowledge of the orientation of prominence features as viewed over several days or from two points of view using the *Solar Terrestrial Relations Observatory (STEREO)* spacecraft.

4.1.1.1 Motions in the Plane of the Sky

Observations of motions in the plane of the sky have the advantage that it is possible to pick out actual moving features, making it straight forward in some cases to track the motion by simply marking off the feature location in successive images.

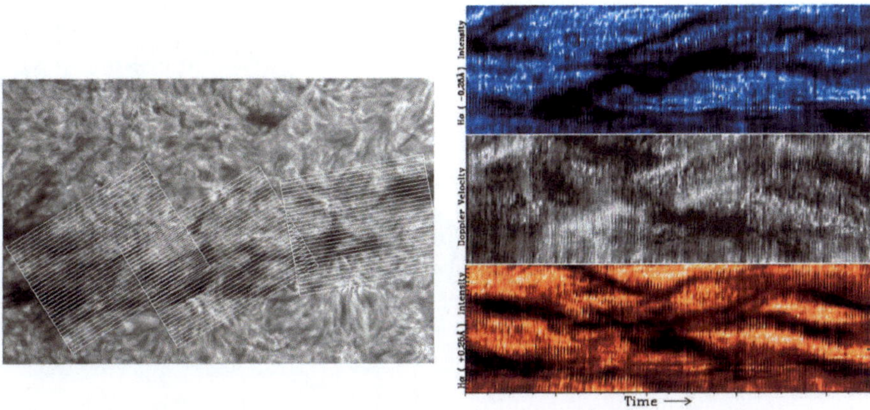

Fig. 4.1 An illustration of the use of time slice diagrams to analyze flows in a Hα filament observed with the SST (Lin et al. 2003). The *left panel* shows the locations of parallel slices used for analyzing velocities in Hα fine structures. At *right* from *top* to *bottom* are the blue wing intensities, Doppler shift, and red wing intensities along a single time slice located in the middle section of the filament. Diagonal features show motions along the slice

There are a number of general things to be careful of, however. The very fact that such measurements are feature driven makes it possible to miss evenly moving flows that vary below the resolution of the data. Also, there is the question of what is actually moving. Often a variation in intensity may indeed be due to actual moving material, but it is also possible it is due to a propagating disturbance in density or temperature. There are cases, especially in optically thin plasmas, in which multiple layers of plasma can produce confusing intensity variations and make the data hard to interpret. It is important to have sufficiently high spatial and temporal resolution to avoid aliasing and be able to reliably identify and track particular features. One should also consider projection effects. For instance, when a prominence is seen along the limb it can be difficult to distinguish vertical motions from horizontal flows curving over the solar limb.

Time-Slice A common variation of feature tracking is to use the time slice method (see Fig. 4.1). The intensity or velocity along a trajectory traced out in consecutive images of the filament is followed versus time, and the slope of the brightening or darkening in this 2D diagram gives a measurement of the velocity of the feature in the plane of the sky. This method can be easily applied to high spatial resolution images, e.g., with the SST or the Dutch Open Telescope (DOT) at La Palma, and has also been used with EUV images. The slices can be linear, but they can also curve to follow the trajectory of a particular feature.

Optical Flow Techniques There has also been some use of automatic tracking to trace prominence flow fields. This can be difficult because of the complex nature of prominences, often including multiple sources that may be optically thin or combine both emission and absorption. The techniques that have been used successfully are

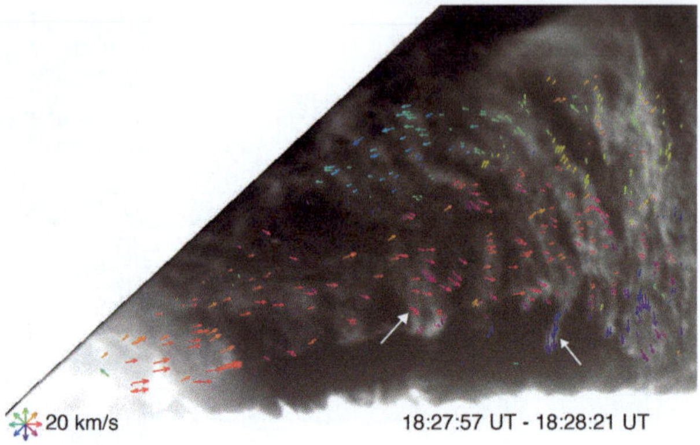

Fig. 4.2 Vertically oriented features seen in a hedgerow prominence observed with *Hinode*/SOT in the Hα line (Chae et al. 2008). They appear to move both horizontally and vertically in the plane of the sky. *White arrows* point out a newly formed feature and one which is falling downwards

best applied to areas that can be described in terms of flow fields rather than discrete moving features.

One of the most commonly used of these methods is Local Correlation Tracking (LCT) (Leese et al. 1970), in which a two dimensional cross-correlation is applied to a pair of images to determine the shifts for different parts of the images so as to maximize the function (November and Simon 1988)

$$C(\delta, x) = \int J_t(\xi - \frac{\delta}{2}) J_{t+\tau}(\xi + \frac{\delta}{2}) W(x - \xi) \partial \xi \tag{4.1}$$

where $J_t(x)$ and $J_{t+\tau}(x)$ are the two images taken at times t and $t + \tau$ respectively, with a vector displacement δ between them. $W(x)$ is a windowing function that controls the size of the region over which the images are compared.

Also used are optical flow techniques originally developed for magnetogram analysis, such as the nonlinear affine velocity estimator (NAVE) (Schuck 2006) which was used to track the flows shown in Fig. 4.2. These algorithms utilize spatial derivatives of the images and allow for more complex local flows (Chae and Sakurai 2008).

4.1.1.2 Along the Line of Sight: Doppler Observations

Doppler measurements more reliably give access to steady flows. If there is sufficient spectral resolution the Doppler velocity is generally determined by fitting the observed spectral line, usually with a Gaussian function, and applying the standard formula for Doppler shifts,

$$v/c = (\lambda - \lambda_0)/\lambda_0 \tag{4.2}$$

where λ_0 is the rest wavelength of the line, λ the line wavelength of the source, v the source velocity, and c the speed of light.

When specifying a Doppler velocity it is important to be clear how the zero velocity point is determined. In many cases one cannot assume that the data one is using has been calibrated with respect to the absolute wavelength. The wavelength scale often shifts in the instrument due to instrumental heating or, for ground based instruments, changes in the refractive index of the atmosphere. Although it is possible for some wavebands to incorporate a wavelength standard into an instrument this is not often done in space-based solar spectrographs.

One method of velocity calibration is to select a feature in the field of view or that has been recently observed and declare it to be at rest so that the velocity values are relative to that feature. For spectral observations in visible wavelengths one should optimally precede or follow one's observations with observations of a quiet area at disk center. The average spectral profile over the region is then used to determine the rest wavelength position. The wavelength scale is also sometimes calibrated by measuring it compared to certain spectral lines that are assumed to be at rest. For instance, for UV or EUV observations of the solar disk absolute wavelength calibrations are often done by assuming that chromospheric spectral lines averaged over a large area are at rest. If the Doppler shifts of interest are to be measured using a spectral line produced at chromospheric or transition region temperatures observed above the limb (as is the case for a prominence), a coronal line at a near-by wavelength (again averaged over a large area) might be used as a standard. However, coronal plasma may not be at rest either. Thus one should be aware of the uncertainties in the chosen method and consider how well the absolute Doppler shift has been measured.

A spectrally resolved profile is generally fit with some function, usually a Gaussian. Often there is more than one Doppler component present. This is especially likely in optically thin plasmas, where the emission is an integration of the LOS velocity of many different features so that the measurement provides a value averaged along the LOS. It is possible, for instance, for some of the plasma in the field of view to be relatively stationary and for other parts of it to be moving. An example is shown in Fig. 4.3. Here, fitting the entire spectral line as a unit could result in a measurement of almost no shift at all. At best one would be left with the impression of a broadened line profile with no information concerning the details of the flow. Instead, the line has been fit by two Gaussian functions, indicating a blue shifted component moving at 12 km s^{-1} and a red shifted component at about 4 km s^{-1} (Orozco Suárez et al. 2012).

A related issue is that of line blending, in which multiple lines are present. If all the lines are from the same source they will move in tandem, but if they are not (if, for instance, they are formed at different temperatures in different regions) fitting the line as a single source will again yield incorrect results. Even if two blended lines are formed from the same ion, a density dependence in their ratio can lead to spurious Doppler signals.

Another aspect to Doppler line fitting is that of the reversed lines, which is particularly relevant to prominences. Some optically thick lines exhibit central reversals. Full modeling of the behavior of such lines is discussed in Heinzel (2014; Chapter 5), but the location of the center of the absorption feature can be an indicator of the relative velocities of different parts of the structure.

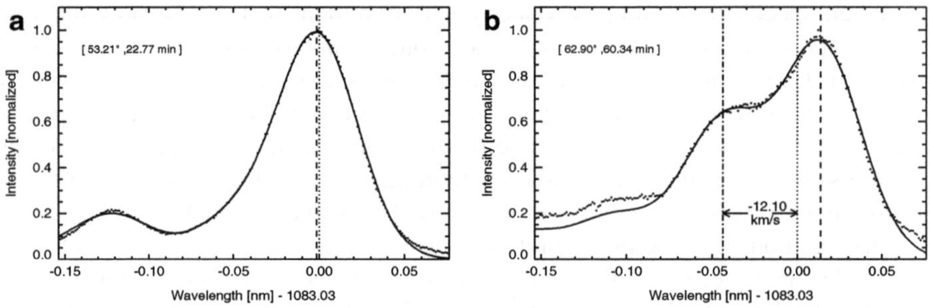

Fig. 4.3 Example of a multi-component Doppler shift in a prominence footpoint in He I 10,830 Å (Orozco Suárez et al. 2012). (**a**) shows a rest profile consisting of a weak line at 10,829.09 Å (due to the $^3S_1$-$^3P_0$ transition) and the main line at 10,830.29 Å ($^2S_1$-$^3P_1$ and $^2S_1$-$^3P_2$). (**b**) shows a spectrum with two Doppler components in the main line, one red shifted and one blue shifted. A spectral fit with a single component might indicate little or no motion along the line of sight. The measurements were made with data from Tenerife Infrared Polarimeter attached to the German Vacuum Tower Telescope. © AAS. Reproduced with permission

For observations of filaments on the disk, the signal coming from the prominence must be disentangled from the chromospheric background. Various techniques have been developed to do this, principally based on cloud model methods (Beckers 1964; Mein et al. 1996; Tziotziou 2007). With such methods, the data is fit to a non-LTE model of the prominence as a "cloud" suspended above the chromosphere utilizing four parameters: the LOS velocity V, the source function S, the optical thickness τ, and the line width $\Delta\lambda_D$. S and τ are strongly coupled but V can be computed. Cloud models are described in more detail in Heinzel (2014; Chapter 5)

For all these caveats and difficulties, there are short-cuts that are taken. Lines are sometimes sampled at only two or three wavelengths. Differences in the intensities of the red and blue wings of the lines are then used to identify general regions of flow or to separate out material flowing in different directions.

Optimally measurement of flows in and transverse to the plane of the sky can be combined to reveal the three dimensional trajectory of the motion. One particular method used to combine the two is to trace motions using an image slice method applied to the red or blue wing of a spectral line. An example of this is shown in Fig. 4.1, in which a time slice analysis is done in the red and blue wings of the Hα line, which show different features (Lin et al. 2003).

4.1.2 Observations of Flows in Prominences

Prominences show a variety of different flows. Here we describe some of the characteristic ones for non-erupting prominences. Eruptions and eruption precursors are discussed in Webb (2014), Gopalswamy (2014; Chapters 15 and 16).

4.1.2.1 Flows in Quiescent Prominences

What are referred to as quiescent prominences exhibit a range of flow behaviors including motions of vertically aligned structures in hedgerow prominences and also the motions in the spine and barb formations of lower latitude prominences in quiet regions. Although traditionally quiescent prominences are thought of as relatively stationary, even the most stable quiescent filaments are dynamic structures exhibiting flows and oscillations.

Spine flows Quiescent prominences not in the polar crown often show a two-part structure of a long spine accompanied by barbs which extend down to the chromosphere (Engvold 2014). Observations in Hα show flows along prominence spines and barbs at speeds of 10–20 km s^{-1}. As shown in the schematic in Fig. 4.4, the flows can go in both directions simultaneously along the spine, a phenomenon known as counter-streaming (Zirker et al. 1998). These observations report individual moving features traced over distances of 10,000–100,000 km. Other observations of spine flows in an intermediate prominence show moving features changing direction, suggesting counter-streaming may be the result of plasma oscillating along the magnetic field (Ahn et al. 2010).

Although polar crown prominences often appear to consist mostly of barbs when observed in Hα, observations in the 304 Å EUV imaging band show flowing material along spines, evidently too hot or optically thin to be observed in Hα but apparent in the more optically thick He II 304 Å line.

He II 304 Å emitting plasma has been observed moving along prominence spines at speeds up to 75 km s^{-1} in the plane of the sky (Wang 1999), much faster than considered typical for Hα flows, and similar observations have been made in other transition region temperature lines in the EUV (Kucera et al. 2003).

Flows in Barbs Although barbs in polar crown prominences often take the form of hedgerows (see below), in lower latitude prominences they are narrower structures, appearing as short outgrowths from the spine against the disk in Hα and as thin dark pillars in EUV images of prominences along the limb. These EUV pillars exhibit swaying motions that can be interpreted as material oscillating back and forth (Panasenco et al. 2014), perhaps in dipped field lines. Alternatively, it has been suggested based on Doppler data that these motions may be due to helical motions around a roughly vertical axis (Orozco Suárez et al. 2012). This seems hard to reconcile with the appearance of these structures from above. From that point of view they appear to consist of relatively straight threads.

Hedgerow Prominence Flows Hedgerow prominences (Engvold 2014) tend to form in the polar crown. As mentioned above, hedgerows as seen in Hα are often interpreted as large barbs, with the prominence spine not strongly visible in Hα but observed in He II 304 Å line.

Hedgerows barbs as seen in Hα often show structures that are aligned perpendicular to the solar limb. These structures have been observed for many years, but recent high resolution observations have renewed focus on these features and related motions.

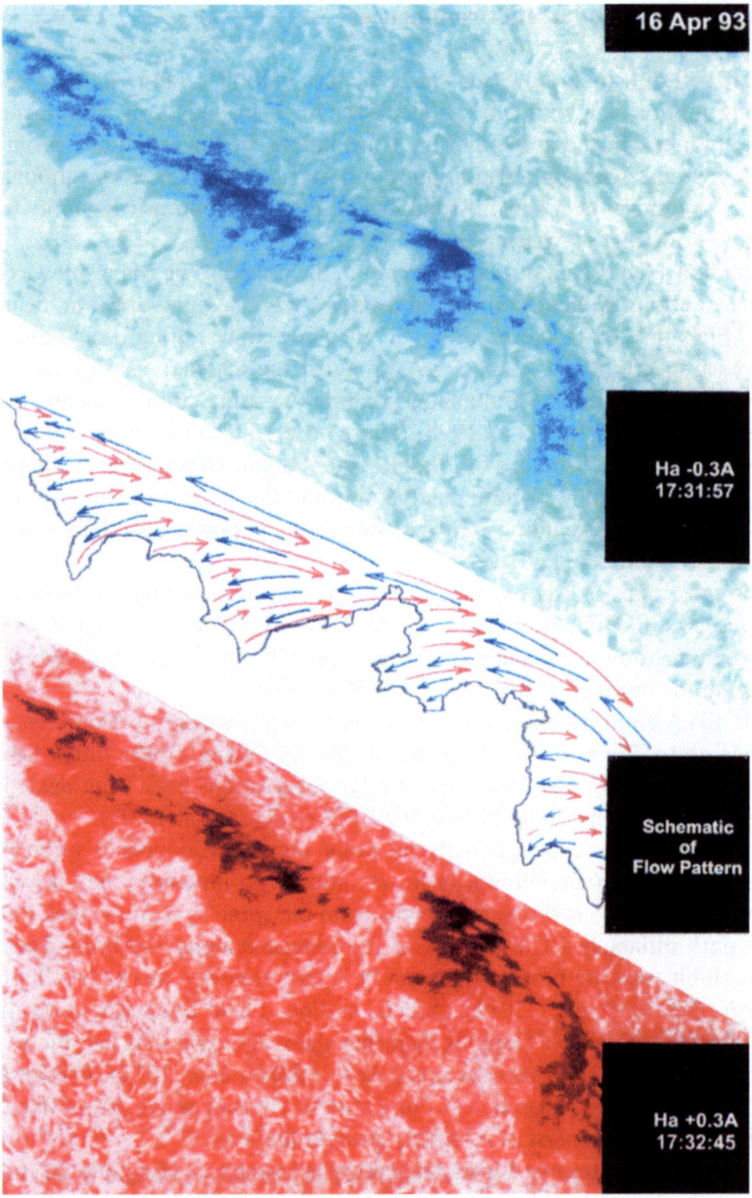

Fig. 4.4 Counter-streaming observed in along a prominence spine and barbs observed at Big Bear Solar Observatory (Zirker et al. 1998). The flows were detected by constructing movies from images in the blue wing (*top*) and red wing (*bottom*) of the Hα line. Reprinted by permission from Macmillan Publishers Ltd: Nature copyright 1998

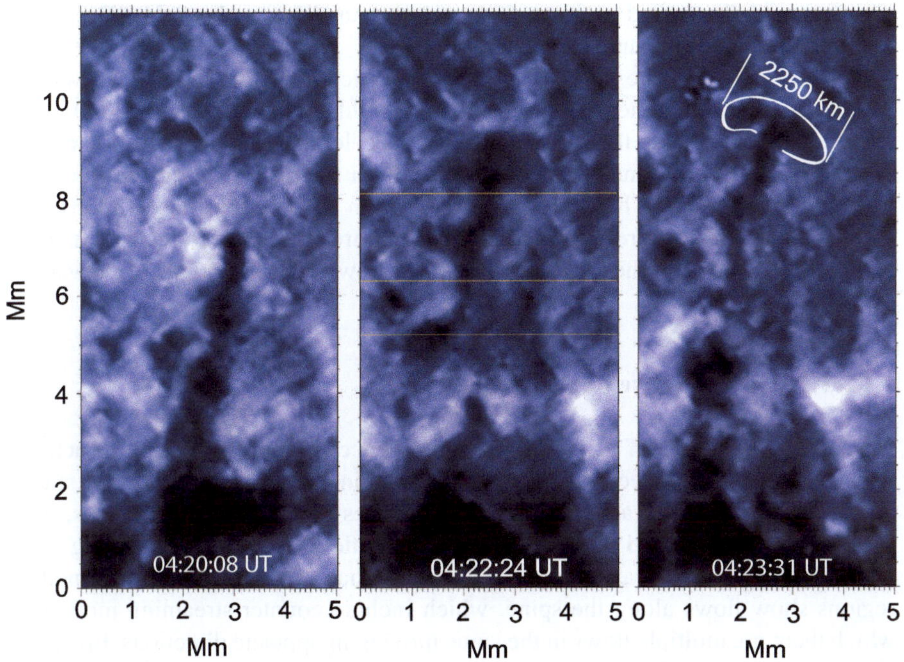

Fig. 4.5 Dark upward moving feature with a "mushroom cap" head observed on the limb with *Hinode*/SOT in the Ca II H (3,968 Å) line, ascending with a mean velocity of 22.9 km s^{-1} in the plane of the sky (Berger et al. 2008). © AAS. Reproduced with permission

The *Hinode* SOT instrument in particular has made possible images of prominences and prominence flows at the limb with unprecedented resolution and stability. Observations of large, quiescent hedgerow prominences in the Ca II H and Hα lines show vertically aligned bright and dark features. An example of the bright features is shown in Fig. 4.2, which shows a series of bright vertical structures exhibiting vortical-type motions. They move horizontally at about 10 km s^{-1} and then move downwards so that individual blobs attain speeds of 35 km s^{-1}. The downwards acceleration of the blobs is 0.015 km s^{-2} or less, significantly below the gravitational acceleration of 0.27 km s^{-2} (Chae et al. 2008). Also observed are turbulent looking dark features moving upwards from the bottom edge of the hedgerow (Fig. 4.5). These plumes have maximum initial speeds in the plane of the sky of 20–30 km s^{-1} and decelerate as they rise (Berger et al. 2010).

A number of possible explanations for the sub-gravitational acceleration in hedgerows have been suggested. This phenomenon is also observed in coronal cloud prominences and coronal rain (see Sect. 4.1.2.3), although it is not clear if the same phenomena is at work. Possible mechanisms include density enhancements in vertical magnetic fields (Mackay and Galsgaard 2001), Lorentz forces associated with small scale horizontal or tangled fields (Low and Petrie 2005; van Ballegooijen and Cranmer 2010), and pressure from waves transverse to the field (Pecseli and Engvold 2000; Antolin and Verwichte 2011).

Doppler data indicate that the apparently vertical structures in hedgerows also have a velocity component along the line of sight (Schmieder et al. 2010). This may be relevant to both the sub-gravitational acceleration and the question of how these features on the limb relate to those seen on the disk. Contrary to the rather turbulent appearance of hedgerows observed from the side, Hα observations from above show relatively direct flows along straight thread-like structures with inclined up- and down-flows in the barbs. It is not clear how these two points of view are to be reconciled.

4.1.2.2 Flows in Active Region Prominences

One of the signatures of active region prominences is their high level of activity, which includes easily seen flows along the prominence axis. A number of studies report the formation of active region prominences via abrupt jet-like flows, often associated with observed activity in the magnetic field. Velocities in forming active region filaments can be as high as 250 km s^{-1} (Chae 2003). Even once formed the regions show flows along the spine, which include counter-streaming motions in which there are multiple flows in the spine moving in opposite directions. Figure 4.6 shows motions of almost 100 km s^{-1} in an active region filament observed in the 193 Å band (Alexander et al. 2013). In addition to these flows observed in cool prominence plasma, active regions filaments are also associated with coronal temperature jet-like features directed parallel to or spiraling around the cooler filament material.

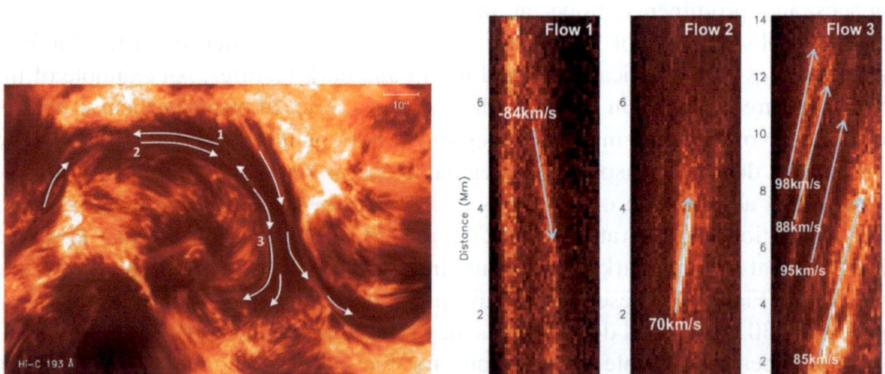

Fig. 4.6 Flows observed in an active region filament with the rocket-borne High-resolution Coronal Imager (Hi-C) instrument in a 193 Å band (Alexander et al. 2013). Flows were observed in both directions along the spine with velocities in the plane of the sky between 70 and 100 km s^{-1}. © AAS. Reproduced with permission

4.1.2.3 Coronal Cloud Prominences and Coronal Rain

Coronal cloud prominences are characterized by drainage along curved trajectories. One study from the 1970s found acceleration in some coronal cloud prominence flows to be consistent with that expected from gravitation (Engvold 1976). More recent observations report accelerations significantly less than gravitational acceleration (Stenborg et al. 2008; Liu et al. 2012). As discussed in Sect. 4.1.2.1, a number of theories have been put forward to explain such sub-gravitational acceleration,

The draining of chromospheric temperature material from the corona is know as "coronal rain." Coronal rain is also commonly observed in unstable loops above active regions. Combined EUV and visible observations of such loops show that the phenomena is due to hot, coronal temperature loops cooling, resulting in the raining down of condensing material (Schrijver 2001). This may be due to the sort of evaporation-condensation process hypothesized for prominences (Karpen 2014) but applied to arched field lines instead of the horizontal or dipped field lines associated with prominence inversion lines. As in coronal cloud prominences, the observed acceleration is less than that expected from gravity. Average speeds are about 70 km s^{-1} (Antolin and Rouppe van der Voort 2012).

4.1.2.4 Flows Observed in the PCTR and Corona

In many ways, observations of the prominence corona transition region (PCTR) and corona in and around prominences are quite limited compared to Hα observations since they rarely reach the combination of high temporal, spatial, and sometimes spectral resolutions available in Hα (although IRIS should offer an improvement on this for some transition region lines). However, the insight they provide with regard to temperature makes them important to our understanding of basic prominence properties.

One key question is how hotter temperature plasma is moving relative to the cool material forming the prominence core. Some observations do show emissions formed at different transition region temperatures from what appear to be the same moving source, especially below about 2.5×10^5 K (Wiik et al. 1993; Kucera et al. 2003; Kucera and Landi 2006, 2008), suggesting a cool core with a PCTR accompanying it. Other studies, especially Doppler and related modeling, have suggested that there may be different threads formed at different temperatures with a range of velocities (de Boer et al. 1998; Cirigliano et al. 2004).

As described above, EUV observations of transition region temperature material in prominences have shown faster moving plasma than what is considered normal in Hα. This may be because they highlight portions of the prominences not observed in Hα. However, even observations of absorption near 195 Å, which should reflect the same plasma as Hα, report quite fast motions (Panasenco et al. 2014). This may be a selection effect resulting from the relatively low spatial resolution of most EUV images as compared to the highest resolution Hα telescopes.

The higher temperature transition region lines and even corona lines highlight the extent to which the prominence is part of a larger magnetic structure. In active prominences bright jets of coronal temperature material spiral around prominence spines. Even quiescent prominences viewed in EUV at the limb along direction of the prominence spine reveal flows along apparently spiral tracks inside prominence cavity, while the cool dense prominence material collects along the bottom. Flows in cavities associated with prominences are discussed in detail in Gibson (2014; Chapter 13).

4.1.2.5 Flows on Time Scales of Days

Prominences evolve on different time scales. For instance, they are been measured to rise and expand with time, especially in the days before eruption (Liu et al. 2012). Such changes are presumably due to evolution of the magnetic field in which the prominence is embedded (Gibson 2014).

Another sort of slow evolution has been reported in stable filaments. Kilper et al. (2009) report that such filaments show a decrease in ratio of He I 10830 Å to Hα emission as a function of height over time, although this variation disappears in more active or erupting filaments. They interpret this as sign of cross field diffusion which would be expected to be much faster for the heavier helium atoms than for hydrogen (Gilbert et al. 2007).

4.2 Prominence Mass

Prominence mass is an important quantity in our understanding of prominence plasma—a basic physical quantity to be accounted for by models of prominence formation and also eruption. Here we will discuss methods used for measuring prominence mass and summarize some of the results.

4.2.1 Continuum Absorption

The most frequently used method for measuring prominence mass is via measurements of continuum absorption. An example of a prominence seen in absorption in EUV is shown in Fig. 4.7. This is absorption due to the photoionization of neutral hydrogen and neutral and once ionized helium, chiefly Lyman-absorption from the ground state, and is proportional to $\exp(-\sigma N)$ where N is the column number of absorbing atoms or ions and σ is the absorption cross section. As shown in Fig. 4.8, the absorption cross section for each of these species has an upper limit in wavelength, 912 Å for H^0, 504 Å for He^0, and 228 Å He^+ below which it decreases. There are also auto-ionization resonances in the neutral helium cross

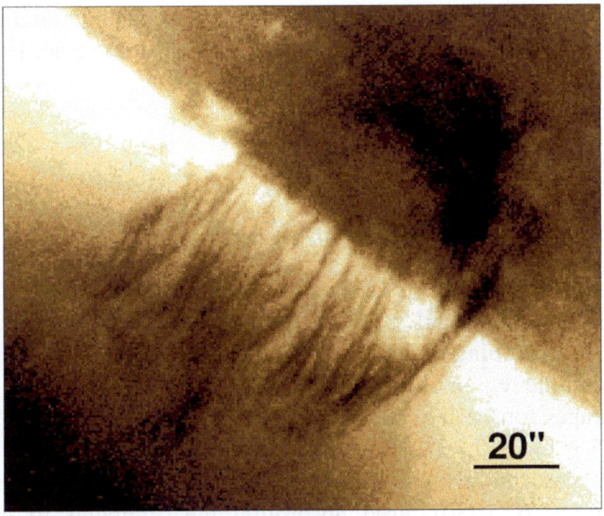

Fig. 4.7 Prominence observed in absorption in the 193 Å band of *SDO*/AIA on 2010 August 11 at 20:00 UT. Credit: NASA/SDO and the AIA science team

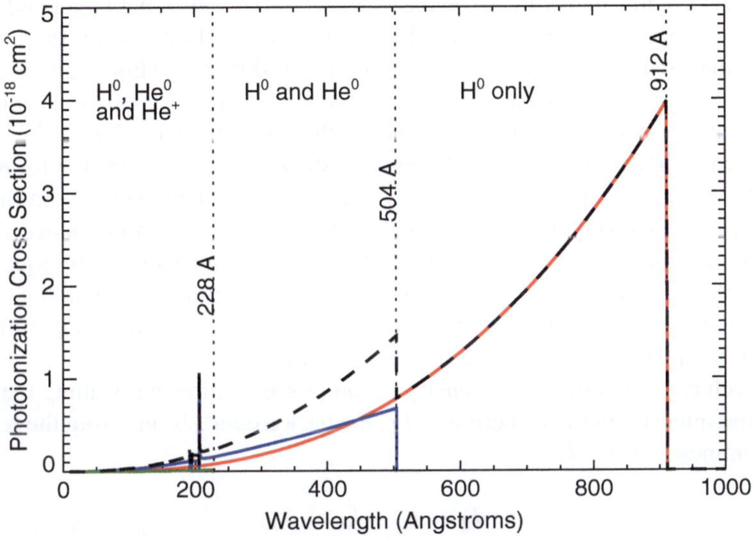

Fig. 4.8 Photoionization cross-section per atom/ion in a prominence of assumed hydrogen fraction 0.9, hydrogen fractional ionization 0.3 and helium ionization fraction 0.1. Neutral hydrogen (Karzas and Latter 1961) (*red*) contributes below 912 Å, He^0 (Fernley et al. 1987) (*blue*) below 504 Å and He^+ (Karzas and Latter 1961) (*green*) below 228 Å. The He I curve shows the locations of the helium auto-ionization features. The total average is shown by the *dashed line*. Figure made by author using code courtesy V. Andretta

section. Formulations for the individual cross sections can be found in a number of sources (Karzas and Latter 1961; Fernley et al. 1987; Rumph et al. 1994; Keady and Kilcrease 2000).

The total cross section, σ is

$$\begin{aligned}\sigma &= N_{H^0}\sigma_H + N_{He^0}\sigma_{He^0} + N_{He^+}\sigma_{He} \\ &= \epsilon_H(1 - x_H)\sigma_H + (1 - \epsilon_H)(1 - x_{He^+} - x_{He^{+2}})\sigma_{He^0} + \epsilon_{He} x_{He^+}\sigma_{He}.\end{aligned} \tag{4.3}$$

Here ϵ_H and ϵ_{He} are the fractional abundances of hydrogen and helium by number ($\epsilon_H + \epsilon_{He} \approx 1$), and x_H, x_{He^+} and $x_{He^{+2}}$ are the ionization fractions of H and He in the absorbing region. We usually assume that all helium is neutral or singly ionized in the region of interest, so that $x_{He^{+2}} = 0$.

Possible ranges for these quantities would be: $0.1 \gtrsim x_H \gtrsim 0.5$, $0.005 \gtrsim x_{He} \gtrsim 0.14$, and $0.85 \gtrsim \epsilon_H \gtrsim 0.95$ (Gilbert et al. 2005).

Continuum absorption in prominences was first noted in *Skylab* data. Orrall and Schmahl (1976, 1980), used absorption measurements to determine column densities of neutral hydrogen. Similar measurements were carried out with *SOHO* (Kucera et al. 1998; Penn 2000) and *TRACE* (Golub et al. 1999), and have been performed using a number of different instruments and combinations thereof. Typical H I column densities are on the order 10^{18}–10^{19} cm^{-2}, but larger and smaller values have been reported. Clearly, the exact values will depend on the particular feature analyzed and its orientation. Gilbert et al. (2005, 2006) first used absorption-based column density measurements to estimate the mass for entire prominences. They found total prominence mass values in the range 8×10^{13}–2×10^{15} g.

The basic geometry of the problem is as shown in Fig. 4.9. In this formulation I_0 is the intensity that would be seen by the observer if there were no prominence, while I_1 is the intensity in the portion of the sky containing the prominence material. In areas not including the prominence in the line of sight these are equal. I_0 in the region containing the prominence is determined by either interpolating spatially using adjacent areas without prominence material or interpolating temporally if the prominence material is moving across the field of view.

In each case the observed intensity is comprised of three parts along the line of sight: the intensity in the foreground, I_f, the background, I_b and from the region of the prominence itself, I_p.

$$I_0 = I_b + I_{0p} + I_f, \tag{4.4}$$

$$I_1 = \alpha I_b + I_{1p} + I_f. \tag{4.5}$$

where the absorption is represented by the extinction factor,

$$\alpha = \exp\left(-\int_0^l n\sigma ds\right), \tag{4.6}$$

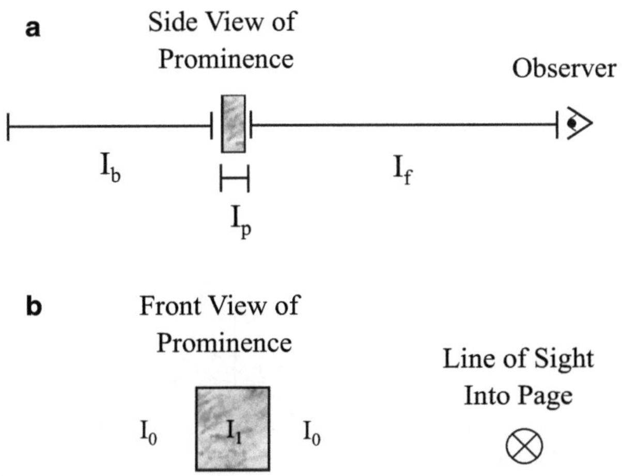

Fig. 4.9 Schematics showing the a prominence observed from (**a**) the side and (**b**) the front (Gilbert et al. 2005). (**a**) I_b and I_f represent the emission in the background and foreground of the prominence while I_p is represents the coronal-temperature radiation from the region of the prominence. (**b**) I_0 represents the observed intensity in areas of the sky without the prominence while I_1 represents the intensity with the prominence. © AAS. Reproduced with permission

in which the product of the number density of particles and the absorption cross section in the source is integrated over the depth of the prominence along the line of sight, l.

The total mass is then

$$M = (4(1 - \epsilon_H) + \epsilon_H)m_H \int\int -\frac{\ln \alpha}{\sigma} da, \tag{4.7}$$

where m_H is the mass of hydrogen and a is the area associated with the prominence absorption.

Thus there are at least five unknowns connected to the geometry alone. There are additional ones associated with the abundance and ionization parameters [Eq. (4.8)] and other issues like unresolved structure and portions of the prominence not detected in absorption. Researchers using continuum absorption techniques to calculate column densities and prominence mass have approached these problems with a variety of assumptions and combinations of data.

Geometry For a prominence sufficiently in front of the plane of the sky (erupting from near disk center, for instance) one might assume $I_f << I_b$ and neglect the foreground radiation all together. It has also been suggested that for a prominence on the limb one might assume $I_f \approx I_b$, although the corona is highly non-uniform, so such an assumption could be problematic.

Gilbert et al. (2005) developed a technique for estimating I_f and I_b by comparing the prominence emission against adjacent regions in which the background but not the foreground emission is presumed to be varying (at the solar limb, for instance).

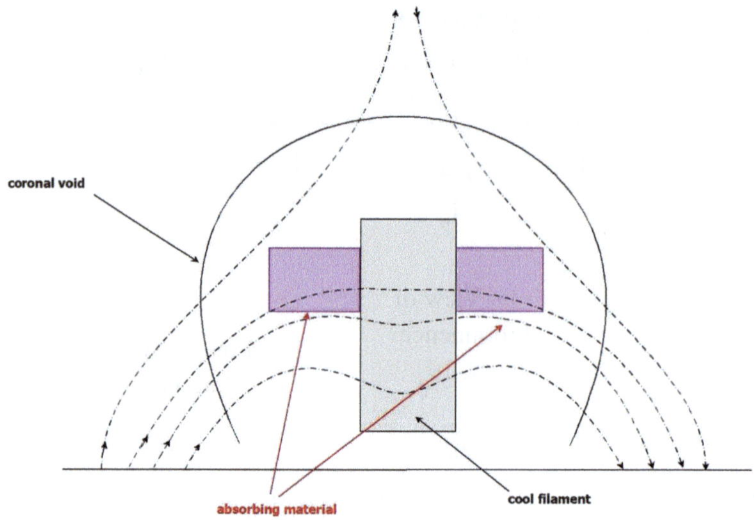

Fig. 4.10 Cartoon based on H-Lyα observations (adapted from Vial et al. 2012) showing the location of the extended EUV filament in a filament channel viewed along the inversion line. The central "cool filament" area shows the location of the prominence as would be seen in Hα. The "absorbing material" shows additional areas with material detected in Lyα. Also shown is the outline of the coronal cavity ("coronal void") and magnetic field lines (*dashed lines*)

The Extended EUV Filament Some EUV observations reveal a more extended filament structure than that often considered in the Hα line. Hα is similar in optical depth to the continuum absorption near about 195 Å (Heinzel et al. 2008). Thus, as can be seen from Fig. 4.8, observations of filaments in wavelengths that are significantly greater than 195 Å but still less than the Lyman cut-off at 912 would be expected to show more tenuous portions of the prominence. It is indeed the case that EUV observations of filament channels, for instance in Mg X 625 Å, show substantially wider filaments than those observed in Hα (Heinzel et al. 2003; Schwartz et al. 2004). Extended filament structure is also observed in hydrogen Lyman-α observations of prominences (Vial et al. 2012; Schwartz et al. 2012) and, as mentioned earlier in this chapter, prominences appear much larger in He II 304 Å than in Hα. Figure 4.10 shows a suggested geometry for the extended filament based on H-Lyα observations. It is not clear in such lines the extent to which we are observing more diffuse cool material or starting to observe the PCTR. It has been estimated that the extra mass due to the filament extension may be equal to 50–100 % the mass derived from the Hα emitting prominence (Aulanier and Schmieder 2002; Heinzel et al. 2003).

Emissivity Blocking Emissivity Blocking (at one time referred to as volume blocking) is a reduction in the observed intensity in the corona due to the fact that cool structures do not emit radiation in lines formed at coronal temperatures (Anzer and Heinzel 2005; Heinzel et al. 2008). This applies both to the situation where the prominence is observable via absorption, and also to coronal line observations at wavelengths not effected by continuum absorption or where such absorption is relatively weak. In these later cases there may be darkening due to emissivity blocking alone. Measurement of emissivity blocking can be performed with a combination of data taken at wavelengths that do not exhibit continuum absorption and ones that do. These can be combined to determine the amount of darkening which occurs simply because of a lack of emission, allowing for a accurate accounting for the I_{0p} and I_{1p} terms in Eqs. (4.4) and (4.5).

Substructure Unresolved sub-structure can also affect masses deduced from continuum absorption (Orrall and Schmahl 1980; Kucera et al. 1998, 2014). One way to account for such sub-structure is through a filling factor describing the fraction of the observing element that is filled with material. This quantity, f_{pos}, is an area filling factor and is distinct from the volume filling factor. Including it changes Eq. (4.6) to:

$$\alpha = f_{pos} \exp\left(-\int_0^l n\sigma ds\right) + (1 - f_{pos}) \tag{4.8}$$

If f_{pos} is low then the amount of material needed to account for a measured amount of absorption, α, will be higher than that needed if there is no unresolved substructure.

Prominence Emission It is also often assumed that $I_{1p} = 0$, i.e., that the emission from the prominence itself is negligible. However, this may not be the case in some wave bands containing contributions from the PCTR. In general when using continuum absorption it is best to select lines or wavebands in which there is no extra emission from the prominence itself and substantial background emission.

Other Uses for Continuum Absorption Measurements in Prominences Uncertainties in ionization fraction and helium abundance present difficulties for mass determinations using Lyman-continuum absorption. The converse of that is that this absorption has potential for measurements of abundances and temperature. Gilbert et al. (2011) attempted to calculate the neutral H/He ratio by using *SOHO* CDS data above and below the 504 Å for neutral helium absorption. The attempt was foiled by optically thick absorption in one of the lines used (Mg X at 625 Å), which is another thing to be careful of when analyzing absorption. However, the measurement may be possible with a more sensitive instrument able to detect fainter coronal lines and fainter features. Landi and Reale (2013) use the dependence of the continuum absorption on ionization fraction and thus temperature to estimate the temperature in an eruptive prominence.

4.2.2 Cloud Modeling

Mass can be determined with spectral lines by comparing observations with the results of non-LTE magneto-hydrostatic models (Mein et al. 1996; Heinzel et al. 1999; Heinzel 2014). By measuring or making assumptions about temperature, bulk velocity and turbulent velocity it is possible use a grid of model results combined with measurements of the filament dimensions to calculate the filament mass. As with the continuum absorption method, uncertainties in ionization fractions result in significant uncertainties in the mass values. Masses calculated in this way give values in the range 2–6$\times 10^{15}$ g, on the order of values obtained using calculations based on continuum absorption (Koutchmy et al. 2008; Grechnev et al. 2014).

4.2.3 *White Light Measurements*

Erupting prominences contribute to the mass calculated for CMEs done using white light coronagraph observations. If the emission observed is purely due to Thompson scattering off electrons then mass is proportional to the electron density with an angle dependence emphasizing the contribution in the plane of the sky. Inversions of the white light intensity to calculate the CME mass have been performed since the 1970s (Stewart et al. 1974).

To calculate CME mass the white light associated with the CME is determined by subtracting off a pre-event image. The mass is then (Vourlidas et al. 2010):

$$M_{CME} = I_{CME}\, C_e\, C_{plasma}, \tag{4.9}$$

where I_{CME} is the excess brightness associated with the CME in units of mean solar brightness and C_{plasma} is the composition of the CME plasma. If one assumes a composition of 10 % helium C_{plasma} is 2×10^{-24} g electron $^{-1}$. C_e is obtained from Thompson scattering theory (Billings 1966; Hundhausen 1993) using the assumption that all electrons lie in the plane of the sky. The mass is determined for each pixel and summed over the area associated with the CME.

It is also usually assumed that the prominence material still present is completely ionized (Vourlidas et al. 2010). If that is not the case it can result in an over estimate of CME mass due to Hα emission in the white light bandpass. Athay and Illing (1986) took into account both white light and Hα to calculate the mass of an erupting prominence, including an analysis of the ionization state of hydrogen, to obtain a mass of 1×10^{16} g.

4.3 Some Outstanding Questions Related to Prominence Flows and Mass

There are numerous questions concerning prominences to which measurements of flows and mass are key. Most of them tie into the basic question of the origins of prominence mass. Clearly we need to bring all available observations of flows and oscillations along with measurements of temperature, mass, density, and magnetic field and, of course, modeling to bear on this larger question and the related questions listed below.

- Where and at what temperature does prominence plasma originate? This is an important distinguisher between different theories of the origin of prominence mass. Some observations, especially in active regions, suggest that prominence material originates as relatively cool plasma ejected from the chromosphere (Wang 1999; Schmieder et al. 2004). Other observations suggest that the cool material condenses from hot material (Berger et al. 2012). These differences may be the result of more than one mechanism at work in different situations, for instance between active region vs. quiescent prominences, but this still is not entirely clear.
- In a related question, what is the role of barbs? Are they the chief conduit of material to and from the chromosphere, important to the existence of prominence plasma, or are they only the result of perturbations on the prominence channel magnetic field with no special role in the processes moving material into the prominence?
- The complex flows in hedgerow prominences offer a fascinating challenge, offering us a look into a regime in which apparently different processes dominate than in lower latitude prominences. Further detailed observations of these motions in three dimensions (spectra with high spatial and temporal resolution combined with high quality imaging) and magnetic field information (Orozco Suárez et al. 2014) should help us understand the processes at work.
- How are the plasmas in the prominence and the larger prominence cavity related? Three dimensional morphological modeling using EUV images and coronal magnetic field measurements could help us understand the prominence cavity system.
- What determines the size and structure of prominence flows at different scales? Substructure may reflect the manner in which the material is inserted into the corona (Kucera et al. 2014), or, alternatively, be a result of processes occurring in the prominence itself, including evolution of the magnetic field or wave based variations (Antolin et al. 2014).

The understanding of flows and mass distributions in prominences is central to these and other important questions about the nature of these complicated and mysterious features of the solar atmosphere.

References

Ahn, K., Chae, J., Cao, W., & Goode, P. R. (2010, September). Patterns of flows in an intermediate prominence observed by hinode. *The Astrophysical Journal*, *721*, 74–79.

Alexander, C. E., Walsh, R. W., Régnier, S., Cirtain, J., Winebarger, A. R., Golub, L., et al. (2013, September). Anti-parallel EUV flows observed along active region filament threads with Hi-C. *The Astrophysical Journal*, *775*, L32.

Antolin, P., & Rouppe van der Voort, L. (2012, February). Observing the fine structure of loops through high-resolution spectroscopic observations of coronal rain with the crisp instrument at the swedish solar telescope. *The Astrophysical Journal*, *745*, 152.

Antolin, P., & Verwichte, E. (2011, August). Transverse oscillations of loops with coronal rain observed by hinode/solar optical telescope. *The Astrophysical Journal, 736*, 121.

Antolin, P., Yokoyama, T., & Van Doorsselaere, T. (2014, June). Fine strand-like structure in the solar corona from magnetohydrodynamic transverse oscillations. *The Astrophysical Journal, 787*, L22.

Anzer, U., & Heinzel, P. (2005, March). On the nature of dark extreme ultraviolet structures seen by SOHO/EIT and TRACE. *The Astrophysical Journal, 622*, 714–721.

Athay, R. G., & Illing, R. M. E. (1986, October). Analysis of the prominence associated with the coronal mass ejection of August 18, 1980. *Journal of Geophysical Research, 91*, 10961–10973.

Aulanier, G., & Schmieder, B. (2002, May). The magnetic nature of wide EUV filament channels and their role in the mass loading of CMEs. *Astronomy and Astrophysics, 386*, 1106–1122.

Ballester, J. L. (2014). Magnetism and dynamics of prominences: MHD waves. In J.-C. Vial, & O. Engvold (Eds.), *Solar prominences, ASSL* (Vol. 415, pp. 257–294). Springer.

Beckers, J. M. (1964, September). *A study of the fine structures in the solar chromosphere*. PhD thesis, Sacramento Peak Observatory, Air Force Cambridge Research Laboratories, Mass.

Berger, T. E., Liu, W., & Low, B. C. (2012). SDO/AIA detection of solar prominence formation within a coronal cavity. *The Astrophysical Journal, 758*, L37.

Berger, T. E., Shine, R. A., Slater, G. L., Tarbell, T. D., Title, A. M., Okamoto, T. J., et al. (2008, March). Hinode SOT observations of solar quiescent prominence dynamics. *The Astrophysical Journal, 676*, L89–L92.

Berger, T. E., Slater, G., Hurlburt, N., Shine, R., Tarbell, T., Title, A., et al. (2010, June). Quiescent prominence dynamics observed with the Hinode Solar Optical Telescope. I. Turbulent upflow plumes. *The Astrophysical Journal, 716*, 1288–1307.

Billings, D. E. (1996). *A guide to the solar corona*, chapter 6B. New York: Academic Press.

Chae, J. (2003, February). The formation of a prominence in noaa active region 8668. II. TRACE observations of jets and eruptions associated with canceling magnetic features. *The Astrophysical Journal, 584*, 1084–1094.

Chae, J., Ahn, K., Lim, E.-K., Choe, G.S., & Sakurai, T. (2008, December). Persistent horizontal flows and magnetic support of vertical threads in a quiescent prominence. *The Astrophysical Journal, 689*, L73–L76.

Chae, J., & Sakurai, T. (2008, December). A test of three optical flow techniques-LCT, DAVE, and NAVE. *The Astrophysical Journal, 689*, 593–612.

Cirigliano, D., Vial, J.-C., & Rovira, M. (2004, September). Prominence corona transition region plasma diagnostics from SOHO observations. *Solar Physics, 223*, 95–118.

de Boer, C. R., Stellmacher, G., & Wiehr, E. (1998, June). The hot prominence periphery in EUV lines. *Astronomy and Astrophysics, 334*, 280–288.

Engvold, O. (1976, August). The fine structure of prominences. I - Observations - H-alpha filtergrams. *Solar Physics, 49*, 283–295.

Engvold, O. (2014) Description and classification of prominences. In J.-C. Vial, & O. Engvold (Eds.), *Solar prominences, ASSL* (Vol. 415, pp. 31–60). Springer.

Fernley, J. A., Seaton, M. J., & Taylor, K. T. (1987, December). Atomic data for opacity calculations. VII - Energy levels, f values and photoionisation cross sections for He-like ions. *Journal of Physics B Atomic Molecular Physics, 20*, 6457–6476.

Gibson, S. (2014). Coronal cavities: observations and implications for the magnetic environment of prominences. In J.-C. Vial, & O. Engvold (Eds.), *Solar prominences, ASSL* (Vol. 415, pp. 321–351). Springer.

Gilbert, H., Kilper, G., & Alexander, D. (2007, December). Observational evidence supporting cross-field diffusion of neutral material in solar filaments. *The Astrophysical Journal, 671*, 978–989.

Gilbert, H., Kilper, G., Alexander, D., & Kucera, T. (2011, January). Comparing spatial distributions of solar prominence mass derived from coronal absorption. *The Astrophysical Journal, 727*, 25.

Gilbert, H. R., Falco, L. E., Holzer, T. E., & MacQueen, R. M. (2006, April). Application of a new technique for deriving prominence mass from SOHO EIT Fe XII (19.5 nm) absorption features. *The Astrophysical Journal, 641*, 606–610.

Gilbert, H. R., Holzer, T. E., & MacQueen, R. M. (2005, January). A new technique for deriving prominence mass from SOHO/EIT Fe XII (19.5 Nanometers) absorption features. *The Astrophysical Journal, 618*, 524–536.

Golub, L., Bookbinder, J., Deluca, E., Karovska, M., Warren, H., Schrijver, C. J., et al. (1999, May). A new view of the solar corona from the transition region and coronal explorer (TRACE). *Physics of Plasmas, 6*, 2205–2216.

Gopalswamy, N. (2014). The dynamic of eruptive prominences. In J.-C. Vial, & O. Engvold (Eds.), *Solar prominences, ASSL* (Vol. 415, pp. 379–408). Springer.

Grechnev, V. V., Uralov, A. M., Slemzin, V. A., Chertok, I. M., Filippov, B. P., Rudenko, G. V. et al. (2014). A challenging solar eruptive event of 18 November 2003 and the causes of the 20 November geomagnetic superstorm. I. Unusual history of an eruptive filament. *Solar Physics, 289*, 289–318.

Heinzel, P. (2014). Radiative transfer in solar prominences. In J.-C. Vial, & O. Engvold (Eds.), *Solar prominences, ASSL* (Vol. 415, pp. 101–128). Springer.

Heinzel, P., Anzer, U., Schmieder, B., & Schwartz, P. (2003, September). EUV-filaments and their mass loading. In A. Wilson (Ed.), *Solar variability as an input to the earth's environment* (vol. 535, pp. 447–457). Noordwijk: ESA Special Publication.

Heinzel, P., Mein, N., & Mein, P. (1999, June). Cloud model with variable source function for solar Hα structures. II. Dynamical models. *Astronomy and Astrophysics, 346*, 322–328.

Heinzel, P., Schmieder, B., Fárník, F., Schwartz, P., Labrosse, N., Kotrč, P., et al. (2008, October) Hinode, TRACE, SOHO, and ground-based observations of a quiescent prominence. *The Astrophysical Journal, 686*, 1383–1396.

Hundhausen, A. J. (1993, August). Sizes and locations of coronal mass ejections - SMM observations from 1980 and 1984–1989. *Journal of Geophysical Research, 98*, 13177.

Karpen, J. (2014). Plasma structure and dynamics. In J.-C. Vial, & O. Engvold (Eds.), *Solar prominences, ASSL* (Vol. 415, pp. 235–255). Springer.

Karzas, W. J., & Latter, R. (1961, May). Electron radiative transitions in a coulomb field. *The Astrophysical Journal, 6*, 167.

Keady, J. J., & Kilcrease, D. P. (2000). Radiation. In A. N. Cox (Ed.), *Allens astrophysical quantities* (pp. 95–120). New York: AIP.

Kilper, G., Gilbert, H., & Alexander, D. (2009, October). Mass composition in pre-eruption quiet sun filaments. *The Astrophysical Journal, 704*, 522–530.

Koutchmy, S., Slemzin, V., Filippov, B., Noens, J.-C., Romeuf, D., & Golub, L. (2008, May). Analysis and interpretation of a fast limb CME with eruptive prominence, C-flare, and EUV dimming. *Astronomy and Astrophysics, 483*, 599–608.

Kucera, T. A., Andretta, V., & Poland, A. I. (1998). Neutral hydrogen column depths in prominences using EUV absorption features. *Solar Physics, 183*, 91.

Kucera, T. A., de Pontieu, B., & Tovar, M. (2003). Prominence motions observed at high cadences in temperatures from 10,000 to 250,000 K. *Solar Physics, 212*, 81.

Kucera, T. A., Gilbert, H. R., & Karpen, J. T. (2014). Mass flows in a prominence spine as observed in EUV. *The Astrophysical Journal, 790*, 68.

Kucera, T. A., & Landi, E. (2006). Ultraviolet observations of prominence activation and cool loop dynamics. *The Astrophysical Journal, 645*, 1525.

Kucera, T. A., & Landi, E. (2008). An observation of low level heating in an erupting prominence. *The Astrophysical Journal, 673*, 611.

Landi, E., & Reale, F. (2013, July). Prominence plasma diagnostics through extreme-ultraviolet absorption. *The Astrophysical Journal, 772*, 71.

Leese, J. A., Novak, C. S., & Taylor, V. R. (1970). The determination of cloud pattern motions from geosynchronous satellite image data. *Pattern Recognition, 2*, 279–280.

Lin, Y., Engvold, O. R., & Wiik, J. E. (2003, September). Counterstreaming in a Large Polar Crown Filament. *Solar Physics, 216*, 109–120.

Liu, K., Wang, Y., Shen, C., & Wang, S. (2012, January). Critical height for the destabilization of solar prominences: Statistical results from STEREO observations. *The Astrophysical Journal, 744*, 168.

Liu, W., Berger, T. E., & Low, B. C. (2012, February). First SDO/AIA observation of solar prominence formation following an eruption: Magnetic dips and sustained condensation and drainage. *The Astrophysical Journal, 745*, L21.

Low, B. C., & Petrie, G. J. D. (2005, June). The internal structures and dynamics of solar quiescent prominences. *The Astrophysical Journal, 626*, 551–562.

Mackay, D. H., & Galsgaard, K. (2001, February). Evolution of a density enhancement in a stratified atmosphere with uniform vertical magnetic field. *Solar Physics, 198*, 289–312.

Mein, N., Mein, P., Heinzel, P., Vial, J.-C., Malherbe, J. M., & Staiger, J. (1996, May). Cloud model with variable source function for solar Hα structures. *Astronomy and Astrophysics, 309*, 275–283.

November, L. J., & Simon, G. W. (1988, October). Precise proper-motion measurement of solar granulation. *The Astrophysical Journal, 333*, 427–442.

Orozco Suárez, D., Asensio Ramos, A., & Trujillo Bueno, J. (2012, December). Evidence for rotational motions in the feet of a quiescent solar prominence. *The Astrophysical Journal, 761*, L25.

Orozco Suárez, D., Díaz, A. J., Asensio Ramos, A., & Trujillo Bueno, J. (2014, April). Time evolution of plasma parameters during the rise of a solar prominence instability. *The Astrophysical Journal, 785*, L10.

Orrall, F. Q., & Schmahl, E. J. (1976). The prominence-corona interface compared with the chromosphere-corona transition region. *Solar Physics, 50*, 365–381.

Orrall, F. Q., & Schmahl, E. J. (1980, September). The H I Lyman continuum in solar prominences and its interpretation in the presence of inhomogeneities. *The Astrophysical Journal, 240*, 908–922.

Panasenco, O., Martin, S. F., & Velli, M. (2014, February). Apparent solar tornado-like prominences. *Solar Physics, 289*, 603–622.

Pécseli, H., & Engvold, O. (2000, May). Modeling of prominence threads in magnetic fields: Levitation by incompressible MHD waves. *Solar Physics, 194*, 73–86.

Penn, M. J. (2000, December). An erupting active region filament: three-dimensional trajectory and hydrogen column density. *Solar Physics, 197*, 313–335.

Rumph, T., Bowyer, S., & Vennes, S. (1994, June). Interstellar medium continuum, autoionization, and line absorption in the extreme ultraviolet. *The Astronomical Journal, 107*, 2108–2114.

Schmieder, B., Chandra, R., Berlicki, A., & Mein, P. (2010, May). Velocity vectors of a quiescent prominence observed by Hinode/SOT and the MSDP (Meudon). *Astronomy and Astrophysics, 514*, A68.

Schmieder, B., Mein, N., Deng, Y., Dumitrache, C., Malherbe, J.-M., Staiger, J., et al. (2004, September). Magnetic changes observed in the formation of two filaments in a complex active region: TRACE and MSDP observations. *Solar Physics, 223*, 119–141.

Schrijver, C. J. (2001, February). Catastrophic cooling and high-speed downflow in quiescent solar coronal loops observed with TRACE. *Solar Physics, 198*, 325–345.

Schuck, P. W. (2006, August). Tracking magnetic footpoints with the magnetic induction equation. *The Astrophysical Journal, 646*, 1358–1391.

Schwartz, P., Heinzel, P., Anzer, U., & Schmieder, B. (2004, July). Determination of the 3D structure of an EUV-filament observed by SoHO/CDS, SoHO/SUMER and VTT/MSDP. *Astronomy and Astrophysics, 421*, 323–338.

Schwartz, P., Schmieder, B., Heinzel, P., & Kotrč, P. (2012, December). Study of an extended EUV filament using SoHO/SUMER observations of the hydrogen lyman lines. II. Lyman α line observed during a multi-wavelength campaign. *Solar Physics, 281*, 707–728.

Stenborg, G., Vourlidas, A., & Howard, R. A. (2008, February). A fresh view of the extreme-ultraviolet corona from the application of a new image-processing technique. *The Astrophysical Journal, 674*, 1201–1206.

Stewart, R. T., McCabe, M. K., Koomen, M. J., Hansen, R. T., & Dulk, G. A. (1974, May). Observations of coronal disturbances from 1 to 9 R_{sun}. I: First event of 1973, January 11. *Solar Physics, 36*, 203–217.
Tziotziou, K. (2007, May). Chromospheric cloud-model inversion techniques. In P. Heinzel, I. Dorotovič, & R. J. Rutten (Eds.), *The physics of chromospheric plasmas. Astronomical Society of the Pacific conference series* (vol. 368, pp. 217). San Francisco: Astronomical Society of the Pacific.
van Ballegooijen, A. A., & Cranmer, S. R. (2010, March). Tangled magnetic fields in solar prominences. *The Astrophysical Journal, 711*, 164–178.
Vial, J.-C., Olivier, K., Philippon, A. A., Vourlidas, A., & Yurchyshyn, V. (2012, May). High spatial resolution VAULT H-Lyα observations and multiwavelength analysis of an active region filament. *Astronomy and Astrophysics, 541*, A108.
Vourlidas, A., Howard, R. A., Esfandiari, E., Patsourakos, S., Yashiro, S., & Michalek, G. (2010, October). Comprehensive analysis of coronal mass ejection mass and energy properties over a full solar cycle. *The Astrophysical Journal, 722*, 1522–1538.
Wang, Y.-M. (1999, July). The jetlike nature of He II λ304 prominences. *The Astrophysical Journal, 520*, L71–L74.
Webb, D. (2014) Eruptive prominences and their association with coronal mass ejections. In J.-C. Vial & O. Engvold (Eds.), *Solar prominences, ASSL* (Vol. 415, pp. 409–430). Springer.
Wiik, J. E., Dere, K., & Schmieder, B. (1993). UV prominences observed with the HRTS: structure and physical properties. *Astronomy and Astrophysics, 273*, 267.
Zirker, J. B., Engvold, O., & Martin, S. F. (1998, December). Counter-streaming gas flows in solar prominences as evidence for vertical magnetic fields. *Nature, 396*, 440–441.

Chapter 5
Radiative Transfer in Solar Prominences

Petr Heinzel

Abstract Critical information about prominence physical conditions can be obtained only by analyzing prominence spectra or high-resolution monochromatic images. Solar prominences are low-density structures and, therefore, the non-equilibrium (NLTE) spectroscopy tools have to be used. We describe the basics of the NLTE radiative-transfer theory applied to typical prominence conditions and demonstrate why the NLTE approach is necessary. Starting from the simplest 1D slab models, we gradually move to more complex problems which include 2D transfer, importance of partial redistribution in the formation of resonance lines and details of multilevel problems including the most up-to-date numerical methods for constructing NLTE models. We also mention some recent achievements in prominence modeling (e.g. magnetic dips, radiative equilibrium models, oscillations), although detailed description of such results is the subject of other chapters.

5.1 Introduction

The radiation emitted by solar prominences or filaments is typically observed in the form of spectra or monochromatic images and has a multiple importance for the physics of prominences. First, it provides us with the diagnostics of the prominence physical state, which means that using rather sophisticated tools discussed below one can determine the basic thermodynamic quantities like temperature, density, pressure, ionization, but also the structure of the magnetic field. In particular the ionization degree, which is mainly determined by the radiation through photoionization-recombination processes, is the critical parameter for our understanding of the coupling between the plasma and magnetic field. Second, having a good knowledge of the spectral line formation, one can quantitatively predict the visibility of prominences and their fine structures, based on various magnetic models (MHD, extrapolations) which contain a realistic mass loading

P. Heinzel (✉)
Astronomical Institute, Academy of Sciences, Fričova 298, 251 65 Ondřejov, Czech Republic
e-mail: pheinzel@asu.cas.cz

J.-C. Vial, O. Engvold (eds.), *Solar Prominences*, Astrophysics and Space Science Library 415, DOI 10.1007/978-3-319-10416-4_5

e.g. in numerous magnetic dips. Finally, the radiation field plays also a crucial role in the global energy budget of prominences and has to be properly considered when evaluating self-consistently the prominence internal structure. Similarly as in the solar chromosphere, hydrogen and some other species (ions) represent most important coolants of the plasma and thus their line and continuum radiation, together with their atomic states, must be precisely determined. We shall now start with the prominence spectral diagnostics and show how the spectral lines and continua can be modeled. The reader should consult the new textbook (Hubeny and Mihalas 2014) (hereafter referred to as HM2014) for an extended review of the astrophysical non-equilibrium spectroscopy and Labrosse et al. (2010) for specific aspects of prominence spectroscopy.

5.2 Radiative Transfer in 1D Slabs

As we can see in many examples in this book, prominences are extremely heterogeneous consisting of highly dynamical small-scale structures, which also have limited lifetimes. This sets significant constraints on the prominence radiative-transfer modeling. Moreover, strong departures from the local thermodynamic equilibrium (LTE) exist under typical prominence conditions and thus even a simple "two-level atom" problem (see below) has to be solved numerically. At the end, fully 3D radiation-magneto-hydrodynamical (RMHD) simulations are desirable to account for the complexity of all observed phenomena. However, even relatively simple 1D models have proven to be very useful for our basic understanding of prominence spectra, and namely those of quiescent prominences as observed in various optical lines. In this case we see mainly cool parts of the prominence body and neglecting the presence of a prominence-corona transition region (PCTR), the prominence can be roughly approximated by a 1D isothermal plasma slab with the gas-pressure distribution being either uniform or satisfying certain equilibrium conditions. Therefore, we will first describe the basics of the radiative-transfer theory applicable to rather simple 1D slabs externally illuminated by the surrounding solar atmosphere, see Fig. 5.1.

The radiative transfer equation (RTE) describes the variations of the specific intensity of radiation $I(\nu)$ along an elementary geometrical path ds due to the absorption and emission processes

$$\frac{dI_\nu}{ds} = -\chi_\nu I_\nu + \eta_\nu \,, \tag{5.1}$$

where χ_ν and η_ν are the absorption and emission coefficients, respectively, and ν is the frequency. Using x as the reference coordinate in a 1D prominence slab (i.e. along the normal to the slab—see Fig. 5.1), we can write

$$\frac{dx}{ds} = \cos\theta \equiv \mu \tag{5.2}$$

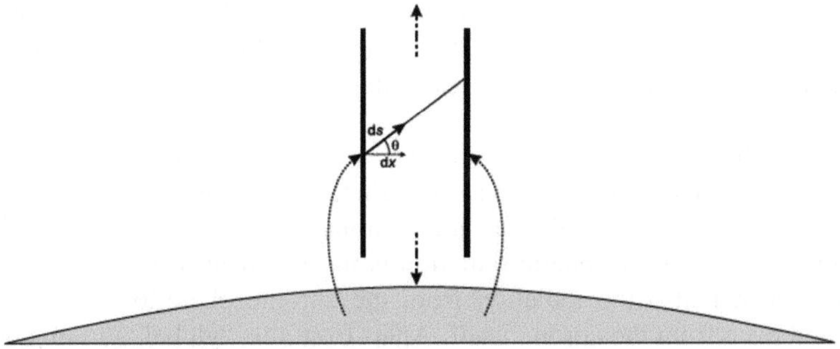

Fig. 5.1 1D prominence slab vertically standing above the solar surface and illuminated on both sides by the solar radiation (see the text)

and defining τ_ν as the optical-depth scale at frequency ν

$$d\tau_\nu = -\chi_\nu dx \tag{5.3}$$

we can finally express the transfer equation in its standard 1D plane-parallel form

$$\mu \frac{dI_\nu}{d\tau_\nu} = I_{\nu\mu} - S_\nu \, . \tag{5.4}$$

The source function S_ν is then defined as the ratio between emission and absorption coefficients

$$S_\nu \equiv \frac{\eta_\nu}{\chi_\nu} \, . \tag{5.5}$$

The mean intensity of the radiation field J_ν is expressed as the average over all directions

$$J_\nu = \frac{1}{2} \int_{-1}^{1} I_{\nu\mu} \mathrm{d}\mu \, . \tag{5.6}$$

Note here that I, J, and S have the same units and we use either erg s^{-1} cm^{-2} sr^{-1} Hz^{-1} or W m^{-2} sr^{-1} Hz^{-1}. Instead of frequency (in Hz), the wavelength units are frequently used (Å) and the conversion factor for the above quantities is $10^8 c/\lambda^2$ where the wavelength λ is in Å, and c is the speed of light in cm s^{-1}.

In cooler parts of prominences where the lines and continua of neutral or singly ionized atoms are formed, the source function strongly depends on the radiation field through the emission coefficient which is the basic ingredient of the NLTE physics dealing with departures from LTE (an alternative notation, non-LTE, is frequently used in the literature). Only after the source function is specified, the integration of RTE is relatively simple. Determination of both the absorption and emission coefficients, and thus the source function, is the principal task of the NLTE radiative-transfer modeling. Note that under the LTE conditions, the source function is simply equal to the Planck function, thanks to strong collisional coupling between atoms and free electrons (high densities). This might be valid in the solar photosphere, but not in prominences where significant departures from LTE exist because of relatively low densities.

RTE is subject to boundary conditions at both slab surfaces, which means that one has to determine the incident radiation coming from the surrounding solar atmosphere. For spectral lines and continua we are mainly interested in, the photospheric and chromospheric radiation is most critical. But in some cases we have to consider also the radiation from the chromosphere-to-corona transition region and even from the corona itself. Apart from this "global" illumination, one can also consider a potential role of atmospheric inhomogeneities surrounding the prominence, like e.g. strong plages or bright active regions when dealing with active or eruptive prominences. The actual conditions depend on the geometry of the problem and on the line or continuum transitions under consideration. The incident radiation strongly affects the source function and thus must be precisely specified for all considered line and continuum frequencies and for all directions. However, once the source function is determined by solving the full NLTE problem, the formal solution of the RTE can be performed along a prescribed line of sight (LOS). Then only the incident radiation along that direction is considered in the boundary conditions (see Sect. 5.4). Observing the prominence on the limb, the LOS directed towards the observer contains the coronal radiation behind the prominence and in front of it. This may be relevant for some UV or EUV transitions (see e.g. discussion on absorption of "hot" coronal radiation by cool prominence material in Kucera 2014), but not for "cool" lines which are not present in the corona. When observing filaments against the solar disk, the situation is different and one has to consider the radiation passing along the LOS from the solar surface below the filament or, in the case of hot lines, take the coronal contributions along the LOS. In the case of optically-thick transitions, this poses a great difficulty because we cannot see the solar surface below the filament and thus it is impossible to determine observationally the incident radiation. This is a typical problem of the so-called cloud model which we will mention later. For hydrogen Lyman lines a solution was proposed in Schwartz et al. (2006).

Depending on the atomic model considered, the incident radiation for computed line and continuum transitions was specified by various authors, mainly on the basis of the observed radiation intensities. Compiled data from various sources have been given e.g. by Heasley et al. (1974); Gouttebroze et al. (1993) (hereafter referred to as GHV) or Heinzel et al. (2005) for hydrogen, in Rudawy and Heinzel (1992) for hydrogen subordinate continua, and in Labrosse and Gouttebroze (2001) for helium. CaII and MgII data were compiled by Gouttebroze and Heinzel (2002) and Heinzel et al. (2014a), respectively. Whenever possible, one should take into account the center-to-limb variations of the incident line and continuum intensities. For EUV irradiation, the hydrogen (912 Å), HeI (504 Å) and HeII (228 Å) resonance continua are important, together with a forest of "hot" coronal emission lines which affects the hydrogen and helium ionization and also provides the radiative heating. The coronal illumination can be synthesized using the CHIANTI code (Landi et al. 2013). Finally, note that one should not neglect the temporal variations of the solar radiation and consider illumination consistent with the dates of the prominence observations. For example the hydrogen Lyman α line irradiation can vary during the solar cycle by tens of %.

5.3 Opacity, Emissivity and the Line Source Function

In order to determine χ_ν and η_ν, we must know the atomic-level populations (number densities of atoms or ions in a given quantum state) and the absorption/emission profiles for the considered spectral line transition between atomic levels i and j (the situation for continua is similar, the interested reader should consult e.g. HM2014). First we introduce the notion of the line absorption profile

$$\phi_\nu = \frac{1}{\sqrt{\pi}\Delta\nu_{\rm D}} H(a, x), \tag{5.7}$$

where $H(a, x)$ is the Voigt function, a is the damping parameter $a = a_i + a_j$ pertinent to the atomic levels i and j and $x = \Delta\nu/\Delta\nu_{\rm D}$ is the frequency displacement from the line centre expressed in units of the Doppler width. Note that x will contain a Doppler shift in cases where the macroscopic velocities are present and the same applies also for the emission profile introduced below. The line profile is normalized

$$\int_0^\infty \phi_\nu \, d\nu = 1 \,. \tag{5.8}$$

Spectral line shapes in prominences are mainly characterized by the Doppler broadening and in the case of optically thick lines, the line-wing broadening is governed by the Voigt profile. Many lines observed in prominences, namely in the optical domain, are optically thin or moderately thick and thus their absorption profile can be considered to be nearly Gaussian.

Using the Einstein coefficients for absorption (B_{ij}), spontaneous emission (A_{ji}) and stimulated emission (B_{ji}), the absorption line coefficient, corrected for stimulated emission, is written as

$$\chi_\nu = n_i B_{ij} \frac{h\nu_{ij}}{4\pi} \phi_\nu - n_j B_{ji} \frac{h\nu_{ij}}{4\pi} \psi_\nu \tag{5.9}$$

and emission coefficient as

$$\eta_\nu = n_j A_{ji} \frac{h\nu_{ij}}{4\pi} \psi_\nu \,. \tag{5.10}$$

Here ψ_ν is the line emission profile (for ψ see Eq. (5.10)).

In these relations n_i and n_j are the atomic level populations and h is the Planck constant. Since both the absorption and stimulated emission are proportional to the intensity of the radiation, it is convenient to combine these two processes into one opacity term (the stimulated emission is also called a negative absorption). The frequency dependence of the absorption and emission processes is given by the profiles ϕ_ν and ψ_ν, respectively. In most practical cases we assume that $\psi_\nu \equiv \phi_\nu$ and this approximation is called complete frequency redistribution—CRD (see Sect. 5.6). The process of absorption and subsequent spontaneous emission is called scattering and in fact most of the prominence radiation is transported by the scattering in spectral lines which we call the diffusion of radiation. Note here that although the absorption and emission profiles can be well approximated by Gaussians even for cases of moderately optically-thick structures, the shape of the spectral line (e.g. the intensity of emission line profile as observed above the limb) can exhibit strong departures from the Gaussian shape, being flat or even somewhat reversed in the line core. In other words, the notions of line absorption or emission profile on one hand and the profile (shape) of the emitted line radiation on the other hand are two principally different things and should not be confused. We will come back to this problem later when discussing the formal solution of RTE.

Finally, from the definition of the source function [Eq. (5.5)] we get

$$S_\nu = \frac{n_j A_{ji} \psi_\nu}{n_i B_{ij} \phi_\nu - n_j B_{ji} \psi_\nu} \simeq \frac{n_j A_{ji}}{n_i B_{ij} - n_j B_{ji}} \rho_{ij}(\nu) \tag{5.11}$$

with $\rho_{ij}(\nu) \equiv \psi_\nu / \phi_\nu$. In CRD case $\rho_{ij}(\nu) = 1$ and the line source function is thus frequency independent. In the case of NLTE, the atomic level populations and $\rho_{ij}(\nu)$ depend on the radiation intensity and thus must be determined consistently with the solution of the transfer equation.

5.4 Formal Solution of the Radiative Transfer Equation

We consider two kinds of 1D slabs having a finite geometrical thickness and oriented either vertically above the solar surface and irradiated symmetrically on both sides (the case of prominences seen on the limb—Fig. 5.1) or parallel to the solar surface and irradiated mainly from below (the case of filaments). More realistic 2D or even 3D geometries will be discussed later.

The so-called formal solution of RTE gives the emergent radiation intensity at the slab surface, for given frequency ν and outgoing direction μ, using the already known source function and the line opacity (i.e. the optical-depth scale)

$$I_{\nu\mu} = I^0_{\nu\mu} \mathrm{e}^{-\tau_\nu/\mu} + \int_0^{\tau_\nu} S(t_\nu) \mathrm{e}^{-t_\nu/\mu} dt_\nu / \mu, \tag{5.12}$$

where $I^0_{\nu\mu}$ is the incident radiation on the opposite side of the slab and τ_ν is the total optical thickness of the slab. This formal solution of transfer equation shows us the basic nature of the line formation in prominence structures (slabs). The emergent spectrum generally consists of the part determined by the depth-dependent variations of the source function and its contribution from each depth is attenuated by the absorption towards the observed surface. The first term plays a role only in case that the LOS crossing the prominence ends up at the prominence surface which is directly illuminated by the background radiation. Then this radiation intensity passes through the prominence slab and is attenuated by a factor $e^{-\tau_\nu/\mu}$. Here we have to stress that even when τ_ν is very large at a given frequency so that we cannot "look" through the slab and the background radiation (if any) cannot penetrate to the visible surface, the emergent line intensity is still non-zero because it will be determined by the source function which, in turn, is mainly driven by the diffusion of radiation scattered from all other directions. This then means that a filament might be better approximated by a narrow vertical 2D slab seen from the top rather than a horizontally infinite 1D slab of finite vertical thickness. The reason is that the illumination of such a 2D slab on both its vertical surfaces may significantly affect the source function due to scattering and this will be detected when looking down from the top. On the other hand, in the case of 1D horizontal slabs, the source function is determined only by the radiation incident on the bottom surface. We have to keep in mind these aspects of the line formation physics when interpreting the spectra arising from prominence structures having various geometrical shapes and being illuminated from various directions with respect to the LOS. Various projection effects may also play a significant role. In Fig. 5.2 we show a nice example of how the prominence radiates being illuminated from the solar surface.

Fig. 5.2 SDO/AIA HeII 304 Å image of an erupting prominence irradiated by the solar disk. We clearly see the effect of the *bottom* illumination—see the text

While the bottom part of the prominence structure is bright as the result of scattering of the incident radiation, the top part is relatively dark because no relevant radiation illuminates it from the corona. Formation of this HeII 304 Å line is discussed in Labrosse (2014).

5.4.1 Depth Variations of the Source Function

A very common approximation used for prominences or even filaments is to assume a uniform source function, i.e. constant through the slab. In the case of limb prominences, the uniformity of the source function e.g. in the Hα line was demonstrated on the grid of models by GHV, who found departures from S = const only for thick and dense slabs. The situation with filaments is more complex as we will see later, but there exist several analyses using the constant source function model called the "cloud model" according to Beckers (1964).

In the case of constant S, the formal solution is written in the form

$$I_{\nu\mu} = I^0_{\nu\mu} \mathrm{e}^{-\tau_\nu/\mu} + S[1 - \mathrm{e}^{-\tau_\nu/\mu}] , \tag{5.13}$$

where τ_ν is again the total optical thickness of the prominence slab at given frequency. Two limiting cases are important, depending on the optical thickness of the slab (we drop the frequency dependence):

$$\tau \ll 1 \Rightarrow S[1 - \mathrm{e}^{-\tau/\mu}] \simeq S\tau/\mu \tag{5.14}$$

$$\tau \gg 1 \Rightarrow S[1 - \mathrm{e}^{-\tau/\mu}] \simeq S . \tag{5.15}$$

These are called the optically-thin and optically-thick cases, respectively.

While a constant source function is a useful approximation for lines which are optically thin or moderately thick in prominences, in a general case the source functions are variable with depth. This behavior is well demonstrated in the GHV grid of models, and namely for the Hα line. In the case of isothermal and isobaric thick slabs, the source function decreases toward both surfaces of a 1D vertical slab. This is even more visible on top of horizontal 1D slabs representing the filaments, because those filament surfaces are not illuminated in such lines from the overlying corona. The Hα line source function in cloud models is thus height variable as shown by Mein et al. (1996). Note that the notion "cloud model" is sometimes also used in connection with general formula Eq. (5.12), where the cloud source function is determined from a detailed solution of the NLTE transfer problem—see e.g. Molowny-Horas et al. (1999) who performed an inversion of the Hα filament profiles using such NLTE cloud models. In optically very thick hydrogen Lyman lines, the decrease of the source function towards the top surface of the filament was first demonstrated by Heinzel et al. (1997) and Schmieder et al. (1998). The actual shape of the Lyman-line source function at the top surface of filaments then depends on the structure of PCTR and this also holds for prominence slabs. A similar situation takes place in other thick lines which are sensitive to temperature/density variations like e.g. the MgII h and k lines (Heinzel et al. 2014a). Here it is worth mentioning the well-known fact that the spatial variations of the source function automatically map into the shape of the optically thick lines and this then results in various line profiles from purely emission ones to those having substantial reversals. But in reality the problem becomes very complex. In the case of hydrogen Lyman α, rather strong line reversal is obtained even for isothermal-isobaric slabs just as the result of partially coherent scattering (Sect. 5.6). On top of that, the line reversal will be modified by the presence of a PCTR.

The mapping of the source function depth-variations into the shape of the emergent line profile is fully described by Eq. (5.12). A simple and useful rule following from this is that the emergent intensity at a given frequency (wavelength) is roughly equal to the value of S at optical depth equal to unity. If we move from the line center toward the wings, we "see" deeper and deeper layers in the line intensity profile. In such a picture, the wavelength dependence of the emergent profile is given by wavelength variations of the line optical depth, while the line source function is in most cases (i.e. under the CRD approximation) wavelength independent. The mapping can be also described in terms of the contribution functions which tell us how much a given depth contributes to the emergent intensity at a given wavelength. For definition of contribution functions and some examples see Heinzel et al. (2005).

5.4.2 *Spectral Visibility of Prominences and Filaments*

The appearance of prominences on the limb and filaments on the disk is different. Above the limb we see prominences in emission against the dark coronal background (exception is a special case of the UV line absorption by resonance continua of hydrogen and helium—see Kucera 2014), on the disk they typically appear as dark filaments against brighter solar disk in all situations.

5.4.2.1 Prominence on the Limb

The spectral line is in emission, $I^0 = 0$ (no background radiation considered from the corona). For $\mu = 1$ we get

$$I = S[1 - \mathrm{e}^{-\tau}] \,. \tag{5.16}$$

In a special case of an optically-thin slab we then have

$$I \simeq S\tau = \eta D, \tag{5.17}$$

where $\tau \ll 1$ and D is the geometrical thickness of the slab. Note again that the intensities I^0 from other directions drive the source function or, equivalently, η. This is the case in central cool parts where the radiation scattering is the dominant process determining the source function. On the other hand, inside the PCTR the temperature is steeply increasing, emission lines become optically-thin and the thermal radiation starts to dominate the scattering process or at least to contribute significantly to the source function

5.4.2.2 Filament on the Disk

The spectral lines are usually in absorption, for $\mu = 1$ we get

$$I = I^0 \mathrm{e}^{-\tau} + S[1 - \mathrm{e}^{-\tau}] . \tag{5.18}$$

As said above, the line source function in central cooler parts is mostly controlled by the photon scattering and thus we can write approximately

$$S \simeq \frac{1}{2} I^0 , \tag{5.19}$$

where 1/2 is the dilution factor by which the incident solar-disk radiation I^0 has to be multiplied because there is roughly only one half of the prominence or filament surrounding from which the incident radiation illuminates it (assuming again no radiation from the corona). For filaments seen against the disk we specify their contrast as

$$\frac{I}{I^0} = \frac{1}{2}[1 + \mathrm{e}^{-\tau}] \tag{5.20}$$

and in two limiting situations we get

$$\begin{aligned} \tau \ll 1 &\Rightarrow \frac{I}{I^0} \simeq 1 \\ \tau \gg 1 &\Rightarrow \frac{I}{I^0} \simeq \frac{1}{2} . \end{aligned} \tag{5.21}$$

For an optically-thin filament, the line-centre contrast approaches unity and in the case of a large optical thickness it becomes 1/2. This is why we can see the filaments as dark structures relative to the background chromosphere. Since τ decreases from the line center towards the line wings, we barely see the filaments when shifting the narrow-band filter out of the line centre. However, the real situation will depend on the spectral line considered. In Hα the filaments are dark and their line-center contrast indeed reaches about 1/2. But for example CaII resonance lines H and K show much lower contrast against the disk and this is related to the behavior of the source function for optically thick lines and also to the fact that CaII lines have different sensitivity to temperature and density compared to Hα. Note that while the Hα line is mostly optically thin or moderately thick, the MgII lines h and k may reach optical thicknesses of the order of 10^3 under the same conditions. The thickest line is the hydrogen Lyman α which can reach the optical thickness $10^5 - 10^6$.

The reason why we observe prominences on the limb in emission and filaments on the disk in absorption is the following. Cool prominence plasma absorbs the incident radiation and scatters it in all directions. Because there is no coronal background in "cool" optical lines, we see on the limb only the scattered radiation and the line is thus in emission. On the other hand, the Hα chromospheric background of the filament is the absorption line which becomes even darker due to filament absorption. The radiation scattered in the direction toward the observer represents only a small fraction of the absorbed one and thus cannot compensate for the absorption. We thus see filaments darker than the background solar surface. Note that although the scattering dominates the line source function in most cases, certain thermal contribution is also possible, namely at higher densities.

5.4.3 *Profiles of Prominence Emission Lines with a Constant Source Function*

Using Eq. (5.13) and taking $\mu = 1$, we get the well-known formula

$$I(\Delta\lambda) = S[1 - \exp(-\tau(\Delta\lambda))]\,, \tag{5.22}$$

where

$$\tau(\Delta\lambda) = \tau_0 \exp(-x^2)\,. \tag{5.23}$$

Note that for practical reasons, we use in this subsection the wavelength scale instead of the frequency one. Now we can express the Doppler width as

$$\Delta\lambda_D = \lambda_0 \frac{v}{c}\,, \tag{5.24}$$

where λ_0 is the line-center wavelength, c is the speed of light and v is the mean velocity corresponding to Maxwellian distribution of atomic and microturbulent motions

$$v^2 = 2kT/m_{\rm A} + v^2_{\rm turb} = 1.65 \times 10^8\, T/A + v^2_{\rm turb}\,, \tag{5.25}$$

where $m_{\rm A}$ is the atomic mass, A the atomic weight, T the kinetic temperature and velocities are in cm s^{-1}. Then the dimensionless wavelength scale is $x = \Delta\lambda/\Delta\lambda_D$, where $\Delta\lambda$ is the wavelength displacement from the line center, i.e. $\Delta\lambda = \lambda - \lambda_0$.

τ has the Gaussian shape according to Eq. (5.23). We can normalize the emission line intensity to its central value as

$$I_{\text{norm}}(\Delta\lambda) = \frac{[1 - \exp(-\tau(\Delta\lambda))]}{[1 - \exp(-\tau_0)]} \quad (5.26)$$

which gives $I_{\text{norm}}(0) = 1$.

For $\tau_0 \ll 1$ we get $\exp(-\tau(\Delta\lambda)) \simeq 1 - \tau(\Delta\lambda)$ and thus Eq. (5.26) reduces to the optically-thin form

$$I_{\text{norm}}(\Delta\lambda) = \exp(-x^2) , \quad (5.27)$$

which represents the Gaussian profile. On the other hand, for thicker cases where $\tau_0 > 1$, the line core starts to be saturated and we get a flat profile there according to Eq. (5.22) or (5.26). Such profile is no longer Gaussian. Note that for Gaussian profiles, FWHM=1.65 $\Delta\lambda_D$. Having observed two optically-thin lines belonging to two species with substantially different atomic weights, one can use the above relations to derive two unknown parameters T and v_{turb}. This simple method was used by several authors, see a recent work by Park et al. (2013).

Integrating the Gaussian line intensity over frequencies or wavelengths, we get the integrated line intensity E

$$E = \sqrt{(\pi)} I_0 \Delta\lambda_{\text{D}} , \quad (5.28)$$

where I_0 is the line-center (peak) intensity. Having the map of line intensities E and I_0, one can directly compute the Doppler width. A generalization of this approach to moderately thick lines assuming a constant source function can be found in Jejčič et al. (2014).

5.4.4 Formation of Millimeter Radio Continua in Prominences

Although this Chapter is dealing primarily with the optical and ultraviolet NLTE spectroscopy, the methods used are also applicable to prominence millimeter and sub-millimeter radio spectra. They have been detected by various radio instruments, but so far only with very low spatial resolution. However, with the newly operating interferometer ALMA (Atacama Large Millimeter Array), the prominences in radio should be seen at unprecedented spatial resolution. A great advantage of these radio spectral continua is that they can be used to measure the prominence kinetic temperature in a more straightforward way than using other spectral features in optical or UV. Here we briefly show how the radio mm and sub-mm continua are formed in solar prominences.

Under characteristic prominence conditions, the dominant source of opacity is the hydrogen free-free continuum for which the absorption coefficient at frequency ν is given as (see Rybicki and Lightman 1979)

$$\chi_\nu(\mathrm{H}) = [3.7 \times 10^8 T^{-1/2} n_\mathrm{e} n_\mathrm{p} \nu^{-3} g_\mathrm{ff}](1 - \mathrm{e}^{-h\nu/kT}), \tag{5.29}$$

where n_e and n_p are the electron and proton densities, respectively, T is the kinetic temperature, and $g_\mathrm{ff} \approx 1$ is the Gaunt factor (cgs units are used). At low temperatures, H^- free-free opacity can also play a role. The last term represents the correction for stimulated emission which is important in the radio domain, contrary to UV.

The synthetic intensity I_ν, emergent from the prominence on the limb, is obtained as the formal solution of the transfer equation

$$I_\nu = \int_0^{\tau_\nu} B_\nu(T)\mathrm{e}^{-t_\nu} dt_\nu = \int_0^{\tau_\nu} \eta_\nu \mathrm{e}^{-t_\nu} ds \quad \eta_\nu = \chi_\nu B_\nu \quad dt_\nu = \chi_\nu ds, \tag{5.30}$$

where $B_\nu(T)$ is the Planck source function, η_ν the emission coefficient, t_ν the optical depth, and s is the geometrical path length along the LOS.

Free-free processes are the collisional processes and thus the corresponding source function is Planckian. But this does not mean that the radio continuum formation is governed by the LTE conditions. This is because the hydrogen ionization and thus n_e and n_p satisfy the NLTE conditions in low-density prominences and not simple Saha equilibrium.

In the radio domain, I_ν and B_ν are directly proportional to the brightness temperature T_b and to the local plasma (kinetic) temperature T, respectively

$$I_\nu = \frac{2\nu^2 k}{c^2} T_\mathrm{b} \quad B_\nu = \frac{2\nu^2 k}{c^2} T, \tag{5.31}$$

where c is the speed of light. Using this Rayleigh–Jeans law, we can write

$$T_\mathrm{b} = \int_0^{\tau_\nu} T\mathrm{e}^{-t_\nu} dt_\nu = \int_0^{\tau_\nu} T\mathrm{e}^{-t_\nu} \chi_\nu ds. \tag{5.32}$$

In the simplest case, assuming a uniform kinetic temperature T along the LOS, we get for a prominence slab of total optical thickness τ_ν

$$T_\mathrm{b} = T(1 - \mathrm{e}^{-\tau_\nu}), \tag{5.33}$$

where $\tau_\nu = \chi_\nu D$ and D is the geometrical thickness of the prominence along the LOS. This is a result analogous to the previously described case of a constant source function. In the optically-thin limit we get simply $T_\mathrm{b} = T\tau_\nu$. We thus see that the kinetic temperature and observed brightness temperature are simply related through the optical thickness at given frequency.

5.5 Statistical Equilibrium for Atomic Level Populations

The NLTE source function for both lines and continua is a priori not known and, moreover, we cannot assume its uniformity in a general case. As we have seen before, the line source function depends on atomic level populations and line profile functions, which must be computed consistently with the radiation field. Since the latter is determined by the solution of RTE, we are faced with a complex NLTE problem of solving a coupled set of equations for level populations, called the equations of statistical equilibrium (ESE), and the radiation field. General form of ESE is written as

$$\frac{dn_i}{dt} = \sum n_j(R_{ji} + C_{ji}) - n_i \sum (R_{ij} + C_{ij}) \tag{5.34}$$

$$\frac{dn_i}{dt} = \frac{\partial n_i}{\partial t} + \frac{\partial n_i v}{\partial x} .$$

R_{ij} are the radiative rates, those for absorption and stimulated emission depend on the line intensity. $C_{ij} = n_e \Omega_{ij}(T)$ are the collisional rates proportional to the electron density n_e and dependent on temperature T through the function $\Omega(T)$. C_{ji} then follow from the detailed balance. The time-derivative on the left-hand side splits into the local temporal variation of n_i (e.g. due to time-dependent processes) and the divergence of the flux of atoms in the state i (v is the macroscopic flow velocity). Other equations to be used are the charge-conservation equation and the state equation for the gas pressure p_g

$$p_g = NkT \tag{5.35}$$

which determines the total particle-number density N (see also HM2014). Finally, knowing N, the electron density and the atomic abundances together with the atomic masses, one can compute the gas density ρ .

To be more specific, we write the line radiative rates in the form $R_{ij} = B_{ij}\overline{J}_{ij}$ for absorption, $R_{ji}(\text{spont}) = A_{ji}$ for spontaneous emission and $R_{ji}(\text{stim}) = B_{ji}\overline{J}_{ij}$ for stimulated emission. Then $R_{ji} = R_{ji}(\text{spont}) + R_{ji}(\text{stim})$.

$$\overline{J}_{ij} = \int_0^\infty J_\nu \phi_\nu \, d\nu \tag{5.36}$$

is the integrated mean intensity weighted by the absorption profile. This quantity tells us how many line photons are actually absorbed from the mean radiation field, owing to the frequency dependence of the absorption coefficient represented by the line profile function ϕ_ν.

Finally, we define the net radiative rates, to be used later, as

$$R_{ij}^{\text{net}} \equiv n_j A_{ji} - (n_i B_{ij} - n_j B_{ji})\overline{J}_{ij} . \tag{5.37}$$

The radiative and collisional rates for continuum transitions are detailed in HM2014.

5.5.1 Two-Level Model Atom

It is illustrative to consider a special case of the so-called "two-level atom" (see Fig. 5.3), where ESE simply reduce to

$$n_1 B_{12} \overline{J}_{12} + n_1 C_{12} = n_2 A_{21} + n_2 B_{21} \overline{J}_{12} + n_2 C_{21} \,. \tag{5.38}$$

Combined with the expression for the line source function Eq. (5.11), we get, after some algebra, the well-known formula

$$S = (1 - \epsilon)\overline{J} + \epsilon B_{\nu_0} \,, \tag{5.39}$$

where ϵ in the typical case of UV resonance line ($h\nu/kT \gg 1$), such as the Lyman-α line of hydrogen, can be expressed as

$$\epsilon \approx \frac{C_{21}}{(C_{21} + A_{21})} \,. \tag{5.40}$$

This represents the probability of a photon destruction. At high densities, ϵ reaches unity and the LTE conditions are achieved. On the other hand, at low densities, ϵ is very small and thus large departures from LTE take place. This is the case of solar prominences. Taking typical temperature $T = 8{,}000$ K and electron density as $n_e = 10^{10}$ cm^{-3}, we get for the hydrogen Lyman α line $\epsilon \simeq 10^{-6}$. This clearly shows how the source function in this strong resonance line is dominated by the scattering term, while the thermal contribution (second term) starts to play a role only in hotter PCTR. However, even when ϵ is extremely small under the prominence conditions, it cannot be neglected because the depth-dependent behavior of the source function depends on it, and namely the thermalization depth at which $S \to B$ (see HM2014). Normally the two-level atom model is not used in practical NLTE modeling, but in cases of strong resonance lines it can serve as a reasonable estimate of the line intensities, see e.g. Paletou et al. (1993) who used it for hydrogen, CaII and MgII line formation in prominences.

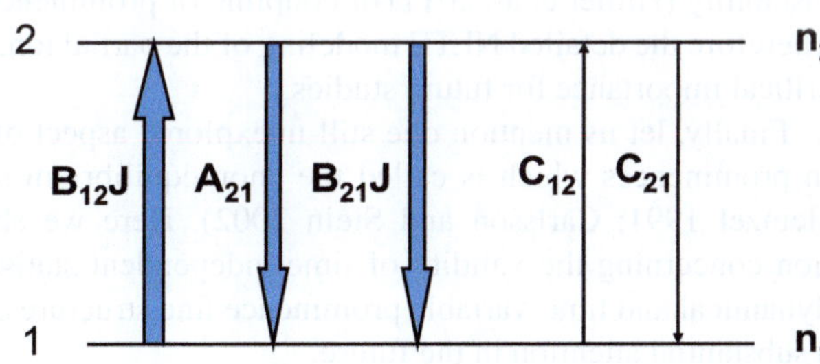

Fig. 5.3 Schematic two-level atom model. For individual quantities see the text

5.5.2 Partial Ionization of Prominence Plasmas

Statistical-equilibrium equations contain both radiative as well as collisional ionization terms, the former being called photoionization. Since the prominence densities are rather low, the photoionization of hydrogen and helium completely determines the prominence electron densities at low temperatures. In central cool parts of prominences or their fine structures, which are visible in optical lines like Hα, also the kinetic temperature is low (mostly below 10^4 K) and thus the plasma is only partially ionized. We define the ionization degree of hydrogen as $i = n_{\rm p}/n_{\rm H}$, where $n_{\rm p}$ is the proton number density and $n_{\rm H}$ the total hydrogen (protons + neutrals) number density. In central cool parts this ionization degree typically varies in a range 0.3–0.7. This means that a significant portion of neutrals exists in the prominence plasma and the question may arise how this cool neutral material is kept at prominence heights against the gravity. This problem is discussed in Gilbert (2014). Here we are just interested which processes dominate the prominence plasma ionization. The photoionization of hydrogen is dominated by the Lyman continuum (head at 912 Å) and by first subordinate continua like Balmer, Paschen, etc. In the statistical-equilibrium equations we need to specify the radiation terms for these continua at each depth of the prominence structure and this is achieved by coupled solution with the RTE, subject to radiative boundary conditions, namely for the Lyman continuum which is usually optically thick in the prominence slabs (see GHV). The other hydrogen continua are very thin and the internal radiation field is entirely determined by external illuminations. The relative role of the Lyman and other continua in hydrogen ionization is thus a complex problem. The general behavior is that optically-thick Lyman continuum radiation ionizes hydrogen predominantly in surface layers, while the Balmer continuum contributes at all depths since the external radiation can freely penetrate everywhere. Helium is a less important contributor to the electron density, in an ideal case of fully ionized hydrogen/helium plasma the helium contributes by 20 % (note that in this case also the collisional ionization may play a role at higher temperatures). Helium ionization properties are further discussed in Labrosse (2014). We note that the overall ionization structure inside the prominences directly affects processes like the thermal conductivity, ambipolar diffusion, wave damping, Rayleigh-Taylor instability (Hillier et al. 2011) or coupling of prominence plasma to magnetic field. Therefore the detailed NLTE modeling of the partial ionization in prominences is of critical importance for future studies.

Finally, let us mention one still unexplored aspect of the statistical equilibrium in prominences which is called the "non-equilibrium ionization" (Engvold 1980; Heinzel 1991; Carlsson and Stein 2002). Here we should rise a general question concerning the validity of time-independent statistical equilibrium in highly dynamical and time-variable prominence fine structures. This problem will deserve a substantial attention in the future.

5.6 Partially-Coherent Scattering in Prominence Plasmas

In a general case of the photon scattering in spectral lines, the emission profile differs from the absorption one and thus ρ_ν in Eq. (5.11) differs from unity. This is the case when the frequencies of the absorbed and re-emitted photons are somehow correlated and thus the photons are no longer completely redistributed over the line, but we speak about partial frequency redistribution (PRD) or partially-coherent scattering. Even more complex is the situation when the angular correlation for the two photons is also taken into account, see e.g. Cram and Vardavas (1978) or Heinzel (1983) for examples dealing with prominences. In the case of frequency redistribution, the line emission profile takes the form

$$\psi_\nu = \frac{\int_0^\infty R_{\nu'\nu} J_{\nu'} d\nu'}{\bar{J}} \,, \tag{5.41}$$

where $R_{\nu'\nu}$ is the redistribution function, i.e. the joint probability that the radiation absorbed at frequency ν' will be re-emitted at frequency ν. This form of ψ_ν now determines the frequency-dependence of the line source function. CRD assumes that $\rho_\nu = 1$ and then $R_{\nu'\nu} = \phi_{\nu'}\phi_\nu$ which automatically leads to relation $\psi_\nu \equiv \phi_\nu$.

For resonance lines (transitions from the ground state) one has

$$R_{\nu'\nu} = \gamma R_{II} + (1-\gamma) R_{III} \,. \tag{5.42}$$

The function R_{II} follows from purely coherent scattering in the atom's frame while R_{III} reflects the complete redistribution in the atom's frame due to elastic collisions having the rate $Q_{\rm E}$ (HM2014). γ is the branching ratio, i.e. the probability that the coherence in the atom's frame is not destroyed by a collisional perturbation of the upper atomic state. $R_{\nu'\nu}$ is the velocity-averaged (i.e. in the observer's or laboratory frame) redistribution function, here also averaged over all directions and thus neglecting the angular dependence of the scattering.

The critical importance of PRD for optically thick resonance lines like the hydrogen Lyman α formed in solar prominences was first clearly demonstrated by Heinzel et al. (1987), who computed theoretical line profiles emergent from isothermal-isobaric 1D slabs illuminated by realistic solar Lyman α radiation. This has been shown to be a critical issue, because the chromospheric Lyman α peaks are quasi-reproduced due to the partially-coherent scattering of the incident radiation in a prominence. On the contrary, if CRD is used, no significant peaks in Lyman α are obtained. This is nicely demonstrated in Fig. 5.4. This example also shows the importance of taking into account the precise shape of the incident line radiation.

1D and 2D NLTE models of prominences and filaments, including their fine structure (threads), have been constructed using the angle-averaged PRD for first two Lyman lines (for higher members of the hydrogen Lyman series the coherence effects become less important and one can use CRD). Within 1D models, PRD was also used for strong resonance lines of helium (Labrosse and Gouttebroze 2001),

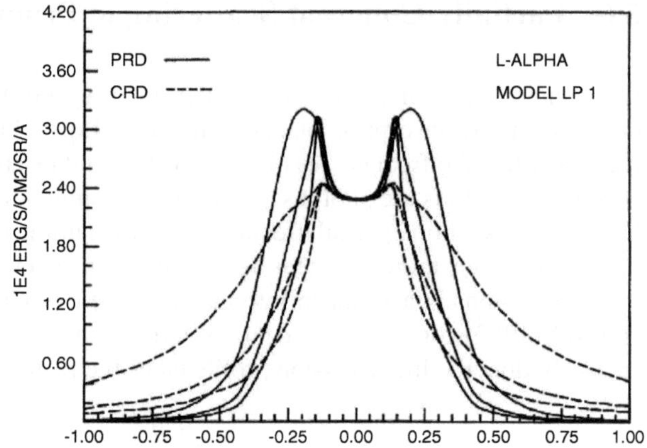

Fig. 5.4 Hydrogen Lyman α line profile emergent from an isobaric-isothermal prominence slab. The difference between PRD and CRD is clearly evidenced. From Heinzel et al. (1987)

for ionized calcium CaII (Gouttebroze and Heinzel 2002) and recently for MgII (Heinzel et al. 2014a). The emergent profiles computed with PRD may significantly differ from those computed assuming CRD and this substantially affects the resonance-line diagnostics. The resulting line profiles do reflect both the PRD scattering physics and the actual shape of the line radiation being scattered. This can be easily understood by inspecting the scattering integral in Eq. (5.41). If the frequency distribution of the radiation field is flat enough over the line absorption profile, then $\psi_\nu = \phi_\nu$ and we get the CRD case—we call this special case "natural excitation". This shows how critical is the real shape of J_ν in combination with the redistribution function. Finally, let us mention that the line-radiation scattering in real heterogeneous prominences is highly anisotropical and thus the angle-dependent redistribution functions should be considered for even more realistic diagnostics. To our knowledge, no detailed study of optically-thick line transfer was performed for prominences using the angle-dependent PRD.

The subordinate lines like the hydrogen Hα which arise between two excited atomic levels which are both broadened can be well described by the CRD approximation—the coherence is partially destroyed by the lower-level broadening (but see Heinzel (1983) for some specific cases).

5.7 General NLTE Problems

5.7.1 Multilevel Atoms

In the most general case of a multilevel atom with continuum, we are dealing with the coupled solution of RTE and ESE using the appropriate geometry of the prominence or its fine structures. The absorption and emission coefficients introduced for a two-level line transition take the general form described in HM2014 (Sect. 18.1).

The continuum formation is dependent on the atomic species considered. For example hydrogen has a strong Lyman continuum below 912 Å which was observed e.g. by SOHO/SUMER (see also the results mentioned in Labrosse 2014) and this resonance continuum is usually optically thick in prominences (GHV). Therefore, the transfer equation must be solved for the hydrogen Lyman continuum, subject to boundary conditions as discussed above. Note that EUV disk continua of HeI below 504 Å and HeII below 228 Å, together with a forest of EUV coronal lines, also affect the formation of the Lyman continuum of hydrogen and the resonance continua of helium. On the other hand, subordinate hydrogen continua like Balmer, Paschen and others are extremely optically thin in prominences, even for large structures (GHV give the optical thickness of the Balmer continuum). Therefore, one can simply evaluate the continuum radiation field at a given position within a prominence and use it to fix the corresponding photoionization rate. A similar situation takes place for helium subordinate continua. However, CaII resonance-continuum formation below 1044 Å is strongly driven by the internal radiation field in hydrogen Lyman continuum and Lyman lines starting from Lyman β. This was found already by Ishizawa (1971a) and Ishizawa (1971b) and further used in Gouttebroze and Heinzel (2002). The photoionization of MgII in the resonance continuum is also almost entirely dependent on the hydrogen-continuum radiation field inside the prominence. These are only some examples of the behavior of the continua in prominences and for any other species one has to take into account carefully the overlap with other important transitions of various species. Accurate evaluation of these continua is critical for all studies which deal with the partial ionization in prominence plasmas, and namely with partial ionization of the hydrogen and helium. The free-free continuum terms are usually negligible in prominences but see the radio continua discussed in Sect. 5.4.4. Thomson scattering on free electrons contributes to the prominence continuum emission in the visible spectrum which, together with the Hα line, causes a pink color of prominences when observed during solar eclipses (see Jejčič and Heinzel 2009). Finally, the photoionization of hydrogen and helium (both HeI and HeII) can be produced also by EUV coronal lines (Anzer and Heinzel 2005), which are then attenuated and we see the prominences on the limb as dark features against the bright corona in various coronal lines—see many illustrative images from SOHO/EIT, TRACE or currently from SDO/AIA. This interesting and relatively new topic is discussed in detail by Kucera (2014).

5.7.2 Numerical Techniques

NLTE models of prominences with various degrees of complexity have been constructed since the seventies and different authors used different numerical techniques. Most popular were complete linearization methods (e.g. Heasley and Mihalas 1976) or methods based on the so-called equivalent two-level atom model approach (Gouttebroze and Labrosse 2000). In some cases, both these techniques have been combined (Heinzel et al. 1987). However, during last decades, new techniques called Accelerated Lambda Iterations (ALI) have been developed for stellar atmospheric modeling and are now routinely used—see HM2014. For prominences they have been first applied in Auer and Paletou (1994) in the frame of a two-level atom and in Heinzel (1995) for a general multilevel case. To describe them briefly, we write the formal solution of RTE at a given depth in terms of the lambda operator as

$$I_{\nu\mu} = \Lambda_{\nu\mu}[S_{\nu\mu}]\,. \tag{5.43}$$

This is used in the so-called lambda iterations (LI) to solve the NLTE problem, i.e. we can write

$$\overline{J}^n_{ij} = \Lambda[S^{n-1}_{ij}]\,, \tag{5.44}$$

where the integrated mean intensity $\overline{J}_{ij}$ which is needed to compute the radiative rates in ESE is obtained at the nth iteration using the source function taken from the previous iteration. This, however, leads to very inefficient and extremely slow convergence rates and thus LI technique is of no practical use. Examples of the LI failure are given in HM2014. Therefore, it has been suggested to use the ALI which are based on the idea of the lambda-operator splitting

$$\Lambda = \Lambda^* + (\Lambda - \Lambda^*)\,, \tag{5.45}$$

where Λ^* is the Approximate Lambda Operator (ALO). Then the iterative scheme can be written as

$$\overline{J}^n_{ij} = \Lambda^*[S^n_{ij}] + (\Lambda - \Lambda^*)[S^{n-1}_{ij}] = \Lambda^*[S^n_{ij}] + \Delta\overline{J}^{n-1}_{ij}\,. \tag{5.46}$$

We can immediately see that in this case $\overline{J}^n_{ij}$ is consistent with the current source function S^n_{ij} and only a correction to it is computed using the lagged source function S^{n-1}_{ij}. Although the coupling between $\overline{J}^n_{ij}$ and S^n_{ij} is only approximate (because of the use of Λ^* instead of the exact operator Λ), we get the exact solution after a limited number of iterations. Further, one can insert this into the ESE and thus precondition them as suggested by Rybicki and Hummer (1991)). Using the formula (5.11) for the line source function, we can express $R^{\rm net}_{ij}$ [Eq. (5.37)] as

$$R^{\rm net}_{ij} = n_j A_{ji}(1 - \Lambda^*) - (n_i B_{ij} - n_j B_{ji})\Delta\overline{J}^{n-1}_{ij}\,, \tag{5.47}$$

where Λ^* and $\Delta\overline{J}_{ij}$ are the angle and frequency-averaged quantities. In this way the net radiative rates no longer depend on the current radiation intensity which represents a great advantage. For multilevel atoms, this approach is called the MALI (Multilevel ALI) method (Rybicki and Hummer 1991). For prominences it was first applied by Heinzel (1995) who solved the multilevel NLTE transfer problem in 1D prominence slabs with CRD. PRD was implemented into prominence ALI modeling by Auer and Paletou (1994) and for MALI by Paletou (1995) who also considered 2D slabs.

5.7.3 Multidimensional Radiative Transfer

So far we have been dealing only with relatively simple 1D geometry in the form of various slabs. This modeling was very successful, providing us with many useful results which are further described in Labrosse (2014). However, it became evident already in the seventies that prominences will require more sophisticated modeling in order to better understand the formation of their spectra. The problem of NLTE modeling in more than one dimension becomes, however, extremely demanding on computing resources even when modern methods of numerical radiative transfer like MALI are used. We can distinguish between two kinds of multidimensional models: (1) models of isolated structures like 2D or 3D slabs, cylinders (loops, threads) or models of magnetic dips, and (2) models of heterogeneous structure of prominences consisting of many fine-structure elements like threads, blobs, plumes etc. which in general interact radiatively among themselves. The first detailed approach to 2D modeling of the free-standing slabs representing prominences was presented by Mihalas et al. (1978) who used simple two-level atom described in subsection 5.5.1, but solved the 2D transfer equation taking into account more realistic illumination of all boundaries. They used the standard Feautrier method (see HM2014) to solve the transfer problem. This pioneering work was continued by Vial (1982) who used the same numerical code. Later on Auer and Paletou (1994) used the more efficient ALI techniques together with the short-characteristics formal solution of RTE and also introduced PRD into the 2D transfer, but still within a two-level approach. In Paletou et al. (1993) the two-level atom solutions and PRD are used for resonance lines of hydrogen, CaII and MgII. Multilevel 2D modeling with PRD for whole prominence slabs was done by Paletou (1995), while Heinzel and Anzer (2001) used the same 2D technique to model the prominence fine-structure threads in magneto-hydrostatic equilibrium (i.e. magnetic dips, see Sect. 5.8). Multi-thread models with velocities, aimed at describing the prominence fine structures, were then developed, see the review by Gunár (2014). All these 2D models are based on the Cartesian geometry. On the other hand, Gouttebroze (2007) (and see the references therein) has developed various 1D and 2D codes using the cylindrical geometry to account for the shape of prominence magnetic loop structures.

5.7.4 *Effect of Prominence Velocities on Line Radiation*

The effect of prominence dynamics on the line radiation is two-fold. First, the macroscopic velocity is responsible for the Doppler shifts of the absorption and emission profiles of spectral lines (for continua, the Doppler-shift effects are quite negligible). In case when we already know the line source function, this Doppler shifted absorption profile is used to perform the formal solution of the RTE. In optically-thick cases and in the presence of velocity gradients, this leads to asymmetrical profiles of the emergent line radiation. The other effect is on atomic level populations, through the statistical equilibrium. In this case, the radiation-field terms in ESE are affected by the relative shift of the absorption profile and the intensity profile. In the case of hydrogen lines, this effect can be neglected for small velocity amplitudes typical for quiescent prominences. The other term in ESE is the divergence of the atomic-population flow on the left-hand side. This can be important e.g. in the case of ambipolar diffusion where the neutral atoms move relative to ions (see Fontenla et al. 1996). The effects of large bulk velocities of eruptive prominences on the line emission are discussed in Labrosse (2014). We call them Doppler brightening or dimming and they are easily accounted for by modifying the boundary conditions which become velocity dependent (Heinzel and Rompolt 1987).

5.8 Coupling of Radiation to Magneto-hydrodynamics

The radiation properties of prominences have an important impact on the prominence magneto-hydrodynamical (MHD) structure, as well as on the energy balance. The dynamics of the prominence plasma in a magnetic field strongly depends on the degree of hydrogen and helium ionization, which in turn is dominated by photoionization processes, mainly in hydrogen continua. Below we briefly mention the magneto-hydrostatic (MHS) models, the radiative equilibrium, and MHD waves and oscillations.

5.8.1 *Models in Magneto-Hydrostatic Equilibrium*

Heasley and Mihalas (1976) and Anzer and Heinzel (1999) used the Kippenhahn-Schlüter (KS) model of prominences (Kippenhahn and Schlüter 1957) and solved the MHS equilibrium problem coupled to NLTE radiative transfer for multilevel hydrogen and helium atoms. Using the column-mass scale $dm = \rho dx$ instead of the geometrical one, they were able to derive the analytical form for the pressure variations within a 1D slab

$$p(m) = 4p_c \frac{m}{M}\left(1 - \frac{m}{M}\right) + p_0 \,, \tag{5.48}$$

where p_0 is the coronal pressure at the slab surfaces and M is the total column mass of the slab. x-component of the magnetic field B_x is constant through the slab and the z-component at the slab surface has the value $B_z \equiv B_{z1}$ which gives

$$M = \frac{B_x B_{z1}}{2\pi g} \,, \tag{5.49}$$

where g is the gravity acceleration at the solar surface. Using this formula, we obtain for p_c

$$p_c = \frac{\pi g^2}{B_x^2}\frac{M^2}{2} = \frac{B_{z1}^2}{8\pi} \,. \tag{5.50}$$

The quantity p_c can be interpreted in the following way: at the slab centre we have the pressure

$$p_{cen} = p(M/2) = p_c + p_0 \,. \tag{5.51}$$

If p_0 would be zero, then $p_{cen} = p_c = B_{z1}^2/8\pi$, which is the magnetic pressure. Therefore, in this case the plasma pressure at the slab centre will be equal to the magnetic pressure calculated with $B = B_{z1}$. Note that by plasma pressure we mean here the gas plus turbulent pressure $p = NkT + \rho v_{turb}^2/2$.

To get the density $\rho(m)$ we use the state equation with the mean molecular mass

$$\overline{m} = \frac{1 + 4\alpha}{1 + \alpha + i} m_H \,, \tag{5.52}$$

where i is the ionization degree of hydrogen $i = n_p/n_H$ (n_p and n_H are the proton and hydrogen densities, respectively), α the helium abundance relative to hydrogen and m_H the hydrogen atom mass. i varies between zero (neutral gas) and unity (fully-ionized hydrogen). Inside the prominence with an arbitrary temperature distribution, the ionization structure results from the complex NLTE calculations described above. The temperature structure is subject to energy-balance conditions, which is represented by the radiative equilibrium in the simplest case. The models of Heasley and Mihalas (1976) were generalized to 2D geometry by Heinzel and Anzer (2001), who coupled the MHS KS-type equilibrium to 2D NLTE radiative transfer in a multilevel hydrogen. This latter work was aimed at complex modeling of quasi-vertical fine-structure threads frequently observed in quiescent prominences. For detailed description see reviews (Heinzel and Anzer 2005; Heinzel 2007), while later development and results are described by Gunár (2014) and in Labrosse (2014). The current modeling aims at 2D or 3D NLTE simulations for arbitrary magnetic dips filled with the plasma in order to predict the visibility of prominence fine structure in various spectral lines (Gunár et al. 2013).

5.8.2 Radiative Equilibrium

Looking again at cartoon in Fig. 5.3, we can easily understand the processes of radiative heating and cooling (other processes appearing in the general equation of energy balance are discussed in Gilbert 2014). In the process of scattering between two atomic levels, the energies of the absorbed and reemitted photons are practically the same, small energy difference corresponds to the broadening of spectral lines and is typically several orders of magnitude smaller than the energy of the line transition itself, i.e. $h\Delta\nu_{\rm D} \ll h\nu_{ij}$. Therefore a negligible energy exchange takes place during the scattering in prominences. However, if the absorbed photon is 'destroyed' by inelastic collisional de-excitation, its energy which initially excited the atom is converted to the thermal pool of the plasma. In this case the radiation heats the plasma. In the opposite situation, the atom is excited by collision with a free electron and the excitation energy is taken out from the thermal pool. When a photon is then emitted from such excited level, it takes this energy out of a given location and we speak about radiative cooling. In both cases we deal with the energy exchange between plasma and radiation field and in a single process described above this energy is roughly equal to the photon energy $h\nu_{ij}$. Locally the problem of radiative heating and cooling reduces to evaluation of the net radiative losses

$$L = 4\pi \int_0^\infty (\eta_\nu - \chi_\nu J_\nu) d\nu = 4\pi \int_0^\infty \chi_\nu (S_\nu - J_\nu) d\nu \, . \tag{5.53}$$

The amount of radiative heating in prominences largely depends on the prominence illumination from surrounding solar atmosphere. A special case when $L = 0$ is called the radiative equilibrium, a well understood situation in stellar atmospheres (HM2014). In the case of solar prominences, Heasley and Mihalas (1976) have argued that the radiative-equilibrium temperatures inside 1D prominence slabs can be as low as 4,600 K, i.e. much lower than what is typically deduced from prominence spectral observations (e.g. Park et al. 2013; Jejčič et al. 2014). Therefore, various authors have attempted to investigate possible sources of prominence heating and this is discussed in Gilbert (2014).

Numerical evaluation of the radiative losses inside the cool prominence parts represents a difficult task because as we see from Eq. (5.53) L depends on the source function and on the radiation field. Moreover, the integration over frequencies includes various lines and continua of species which are supposed to be important coolants of the prominence plasmas. In prominences, where the thermodynamic conditions are similar to those in the solar chromosphere, the most important contributors to net radiative losses are the hydrogen, CaII, and MgII and some other ions. This was clearly demonstrated by Heinzel and Anzer (2012) and Heinzel et al. (2014a), who constructed radiative-equilibrium models by considering the relaxation of the 1D slabs to their radiative-equilibrium state described by $L = 0$. This kind of relaxation, first studied by Gouttebroze (2007) for purely hydrogen plasma contained in 1D cylinders, depends on the gas pressure,

Table 5.1 Radiative-equilibrium temperatures (in K) at 1D slab center and at its surface (in parentheses)

p[dyn cm^{-2}]	0.01	0.1	0.5	Net losses
$D = 200$	9750 (9760)	7990 (8180)	7010 (7560)	HI only
$D = 1{,}000$	9700 (9710)	7400 (8050)	6620 (7550)	
$D = 5{,}000$	9480 (9570)	6780 (8020)	6340 (7550)	
$D = 200$	8280 (8280)	6080 (6530)	4960 (5720)	HI, CaII and MgII
$D = 1{,}000$	8140 (8190)	5260 (6370)	4680 (5710)	
$D = 5{,}000$	7650 (7920)	4880 (6360)	4430 (5690)	

D is the slab thickness in km

Two sets of models are shown: with only hydrogen net losses (consistent with Gouttebroze 2007), and with hydrogen, CaII and MgII net losses—we see a significant effect of additional losses, namely in the case of thick and dense slabs. Adapted from Heinzel et al. (2014a)

geometrical thickness of the slab and on radiative boundary conditions. In most cases, adding CaII and MgII leads to significant cooling in addition to that caused by hydrogen. However, it is interesting to see that for low pressures which are found in quiescent prominences and for small geometrical extensions simulating narrow threads, the radiative-equilibrium temperatures are mostly above the lower limit of common observational determinations. This suggests that under some conditions, the prominence fine structures will not need any additional heating. We show these results in Table 5.1 adapted from Heinzel et al. (2014a).

5.8.3 Synthetic Spectra of Prominence MHD Oscillations

In a recent study by Heinzel et al. (2014b), a first attempt was made to synthesize the time-dependent hydrogen spectra in the case of prominence oscillations which are caused by linear perturbations to 1D-slab MHD equilibria (see Ballester 2014). For short oscillation periods where the period is comparable to radiative-relaxation times of the hydrogen plasma, a fully time-dependent solution of the NLTE problem is needed using the general form of ESE [Eq. (5.34)]. However, when the periods are large enough, one can solve the NLTE problem for a series of stationary snapshots assuming the statistical equilibrium for atomic level populations and ion densities to be time independent. Another approximation used in Heinzel et al. (2014b) is that for very small amplitudes of velocities (up to 2 km s^{-1} in their oscillatory models), one can solve the NLTE transfer problem in two subsequent steps. First, the static NLTE model is computed, and then by using precomputed atomic-level populations and electron densities, one performs the formal solution of RTE, including the LOS velocity distribution into the line opacities and emissivities, modifying the respective profiles by accounting for the Doppler shifts. Using the prescribed time-dependent MHD model of oscillations, Heinzel et al. (2014b) performed the NLTE transfer modeling for a hydrogen model atom having five bound levels and continuum. As a result, they obtained the emergent intensity profiles of the studied spectral lines for each time step (snapshot). Such profiles are in general asymmetrical due to velocity gradients and can be Doppler shifted.

Conclusions and Future Prospects

Radiative transfer applied to solar prominences provides the basis for the non-equilibrium spectroscopy of these fascinating objects. Further developments in this field will focus on fully 3D radiative-transfer modeling, taking into account the fine structure and its dynamics on spatial scales of tens of km and time scales of tens of sec. Under such conditions, the time-dependent ionization will play the role in the statistical equilibrium, similarly to situation we meet in the chromosphere. Sometimes the question is posed whether we cannot do 3D time-dependent simulations of prominence fine-structure dynamics as in the case of the quiet chromosphere. The main problem is largely unknown structure of the magnetic field on very small spatial scales and its dynamics. Only recently some studies have been initiated to explore the complexity of the dipped magnetic field and reliable visualizations of the prominence in various spectral lines. The realistic field topology is still static, while the dynamics, including the prominence condensation process, was so far simulated in 1D only (see Karpen 2014). With the advent of new ground-based and space instrumentation (4m class telescopes, UV imagers and spectrometers in space like those on Solar Orbiter and Solar-C) will generate a vast amount of multiwavelength data which must be analyzed in terms of the non-equilibrium spectroscopy outlined in this chapter. It will be a great challenge to interpret spectra and images from such high-resolution instruments of new generation. The radiative transfer theory has now solid grounds (for a review we again refer to HM2014), and it has to be properly applied to complex numerical simulations of the prominence structure and dynamics. The use of extensive parallel simulations is unavoidable. In this chapter we described general methods which are used to study solar prominences, but we concentrated mostly on cool structures typically emitting the lines like hydrogen Hα or similar. However, relatively cool structures are visible also in the cores of CMEs, frequently bringing the form of an eruptive prominence visible up to large altitudes. These structures show up temperatures of the order of 10^5 K and very low electron densities, implying very low gas pressures of the order of 10^{-3} dyn cm^{-2} or even lower. These structures move with large velocities and their spectra are formed under the extreme conditions, including the angle-dependent PRD scattering processes and Doppler dimming effects (see e.g. Ciaravella et al. 2003). CMEs in the hydrogen Lyman α line and the visible light will be detected by METIS coronagraph on board of ESA's Solar Orbiter mission.

References

Anzer, U., & Heinzel, P. (1999). *Astronomy and Astrophysics, 349*, 974.

Anzer, U., & Heinzel, P. (2005). *The Astrophysical Journal, 622*, 714. doi:10.1086/427817.

Auer, L. H., & Paletou, F. (1994). *Astronomy and Astrophysics, 285*, 675.

Ballester, J. (2014). In J. C. Vial, & O. Engvold (Eds.), *Solar prominences, ASSL* (Vol. 415, pp. 257–294). Springer.

Beckers, J. M. (1964). A study of the fine structures in the solar chromosphere. Ph.D. thesis, Sacramento Peak Observatory, Air Force Cambridge Research Laboratories, Mass.

Carlsson, M., & Stein, R. F. (2002). *The Astrophysical Journal, 572*, 626. doi:10.1086/340293.

Ciaravella, A., Raymond, J. C., van Ballegooijen, A., Strachan, L., Vourlidas, A., Li, J., et al. (2003). *The Astrophysical Journal, 597*, 1118. doi:10.1086/381220.

Cram, L. E., & Vardavas, I. M. (1978). *Solar Physics, 57*, 27. doi:10.1007/BF00152041.

Engvold, O. (1980). *Solar Physics, 67*, 351. doi:10.1007/BF00149812.

Fontenla, J. M., Rovira, M., Vial, J. C., & Gouttebroze, P. (1996). *The Astrophysical Journal, 466*, 496. doi:10.1086/177527.

Gilbert, H. (2014). In J. C. Vial, & O. Engvold (Eds.), *Solar prominences, ASSL* (Vol. 415, pp. 155–176). Springer.

Gouttebroze, P. (2007). *Astronomy and Astrophysics, 465*, 1041. doi:10.1051/0004-6361:20066636.

Gouttebroze, P., & Heinzel, P. (2002). *Astronomy and Astrophysics, 385*, 273. doi:10.1051/0004-6361:20020142.

Gouttebroze, P., Heinzel, P., & Vial, J. C. (1993). *Astronomy and Astrophysics Supplement, 99*, 513.

Gouttebroze, P., & Labrosse, N. (2000). *Solar Physics, 196*, 349 doi:10.1023/A:1005229231465.

Gunár, S. (2014). In B. Schmieder, J. M. Malherbe, & S. T. Wu (Eds.), *IAU Symposium* (Vol. 300, pp. 59–68). doi:10.1017/S1743921313010752.

Gunár, S., Mackay, D. H., Anzer, U., & Heinzel, P. (2013). *Astronomy and Astrophysics, 551*, A3. doi:10.1051/0004-6361/201220597.

Heasley, J. N., & Mihalas, D. (1976). *The Astrophysical Journal, 205*, 273. doi:10.1086/154273.

Heasley, J. N., Mihalas, D., & Poland, A. I. (1974). *The Astrophysical Journal, 192*, 181. doi:10.1086/153049.

Heinzel, P. (1983). *Bulletin of the Astronomical Institutes of Czechoslovakia, 34*, 1.

Heinzel, P. (1991). *Solar Physics, 135*, 65. doi:10.1007/BF00146699.

Heinzel, P. (1995). *Astronomy and Astrophysics, 299*, 563.

Heinzel, P. (2007). In P. Heinzel, I. Dorotovič, & R.J. Rutten (Eds.), *The Physics of Chromospheric Plasmas. Astronomical Society of the Pacific Conference Series* (Vol. 368, p. 271). Astronomical Society of the Pacific.

Heinzel, P., & Anzer, U. (2001). *Astronomy and Astrophysics, 375*, 1082. doi:10.1051/0004-6361:20010926.

Heinzel, P., & Anzer, U. (2005). In A. Hanslmeier, A. Veronig, & M. Messerotti (Eds.), *Solar Magnetic Phenomena. Astrophysics and Space Science Library* (Vol. 320, pp. 115–138). Netherlands: Springer.

Heinzel, P., & Anzer, U. (2012). *Astronomy and Astrophysics, 539*, A49. doi:10.1051/0004-6361/200913537.

Heinzel, P., Anzer, U., & Gunár, S. (2005). *Astronomy and Astrophysics, 442*, 331. doi:10.1051/0004-6361:20053360

Heinzel, P., Gouttebroze, P., & Vial, J. C. (1987). *Astronomy and Astrophysics, 183*, 351.

Heinzel, P., & Rompolt, B. (1987). *Solar Physics, 110*, 171. doi:10.1007/BF00148210.

Heinzel, P., Schmieder, B., & Vial, J. C. (1997). In A. Wilson (Ed.), *Fifth SOHO Workshop: The Corona and Solar Wind Near Minimum Activity* (Vol. 404, p. 427). Oslo: ESA Special Publication.

Heinzel, P., Vial, J. C., & Anzer, U. (2014a). *Astronomy and Astrophysics, 564*, A132. doi:10.1051/0004-6361/201322886.
Heinzel, P., Zapiór, M., Oliver, R., & Ballester, J. L. (2014b). *Astronomy and Astrophysics, 562*, A103. doi:10.1051/0004-6361/201322346.
Hillier, A., Isobe, H., Shibata, K., & Berger, T. (2011). *The Astrophysical Journal Letters, 736*, L1. doi:10.1088/2041-8205/736/1/L1.
Hubeny, I., & Mihalas, D. (2014). *Physics of stellar atmospheres*. Princeton Univ. Press.
Ishizawa, T. (1971a). *Publications of the Astronomical Society of Japan, 23*, 75.
Ishizawa, T. (1971b). *Publications of the Astronomical Society of Japan, 23*, 121.
Jejčič, S., & Heinzel, P. (2009). *Solar Physics, 254*, 89. doi:10.1007/s11207-008-9289-2.
Jejčič, S., Heinzel, P., Zapiór, M., Druckmüller, M., Gunár, S., & Kotrč, P. (2014). *Solar Physics, 289*, 2487. doi:10.1007/s11207-014-0482-1.
Karpen, J. (2014). In J. C. Vial, & O. Engvold (Eds.), *Solar prominences, ASSL* (Vol. 415, pp. 235–255). Springer.
Kippenhahn, R., & Schlüter, A. (1957). *Z. für Astrophysics, 43*, 36.
Kucera, T. (2014). In J. C. Vial, & O. Engvold (Eds.), *Solar prominences, ASSL* (Vol. 415, pp. 77–99). Springer.
Labrosse, P. (2014). In J. C. Vial, & O. Engvold (Eds.), *Solar prominences, ASSL* (Vol. 415, pp. 431–451). Springer.
Labrosse, N., & Gouttebroze, P. (2001). *Astronomy and Astrophysics, 380*, 323. doi:10.1051/0004-6361:20011395.
Labrosse, N., Heinzel, P., Vial, J. C., Kucera, T., Parenti, S., Gunár, S., et al. (2010). *Space Science Reviews, 151*, 243. doi:10.1007/s11214-010-9630-6.
Landi, E., Young, P. R., Dere, K. P., Del Zanna, G., & Mason, H. E. (2013). *The Astrophysical Journal, 763*, 86. doi:10.1088/0004-637X/763/2/86.
Mein, N., Mein, P., Heinzel, P., Vial, J. C., Malherbe, J. M., & Staiger, J. (1996). *Astronomy and Astrophysics, 309*, 275.
Mihalas, D., Auer, L. H., & Mihalas, B. R. (1978). *The Astrophysical Journal, 220*, 1001. doi:10.1086/155988.
Molowny-Horas, R., Heinzel, P., Mein, P., & Mein, N. (1999). *Astronomy and Astrophysics, 345*, 618.
Paletou, F. (1995). *Astronomy and Astrophysics, 302*, 587.
Paletou, F., Vial, J. C., & Auer, L. H. (1993). *Astronomy and Astrophysics, 274*, 571.
Park, H., Chae, J., Song, D., Maurya, R. A., Yang, H., Park, Y. D., et al. (2013). *Solar Physics, 288*, 105. doi:10.1007/s11207-013-0271-2.
Rudawy, P., & Heinzel, P. (1992). *Solar Physics, 138*, 123. doi:10.1007/BF00146200.
Rybicki, G. B., & Hummer, D. G. (1991). *Astronomy and Astrophysics, 245*, 171.
Rybicki, G. B., & Lightman, A. P. (1979). *Radiative processes in astrophysics*. New York: Wiley.
Schmieder, B., Heinzel, P., Kucera, T., & Vial, J. C. (1998). *Solar Physics, 181*, 309. doi:10.1023/A:1005022003293.
Schwartz, P., Heinzel, P., Schmieder, B., & Anzer, U. (2006). *Astronomy and Astrophysics, 459*, 651. doi:10.1051/0004-6361:20065619.
Vial, J. C. (1982). *The Astrophysical Journal, 254*, 780. doi:10.1086/159789.

Chapter 6
Derivation of the Major Properties of Prominences Using NLTE Modelling

Nicolas Labrosse

Abstract I introduce techniques to derive the major properties of prominences based on NLTE modelling. The main results of one- and two-dimensional models of the prominences and their fine-structures are presented. Modelling the radiative transfer processes out of local thermodynamic equilibrium allows one to retrieve the prominence plasma parameters (temperature, density, pressure, ionisation degree) as well as the prominence's mass.

6.1 Introduction

The previous chapter has introduced the basics of the radiative transfer theory as applied to solar prominences, and a detailed account of how NLTE[1] radiative transfer modelling can be implemented in this context. It is clear that this type of heavy modelling is unavoidable in view of the very complex processes of line and continua formation. The atomic level populations are non-linearly coupled to the radiation field which has a non-local character extending to the whole medium. A well-known example of this lies in the properties of the Hα line which may be considered as optically thin in some cases, but is inevitably linked to the Lα and Lβ transitions which are definitely optically thick in prominences. Another example can be found in the coupling between singlet and triplet states within the He I atom, linking lines such as the He I triplet 5,876 Å (D3) and 10,830 Å lines to the optically thick resonance lines found in the singlet system (Labrosse and Gouttebroze 2004). Part of this coupling comes from the photoionisation-recombination mechanism, where the He I atom is ionised from the ground state (singlet), which is then followed by recombination from the ground state of ionised helium preferably towards the triplet states. Hence the coupling between neutral and ionised helium is important and must also be taken into account in the NLTE modelling, even

[1] NLTE stands for Non-LTE, i.e. departures from LTE (Local Thermodynamic Equilibrium).

N. Labrosse (✉)
SUPA, School of Physics and Astronomy, University of Glasgow, Glasgow, Scotland
e-mail: Nicolas.Labrosse@glasgow.ac.uk

J.-C. Vial, O. Engvold (eds.), *Solar Prominences*, Astrophysics and Space Science Library 415, DOI 10.1007/978-3-319-10416-4_6

if one is only interested in the neutral lines. A third, equally important example, concerns the impact that the hydrogen Lyman lines have on the ionisation state of other elements such as Ca II (Gouttebroze and Heinzel 2002) and Mg II (Heinzel et al. 2014a).

Hence it is necessary to develop comprehensive codes to compute the spectra emitted by atoms and ions in solar prominences by solving the coupled equations of radiative transfer and statistical equilibrium. In this chapter I show how the theoretical ideas relevant to the above discussion (and to the previous chapter) are put in use in order to derive the major properties of solar prominences. An extensive discussion of the results obtained in this area can be found in Labrosse et al. (2010), and recent contributions to this field can be found in Schmieder et al. (2014).

This chapter starts with a brief summary of how NLTE radiative transfer modelling is used in the context of solar prominence studies to infer some of their properties. Then, Sect. 6.3 is devoted to the core topic of the determination of the plasma parameters in quiescent, static prominences, first from the study of the hydrogen spectrum, then helium, and trace elements such as ionised calcium and magnesium. I will present the basic results from one-dimensional (1D) and two-dimensional (2D) NLTE radiative transfer computations. The same techniques can be used to derive the properties of active and eruptive prominences exhibiting plasma flows, and Sect. 6.4 shows the results obtained from the adoption of a velocity-dependent incident radiation field in these models. Finally, some conclusions are given in the section "Conclusions".

6.2 What Is NLTE Modelling and How Does It Work?

The principles of NLTE modelling have been described in the previous chapter and the interested reader is advised to refer to it (Heinzel 2014). I will just highlight the basic philosophy here for the reader who is mostly interested in learning about NLTE diagnostics without going into the details of how it is done.

NLTE modelling represents a forward method. This means that starting from a given prominence atmosphere model (spatial distribution of temperature, pressure, gas density), one evaluates the excitation and ionisation balance for given species, determines the opacities and emissivities, and finally solves the transfer equation along the line-of-sight (LOS) to get the emergent synthetic spectrum. The latter is then compared to the observed spectrum. In this way, one can adjust the initial model in order to get an optimum agreement with the observations. This procedure is iterative and the final models are called semi-empirical models, provided that they are at least partially data-driven.

6.2.1 A Note on Optically Thick Lines as a Diagnostic Tool for the Prominence Plasma

The importance of studying the optically thick lines and continua emitted from various elements must be emphasized. The following discussion focuses on the hydrogen Lα line as the best example of an optically thick line emitted by a solar prominence under NLTE conditions, however it is relevant to any other line or continuum in the optically thick regime.

The prominence Lα[2] intensity is roughly given by the quiet Sun intensity in that line multiplied by a factor of 0.3–0.4, about the value of the dilution factor. Indeed, as explained in Chap. 5, the intensity of the radiation incident on a point at the prominence surface (which is used as the boundary condition for the solution of the radiative transfer equation and so is crucially important) is given by the intensity of the line coming from the disc multiplied by this dilution factor which is ≤ 0.5 (assuming no illumination from the corona). The dilution factor is wavelength dependent if the disc incident radiation exhibits centre-to-limb variations. If there are no centre-to-limb variations, it is simply given by:

$$W = \frac{1}{2}\left(1 - \sqrt{1 - \left(\frac{R_s}{h + R_s}\right)^2}\right) , \tag{6.1}$$

where R_s is the solar radius and h is the altitude of the prominence.

When the plasma is very optically thick in some line, one mostly sees the external layer of the prominence which is scattering the line from the disc. Generally speaking, the large opacity in UV and EUV resonance lines, combined with the strength of the incident radiation as compared to the local (thermal) radiation field, is a natural explanation for the bright appearance of prominences in these lines. Hence at first sight it could seem that, no matter the physical conditions in the prominence, the intensity of very optically thick lines is primarily determined by the amount of incident radiation from the Sun.

Of course, the reality is more subtle. One has to take into account first that different parts of the line profile allow one to probe different parts of the prominence structure. The core of the line, which is where the optical thickness is largest, is emitted from a layer close to the surface (which in turn may be more sensitive to the incident radiation), while photons in the line wings will come from deeper regions inside the prominence. As an illustration, it has been noted in many studies that the effect of (e.g. in a 1D plane-parallel model) the slab width is usually

[2] In this chapter, I refer to the Lyman hydrogen spectrum (lines and continua) simply as the Lyman spectrum, unless otherwise stated.

negligible on the optically thick resonance lines: when the line centre is saturated, only the optically thin wings of the line profile may become brighter for larger slab thickness. In addition, the relative contribution of thermal processes in the excitation mechanisms of the line may be important (depending on the physical conditions of the plasma), and in such case the line will also reflect somehow these conditions (albeit not so straightforwardly than for an optically thin line with a gaussian profile).

6.3 Determination of the Plasma Parameters

Because of the nature of prominence diagnostics based on NLTE modelling, it is not suitable to describe separately how specific plasma parameters are obtained. Rather, the process is to identify a set of atmosphere parameters from the models, which yield computed spectra in agreement with observations. Hence, the model[3] which best matches observed properties provides all the desired parameters: temperature, densities, column mass and so forth. Therefore in this section, I will discuss what we can learn from the modelling of spectra emitted by various elements, starting with hydrogen.

The intention here is not to present an exhaustive review of the full body of work that has been done in this area, but to focus on some particularly important and enlightening studies which have enabled the determination of the plasma parameters in prominences. The 1970s saw important first steps made by people like Mihalas, Heasley, Milkey, Poland, Morozhenko, and others to lay out the foundations for NLTE radiative transfer modelling of solar prominences. Their pioneering work built prominence models on the basis of integrated intensities (only), with the assumption of a flat incident spectrum, and did not provide the spectral signatures in terms of line profiles of the various models used. The reasons are that the incident (chromospheric and coronal) line profiles were not well known, and the proper treatment of the line radiation scattering was not easy. Moreover, there were only a few observed prominence spectra to compare model products with. Their early models were able to provide a first glimpse at the complexity of the physical processes responsible for the radiation emitted by these structures. However, more complex codes have been developed since then, while more detailed observations have become available, and this is what we are going to discuss now.

[3]This is assuming that a unique solution can be identified. For the modelling to be considered successful when a unique solution cannot be identified from the comparison with observations, there needs to be a small set of models with close enough physical parameters which match the observations.

6.3.1 *Hydrogen Lines and Continua*

The first prominence spectral observations in Lα along with Lβ, Ca II, and Mg II lines came from OSO-8 (Vial 1982a). Starting from a 2D code solving the radiative transfer problem in an externally illuminated structure (Mihalas et al. 1978), Vial (Vial 1982) was able to compare Lα, Mg II h and k, and Ca II H and K computed profiles to observed ones. A two-level atom was assumed and the ionisation was treated with the assumption of photo-ionisation only, but the OSO-8 incident radiation was properly taken into account. This allowed the author to build a reasonable model, with a temperature of 8,000 K and an electron density of $2 \times 10^{10}\,\mathrm{cm}^{-3}$. Much of the subsequent modelling work was carried out in 1D. The progress made in the frame of 1D allows us to shed some light on the continuous improvement of modelling in relation with the physics of the radiation, e.g., partial redistribution in frequency (PRD—see Chap. 5) *vs* complete redistribution (CRD), and the availability of observed line profiles.

6.3.1.1 1D Isothermal and Isobaric Models

A series of modelling efforts in 1D culminated around the GHV paper (Gouttebroze et al. 1993) which, among other observables, provided for the first time an extensive range of realistic emergent line profiles (see Fig. 6.1) comparable with available observed profiles. This was done with a proper account of the incident radiation profiles, and a rigorous treatment of the diffusion within the line profiles. An important feature of this modelling was the use of a 20-level plus continuum atom, which allowed to predict a set of Lyman profiles more complete than what could be actually observed with the then available spectrometers. It is immediately noticeable that the shape of the three Lyman lines shown in Fig. 6.1 are not Gaussian and exhibit a strong self-reversal at line centre (Lα), or a peak in the wings

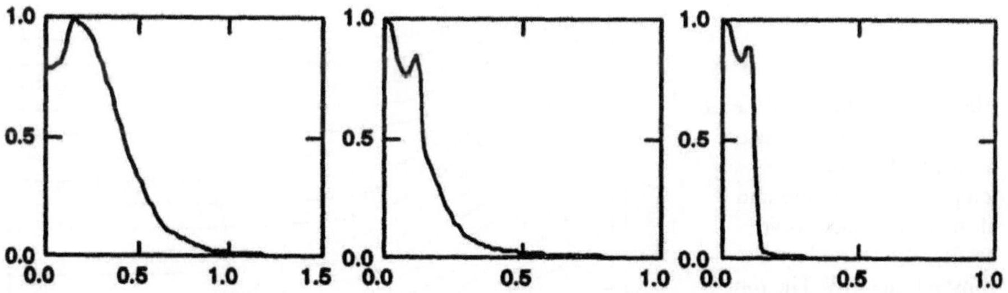

Fig. 6.1 Lα, Lβ and Lγ line half-profiles emergent from a 1D prominence model characterised by its temperature (6,000 K), its pressure (0.05 dyn cm^{-2}), its geometrical thickness (5,000 km) and microturbulence (5 km s^{-1}). Abscissae are in Å. The line profiles are normalised to the value of the maximum intensity, given in a table by GHV. Credit: Gouttebroze et al. (1993), reproduced with permission © ESO

(Lβ and Lγ). These complex line shapes illustrate the necessity of performing NLTE modelling to compute the line profiles and treat the complex line formation process realistically. The variation of the opacity across the line profile, itself determined partly by the temperature and pressure variations inside the prominence, the presence or not of significant thermal excitation, the detailed line shape of the incident line profile, and the effects of coherent scattering, all combine to give the type of profiles seen in this figure.

Correlations connecting the observables to the thermodynamic parameters were derived. One example is the colour temperature of the Lyman continuum (a measure of the slope of the continuum), which was found representative (at not too high temperatures) of the electron temperature, a confirmation of the result of Heasley and Milkey (1983). These correlations, further explored in Heinzel et al. (1994), provided a valuable tool for the interpretation of combined multi-wavelength observations.

These basic 1D models describe the central cool parts of the prominence or its fine structures reasonably well. A prominence whose physical properties vary with altitude may be schematically simulated by a sequence of individual models. For instance, Gouttebroze and Labrosse (2000) computed 20 models to simulate a prominence with a vertical extension of 10^5 km, a temperature increasing from 5,000 K at the bottom to 10,000 K at the top, and a pressure decreasing from 0.2 to 0.02 dyn cm^{-2}. Figure 6.2 shows how the hydrogen Lα line profile varies from the bottom to the top of the prominence. Near the bottom, the prominence is very optically thick in Lα and the emission process is dominated by the scattering of the incident radiation. As the temperature increases and the pressure decreases towards

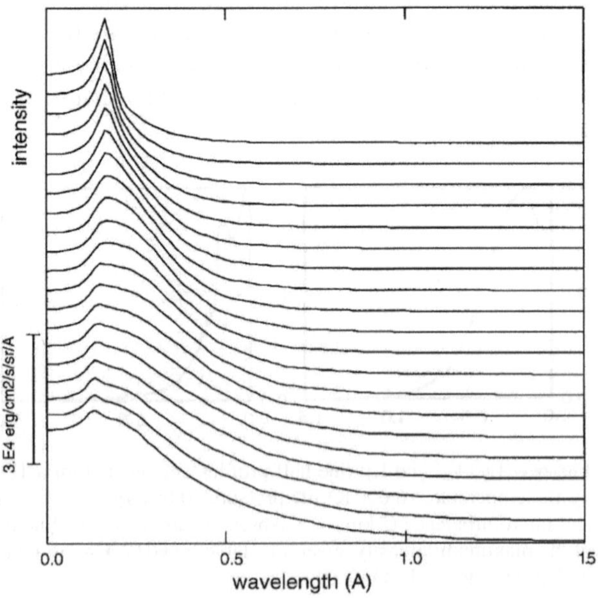

Fig. 6.2 Variations of the Lα profile along a prominence model with variable temperature, pressure and altitude. The successive profiles are shifted by a constant quantity. The *top curve* corresponds to the *top* of the prominence. Credit: Gouttebroze and Labrosse (2000). With kind permission of Springer Science+Business Media

the top, the optical thickness decreases. The incident radiation also decreases with altitude. Therefore, the scattering decreases and the thermal emission of the slab becomes more important at the top of the prominence, relative to the emission due to scattering of the incident radiation. As a consequence, the line profile is relatively broad and flat at the bottom of the prominence (similar to the incident profile). As altitude increases, the profile is transformed with sharp peaks and an intensity at line centre which is larger than the value of the diluted incident intensity (this is due to thermal excitation), and low wings resulting from the decreased scattering. It is worth stressing that as the altitude of the prominence is changed, the boundary conditions of radiative transfer are changed, since the incident radiation is diluted with height as can be seen in Eq. (6.1).

There is a long list of papers who have used the results of GHV and their subsequent paper on theoretical correlations (Heinzel et al. 1994) to infer plasma parameters in prominences. For example, Heinzel et al. (1996) compared Hα observations with results from GHV and derived electron densities in very good agreement with measurements from 2.5×10^9–$6.3 \times 10^{10}\,\mathrm{cm}^{-3}$ (Bommier et al. 1994). This enabled them to obtain values for the geometrical thickness ranging between a few hundreds km up to a few 10^4 km in different prominences, corresponding to a fairly constant total column mass of $10^{-5}\,\mathrm{g\,cm}^{-2}$.

The Lyman Continuum and the Electron Temperature

The hydrogen Lyman continuum in solar prominences is mostly produced through photo-ionisation due to the chromospheric emission, followed by radiative recombination. This emission is related to regions affected by the penetration of the incident Lyman continuum radiation. Under the assumption that the electron temperature T is constant in the emitting region, and that the source function is constant with depth and position, the continuum intensity can be written as:

$$I_\lambda = \frac{2hc^2}{b_1 \lambda^5} \exp\left(-\frac{hc}{\lambda k T}\right) . \qquad (6.2)$$

Here, h, c, and k are the Planck constant, the speed of light, and the Boltzmann constant, respectively. $b_1 = n_1/n_1^\star$ (the ratio of the population of the ground state and its LTE population) is the so-called LTE departure coefficient of the ground state. T and b_1 are then obtained by fitting (6.2) to the observed continuum intensity. Note that this technique is only valid well below 15,000 K (otherwise the formation of the Lyman continuum will be different), and that it underestimates T if the layer is optically thin in the Lyman continuum (in which case Eq. (6.2) cannot be used).

Only a few measurements have been reported. Ofman et al. (1998) found a large range of temperatures (from 5,000 K to 15,000 K), the largest values being probably related to the disappearance (and associated heating) of the prominence. Parenti et al. (2005a) found different temperatures (8,280–7,560 K) in different parts of the same prominence.

Slab Thickness, Column Densities, and Prominence Mass

NLTE calculations can be used also to obtain quantities such as the thickness of the observed prominence, or the column density of the emitter. For example, by combining polarimetric data in the Hα and He I D3 lines on one hand, and observed Hα intensities (compared to theoretical ones) on the other hand, Heinzel et al. (1996) obtained values for the geometrical thickness ranging between a few hundreds km up to a few 10^4 km for 18 measurements in different prominences. They then derived a corresponding total column mass of 10^{-5} g cm^{-2}.

If one wants to derive e.g., the total mass of the prominence or the gas pressure, one has to measure the density of neutral hydrogen (along with minor atoms and ions such as He). This raises the issue of the ionisation degree of the hydrogen plasma (defined in Sect. 5.4 of Chap. 5). It is difficult to determine this from observations as this requires a simultaneous derivation of the electron density and the neutral hydrogen density. A comparison between a set of observed hydrogen line profiles and the predictions of NLTE modelling can provide an adequate answer for this important parameter.

An other approach to derive physical parameters of prominences has been developed based on the absorption mechanism of coronal radiation by H, He I, and He II continua (see Sect. 6.3.4). Because the prominence plasma may be optically thick at these wavelengths and is absorbing the radiation coming from the corona behind the prominence, NLTE radiative transfer effects can be used to relate the observed absorption to the prominence plasma parameters. The opacity derived from observations of Hα and coronal EUV lines (Heinzel et al. 2008) is consistent with models of prominences having an electron density of the order of 10^{11} cm^{-3} with typical parameters for the temperature (6,000–8,000 K) and a thickness of 1–5$\times 10^3$ km. These values yield hydrogen densities in prominences in the range 10^{10}–10^{11} cm^{-3}.

Prominence Oscillations

Prominence oscillations are often observed, and their characteristics are the focus of many observational and theoretical investigations (Chap. 11 - Ballester 2014). An exploratory study of the effect of global oscillations on the emergent Hα and Hβ lines (Heinzel et al. 2014b) shows that under typical temperature and pressure conditions, the fundamental slow mode and its first overtone, and the fundamental fast mode, produce detectable variations in the intensity and width of the lines—but the fast overtone does not. This is an important step in the diagnostics of these oscillating structures.

6.3.1.2 1D Models with a Prominence-to-Corona Transition Region

Prominence models with a prominence-to-corona transition region (PCTR) belong to the category of semi-empirical atmospheric models. Their temperature structure is determined empirically to reach agreement between the synthetic and observed spectra. However, NLTE prominence models are so far either isobaric, or consider certain magneto-hydrostatic equilibrium (MHS – see e.g. Anzer and Heinzel 1999).

Heinzel et al. (2001a) observed and studied three quiescent prominences from Lα to L9 and in the Hα line. Two classes of Lyman line profiles were obtained: deeply reversed and unreversed ones. While isothermal and isobaric models proved to be unsuccessful in reproducing the two classes of observed profiles, a 1D code including a PCTR with the pressure prescribed by the MHS equilibrium allowed the authors to derive a constant gas pressure around 0.2 dyn cm^{-2} for reversed profiles and a variation from 0.12 to 0.04 dyn cm^{-2} for non reversed profiles. This proved to be decisive in interpreting the two types of profiles with respect to the PCTR structure.

> The PCTR across the magnetic field is very thin because of the strongly reduced perpendicular conduction, while the PCTR along the magnetic field, being governed by strong parallel conduction, is rather extended. Unreversed profiles correspond to observations along the field lines where the PCTR provides line centre photons. On the contrary, deeply reversed profiles correspond to observations across the field lines where the cool prominence core is the only contribution.

6.3.1.3 1D Models Addressing the Fine Structure of Prominences

The 1D monolithic slab models discussed so far are, unfortunately, unable to reproduce correctly the observed properties of the Lα and Lβ lines *simultaneously*. In order to overcome this, several solutions were proposed, most dealing with the concept of fine structuring of prominences well supported by observations (see Chaps. 1 and 2: Vial 2014 and Engvold 2014). The Lyman series profiles available in prominences and filaments were useful in this context, too. Schmieder et al. (1999) compared the observed L4 to L9 profiles in a quiescent prominence with the profiles provided by three different classes of models: the (isobaric, isothermal) GHV models, the filamentary models of Fontenla et al. (1996), and a superposition of GHV-type models with very small thickness which represent an elementary filamentary structure. The two main conclusions of the paper are that (1) all Lyman lines seem to be formed at the base of the PCTR, and (2) some temperature gradient corresponding to a PCTR is needed to explain the behaviour of several higher Lyman lines. From the comparison of the different spatial behaviours of Lα and Lβ profiles with predictions of 1D and multi-thread modelling, it was found that the observed Lα/Lβ ratio never goes higher than 180 while the thread modelling leads to much lower values (Vial et al. 2007).

The issue of the number of threads requires more complex computations than the 1D modelling discussed so far. It is also related to observational constraints. High-resolution observations seem to indicate that the fine structure of prominences is not yet fully resolved at this time.

6.3.1.4 2D Models

Although the 1D models discussed above still represent a useful and computationally efficient approach for a number of situations, they do not allow us to consistently study the variation of the radiation and the plasma parameters in two dimensions, e.g., along and across the magnetic field lines. Indeed, observations carried out both from space and from ground may reveal the structure of prominences at various angles with respect to the magnetic field orientation. Proper interpretation of such observations calls for more general multi-dimensional prominence models.

The modelling efforts following Mihalas et al. (1978) and Vial (1982) led to new codes which were more robust, could run faster, and included a more refined treatment of radiation scattering. For example, the study by Paletou et al. (1993) presented computed synthetic profiles of H I (Lα), Ca II (H & K), and Mg II (h & k) lines in a 2D horizontally infinite slab model using the two-level atom approximation. The authors confirmed a strong influence of PRD on synthetic Lα profiles. Later on, a method for multi-level radiative transfer was implemented into 2D horizontally infinite slab geometry by Auer et al. (1994) and by Paletou (1995), both for CRD and PRD. As in 1D, the proper consideration of the incident radiation is critical. Gouttebroze (2006) incorporated the effect of an anisotropic radiation incident on a 2D cylindrical structure, and showed how the hydrogen intensities vary depending on how a particular section of the cylinder is illuminated.

In order to address the high-resolution prominence observations which reveal a variety of fine structures (in particular, elongated thread-like features and knots of plasma), Heinzel and Anzer (2001) described the MHS equilibrium of vertical 2D threads hanging in magnetic dips of the horizontal magnetic field. The 2D temperature variation (Fig. 6.3), prescribed semi-empirically, accounts for two different PCTRs. The PCTR across the magnetic field lines is very narrow (Fig. 6.3B), with a steep temperature gradient from the cool central part of the thread towards its boundaries. The PCTR along the magnetic field lines is much more extended

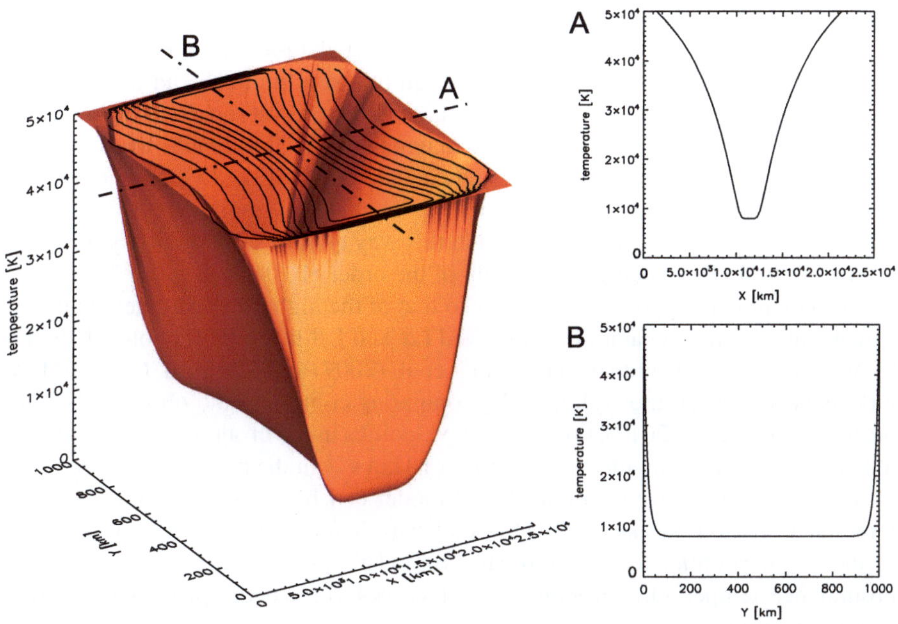

Fig. 6.3 Semi-empirical variation of the temperature within the cross-section of the 2D vertical prominence fine-structure thread. The x and y axis, representing geometrical dimensions of the thread, are not drawn to the same scale. Iso-contours of the temperature are shown. This plot clearly shows two different shapes of the PCTR, with (*A*) a gentle rise of the temperature along the x-axis (along the field lines), and (*B*) a steep gradient of the temperature along the y-axis (across the magnetic field lines). Credit: Labrosse et al. (2010). With kind permission of Springer Science+Business Media

(Fig. 6.3A), with a shallow rise of the temperature. Synthetic profiles of the hydrogen Lyman lines obtained along and across the magnetic field lines show significant differences with considerably reversed profiles obtained across the magnetic field and unreversed profiles obtained along the field. Such a behaviour is in agreement with observations presented by Heinzel et al. (2001a). Using the same code, Gunár et al. (2007a) investigated the Lyman continuum behaviour and showed that a comparison between synthetic and observed Lyman continuum intensities produces useful constraints on the temperature variation of the prominence fine structures.

Attempts to reproduce the observed filamentary fine structure led to the development of a full 2D magneto-hydrostatic model of prominence fine structure (Heinzel et al. 2005), which confirmed the sensitivity of the line profile shapes to the direction of the LOS and the magnetic field. This was successfully tested through the continuous (3 days) observations of a filament/prominence close to the limb (Schmieder et al. 2007) from L3 to L7 and in Hα. With the filament being circular in shape, the authors managed to interpret the daily differences of profiles as the result of the changing angle of the LOS with the magnetic field direction.

The prominence fine-structure properties obtained from these models are in good agreement with expected values. However, Gunár et al. (2007) find that the observed Lyman line profiles can be better reproduced by using multi-thread fine-structure models consisting of a set of identical 2D threads hanging on separate magnetic field lines with the LOS perpendicular to the magnetic field. These multi-thread models were used by Gunár et al. (2008) to study asymmetries of the Lyman line profiles. Each thread of the multi-thread model has a randomly assigned LOS velocity. Synthetic Lyman line profiles obtained in this way exhibit substantial asymmetries, even with the LOS velocities being only of the order of 10 km s^{-1}. The Lα profiles can also exhibit an opposite asymmetry to that of the higher Lyman lines. This is in agreement with the behaviour of observed Lα and Lβ line profiles noticed already by Vial (1982a), and confirmed by Gunár et al. (2007) and Vial et al. (2007). These models have been further validated by comparing computed and observed Hα line profiles (Gunár et al. 2012). Overall LOS velocities in the observed prominence did not exceed 15 km s^{-1}, with values close to 10 km s^{-1} in the prominence core.

The information from a multi-thread model can be used to compute the Differential Emission Measure (DEM—see Chap. 3 for a discussion, Parenti 2014) of the whole prominence structure (Fig. 6.4). This provides a new insight on the prominence temperature structure from the cool core to the prominence-corona transition region.

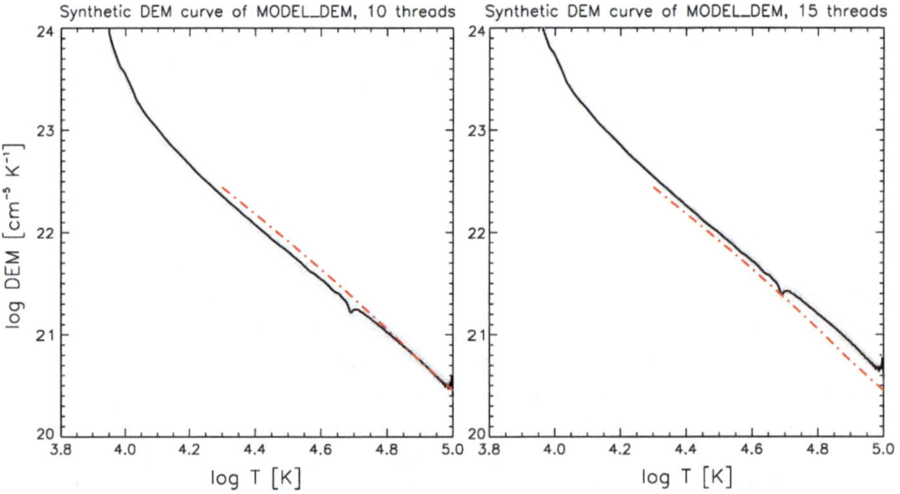

Fig. 6.4 Synthetic DEM curves of a multi-thread configuration with 10 (*left panel*) and 15 threads (*right panel*). *Gray solid lines* represent 100 random realizations of the multi-thread model and the *black solid line* gives the average of these 100 realizations. *Red dash-dotted line* represents the observed DEM curve. Credit: Gunár et al. (2011), reproduced with permission © ESO

We have learnt a great deal by comparing observed and computed (in 1D and 2D) hydrogen spectra. However, the ranges of hydrogen line profiles and inferred physical parameters remain very large. Moreover, the number of threads assumed in the various multi-thread models (larger than 10 along the LOS) implies that observations are insufficient for resolving them, in spite of progress in spatial resolution. This means that for a given set of observations, there might be more than one solution. One way to go around this is to combine the analysis of hydrogen lines with lines emitted by other elements.

6.3.2 Helium Lines and Continua

There are at least two good reasons to include helium in the NLTE modelling of solar prominences. First, this enables us to probe a wider region of the prominence plasma, hence gaining a greater understanding of the physical conditions in these structures. Secondly, this should provide additional constraints to measure the helium-to-hydrogen abundance ratio.

6.3.2.1 1D Isothermal and Isobaric Models

The study of the effects of the slab temperature, pressure, and width, as well as of the He abundance, on the mean population densities and on He I and He II line profiles and integrated intensities (Labrosse and Gouttebroze 2001) illustrates the complexity of the coupling existing between the different transitions. The results can be briefly summarised as follows: at low temperatures/pressures, the main mechanism of formation of the He resonance lines is the scattering of the incident radiation. A substantial amount of ionized helium can be found at the boundary between the prominence and the corona. This ionized helium is produced via the photo-ionisation of neutral helium by the incident radiation, and can scatter the incident radiation at 304 Å. Consequently, even *cool* prominences can in principle emit in the He II line at 304 Å, although the formation temperature of this line is around 80,000 K. This study also shows that the different sensitivities of line intensities (optically thick vs. optically thin, singlet vs. triplet) could be used in conjunction with hydrogen lines to improve the diagnostics of the helium abundance in prominences. The importance of PRD in the formation of the resonance lines of H and He was demonstrated. PRD for the radiative transfer calculations in hydrogen lines and continua is important for the subsequent modelling of helium lines and continua, since the electron density structure inside the prominence slab changes between CRD and PRD.

A comparison between observed and computed H and He I line profiles was presented in Labrosse et al. (2006). Profiles of the Lβ, L5 and He I 584 Å lines were obtained. The agreement was satisfactory, and it was concluded that the temperature of the prominence central part was $\sim 8,600$ K and the pressure 0.03 dyn cm^{-2}.

6.3.2.2 1D Models with a Prominence-to-Corona Transition Region

The presence of the PCTR affects H and He line profiles in different ways, depending on the optical depth and the region of formation of the spectral lines (Labrosse et al. 2002). For example, the inclusion of the PCTR has a dramatic impact on the resulting emergent profiles of hydrogen and helium resonance lines. However, through the radiative coupling between optically thick and optically thin lines, the intensities of the latter are also affected by the inclusion of the PCTR in the models. Similarly, Labrosse and Gouttebroze (2004) found that the presence of the PCTR affects the emitted intensities of the triplet lines by reducing the impact of collisional excitations at high temperatures in comparison with the isothermal and isobaric case. A simple study of helium energy level populations demonstrated how statistical equilibrium is changed when a transition region is present, pointing to the necessity of including an interface between the prominence body and the corona to predict all emergent intensities whatever the region of formation of the radiation is, due to the non-local nature of the coupling between the radiative transfer and statistical equilibrium equations. Labrosse and Gouttebroze (2004) also found a correlation between most of the He I triplet line ratios and the altitude of the model prominence. This allowed the authors to solve some long-standing discrepancies in comparisons between predicted triplet line intensity ratios and observations by extrapolating the computations to higher altitudes. Labrosse et al. (2011) compared computed integrated intensities in the He II 256 Å resonance lines with observations and found a central temperature of 8,700 K, a central pressure of 0.33 dyn cm^{-2}, and a relatively large column mass of 2.5×10^4 g cm^{-2}.

6.3.2.3 1D Models Addressing the Fine Structure of Prominences

Despite its importance, the problem of the fine structure threads has not been addressed so much when interpreting observations in helium lines with 1D models. The importance of multi-threads effects on the population of the triplet levels of He I was first shown with a simplified modelling by Morozhenko (1984). Gouttebroze et al. (2002) considered the superposition of several slabs along the LOS (without radiative interaction) and found that it improves the agreement between observed and computed integrated intensity ratios in some cases (e.g. for the ratio $E(\mathrm{D3})/E(\mathrm{H}\beta)$).

6.3.2.4 2D Models

Léger and Paletou (2009) computed neutral helium line profiles (including the He I 10,830 Å and D3 multiplets) using 2D horizontal prominence fine structure models in both single-thread and multi-thread configurations. They confirmed the importance of the multi-thread approach for the modelling of the prominence fine structures in order to reproduce the observed intensity ratio between individual components of the multiplets for the 10,830 Å and D3 lines. A more complete He I-II-III system was modelled by Gouttebroze and Labrosse (2009) in 2D cylindrical models. They considered the contribution to the electron density from ionised helium: this produces regions where the electron density can be larger than the hydrogen density (see Fig. 6.5). They showed that the ionisation ratios[4] of H, He I and He II are principally controlled by temperature (Fig. 6.5). The consideration of the helium contribution to the electron density in these models should have an impact on the determination of the helium abundance in solar prominences.

6.3.3 Prominence Diagnostics from NLTE Modelling of Trace Elements

6.3.3.1 Ca II Spectra

Other than the hydrogen and helium lines, the optical and IR spectral regions contain, among others, five lines of ionized calcium which are frequently observed in prominences. These five lines are the UV Ca II resonance lines H and K plus three IR lines (triplet).

Heasley and Milkey (1978) computed the ratio r of integrated intensities of the Ca II IR line at 8,542 Å to the hydrogen Hβ line. They found that in the temperature range $6,500 - 9,000$ K, this ratio is relatively insensitive to temperature, practically independent of column mass, and decreasing with gas pressure. This theoretical relation was then used to derive the gas pressure from observations. The method has yielded a range of pressures from low $0.01 - 0.04\,\mathrm{dyn\,cm^{-2}}$ to high values of $0.3 - 0.5\,\mathrm{dyn\,cm^{-2}}$. Gouttebroze and Heinzel (2002) computed a larger grid of models of the GHV type, up to $p = 1\,\mathrm{dyn\,cm^{-2}}$. Their principal result is that at pressures higher than $0.1\,\mathrm{dyn\,cm^{-2}}$, the ratio r is strongly dependent on the temperature: r starts to increase for temperatures below 8,000 K, and decreases at

[4]The ionisation ratio is defined as the population of an ion divided by the total population of the element (i.e. summed over all ionisation stages).

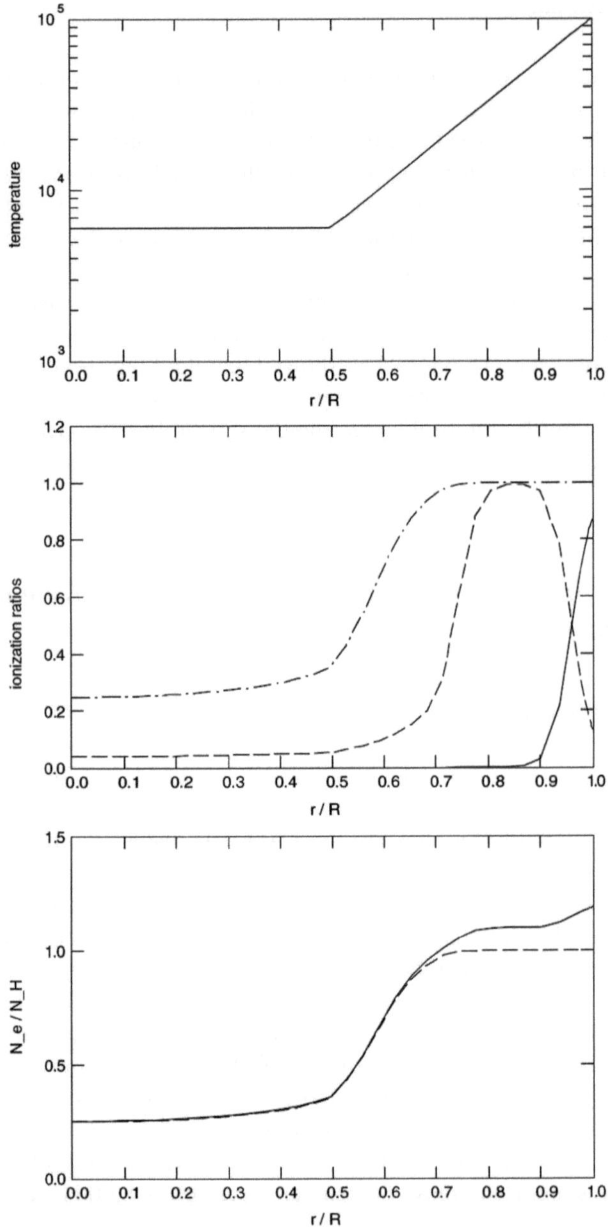

Fig. 6.5 Variations of temperature and population ratios with the distance to the axis (r) at the foot of the loop for a model with varying temperature and a constant pressure fixed at 0.1 dyn cm^{-2}. *Abscissae*: distance to axis relative to the total radius R. *Top*: simple temperature profile adopted in the models. *Middle*: ionisation ratios for hydrogen (*dot-dashed line*), neutral helium (*dashed line*), and ionised helium (*continuous line*). *Bottom*: electron-to-hydrogen ratio (*dashed line*: model assuming neutral helium; *continuous line*: model with both hydrogen and helium ionisation). Note how the ionisation and electron-to-hydrogen ratios increase with the temperature in the transition region interface. Credit: Gouttebroze and Labrosse (2009), reproduced with permission © ESO

higher temperatures. This means that there exists a temperature bifurcation of the ratio r for high pressures. This can explain higher measured values of r, say up to 0.8, which indicate lower temperatures, and gas pressures higher than 0.1 dyn cm^{-2}. This is significant since higher gas pressures mean larger values of the plasma-β parameter. Although the values vary greatly from prominence to prominence, the ratio (and so the pressure) tends to be fairly uniform for a given prominence.

Maps of the temperature, LOS velocity, and microturbulence were built inside a quiescent filament by Tziotziou et al. (2001) using a cloud model.[5] The temperature was found to peak at 8,500 K, the velocity indicated an excess of blueshift (material moving upwards), and the microturbulence peaked at 5 km s^{-1}.

6.3.3.2 Mg II Spectra

Mg II resonance lines have long been predicted to have a large opacity. This means that observations in these lines can be used to investigate the individual fine structures in prominences. The observations obtained at the end of the 1970s (Vial 1982a) were compared to the results of 2D models (Vial 1982; Paletou et al. 1993; Paletou 1995). More recent observations can now be compared with new models from Heinzel et al. (2014a). Their results showed that the line shapes (in particular the presence or not of a self-reversal) are sensitive to the plasma parameters. The line core is sensitive to the presence (or absence) of a PCTR, and of its structure (gradient, thickness). The conclusion is that the NLTE modelling of Mg II line spectra promises to deliver new insight about the fine structures of prominences and will be a useful complement to H and He observations.

6.3.4 Other Uses of NLTE Models for Prominence Diagnostics

Most of the NLTE models discussed so far were used to predict emitted intensities in some lines and continua which may be directly compared with observations. One can make indirect use of these calculations to study other aspects of the physics of the prominence plasma, such as the absorption of coronal radiation by the cool parts of the prominence plasma at wavelengths below 912 Å marking the ionisation edge of hydrogen (see also Chap. 4, Kucera 2014), or the ionisation structure of the prominence.

[5]This method works only if the medium is not too optically thick in the observed line (otherwise the incident radiation component of the model is not visible).

Prominences are observed in EUV coronal lines as dark structures embedded in the brighter background of coronal radiation. Since the prominence material is much cooler than the corona (where such lines are formed), the usually complicated combination of atomic processes is reduced to the relatively simple process of absorption, which is the absorption by the photo-ionisation continuum. For prominences in the EUV, the main absorbers are neutral hydrogen, neutral helium, and singly-ionized helium. Using this method, column densities of neutral hydrogen have been found to be of the order of $10^{18} - 10^{19}\,\mathrm{cm}^{-2}$, and total mass values ranging from 1×10^{14} g to 2×10^{15} g.

As far as the ionisation degree is concerned, an interesting method has been developed (Jejčič et al. 2014) based on the use of eclipse images to infer the electron density and gas pressure inside the prominence. This information is then used to construct 2D plane-parallel models of the prominence using realistic incident radiation to investigate the ionisation degree inside the slab. Figure 6.6 shows the result obtained with this approach. Based on a uniform temperature of 8,000 K and gas pressure of $0.05\,\mathrm{dyn\,cm}^{-2}$, this model shows a noticeable increase of the electron density towards the lower part of the slab (closer to the Sun's surface) due to the stronger ionising Lyman continuum incident radiation. This effect is also seen on the ionisation degree structure.

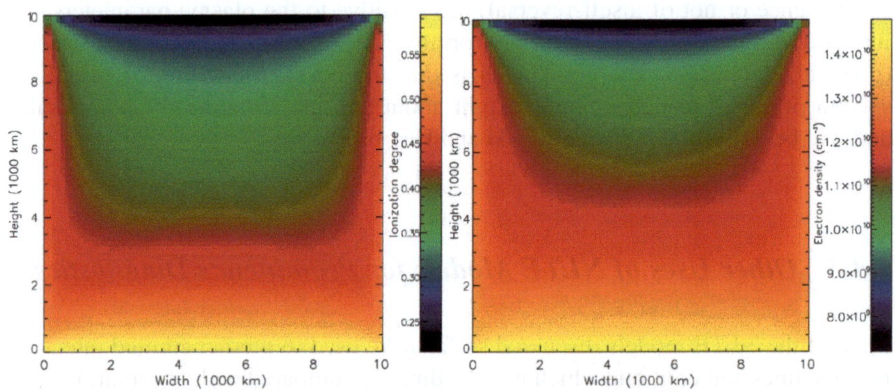

Fig. 6.6 2D model of the electron density (*left*) and ionisation degree structure (*right*) based on plasma parameters that correspond to a prominence observed during the total solar eclipse on 1 August 2008. Credit: Jejčič et al. (2014). With kind permission of Springer Science+Business Media

6.4 Motions in Active and Eruptive Prominences and Their Effects on the Emergent Radiation

When some plasma element of the prominence moves radially outwards, the incident radiation coming from the solar disk and illuminating the structure is shifted to lower frequencies. This is significant for the formation of optically thick resonance lines. Therefore, if mass motions are taking place, the NLTE diagnostic of the prominence is even more complex. In order to improve our understanding of dynamic structures, it is necessary to study the effect of motions in prominences on the relevant lines. Most of this modelling has been done in 1D.

Let us start with a simple description of the effect of radial motions on the emitted spectrum. Consider a simple two-level atom whose upper level is excited by the radiation coming from the Sun: the absorption profile of the radiative transition between the two levels gets out of resonance with the incident radiation when this atom is moving away from (or towards) the Sun, due to the Doppler effect. We namely have a Doppler dimming effect when the incident line is in emission, or a Doppler brightening effect if the incident line is in absorption (Hyder and Lites 1970). In a more realistic situation, an atom has more than two energy levels. Consequently, coupling effects take place between the atomic levels. For that reason, a combination of Doppler dimming and brightening can occur. The maximum effect is achieved when radiative excitation dominates the collisional one in the formation of a line.

The usual approach is to use a velocity-dependent incident radiation illuminating both sides of the prominence slab. It is calculated at a given height, taking into account the centre-to-limb variations (if any) of the incident radiation. The Doppler effect induces a shift in frequency of the incident line profile, and the variation of the Doppler shift with the direction of the incident photons induces a distortion of the incident profile. Figure 6.7 shows this effect for the He I line at 584 Å. Several studies have confirmed the importance of using PRD in the computation of the detailed shape of optically thick line profiles. As expected, the effect of the PCTR on the H and He lines emitted by moving prominence structures is also important.

Labrosse et al. (2008) showed that the He II 304 Å line is strongly dependent on the radial velocity, which can be explained by the dominant role of scattering in the line formation. Other lines such as He I 584 Å and the hydrogen Hα line are also sensitive to the Doppler dimming / brightening due to the radial motion of the plasma at velocities up to $\sim$ 100 km s^{-1}. At higher velocities, the absorption profile of the transition gets significantly out of resonance with the incident radiation, and the resulting variation of the emergent intensity with the radial velocity then depends more strongly (albeit indirectly) on other parameters such as the strength of collisional excitations, or the coupling with other transitions. A similar study was carried out to investigate the Doppler dimming effect on Mg II resonance lines (Heinzel et al. 2014a). In the case of a cool isothermal isobaric 1D prominence slab model ($T = 8{,}000$ K and $p = 0.1$ dyn cm^{-2}), the Mg II k line intensity is found to be sensitive to Doppler dimming.

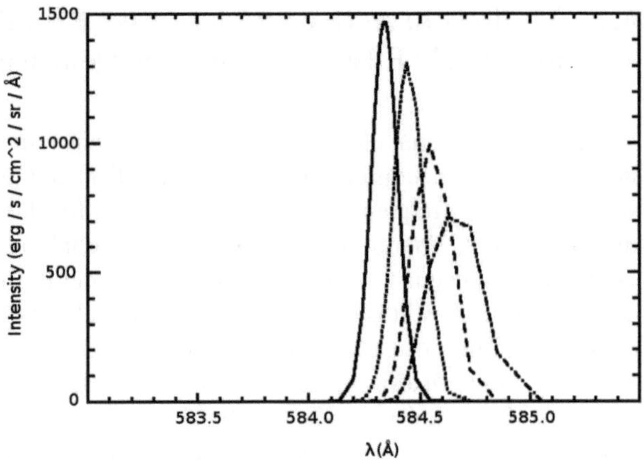

Fig. 6.7 Mean intensity of the incident He I 584 Å emission profile as seen by a moving prominence at the height of 50,000 km. The velocities range from 0 (*solid line*) to 240 km s^{-1} with a step of 80 km s^{-1}. Credit: (Labrosse et al. 2007), reproduced with permission © ESO

Note that Gouttebroze (2008) developed 2D cylindrical models which take into account the Doppler effects produced by 3D velocity fields, i.e. corresponding to a phase of expansion during a process of radial oscillation, to a solid rotational motion of the cylinder around its axis, or to longitudinal flows. The sensitivity of the line profiles and intensities varies with the spectral line and type of motion considered, but there is a rich potential in this type of modelling to study not only the detailed dynamics of individual fine prominence threads, but also that of loop structures.

The study of the variation of the resonance line intensities with the radial velocity of the plasma indicates that the Doppler dimming effects are essentially present when the relative contribution of the thermal emission compared to the scattering of incident radiation in the lines studied is low.

It is necessary to take into account the fact that the prominence plasma is heated during an eruption. In this situation, the Doppler dimming effect may be moderated by an enhanced contribution of collisional excitation. This is one of the major challenges in this type of study: it is essential to make a distinction between a change in the thermodynamic state of the plasma (e.g., an increase in temperature and density) and a change of the radial velocity. A first step in this direction was taken in Labrosse and McGlinchey (2012). The variation of intensity of arbitrarily selected features in a small sample of observations of prominence eruptions as a function of velocity in the plane of the sky is compared with NLTE radiative transfer calculations of the intensity of the He II 304 Å resonance line. It is found that the variation of the plasma parameters during the eruption must be considered to reproduce qualitatively the observations. These theoretical predictions can then be compared with observations. As an example, it was shown that the gas pressure at the surface of an eruptive prominence decreases with altitude (Labrosse 2014).

Another mean of distinguishing between variations in intensity due to a change of the plasma state or to the radial velocity is to use the full line profiles when available. The radial motion of the prominence induces asymmetries in the line profile which cannot be attributed to temperature or pressure effects.

Conclusions

The very first ingredient of a successful modelling of the NLTE radiative transfer problem in solar prominences is the proper consideration of the detailed incident spectrum coming from the solar disc and the surrounding atmosphere. This gives a very good first idea of what the emergent spectrum should look like. The wonderful sophistication of the models comes in next to reach a higher level of realism and accuracy when computing the radiative signatures of prominences: multi-level atoms, coupling between transitions, pressure and temperature variations, fine structure, multi-dimensional radiation transfer, ... It is important to include realistic temperature and pressure variations in the model prominence if we want to understand how these cool structures can exist in the hot corona. Indeed, progress in observations and NLTE modelling of the hydrogen and helium spectra has shown that it is necessary to consider a PCTR where the resonance lines are (partly) formed at various temperatures. The range of global plasma parameters for prominences derived from the NLTE modelling of H and He spectra is in good agreement with those expected from other measurements (Engvold et al. 1990; Labrosse et al. 2010). The multi-dimensional, multi-thread modelling in all optically thick lines available to observations is what is now required in order to learn more about individual fine structures. Multithread modelling does not so far take into account the radiative interactions between threads, and a new picture of prominence fine structures could emerge once the models are able to take this into account.

The precise measurement of abundances could be a precious source of information on the origin of prominence material, and hence on the prominence formation mechanisms. Establishing whether the prominence plasma more closely resembles the photospheric or chromospheric plasma tells us something about its formation. We refer the reader to Chap. 10 (Karpen 2014) for a discussion of the different models of prominence formation. NLTE modelling techniques can be used to study abundances. For instance, Heasley and Milkey (1978) inferred a helium abundance of 0.100 ± 0.025 by number with respect to hydrogen, while Labrosse and Gouttebroze (2001) showed that helium line intensities could be used together with hydrogen line intensities to improve the diagnostics of the helium abundance in prominences. This type of studies should be extended and developed to refine abundance estimates in solar prominences.

(continued)

Progress in the determination of the properties of prominences will come from, ultimately, 3D time-dependent NLTE radiation magneto-hydrodynamics modelling. This is still a very ambitious goal, but a goal we should not lose sight of.

References

Anzer, U., & Heinzel, P. (1999). The energy balance in solar prominences. *Astronomy and Astrophysics*, *349*, 974–984.

Auer, L., Bendicho, P. F., & Trujillo Bueno, J. (1994). Multidimensional radiative transfer with multilevel atoms. 1: ALI method with preconditioning of the rate equations. *Astronomy and Astrophysics*, *292*, 599–615.

Ballester, J. L. (2014). Magnetism and dynamics of prominences: MHD waves. In J.-C. Vial, & O. Engvold (Eds.), *Solar Prominences, ASSL* (Vol. 415, pp. 257–296). Springer.

Bommier, V., Landi Degl'Innocenti, E., Leroy, J.L., & Sahal-Brechot, S. (1994). Complete determination of the magnetic field vector and of the electron density in 14 prominences from linear polarization measurements in the HeI D3 and H-alpha lines. *Solar Physics*, *154*, 231–260. doi:10.1007/BF00681098.

Engvold, O. (2014). Description and classification of prominences. In J.-C. Vial, & O. Engvold (Eds.), *Solar prominences, ASSL* (Vol. 415, pp. 31–60). Springer.

Engvold, O., Hirayama, T., Leroy, J. L., Priest, E.R., & Tandberg-Hanssen, E.(1990). Hvar reference atmosphere of quiescent prominences. In: V. Ruzdjak, & E. Tandberg-Hanssen (Eds.), *IAU Colloq. 117: Dynamics of quiescent prominences*. Lecture notes in physics (Vol. 363, pp. 294–+). Berlin: Springer. doi:10.1007/BFb0025640.

Fontenla, J. M., Rovira, M., Vial, J. C., & Gouttebroze, P. (1996). Prominence thread models including ambipolar diffusion. *Astrophysical Journal*, *466*, 496–+. doi:10.1086/177527.

Gouttebroze, P. (2006). Radiative transfer in cylindrical threads with incident radiation. III. Hydrogen spectrum. *Astronomy and Astrophysics*, *448*, 367–374. doi:10.1051/0004-6361:20054139.

Gouttebroze, P. (2008). Radiative transfer in cylindrical threads with incident radiation. V. 2D transfer with 3D velocity fields. *Astronomy and Astrophysics*, *487*, 805–813. doi:10.1051/0004-6361:20079272.

Gouttebroze, P., & Heinzel, P. (2002). Calcium to hydrogen line ratios in solar prominences. *Astronomy and Astrophysics*, *385*, 273–280. doi:10.1051/0004-6361:20020142.

Gouttebroze, P., & Labrosse, N. (2000). A ready-made code for the computation of prominence NLTE models. *Solar Physics*, *196*, 349–355.

Gouttebroze, P., & Labrosse, N. (2009). Radiative transfer in cylindrical threads with incident radiation. VI. A hydrogen plus helium system. *Astronomy and Astrophysics*, *503*, 663–671. doi:10.1051/0004-6361/200811483, 0905.3466.

Gouttebroze, P., Heinzel, P., & Vial, J. C. (1993). The hydrogen spectrum of model prominences. *Astronomy and Astrophysics Supplement Series*, *99*, 513–543.

Gouttebroze, P., Labrosse, N., Heinzel, P., & Vial, J. C. (2002). Prediction of line intensity ratios in solar prominences. In H. Sawaya-Lacoste (Ed.), *SOLMAG 2002. Proceedings of the magnetic coupling of the solar atmosphere Euroconference* (Vol. 505, pp. 421–424). Santorini: ESA Special Publication.

Gunár, S., Heinzel, P., Schmieder, B., Schwartz, P., & Anzer, U. (2007). Properties of prominence fine-structure threads derived from SOHO/SUMER hydrogen Lyman lines. *Astronomy and Astrophysics*, *472*, 929–936. doi:10.1051/0004-6361:20077785.

Gunár, S., Heinzel, P., & Anzer, U. (2007a). Prominence fine structures in a magnetic equilibrium. III. Lyman continuum in 2D configurations. *Astronomy and Astrophysics*, *463*, 737–743. doi:10.1051/0004-6361:20066142.

Gunár, S., Heinzel, P., Anzer, U., & Schmieder, B. (2008). On Lyman-line asymmetries in quiescent prominences. *Astronomy and Astrophysics*, *490*, 307–313. doi:10.1051/0004-6361:200810127.

Gunár, S., Parenti, S., Anzer, U., Heinzel, P., & Vial, J. C. (2011). Synthetic differential emission measure curves of prominence fine structures. II. The SoHO/SUMER prominence of 8 June 2004. *Astronomy and Astrophysics*, *535*, A122. doi:10.1051/0004-6361/201117429.

Gunár, S., Mein, P., Schmieder, B., Heinzel, P., & Mein, N. (2012). Dynamics of quiescent prominence fine structures analyzed by 2D non-LTE modelling of the Hα line. *Astronomy and Astrophysics 543*, A93. doi:10.1051/0004-6361/201218940.

Heasley, J. N., & Milkey, R. W. (1978). Structure and spectrum of quiescent prominences. III - Application of theoretical models in helium abundance determinations. *Astrophysical Journal*, *221*, 677–688. doi:10.1086/156072.

Heasley, J. N., & Milkey, R. W. (1983) Structure and spectrum of quiescent prominences. IV - The ultraviolet ionization continua of hydrogen and helium. *Astrophysical Journal*, *268*, 398–402. doi:10.1086/160965.

Heinzel, P. (2014). Radiative transfer in solar prominences. In J.-C. Vial, & O. Engvold (Eds.), *Solar prominences, ASSL* (Vol. 415, pp. 101–128). Springer.

Heinzel, P., & Anzer, U. (2001). Prominence fine structures in a magnetic equilibrium: Two-dimensional models with multilevel radiative transfer. *Astronomy and Astrophysics 375*, 1082–1090. doi:10.1051/0004-6361:20010926.

Heinzel, P., Gouttebroze, P., & Vial, J. C. (1994). Theoretical correlations between prominence plasma parameters and the emitted radiation. *Astronomy and Astrophysics*, *292*, 656–668.

Heinzel, P., Bommier, V., & Vial, J. C. (1996). A Complex Diagnostic of Solar Prominences. *Solar Physics*, *164*, 211–222. doi:10.1007/BF00146635.

Heinzel, P., Schmieder, B., Vial, J. C., & Kotrč, P. (2001a). SOHO/SUMER observations and analysis of the hydrogen Lyman spectrum in solar prominences. *Astronomy and Astrophysics*, *370*, 281–297. doi:10.1051/0004-6361:20010265.

Heinzel, P., Anzer, U., & Gunár, S. (2005). Prominence fine structures in a magnetic equilibrium. II. A grid of two-dimensional models. *Astronomy and Astrophysics*, *442*, 331–343. doi:10.1051/0004-6361:20053360.

Heinzel, P., Schmieder, B., Fárník, F., Schwartz, P., Labrosse, N., Kotrč, P., et al. (2008). Hinode, TRACE, SOHO, and ground-based observations of a quiescent prominence. *Astrophysical Journal*, *686*, 1383–1396. doi:10.1086/591018.

Heinzel, P., Vial, J. C., & Anzer, U. (2014a). On the formation of Mg ii h and k lines in solar prominences. *Astronomy and Astrophysics*, *564*, A132. doi:10.1051/0004-6361/201322886.

Heinzel, P., Zapiór, M., Oliver, R., & Ballester, J. L. (2014b). Synthetic hydrogen spectra of prominence oscillations. *Astronomy and Astrophysics*, *562*, A103. doi:10.1051/0004-6361/201322346, 1401.2131.

Hyder, C. L., & Lites, B. W. (1970). Hα doppler brightening and Lyman-α doppler dimming in moving Hα prominences. *Solar Physics*, *14*, 147–156. doi:10.1007/BF00240170.

Jejčič, S., Heinzel, P., Zapiór, M., Druckmüller, M., Gunár, S., & Kotrč, P. (2014). Multi-wavelength eclipse observations of a quiescent prominence. Solar Physics *289*, 2487–2501. doi:10.1007/s11207-014-0482-1.

Karpen, J. (2014). Plasma structure and dynamics. In J.-C. Vial, & O. Engvold (Eds.), *Solar prominences, ASSL* (Vol. 415, pp. 235–255). Springer.

Kucera, T. (2014). Derivations and observations of prominence bulk motions and mass. In J.-C. Vial, & O. Engvold (Eds.), *Solar prominences, ASSL* (Vol. 415, pp. 77–99). Springer.

Labrosse, N. (2014). Plasma properties in eruptive prominences. In: IAU Symposium (Vol. 300, pp. 79–84). doi:10.1017/S1743921313010776.

Labrosse, N., & Gouttebroze, P. (2001). Formation of helium spectrum in solar quiescent prominences. *Astronomy and Astrophysics*, *380*, 323–340. doi:10.1051/0004-6361:20011395.

Labrosse, N., & Gouttebroze, P. (2004). Non-LTE radiative transfer in model prominences. I. Integrated intensities of He I triplet lines. *Astrophysical Journal, 617*, 614–622. doi:10.1086/425168.

Labrosse, N., & McGlinchey, K. (2012). Plasma diagnostic in eruptive prominences from SDO/AIA observations at 304 Å. *Astronomy and Astrophysics, 537*, A100. doi:10.1051/0004-6361/201117801, 1111.4847.

Labrosse, N., Gouttebroze, P., Heinzel, P., & Vial, J. C. (2002). Line profiles and intensity ratios in prominence models with a prominence to corona interface. In: J. Kuijpers (Ed.), *Solar variability: From core to outer frontiers* (Vol. 506, pp. 451–454). Prague: ESA Special Publication.

Labrosse, N., Vial, J. C., & Gouttebroze, P. (2006). Plasma diagnostic of a solar prominence from hydrogen and helium resonance lines. In: D. Barret, F. Casoli, G. Lagache, A. Lecavelier, L. Pagani (Eds.) *SF2A-2006: Semaine de l'Astrophysique Francaise* (pp 549+).

Labrosse, N., Gouttebroze, P., & Vial, J. C. (2007). Effect of motions in prominences on the helium resonance lines in the extreme ultraviolet. *Astronomy and Astrophysics, 463*, 1171–1179. doi:10.1051/0004-6361:20065775, arXiv:astro-ph/0608221.

Labrosse, N., Vial, J. C., & Gouttebroze, P. (2008). Diagnostics of active and eruptive prominences through hydrogen and helium lines modelling. *Annales Geophysicae, 26*, 2961–2965, 0804. 4625.

Labrosse, N., Heinzel, P., Vial, J., Kucera, T., Parenti, S., Gunár, S., et al. (2010). Physics of solar prominences: I–Spectral diagnostics and non-LTE modelling. *Space Science Reviews, 151*, 243–332. doi:10.1007/s11214-010-9630-6, 1001.1620.

Labrosse, N., Schmieder, B., Heinzel, P., & Watanabe, T. (2011). EUV lines observed with EIS/Hinode in a solar prominence. *Astronomy and Astrophysics, 531*, A69. doi:10.1051/0004-6361/201015064, 1105.1400.

Léger, L., & Paletou, F. (2009), 2D non-LTE radiative modelling of He I spectral lines formed in solar prominences. *Astronomy and Astrophysics, 498*, 869–875. doi:10.1051/0004-6361/200810296, 0811.4753.

Mihalas, D., Auer, L. H., & Mihalas, B. R. (1978). Two-dimensional radiative transfer. I - Planar geometry. *Astrophysical Journal, 220*, 1001–1023. doi:10.1086/155988.

Morozhenko, N. N. (1984). On the excitation of lower levels of singlet helium in quiescent prominences. *Solar Physics 92*, 153–160. doi:10.1007/BF00157242.

Ofman, L., Kucera, T. A., Mouradian, Z., & Poland, A. I. (1998). SUMER Observations of the Evolution and the Disappearance of a Solar Prominence. *Solar Physics, 183*, 97–106.

Paletou, F. (1995). Two-dimensional multilevel radiative transfer with standard partial frequency redistribution in isolated solar atmospheric structures. *Astronomy and Astrophysics, 302*, 587+

Paletou, F., Vial, J. C., & Auer, L. H. (1993). Two-dimensional radiative transfer with partial frequency redistribution. II. Application to resonance lines in quiescent prominences. *Astronomy and Astrophysics, 274*, 571+

Parenti, S. (2014). Spectral diagnostics of cool prominence and PCTR optically thin plasmas. In J.-C. Vial, & O. Engvold (Eds.), *Solar prominences, ASSL* (Vol.415, pp. 61–76). Springer.

Parenti, S., Lemaire, P., & Vial, J. C. (2005a). Solar hydrogen-Lyman continuum observations with SOHO/SUMER. *Astronomy and Astrophysics, 443*, 685–689. doi:10.1051/0004-6361:20053431.

Schmieder, B., Heinzel, P., Vial, J. C., & Rudawy, P. (1999). SOHO/SUMER observations and analysis of hydrogen Lyman lines in a quiescent prominence. *Solar Physics, 189*, 109–127.

Schmieder, B., Gunár, S., Heinzel, P., & Anzer, U. (2007). Spectral diagnostics of the magnetic field orientation in a prominence observed with SOHO/SUMER. *Solar Physics, 241*, 53–66. doi:10.1007/s11207-007-0251-5.

Schmieder, B., Malherbe, J. M., & Wu, S. T. (Eds.) (2014). Nature of prominences and their role in space weather. In *IAU Symposium* (Vol. 300).

Tziotziou, K., Heinzel, P., Mein, P., & Mein, N. (2001). Non-LTE inversion of chromospheric {\Ca Ii} cloud-like features. *Astronomy and Astrophysics, 366*, 686–698. doi:10.1051/0004-6361:20000257.

Vial, J. C. (1982). Two-dimensional nonlocal thermodynamic equilibrium transfer computations of resonance lines in quiescent prominences. *Astrophysical Journal, 254*, 780–795. doi:10.1086/159789.

Vial, J. C. (1982a). Optically thick lines in a quiescent prominence - Profiles of Lyman-alpha, Lyman-beta /H I/, K and H /Mg II/, and K and H /Ca II/ lines with the OSO 8 LPSP instrument. *Astrophysical Journal, 253*, 330–352. doi:10.1086/159639.

Vial, J.-C. (2014). Historical background and introduction. In J.-C. Vial, & O. Engvold (Eds.), *Solar prominences, ASSL* (Vol. 415, pp. 1–29). Springer.

Vial, J. C., Ebadi, H., & Ajabshirizadeh, A. (2007). The Ly α and Ly β profiles in solar prominences and prominence fine structure. *Solar Physics, 246*, 327–338. doi:10.1007/s11207-007-9080-9, 0710.1433.

Chapter 7
Energy Balance

Holly Gilbert

Abstract The complexity of prominence formation and structure is intimately related to energy balance. Fundamental properties of these structures are still being investigated and understanding the processes involved with heating and cooling of prominence material, which is partially ionized, is a critical piece of the puzzle. It is important to understand the nature of the chromosphere–corona transition region (CCTR) and, more specifically, the interplay among mechanical heating, radiative cooling, radiative heating, and thermal conduction that determines the location and structure of this transition region. For prominences to exist they need mechanical equilibrium (which is described by the equations of magneto-hydrostatics) and detailed energy balance, in which steady radiative cooling is balanced by heating mechanisms. Aspects of mechanical and energy balance have been thoroughly studied in the past, but models have difficulty accounting for both of these equilibria self-consistently on scales ranging from the central cool parts of the prominence into the corona.

7.1 Introduction

Examining the energy equation and how it relates to prominences requires an understanding of how the chromosphere is heated and cooled. It is useful to begin by considering first the temperature structure of a radiative-equilibrium solar atmosphere and then the observationally inferred temperature structure of the real solar atmosphere. Energy from the center of the Sun is transported outward through the solar interior and beyond the solar surface by a flux of electromagnetic radiation. As in the more familiar case of heat transport by thermal conduction, the outward transport of energy by radiation in the solar interior is associated with a temperature that declines outward from sun center (in the direction of the energy transport). If there were no non-radiative processes operating in the solar atmosphere, then the temperature of the atmosphere would continue its radially

H. Gilbert (✉)
NASA GSFC, Greenbelt, MD, USA
e-mail: holly.r.gilbert@nasa.gov

J.-C. Vial, O. Engvold (eds.), *Solar Prominences*, Astrophysics and Space Science Library 415, DOI 10.1007/978-3-319-10416-4_7

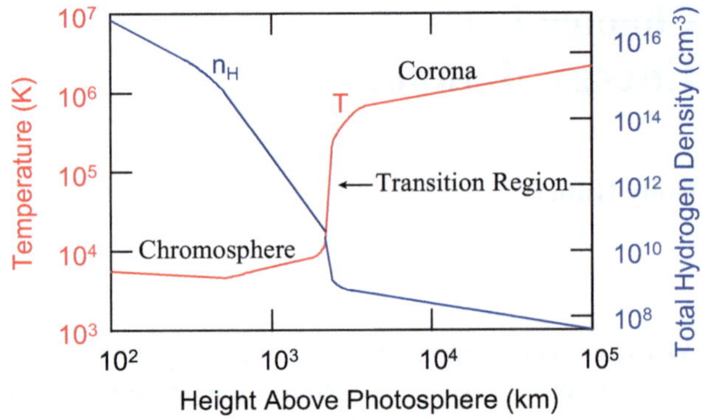

Fig. 7.1 A phenomenological model of the average solar atmosphere, with the temperature and total hydrogen density of the atmosphere plotted as functions of height above the photosphere. Note the extremely steep temperature rise from the chromosphere to the corona

outward decline. Observations of radiation emitted by the solar atmosphere lead to the phenomenological model of the density and temperature structure of the atmosphere shown schematically in Fig. 7.1. The temperature rise through the chromosphere, transition region, and corona leads to the conclusion that some non-radiative process is adding energy to the atmosphere, and the term mechanical heating (as distinct from radiative heating) is normally used to describe this unspecified process.

Generally speaking, the chromosphere acts as a natural thermostat and is essentially isothermal. Dissipated mechanical energy is taken up by latent heat of ionization and is rapidly lost by radiation. Since prominences are at chromospheric temperatures we often refer to the chromosphere while discussing the physics of heating and cooling in this chapter. When possible or relevant, differences between a prominence environment and the general chromosphere are called out. For example, we have to be cognizant that prominences and the upper chromosphere are exposed to different radiative environments and have different dynamical histories. Therefore, calculations of the mean chromosphere may not provide an accurate depiction of prominence characteristics, even if we are able to measure the prominence temperature. Knowing the thermal structure of prominences is essential for understanding energy balance.

The details of chromospheric heating are complicated and still being debated. Heating, which must account for the radiative losses, comes from a combination of collisional effects (Joule, frictional, and viscous heating) and compressional heating. The processes involved with radiative cooling are understood, but the balancing of the cooling with the heating is not well understood. When studying these types of processes in prominences many things need to be considered. The cool part of the prominence (i.e., the interior of the material) is a different environment than the surrounding hotter shell called the prominence–corona transition region (PCTR). These two regions have to be considered separately in some cases. The temperature and pressure structure of the interface between a prominence and the corona (PCTR) is the result of energy balance. A strong temperature gradient in the PCTR was identified decades ago in Skylab data. That data also indicated a lower pressure and thinner transition region for prominences than for the chromosphere–corona TR (Vial 1990). A sophisticated technique, the differential emission measure (DEM) (see Chapter 3 for details: Parenti 2014) provides insight into the temperature gradient in the PCTR. In addition to separating the PCTR from the cooler region, we have to consider how ion-neutral coupling (which involves non-thermal processes) in the partially ionized prominence plasma affects the various physics we are discussing. Lastly, prominences exhibit fine structure, having visible smaller threads that make up larger, sheet-like structures. This fine structure adds to the complexity, especially because many models consider "slabs" of material. All of these factors will be considered as we delve into the various physical processes involved with energy balance in the subsequent sections.

It is worth noting that non-thermal energy balance is extremely important in the context of magnetic forces involved with erupting prominences. Energy conversion in the destabilization of larger magnetic structures is discussed in the context of the storage and release of magnetic energy (Chapter 12: Fan 2014). The present chapter focuses on quiescent prominences: those located in quiet regions on the solar surface, and we limit the discussion to the concept of thermal energy in non-erupting structures. For simplicity, we begin with a general energy balance equation and subsequently study each of the relevant terms in further discussing details of the physics.

7.2 Energy Balance

For solar phenomena that remain in a steady, essentially static state for long periods of time, the relevant physics may be approximated by a magneto-hydrostatic equilibrium solution to the magnetohydrodynamics (MHD) equations (see Chapter 11: Ballester 2014). The relevant equation for this chapter is the energy balance equation, which can be expressed in many forms. Starting from first principles, we begin by examining the heat flow in the total energy balance of a gas. The first law of thermodynamics relates the change in internal energy (U) to the gain of heat (Q) and the work (W) performed:

$$\Delta U = \Delta Q + \Delta W \tag{7.1}$$

which can be written as:

$$\Delta U = \Delta Q - PdV \tag{7.2}$$

P = pressure, and V = volume. The above equation applies if work is performed only by the mechanical compressions and expansions produced as the fluid moves. We consider enthalpy, $H = U + PV$ to account for the work done in compression and expansion of the gas (i.e., by the pressure force), and take the time derivative in terms of directly measured quantities (ρ = mass density and T = temperature) so that we have:

$$\rho\frac{dH}{dt} = c_p\rho\frac{dT}{dt} = \rho\frac{dQ}{dt} \tag{7.3}$$

The above equation is the heat balance for a unit volume of isobaric fluid (c_p is the specific heat at constant pressure).

If we use a one-fluid description of the plasma in the solar atmosphere (i.e., if we sum the conservation laws for all the plasma constituents), then energy balance can be described by an equation written in the familiar form of a conservation law:

$$\frac{\partial \mathrm{E}}{\partial t} + \nabla\cdot\vec{\mathbf{F}} = \mathrm{Q} - \mathrm{L} \tag{7.4}$$

where E is the energy density, $\vec{\mathbf{F}}$ is the energy flux density, Q is the mechanical heating rate, and L is the radiative cooling rate. With these definitions, we have implicitly included any work done against the gravitational field in the second term on the left side of (7.4), so we can write (in the absence of viscous effects)

$$\mathrm{E} = \tfrac{1}{2}\rho\, u^2 + \tfrac{3}{2}p \tag{7.5}$$

$$\vec{\mathbf{F}} = \rho\vec{\mathbf{u}}\left(\tfrac{1}{2}u^2 + \tfrac{5}{2}p/\rho - \tfrac{1}{2}v_g^2\right) + \vec{\mathbf{q}} \tag{7.6}$$

where ρ is the mass density, $\vec{\mathbf{u}}$ is the flow velocity, p is the pressure, v_g is the gravitational escape speed, and $\vec{\mathbf{q}}$ is the heat flux density (associated with thermal conduction), which can be written in the form

$$\vec{\mathbf{q}} = -\kappa\, T^{5/2}\nabla T \tag{7.7}$$

The conduction coefficient, κ, is a constant. It is sometimes more useful to describe energy balance using an equation written in terms of the temperature rather than in terms of the energy density and energy flux density. Such an equation takes the following form

$$3nk\left(\frac{\partial T}{\partial t} + \vec{\mathbf{u}}\cdot\nabla T\right) + p\nabla\cdot\vec{\mathbf{u}} + \nabla\cdot\vec{\mathbf{q}} = \mathrm{Q} - \mathrm{L} \tag{7.8}$$

where k is the Boltzmann constant.

The equations describing fluid motions, which include the equation for mass continuity, the momentum equation, and the energy equation, are completed with the equation of state for a perfect gas (applicable for a fully ionized plasma):

$$P = 2\rho RT = 2nkT \tag{7.9}$$

with $R = 8.32 \times 10^7$ ergs K^{-1} mol^{-1}, which is the gas constant, and ρ = number density = nm_p, m_p = mass of a proton.

Although the energy equation can be written in many forms, the main components we are concerned with in the solar atmosphere are energy losses or gains through thermal conduction, energy losses through radiation, energy gained through heating, and energy losses or gains through advection. If we assume for the moment that the terms in (7.8) associated with flow and thermal conduction are negligible, which is not an unreasonable expectation in the low-temperature chromosphere, then we can rewrite (7.8) in the form:

$$3nk\frac{\partial T}{\partial t} = \mathrm{Q} - \mathrm{L} \tag{7.10}$$

The radiation which prominences emit is important in the fundamental physics, providing diagnostics of the prominence structure, but it also plays a critical role in the global energy budget of prominence structures via the radiative losses or gains. If the net radiative energy emitted at a given point is equal to that absorbed, we have radiative equilibrium, which simply means that the radiative flux integrated over all frequencies is conserved:

$$\frac{dF_r}{dx} = 0 \left(with\ F_r = \int_0^\infty F_\upsilon d \right) \tag{7.11}$$

$$F_\upsilon = \frac{1}{2}\int_{-1}^{1} I_{\upsilon,\mu}\mu d\mu \tag{7.12}$$

Here x is a coordinate in a simple 1D prominence model, ν is the frequency, and μ is cosΦ (Φ is the polar angle in spherical geometry). The integration over frequencies includes various lines and continua of species which act as important coolants of the prominence plasma.

To approach the problem of energy balance, a first step is the estimation of the rate of energy losses and thus the required input. Radiative losses can be computed from a model chromosphere, again noting that in a steady state these net losses must be balanced locally by some form of heating. The next step involves identifying chromospheric heating mechanisms, which are not as well understood as the radiative losses.

In applying Eq. (7.8) to prominences it is beneficial to split them into two different regions. For example, in the cool region of prominences (i.e., below 10,000 K) energy transport by thermal conduction can be neglected because of its strong temperature dependence (see Sect. 7.3). However, the region of cool plasma is complicated because it is optically thick in several atomic transitions, requiring the non-LTE (Chapter 5: Heinzel 2014) radiative transfer problem to be solved. The plasma ionization state is dependent on the radiation field and enters into the equation of state of the plasma. Therefore, the relation between the gas pressure and density depends on the degree of ionization and the global radiation magento-hydrodynamical modelling must take the non-LTE physics into consideration. See Chapters 5 (Heinzel 2014) and 6 (Labrosse 2014) for detailed descriptions of radiative transfer and non-LTE modeling in prominences. In the hotter PTCR, thermal conduction becomes very important but since the plasma is optically thin here, the treatment of radiation is simplified and the non-LTE radiative transfer problem is not required. A closer inspection of the terms in the energy balance equation is provided following a discussion on the thermal instability and its consequences for prominence formation and the existence of the transition region.

7.2.1 When Balance Fails and the Transition Region

Although the existence and stability of prominences requires energy balance, the physical processes involved with their formation involve an imbalance in the heating and cooling, or the radiative thermal instability (Malherbe 1989; Soler et al. 2011). A condensation can form when a density perturbation causes radiative cooling (which increases as density squared) to dominate mechanical heating along magnetic structures within a prominence channel. The perturbed region has a net cooling which results in a temperature and pressure decrease followed by an inflow of material causing the density to increase (examples include the thermal nonequilibrium models of Antiochos and Klimchuk (1991) and Karpen et al. (2006)). The condensation grows until a balance between heating and cooling is found and pressure balance is regained. Soler et al. (2012) studied the stability of thermal modes in prominence plasma, emphasizing the importance of having an accurate description of the radiative loss function. They found that thermal modes may be unstable in prominences for temperatures as low as 15,000 K. This instability can potentially have a large impact on prominence dynamics and evolution as well as formation.

This can be understood more fundamentally by looking at (7.10), where we are neglecting the terms associated with flow and thermal conduction. In (7.10), if the mechanical heating rate exceeds the radiative cooling rate (i.e., if $\mathtt{Q} > \mathtt{L}$), then the local temperature increases with time, whereas if the radiative cooling rate exceeds the mechanical heating (i.e., if $\mathtt{Q} < \mathtt{L}$), then the local temperature decreases with time. This is just what we would expect, and it leads to a well-behaved (i.e., thermally stable) atmosphere, as long as the radiative cooling rate increases with increasing temperature (i.e., $\partial\mathtt{L}/\partial T > 0$). This is because any mismatch in the heating and cooling rates will tend to drive the atmosphere back toward equilibrium (i.e., $\mathtt{Q} = \mathtt{L}$). Referring to the constant-pressure radiative cooling curve in Fig. 7.2, we see that at temperatures less than the Ly-α cooling peak temperature ($\approx 2 \times 10^4$ K), the equilibrium with heating balancing cooling (i.e., $\mathtt{Q} = \mathtt{L}$) is stable, because $\partial\mathtt{L}/\partial T > 0$. In contrast, at temperatures greater than the Ly-α cooling peak temperature ($\approx 2 \times 10^4$ K), the equilibrium with heating balancing cooling (i.e., $\mathtt{Q} = \mathtt{L}$) is unstable, because $\partial\mathtt{L}/\partial T < 0$. The consequences of these thermally stable and thermally unstable regions can be explored by carrying out a thought experiment that will lead to an understanding of why there exists a chromosphere–corona

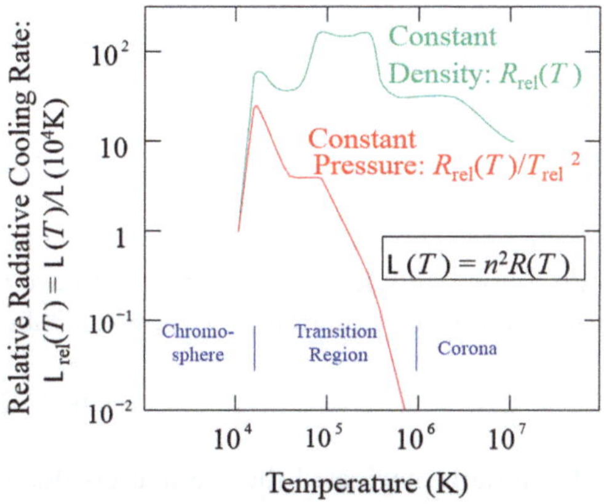

Fig. 7.2 Variation of the optically thin radiative cooling rate with temperature (adapted from Raymond 1976)

transition region and what determines its nature and location, which is applicable to the PCTR that forms on either side of a condensing prominence.

As mentioned above, the equilibrium form of (7.10) is $\mathtt{Q} = \mathtt{L}$, and making use of the equation for the radiative cooling rate (7.23) discussed in Sect. 7.5, we see that in this equilibrium

$$R(T) = \mathtt{Q}/n^2 \tag{7.13}$$

Now let us consider the region of the chromosphere below the Ly-α cooling peak temperature ($\approx 2 \times 10^4$ K). If we have a balance between heating and cooling, then if $\mathtt{Q}$ decreases over a scale height that is smaller than the scale height over which n^2 declines, the atmospheric temperature will decrease with increasing height. If this situation (i.e., $\mathtt{Q}$ decreasing more rapidly with height than n^2) were to be true at all heights above the photosphere, then the solar atmosphere would exhibit a chromosphere with monotonically decreasing temperature and would have no transition region or corona. In contrast, if the scale height of $\mathtt{Q}$ is larger than that of n^2, then the atmospheric temperature will increase with increasing height, and eventually the temperature will reach that of the Ly-α cooling maximum. Beyond this point, we enter the unstable region of the constant-pressure radiative cooling curve, and according to (7.10), the temperature should increase without bound. This situation is clearly unphysical, and we must return to the more general Eq. (7.8) and ask what physical effects we have neglected that become important in this thermally unstable region.

If substantial flows are allowed (i.e., more substantial than generally seen in prominences, such as those observed in magnetically open regions like coronal holes) then two different physical effects become important in the description provided by (7.8): advection [involving the second and third terms in (7.8)], and thermal conduction [involving the fourth term in (7.8)]. In the case of prominences, the magnetic structure holding the mass is closed at low altitudes, and advection can only be important during a transient adjustment to a new equilibrium. Hence, for this thought experiment we consider only the effects of thermal conduction in the modification of (7.10), so we can write

$$3nk\frac{\partial T}{\partial t} + \nabla \cdot \overrightarrow{\mathbf{q}} = \mathtt{Q} - \mathtt{L} \tag{7.14}$$

for the general situation in which we can have transient adjustments (but in which advection is not important). Thus, we have a new equilibrium equation (to replace the equilibrium equation $\mathtt{Q} = \mathtt{L}$), which takes the form

$$\nabla \cdot \overrightarrow{\mathbf{q}} = \mathtt{Q} - \mathtt{L} \tag{7.15}$$

We initially neglected the thermal conduction term that we are now including in (7.14) and (7.15) because of the extremely strong temperature dependence of the heat flux, which is manifest in Eq. (7.7). Indeed, a plasma at temperatures below 5×10^4 K is a very good thermal insulator. It follows [see (7.7)] that the only way thermal conduction can mitigate the thermal instability at these low temperatures is with an extremely strong temperature gradient. It is for this reason that the temperature rises so steeply in the lower transition region and levels off in the corona (see Figs. 7.1 and 7.2), where the plasma has become a better thermal conductor than copper, so that only a mild temperature gradient produces a significant heat flux. Describing how the chromosphere–corona transition region forms helps our fundamental understanding of how the PCTR forms on either side of a prominence.

Although the PCTR is very thin and in general similar to the CCTR (Rabin 1986) some differences between the two transition regions have been discovered. Schmahl (1979) and Engvold et al. (1987) found the rise of temperature in the prominence–corona interface is less steep than the chromosphere–corona transition. Parenti and Vial (2007) estimated that the emission of the PCTR is 2.5–10 times less than the CCTR, and analysis of UV emission lines formed in the P-C transition sheath give electron pressures between 1/2 and 1/5 times the pressure in the CCTR (Noyes et al. 1972; Mariska et al. 1979). Lastly, Chiuderi and Chiuderi Drago (1991) discovered a variation in the PCTR thickness with the angle between the magnetic field direction and the temperature gradient. More specifically, they claim a thinner PCTR when the angle approaches 90°.

7.3 Conductive Flux

Heat conduction is not a source of energy in the corona, but rather a means of transporting it. Simply put, thermal conduction serves to redistribute the energy deposited by mechanical heating to a region where radiative cooling can remove the energy from the atmosphere. In particular, in the simple case of the mean solar atmosphere of Fig. 7.1, thermal conduction carries the bulk of the energy (that is the energy not lost through local radiative cooling) deposited above the chromosphere (through mechanical heating) back downward to the chromosphere, where it can be radiated away. In the case of prominences, thermal conduction carries energy from the corona on the each side of the structure, through the PCTR and into the cooler regions. The redistribution of the energy by thermal conduction leads to an enhanced heating, and thus an enhanced density and pressure at the base of the transition region. Since the transition region is so narrow (in both the chromosphere and in prominences), the pressure remains nearly constant throughout it, and the net effect of the downward thermal conduction is to increase the pressure (and the density) at the base of the corona (i.e., at about 10^6 K).

Heat flux due to thermal conduction in a gas is given by

$$F_{cond} = -\kappa \nabla T \tag{7.16}$$

where κ is the thermal conductivity given by (see Mariska 1992):

$$\kappa = 10^{-6} \times T^{5/2} \text{ erg s}^{-1} \text{ cm}^{-1} \text{ K}^{-1} \tag{7.17}$$

This is for a fully ionized plasma where the direction of the conductivity is parallel to the magnetic field lines. Conduction parallel to the magnetic field is carried by electrons, while cross-field conduction is due mainly to protons. The ion conductivity coefficient is much smaller than the electron conductivity coefficient and thus can be ignored [Braginskii (1965) found the ion coefficient is approximately 4 % of the electron coefficient]. Typically, the Spitzer (1962) value of the perpendicular thermal conductivity is many orders of magnitude smaller than the corresponding parallel conductivity, because the gyroradius of electrons is much shorter than their mean free path. Under these conditions, the dominant effect of **B** on thermal conduction is to decrease the path-length over which heat-flux-carrying electrons travel (Rosner and Tucker 1989).

The net energy deposited (or extracted) per unit volume requires calculation of the divergence of the conductive flux, which can be expressed as

$$\nabla \cdot F_{cond} = -\nabla \cdot (\kappa \, \nabla T) \tag{7.18}$$

Given the environment we are concerned about (i.e., the PCTR), to a good approximation, this is given by:

$$\nabla \cdot (\kappa \, \nabla T) = B \cdot \nabla \left(\frac{\kappa_{par}}{B^2} B \cdot \nabla T \right) + \nabla \cdot \left(\kappa_{perp} \nabla T \right) \tag{7.19}$$

with $\kappa_{par} = \kappa_0 T^{5/2}$ and $\kappa_0 = 10^{-11}$

When thermal conduction is the dominant term in the energy balance (which occurs in the high temperatures of the transition region where T > 100,000 K), a constant conductive flux can be assumed, so that

$$\nabla \cdot F_{cond} = -\kappa \, \nabla^2 T \tag{7.20}$$

Conductivity is strongly temperature dependent and the $T^{5/2}$ dependence makes energy transport by thermal conduction very important in the outer part of the transition region, but negligible at relatively low temperatures. In fact, it can be neglected in the inner, cool region of prominences and in the relatively low chromosphere. For example, Ulmschneider (1970) calculated that for heights $<$2,000 km the amplitude of the conductive flux ranges from only 0.4 to 6.0 ergs cm^{-2} s^{-1}, while the models of Alvarez (1980) produced a conductive flux of 9.47×10^4 ergs cm^{-2} s^{-1} for their "quiet region" transition region ($\sim$400,000 K). Yang et al. (1975) found an averaged conductive flux over the PCTR on the order of 10^5–10^6 ergs cm^{-2} s^{-1} (for temperatures of $\sim$40,000–100,000 K). Thermal conduction is clearly important in the PCTR, and in this layer the efficiency of thermal conduction can be enhanced due to ambipolar diffusion, or the decoupling of neutral particles from the ionized component of the plasma, which affects the energy balance and the hydrogen ionization. The importance of these effects in the CCTR was explored in calculations of Fontenla et al. (1990, 1991, 1993), but Fontenla et al. (1996) considered the effects of ambipolar diffusion on energy balance in prominences. Interestingly, in addition to enhancing thermal conduction in the PCTR, ambipolar diffusion has other effects on solar prominence threads. When Fontenla et al. (1996) considered ambipolar diffusion in the hydrogen ionization by introducing transport of neutral hydrogen toward the corona (and transport of protons toward the prominence slab center), they found that ambipolar diffusion affects the threads in prominences by increasing the thread geometrical and optical thicknesses.

In considering thermal conduction, the angle between the magnetic field and the temperature gradient becomes very important. This angle greatly influences the heat conduction and the thickness of the PCTR (Chiuderi and Chiuderi Drago 1991; Chiuderi Drago and Landi 2002). Thermal conduction presents a difficulty for a two-dimensional prominence model since the magnetic field lines in such a configuration lie wholly in the plane of the calculation, where the gradient of temperature must lie. Conduction along the field lines from the hot corona into the cooling condensation is so efficient in this case that the model condensation quickly disappears (Oster and Sofia 1966). On average, magnetic fields threading a three-dimensional prominence have a large component along the long axis of the prominence. The conductive heat flux decreases more rapidly than the square of the cosine of the angle between the magnetic field and the thermal gradient. Therefore, the axial field serves to shield the prominence from thermal conduction. If the angle between the field and the thermal gradient is not accounted for (i.e., assuming **B** parallel to ∇T), a much lower pressure must be assumed in the PCTR at the prominence top to account for the observed data.

7.4 Enthalpy Flux

A fluid that is moving carries energy in the form of kinetic energy (KE) flux

$$\frac{1}{2}\rho u^3 \tag{7.21}$$

and an enthalpy flux:

$$\gamma P v / (\gamma - 1) \tag{7.22}$$

where γ (adiabatic index) is the ratio of the specific heat at constant pressure to the specific heat at constant volume, P = pressure, u is the flow velocity, and v = thermal velocity.

We consider fluxes due to kinetic energy since prominences have plasma that move in the Sun's gravitational field. At transition region densities and velocities, the KE flux is small, but the KE flux associated with prominence flows of at least 20 km s^{-1} is on the order of a few times 10^5 ergs s^{-1} cm^{-2} (Kucera et al. 2014).

The enthalpy flux is composed of two parts: the flux of internal energy transported by the flow and the work done by the pressure force. Considering that prominences are suspended in the hotter corona, they are exposed to advection of heat by downward flowing coronal material from above and upward flowing coronal material from below. For non-static prominences, enthalpy inflow can play a major role in heating. For example, in their siphon model of prominence formation, Poland and Mariska (1986) found that the hot material flowing into the "well" in their magnetic configuration causes an increase in the energy transported by the enthalpy flux, leading to hotter prominence material than the expected value of about 10,000 K. They require that the siphon-like flow of chromospheric material into their seed condensation be slow as the formation occurs to mitigate the heating effects of the enthalpy flux. At transition region densities and velocities, the enthalpy flux can be substantial. For a pressure of 0.2 dyn cm^{-2} s^{-1} and a velocity of 8 km s^{-1} (and assuming the ratio of specific heats is 5/3) the enthalpy flux is 4×10^5 erg cm^{-2} s^{-1} (Mariska 1992) which is comparable to the conductive flux in this region (see Sect. 7.3). Flows will dominate the plasma if the enthalpy flux is comparable to or exceeds the conductive flux.

Anzer and Heinzel (2000) considered the inflow of enthalpy and ionization energy to study prominence heating associated with mass inflow into the cool parts in various 1D models. This type of analysis accounts for the fact that the inflow is occurring on both sides of a prominence, which is different from the chromosphere. The resulting energy gains were compared with integrated radiative losses obtained for such slabs by Anzer and Heinzel (1999) and many of the models they considered were in energy equilibrium as long as reasonable flows were assumed (inflow velocities of 1.5 km s^{-1} for temperatures of 30,000 K were considered). We now explore in more detail the radiative loss and other sources of heating in the solar chromosphere and prominences.

7.5 Radiative Losses

Radiative cooling results from collisions between particles, during which a part of the particle kinetic energy is converted to photon energy, and the resulting photon travels out of the region in which it was produced. The most common such collisional process is a collision between an electron and an atom (which may be partially ionized), in which the electron gives up part of its energy to raising the atom to an excited state. Relaxation from this excited state then produces a photon, which we assume escapes the atmosphere. The rate of such photon production (and thus the rate of radiative cooling) associated with this particular type of collision is going to be proportional to the electron density and to the atom density. Each of these densities can be considered directly proportional to the total hydrogen density, multiplied by proportionality factors accounting for the abundance of the atomic species and the ionization state of the medium. The total correction factor will generally be dependent on temperature, so it follows that we can write the optically thin radiative cooling rate (associated with all the types of collisions leading to photon production) in the form

$$\mathrm{L} = n^2 R(T) \tag{7.23}$$

where n is the total hydrogen density, and the radiative cooling function $R(T)$ is a complex function of temperature.

The optically thin radiative cooling rate, the variation of which has be calculated by various people using different parameters, is important for the PCTR. We show an example in Fig. 7.2, where L is measured relative to the cooling rate at $T = 10^4$ K. Two cases of particular interest are shown: the radiative cooling rate when the density is held fixed as the temperature varies and the rate when the pressure is held fixed as the temperature varies. In the former case (constant density), it follows from (7.23) that

$$\mathrm{L}_{\mathrm{rel}}(T) = \mathrm{L}(T)/\mathrm{L}\left(10^4\ \mathrm{K}\right) = \left[n^2 R(T)\right] / \left[n^2 R\left(10^4\ \mathrm{K}\right)\right] = R_{\mathrm{rel}}(T) \tag{7.24}$$

while in the latter case (constant pressure) we have $n \propto 1/T$, and from (7.23) it follows that

$$\mathrm{L}_{\mathrm{rel}} = \frac{R(T)/R\left(10^4\ \mathrm{K}\right)}{T^{\,2}/\left(10^4\ \mathrm{K}\right)^2} = R_{\mathrm{rel}}(T)/T_{\mathrm{rel}}^2 \tag{7.25}$$

where the 'rel' subscript refers to a quantity measured relative to its value at $T = 10^4$ K.

The peak in the radiative cooling curve that occurs near $T = 2 \times 10^4$ K (a relative maximum in the constant-density curve and an absolute maximum in the constant-pressure curve) is associated with cooling through the production of hydrogen Ly-α photons. Ly-α cooling becomes particularly efficient near $T = 10^4$ K, where the

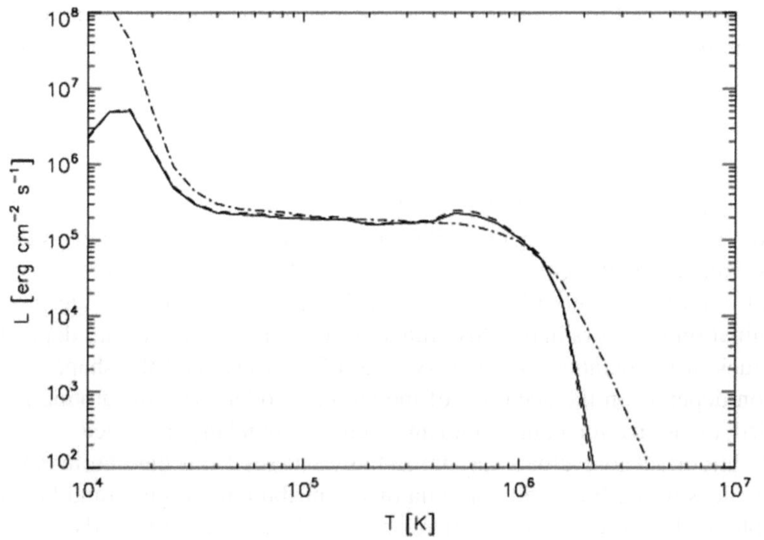

Fig. 7.3 Integrated radiative loss rates for the QS (*dash-dotted*) and the prominence in two different spatial locations (*dashed* and *solid*) from Parenti and Vial (2007)

electron thermal energy becomes relatively large and a significant fraction of the hydrogen atoms remains neutral (so that the atoms are subject to excitation). Ly-α cooling is the dominant radiative cooling mechanism in this temperature range. Yet, as the temperature begins to increase further, hydrogen becomes rapidly nearly fully ionized, and the atoms that are available to be excited to produce Ly-α photons become very rare. Thus, this natural thermostat stops operating. These two effects—increased excitation efficiency and rapid hydrogen ionization—lead to a sharp peak in the radiative cooling curve. At temperatures below the peak temperature ($T = 2 \times 10^4$ K), the relative radiative cooling rate is very similar for the constant-density and the constant-pressure cases, but above the peak, the two curves diverge significantly. If we are primarily concerned with the very narrow transition region (which is nearly a constant pressure region owing to its spatial narrowness) when considering radiative cooling in the temperature range 2×10^4 K $< T < 10^6$ K, the constant-pressure radiative cooling curve is the relevant one. This also leads to the temperature plateau between 6,000 and 7,000 K in Fig. 7.1.

Another loss curve is shown in Fig. 7.3, which was calculated by Parenti and Vial (2007) by using the differential emission measure in the quiet-Sun and in a prominence. The shape of the constant pressure cooling rate in Fig. 7.2 is similar to the one given in Fig. 7.3.

Above 10^5 K the plasma is optically thin and the radiative loss rate calculation only requires the solution to the ionization and excitation balance for each major element in the plasma. This loss rate is the sum of the emitted radiation from bound–bound, bound–free, and free–free transitions at a certain temperature. Therefore, for the PCTR, simplifying assumptions can be made. Namely, the optically thin approximation, and the assumption that the plasma is in ionization and excitation equilibrium. For example, Anzer and Heinzel (1999) use the power laws of Hildner (1974) to get:

$$L = p^2 \chi_i T^{\alpha_i - 2} \tag{7.26}$$

(χ_i and α_i are tabulated by Hildner (1974))

Radiative losses also occur in the chromosphere, which is more relevant to energy balance in the core of prominences, but the chromosphere has a higher level of complexity with all radiative transfer collisional rates playing a role. The determination of the radiative loss function in prominence plasma, depending on the values of temperature and density, is a difficult task and the shape of the loss function depends on the accuracy of the atomic model used, the atomic processes included, the ionization equilibrium and element abundance assumed.

It is important to evaluate the net radiative losses from the chromosphere and prominences in the lines and continua that contribute to cooling (and heating) the atmosphere. In these regimes, where increased mass density leads to increased probability of absorption of photons and collisional de-excitation, not all photons will escape the atmosphere. However, in strong lines (e.g., Hα and Ca II K) most of the photons emitted are immediately absorbed in the same transition but the collisional destruction probability is often very low and the photon may escape after a very large number of scattering events. There may also be photons absorbed in one line (e.g., Lyβ) escaping in another transition (e.g., Hα). As mentioned above, a full description of these processes entails solving the transfer equations coupled with the non-LTE rate equations (or statistical equilibrium equations if rates are assumed to be instantaneous) (see Chapter 5: Heinzel 2014).

Prominence models have to account for the optically thin radiative loss rate, the probability of energy escape from the atmosphere, and the fraction of atoms in the given ionization state. Radiative losses can be computed from a model chromosphere or prominence, but evaluation of these losses inside the cool part of the prominence material is difficult because the radiative cooling function depends on the source function and on the radiation field.

Labrosse et al. (2010) gives the general formula for radiation-flux divergence:

$$L = 4\pi \int_0^\infty [\eta(\upsilon) - \chi(\nu) J(\upsilon)]\, d\upsilon = 4\pi \int_0^\infty \chi(\eta) [S(\upsilon) - J(\upsilon)]\, d\upsilon \tag{7.27}$$

$\eta(\upsilon)$ and $\chi(\nu)$ are total emissivities and opacities, $J(\upsilon)$ is the mean intensity of the radiation field, and $S(\upsilon)$ is the source function.

The most important coolants in the chromosphere and in prominences are hydrogen, calcium and magnesium. Decades ago it was argued that radiative losses in the low chromosphere are mainly due to the H- ion and the Balmer series of hydrogen (Athay 1966; Ulmschneider 1974). However, more recent studies have questioned the validity of the earlier work. Vernazza et al. (1981) computed the net radiative cooling rates, using a model chromosphere, for several atoms and ions, and found that the H- cooling rate is smaller than that calculated previously (by a factor of 10). For the average quiet Sun, they found the largest integrated cooling rates are due to Ca II infrared-triplet and resonance lines, the Mg II resonance lines, H-, and Lα (in order of decreasing importance).

Carlsson and Leenaarts (2012) discuss the various contributions to radiative cooling in hydrogen and find that the Lyα transition dominates (20 % or less comes from other transitions, with Lyβ and Lyman continuum as the most important). Below temperatures of 32,000 K hydrogen optically thin radiation dominates over contributions from other elements, and below temperatures of 7,000 K Lyman continuum starts to contribute and Hα begins to dominate.

For prominences, comprehensive studies have been performed for radiation losses due to hydrogen. Poland and Anzer (1971) made a crude estimate of Ca II h and k line losses and indicate that the energy loss for those lines is relatively unimportant when compared with the hydrogen radiation. More recently, detailed transfer modelling of Anzer and Heinzel (1999) have shown that calcium losses are negligible for prominence slabs. Mg II losses were studied by Heinzel et al. (2014) where they extended the work on radiative equilibrium in prominences (Heinzel and Anzer 2012). In their latest work they include an additional cooling term which is due to the presence of the Mg II lines and find that the effect on the resulting radiative equilibrium temperatures can be important but only in the case of very low pressures. They conclude that role of Mg II h and k radiation losses on prominence energy balance is dominant at lower pressures, but the losses due to hydrogen and Ca II dominate at higher pressures.

Examples of calculated cooling rates include the following: Parenti and Vial (2007) found that for a prominence the integral over the temperature range of 10^4–10^6 K yielded losses of 3×10^6 ergs cm^{-2} s^{-1}. This is similar to the radiative losses given in Withbroe and Noyes (1977) for chromospheric (quiet Sun), which range from 3×10^5 ergs cm^{-2} s^{-1} (upper chromosphere) to 2×10^6 ergs cm^{-2} s^{-1} (low chromosphere), with a total chromospheric loss of 4×10^6 ergs cm^{-2} s^{-1}. For the low chromosphere Vernazza et al. (1981) found that the H- cooling rate is much less than the contributions due to the Ca II lines, and the H- bound-free cooling rate never exceeds 0.01 ergs cm^{-3} s^{-1}. They found integrated net cooling rates for Ca II k (7×10^5 ergs cm^{-2} s^{-1}), for H- (2×10^5 ergs cm^{-2} s^{-1}), and Mg II (4–5×10^5 ergs cm^{-2} s^{-1}). Vernazza also found that the Lyα cooling rate is the dominant one in the transition region at temperatures of 20,000–30,000 K. There can be significant differences between the rates obtained from observed emission fluxes and those calculated from an atmospheric model.

Although useful in the energy balance problem, it is worth noting that detailed calculations of radiative losses from the chromosphere are of limited value unless the other contributions to the energy balance can also be calculated with a similar level of accuracy.

7.6 Other Heating Mechanisms

To address the mechanical heating term in the energy balance equation, we note that there is a "zoo" of possible chromospheric heating mechanisms. We've already discussed thermal conduction, kinetic energy, and enthalpy, but there are a few more that might play a role in prominence heating and energy balance. Not all of these mechanisms will apply to prominences, but for completeness we address some of the more important ones for chromospheric heating. It is worth noting that prominences have a different geometry and location relative to the photosphere than the chromospheric layer that sits above the photosphere, and these differences need to be considered in the context of heating. Generally speaking, the bulk of the chromosphere is a few thousand degrees hotter than the underlying photosphere and far cooler than the corona. The chromosphere is so dense relative to the corona that it requires roughly ten times more heat input (when measured as a height-integrated rate of change of energy density) to maintain its elevated temperature (Leake et al. 2014). As discussed earlier, the increase in temperature through the chromosphere and up into the transition region and corona must be explained by some non-radiative form of energy. Otherwise the sum of all the net radiative gains and losses from the solar atmosphere would equal zero. This is also true in the case of prominences. In exploring various heating mechanisms, it is useful to determine the lowest atmospheric level at which mechanical heating is required. To determine this level, we can compare the level at which the observed atmosphere becomes hotter than a model of a theoretical radiative–convective atmosphere in which no mechanical heating is included. In doing so, one has to be cognizant of the large uncertainties in both model and observed atmospheres, but if temperatures rise above the values of the radiative–convective model by a significant amount, it can be concluded that mechanical heating must be present. Similarly, in the context of prominences, models can be compared to observations to determine whether heating sources must be present. Heasley and Mihalas (1976) demonstrated that the radiative equilibrium temperature inside one-dimensional prominence slabs is much lower than what is typically deduced from spectral observations. This led to investigations into possible prominence heating sources [e.g., vertical downflows in Anzer and Heinzel (2000)].

Returning to the general chromospheric heating problem, many sources have been proposed as important. Two main theories come to the front as viable options to solve this problem: mechanical heating by upward-propagating waves (Alfven 1947) and Joule heating associated with magnetic field reconnection and the resistive dissipation of electric currents (Rabin and Moore 1984; Parker 1988). Since Joule and viscous heating mechanisms become relatively more important on progressively smaller scales, such as those associated with current sheets, shocks, or wave motions, it is not clear whether these should be considered in the context of prominences. We explore wave and radiative heating in a bit more detail as potentially relevant to solar prominences.

7.6.1 Wave Heating

The turbulent convection zone below the solar surface is capable of supplying a flux of wave energy into the chromosphere, which leads to one of the large classes of potential heating mechanisms. In this class of mechanisms, acoustic waves that are generated in the convection zone travel into the atmospheric layers above where they steepen into shocks and dissipate. However, it is not clear that acoustic waves are able to propagate high enough to deposit energy in the chromosphere (Ulmschneider 1990; Fossum and Carlsson 2005, 2006; Kalkofen 2007), and especially in solar prominences suspended in the corona. Kalkofen (2008) determined that the observational estimates of the acoustic flux in the photosphere are significantly below that needed to match the energy radiated by the chromosphere. Acoustic waves are unlikely a dominant source of heating in the chromosphere or solar prominences, but magnetic waves (e.g., Alfven) are potential candidate sources.

Prominences are very dynamic with ubiquitous transverse thread oscillations and propagating waves along the threads, which have been interpreted in terms of MHD waves (see the reviews by Ballester 2006; Oliver 2009). MHD waves have recently been investigated in the chromosphere; particularly Alfven waves as they can propagate upwards along magnetic field lines (e.g., De Pontieu et al. 2007; Tomczyk et al. 2007). These waves lose energy as they travel along the magnetic field and thus the heating will decrease along the field. The situation becomes particularly complex for prominences due to the almost discontinuous density change from the corona to the prominence, causing wave reflection and a reduced wave heating inside the prominence. There are a variety of wave energy dissipative mechanisms (e.g., viscosity, thermal and electric conductivity) which cause the wave amplitude to become progressively attenuated. The frictional damping of magnetoacoustic waves in a partially ionized plasma is much stronger than in a fully ionized plasma because the presence of neutral atoms causes the Joule dissipation to increase as a result of collisions of electrons with neutrals and ions and, more importantly, of collisions of ions with neutrals (Khodachenko et al. 2004). A comparative study of the role of ion-neutral damping of MHD waves and their damping due to viscosity and thermal conductivity was made by Khodachenko et al. (2004) and Khodachenko et al. (2006) and it was found that collisional damping is dominant.

Parenti and Vial (2007) considered energy fluxes carried by both Alfven and sound waves in solar prominences:

$$F_A = \rho \xi^2 V_A \qquad (7.28)$$

$$F_s = \rho \xi^2 C_s \qquad (7.29)$$

with the Alfven speed given by $V_A = \sqrt{\left(B/_{4\pi\rho}\right)}$ and the sound speed, $C_S = \sqrt{\left(\gamma p/_{\rho}\right)}$

In this work the measured non-thermal velocities in prominences were interpreted as the averaged value of the transverse wave propagation. The two different prominence regions considered (i.e., cooler and PCTR) showed different results regarding whether the waves are efficient enough for transporting the energy needed to compensate for the radiative losses. For the cooler region neither Alfven nor sound waves could transport enough energy to account for the radiative losses, but in the hotter region, the Alfven wave flux was sufficient for temperatures higher than 7×10^4 K and the sound wave flux became important for temperatures higher than 2×10^5 K.

Another factor to consider in the damping of Alfven waves is the effect of neutral helium. Zaqarashvili et al. (2013) found that neutral helium atoms can significantly enhance the damping at certain temperatures (10,000–40,000 K), where the ratio of neutral helium and neutral hydrogen atoms is increased, which is important in the PCTR. Soler (2010, PhD thesis) suggested that the neutral helium does not have a significant influence on the damping rate in the prominence cores (i.e., below 8,000 K temperatures). By investigating Alfven wave damping in cool prominence cores they found that the damping is not very efficient because the theoretical damping times are between one and two orders of magnitude larger than the corresponding periods (applicable to both individual and collective oscillations of prominence threads).

7.6.2 *Radiative Heating*

The chromosphere is heated by radiation originating in the photosphere and in the corona above, so the contribution from these two sources must be subtracted. However, prominences, which are suspended in the corona, are more thermally isolated from the photosphere and have a different magnetic geometry than the chromosphere. The radiation from the photosphere and corona mostly travels across the prominence magnetic field compared to being aligned with the field (as in the chromosphere). The optically thin radiative losses from the corona emit energy towards the Sun which is absorbed in the chromosphere where it contributes to radiative heating. For prominences suspended fairly high in the atmosphere, coronal radiation is absorbed from below and above the structure since it is surrounded by the corona. As discussed earlier, cooling in strong lines and continua play a role in energy balance, but those same transitions may also provide heating of cool pockets in the chromosphere. Carlsson and Leenaarts (2012) suggest that most of this radiation is absorbed in the continua of neutral helium and neutral hydrogen, and that heating in hydrogen transitions is mainly due to absorption of Ly-α photons produced by a nearby source. In that work, radiative heating in the h and k lines of Mg II is negligible, and small for the lines of Ca II. They also demonstrate that radiative heating can be approximated with the same functional form as radiative cooling. However, the heating is strongly dependent on the details of the local radiation field and absorption profile, and because of this they conclude the accuracy is better for cooling.

7.7 Concluding Remarks

Much work still needs to be done in solving the energy balance problem in solar prominences. Models often have to rely on assumptions concerning the fine structure and plasma conditions. As we gain more knowledge about the conditions, more non-LTE modelling efforts need to be made, taking into account the fine structure. We also need to address whether ionization equilibrium is always a valid assumption. Ionization of hydrogen in the chromosphere and PCTR does not obey LTE or instantaneous statistical equilibrium because the timescale is long relative to hydrodynamical timescales (Carlsson and Stein 2002), and the ionization state of hydrogen strongly influences the temperature, pressure and electron density. Radiation magnetohydrodynamic (MHD) simulations of the atmosphere must therefore account properly for hydrogen ionization (Leenaarts et al. 2007; Bradshaw 2009). In general, time-dependent numerical codes need to be developed to properly evaluate ionization and recombination times. Moreover, a coupling of the non-LTE radiative transfer calculations to other types of numerical modelling that require a precise determination of the energy losses due to emitted radiation would also be beneficial.

Heating mechanisms need to be studied more carefully, especially those involving waves, and the role of ambipolar diffusion requires further attention. It is not clear on which spatial or temporal scales prominence heating takes place, and the properties of the PCTR remain uncertain. A more detailed thermal structure through the PCTR would also be extremely useful.

Thanks to improved prominence observations and modeling, we have come a long way in understanding basic physical properties of these solar phenomena. However, it is critical to continue improving both modeling and observations, and absolutely necessary to utilize them in conjunction with each other.

References

Alfven, H. (1947). Magneto hydrodynamic waves, and the heating of the solar corona. *Monthly Notices of the Royal Astronomical Society, 107*, 211.

Alvarez, M. (1980). Energy balance from the chromosphere–corona transition region. *Astrophysical Journal, 240*, 322.

Antiochos, S. K., & Klimchuk, J. A. (1991). A model for the formation of solar prominences. *Astrophysical Journal, 378*, 372.

Anzer, U., & Heinzel, P. (1999). The energy balance in solar prominences. *Astronomy and Astrophysics, 349*, 974.

Anzer, U., & Heinzel, P. (2000). Energy considerations for solar prominences with mass inflow. *Astronomy and Astrophysics, 358*, L75.

Athay, R. G. (1966). Radiative energy loss from the solar chromosphere and corona. *Astrophysical Journal, 146*, 223.

Ballester, J. L. (2006). Seismology of prominence-fine structures: Observations and theory. *Space Science Reviews, 122*, 129.

Ballester, J. L. (2014). Magnetism and dynamics of prominences: MHD waves. In J.-C. Vial & O. Engvold (Eds.), *Solar prominences, ASSL* (Vol. 415, pp. 257–294). New York: Springer.

Bradshaw, S. J. (2009). A numerical tool for the calculation of non-equilibrium ionisation states in the solar corona and other astrophysical plasma environments. *Astronomy and Astrophysics, 502*, 409.

Braginskii, S. I. (1965). Transport processes in plasmas. In M. A. Leontovich (Ed.), *Reviews of plasma physics* (Vol. I, p. 205). New York: Consultants Bureau.

Carlsson, M., & Leenaarts, J. (2012). Approximations for radiative cooling and heating in the solar chromosphere. *Astronomy and Astrophysics, 539*, 10.

Carlsson, M., & Stein, R. F. (2002). Dynamic hydrogen ionization. *Astrophysical Journal, 572*, 626.

Chiuderi Drago, F., & Landi, E. (2002). The prominence–corona and the filament–corona transition region: Is there any difference? *Solar Physics, 206*, 315.

Chiuderi, C., & Chiuderi Drago, F. (1991). Energy balance in the prominence–corona transition region. *Solar Physics, 132*, 81–94.

De Pontieu, B., et al. (2007). Chromospheric Alfvénic waves strong enough to power the solar wind. *Science, 318*, 1574.

Engvold, O., Kjeldseth-Moe, O., Bartoe, J.-D. F., & Brueckner, G. E. (1987). Observations and modeling of the prominence/corona transition region, In *ESA, proceedings of the 21st ESLAB symposium on small scale plasma processes in the solar chromosphere/corona, interplanetary medium and planetary magnetospheres* (p. 21).

Fan, Y. (2014). MHD equilibria and triggers for eruption. In J.-C. Vial & O. Engvold (Eds.), *Solar prominences, ASSL* (Vol. 415, pp. 295–320). New York: Springer.

Fontenla, J. M., Avrett, E. H., & Loeser, R. (1990). Energy balance in the solar transition region. I – Hydrostatic thermal models with ambipolar diffusion. *Astrophysical Journal, 355*, 700.

Fontenla, J. M., Avrett, E. H., & Loeser, R. (1991). Energy balance in the solar transition region. II – Effects of pressure and energy input on hydrostatic models. *Astrophysical Journal, 377*, 712.

Fontenla, J. M., Avrett, E. H., & Loeser, R. (1993). Energy balance in the solar transition region. III – Helium emission in hydrostatic, constant-abundance models with diffusion. *Astrophysical Journal, 406*, 319.

Fontenla, J. M., Rovira, M., Vial, J.-C., & Gouttebroze, P. (1996). Prominence thread models including ambipolar diffusion. *Astrophysical Journal, 466*, 496.

Fossum, A., & Carlsson, M. (2005). High-frequency acoustic waves are not sufficient to heat the solar chromosphere. *Nature, 435*, 919.

Fossum, A., & Carlsson, M. (2006). Determination of the acoustic wave flux in the lower solar chromosphere. *Astrophysical Journal, 646*, 579.

Heasley, J. N., & Mihalas, D. (1976). Structure and spectrum of quiescent prominences – Energy balance and hydrogen spectrum. *Astrophysical Journal, 205*, 273.

Heinzel, P. (2014). Radiative transfer in solar prominences. In J.-C. Vial & O. Engvold (Eds.), *Solar prominences, ASSL* (Vol. 415, pp. 101–128). New York: Springer.

Heinzel, P., & Anzer, U. (2012). Radiative equilibrium in solar prominences reconsidered. *Astronomy and Astrophysics, 539*, 6.

Heinzel, P., Vial, J.-C., & Anzer, U. (2014). On the formation of Mg II h and k lines in solar prominences. *Astronomy and Astrophysics, 564*, A132.
Hildner, E. (1974). The formation of solar quiescent prominences by condensation. *Solar Physics, 35*, 123.
Kalkofen, W. (2007). Is the solar chromosphere heated by acoustic waves? *Astrophysical Journal, 671*, 2154.
Kalkofen, W. (2008). Wave heating of the solar chromosphere. *Journal of Astrophysics and Astronomy, 29*, 163.
Karpen, J. T., Antiochos, S. K., & Klimchuk, J. A. (2006). The origin of high-speed motions and threads in prominences. *Astrophysical Journal, 637*, 531.
Khodachenko, M. L., Arber, T. D., Rucker, H. O., & Hanslmeier, A. (2004). Collisional and viscous damping of MHD waves in partially ionized plasmas of the solar atmosphere. *Astronomy and Astrophysics, 422*, 1073.
Khodachenko, M. L., Rucker, H. O., Oliver, R., Arber, T. D., & Hanslmeier, A. (2006). On the mechanisms of MHD wave damping in the partially ionized solar plasmas. *Advances in Space Research, 37*, 447.
Kucera, T., Gilbert, H. & Karpen, J. (2014). Mass flows in a prominence spine as observed in EUV. *Astrophysical Journal, 790*, 68.
Labrosse, N. (2014). Derivation of the major properties of prominences using non-LTE modeling. In J.-C. Vial & O. Engvold (Eds.), *Solar prominences*. New York: Springer.
Labrosse, N., Heinzel, P., Vial, J.-C., Kucera, T., Parenti, S., Gunar, S., Schmieder, B., & Kilper, G. (2010). Physics of solar prominences: I—Spectral diagnostics and non-LTE modelling. *Space Science Reviews, 151*(4), 243–332.
Leake, J. E., DeVore, C. R., Thayer, J. P., Burns, A., Crowley, G., Gilbert, H. R., Huba, J. D., Judge, P., Krall, J., Linton, M. G., Lukin, V. S., Rodrigues, F., & Wang, W. (2014). Ionized plasma and neutral gas coupling in the sun's chromosphere and earth's ionosphere/thermosphere. *Space Science Reviews* (184).
Leenaarts, J., Carlsson, M., Hansteen, V., & Rutten, R. J. (2007). Non-equilibrium hydrogen ionization in 2D simulations of the solar atmosphere. *Astronomy and Astrophysics, 473*, 625.
Malherbe, J.-M. (1989). The formation of solar prominences. In *Dynamics and structures of quiescent solar prominences: Proceedings of the workshop, Palma de Mallorca, Spain, November 1987* (A89-51201 22–92) (pp. 115–141). Dordrecht: Kluwer Academic Publishers.
Mariska, J. T. (1992). Model equations: The solar transition region. In R. F. Carswell, D. N. C. Lin & J. E. Pringle (Eds.), (p. 201). New York: Cambridge University Press.
Mariska, J. T., Doschek, G. A., & Feldman, U. (1979). Extreme-ultraviolet limb spectra of a prominence observed from SKYLAB. *Astrophysical Journal, 232*, 929.
Noyes, R. W., Dupree, A. K., Huber, M. C. E., Parkinson, W. H., Reeves, E. M., & Withbroe, G. L. (1972). Extreme-ultraviolet emission from solar prominences. *Astrophysical Journal, 178*, 515.
Oliver, R. (2009). Prominence seismology using small amplitude oscillations. *Space Science Reviews, 149*, 175.
Oster, L., & Sofia, S. (1966). Thermal dissipation and its application to flare phenomena. *Astrophysical Journal, 143*, 944.
Parenti, S. (2014). Spectral diagnostics of cool and PCTR optically thin plasma. In J.-C. Vial & O. Engvold (Eds.), *Solar prominences, ASSL* (Vol. 415, pp. 61–76). New York: Springer.
Parenti, S., & Vial, J.-C. (2007). Prominence and quiet-Sun plasma parameters derived from FUV spectral emission. *Astronomy and Astrophysics, 469*, 1109.
Parker, E. N. (1988). Nanoflares and the solar X-ray corona. *Astrophysical Journal, 330*, 474.
Poland, A., & Anzer, U. (1971). Energy balance in cool quiescent prominences. *Solar Physics, 19*(2), 401.
Poland, A. I., & Mariska, J. T. (1986). A siphon mechanism for supplying prominence mass. *Solar Physics, 104*, 303.

Rabin, D. (1986). The prominence–corona interface and its relationship to the chromosphere–corona transition. In *NASA Goddard Space Flight Center Coronal and Prominence Plasmas* (pp. 135–142) (SEE N87-20871 13-92).
Rabin, D., & Moore, R. (1984). Heating the sun's lower transition region with fine-scale electric currents. *Astrophysical Journal, 285*, 359.
Raymond, J. C., Cox, D. P., & Smith, B. W. (1976). Radiative cooling of a low-density plasma. *Astrophysical Journal, 204*, 290.
Rosner, R., & Tucker, W. H. (1989). On magnetic fields, heating and thermal conduction in halos, and the suppression of cooling flows. *Astrophysical Journal, 338*, 761.
Schmahl, E. J. (1979). The prominence–corona interface – A review. In *Physics of Solar Prominences: Proceedings of the Colloquium, Oslo, Norway* (pp. 102–120), August 14–18, 1978 (A79-46076 20–92). Oslo: Universitetet i Oslo.
Soler, R. (2010). *Damping of magnetohydrodynamic waves in solar prominence fine structures.* PhD thesis, Universitat de les Illes Balears.
Soler, R., Ballester, J. L., & Goossens, M. (2011). The thermal instability of solar prominence threads. *Astrophysical Journal, 731*, 39.
Soler, R., Ballester, J. L., & Parenti, S. (2012). Stability of thermal modes in cool prominence plasmas. *Astronomy and Astrophysics, 540*, 176–181.
Spitzer, L. (1962). *Physics of fully ionized gases* (2nd ed.). New York: Interscience.
Tomczyk, S., McIntosh, S. W., Keil, S. L., Judge, P. G., Schad, T., Seeley, D. H., & Edmondson, J. (2007). Alfvén waves in the solar corona. *Science, 317*, 1192.
Ulmschneider, P. (1970). Thermal conductivity in stellar atmospheres I. Without magnetic field. *Astronomy and Astrophysics, 4*, 144.
Ulmschneider, P. (1974). Radiation loss and mechanical heating in the solar chromosphere. *Solar Physics, 39*, 327.
Ulmschneider, P. (1990). Recent advances in acoustic heating. In *Basic plasma processes on the sun* (A92-30901 12–92) (pp. 231–234). Dordrecht, Netherlands: Kluwer Academic Publishers.
Vernazza, J. E., Avrett, E. H., & Loeser, R. (1981). Structure of the solar chromosphere. III. Models of the EUV brightness components of the quiet Sun. *Astrophysical Journal, 45*, 635.
Vial, J. C. (1990). The prominence–corona interface. In V. Ruždjak & E. Tandberg-Hanssen (Eds.), *Dynamics of Quiescent Prominences, Proceedings of the No. 117 Colloquium of the International Astronomical Union Hvar, SR Croatia, Yugoslavia 1989* (Vol. 363, pp 106–119).
Withbroe, G. L., & Noyes, R. W. (1977). Mass and energy flow in the solar chromosphere and corona. *Annual Review of Astronomy and Astrophysics, 15*, 363.
Yang, C. Y., Nicholls, R. W., & Morgan, F. J. (1975). Studies of the prominence–corona transition zone from rocket ultraviolet spectra of the March 1970 eclipse. *Solar Physics, 45*, 351.
Zaqarashvili, T. V., Khodachenko, M. L., & Soler, R. (2013). Torsional Alfvén waves in partially ionized solar plasma: Effects of neutral helium and stratification. *Astronomy and Astrophysics, 549*, 9.

Chapter 8
Magnetometry of Prominences

Arturo López Ariste

Abstract We describe the measurement of magnetic fields in prominences. Using the He D_3 line as example we describe and illustrate the computation of the polarization emitted by He atoms in the presence of magnetic fields. The relatively weak magnetic fields expected in prominences require taking into consideration a long series of quantum coherences between the atomic levels of the He atom, coherences that are critical for the sensitivity of the emitted radiation and its polarization to the magnetic fields. But solving that quantum problem is only half the task: the observed polarized profiles need to be compared to the computed ones until a match is found. This inference or inversion can only work when the appropriate numerical technicals are put to work, techniques that help identify what magnetic field is the best solution for an observed profile, that can make use of all the available observables while being robust in front of noise and the low brightness of prominences and filaments respect to the entouring photosphere. These difficulties can be tackled but at the prixe of some approximations that have to be kept in mind in the analysis of prominence magnetic fields at the risk of serious mistakes on the inferred magnetic fields. Improving upon those approximations marks also the path for the future, with which description we will conclude this chapter.

8.1 Methods to Remotely Measure Magnetic Fields: A Fly-Over

Since our main source of physical information on astrophysical sources is radiation and, in particular, spectral lines, it is not surprising that the measurement of magnetic fields relies on those modifications in the emitted or absorbed spectra induced by those magnetic fields. One big difference, and source of no small difficulties, is that those modifications of the spectra appear mostly in polarization. Prominences are no exception to these general statements and the measurement of magnetic

A. López Ariste (✉)
IRAP - THEMIS, CNRS

J.-C. Vial, O. Engvold (eds.), *Solar Prominences*, Astrophysics and Space Science Library 415, DOI 10.1007/978-3-319-10416-4_8

fields in prominences has traditionally relied on the polarimetry of the several atomic lines emitted or absorbed in prominences and filaments. Two atomic effects are responsible of the changes in the polarization of the spectral lines due to the magnetic fields: the Zeeman effect and the Hanle effect.

The Zeeman effect is better known in solar physics, because of its relative simplicity and because of its generalized use in the measurement of magnetic fields in the solar photosphere. If an atomic level is characterized by a set of quantum numbers N and L, a series of Zeeman sublevels labelled with the quantum numbers M varying from $-L$ to $+L$ share the same energy: they are degenerated. A magnetic field breaks that degeneracy and the Zeeman sublevels split apart into $2L + 1$ levels of different energy where before only one level was found. The larger the field the larger the energy separation between the magnetic sublevels. If that level $|N, L >$ was the origin of a radiative transition resulting in a spectral line, now $2L + 1$ transitions appear: the spectral line is split in $2L + 1$ transitions. The spectral separation of these split lines is a function of the magnetic field, that can be thus measured. The sublevels $|N, L, M >$ not only have different energies for different values of M, they also have different angular momentum. This is so because the actual meaning of the quantum number M is the projection of the orbital angular momentum L onto the magnetic field vector. One can see M as the projection of L on magnetic field direction. This projection is quantized, its possible values being the different values of M.

A radiative transition between two sublevels $|N, L, M >$ and $|N', L', M' >$ will produce a photon with total angular momentum unity, but whose spin (or helicity) will be the same as the difference $M - M'$. This spin of the photon is nothing else than its polarization. Thus the polarization of the photon will be $+1$, 0 or -1 if it came from sublevels with $M - M' = +1, 0$ or -1, other values being forbidden by the selection rules of radiative transitions. These values are to be interpreted in the natural polar basis of the spin, for which ± 1 represent states of circular polarization of different sense, and 0 is linear polarization.

Summarizing, as the atomic level $|N, L >$ splits in the presence of a magnetic field, the emitted atomic line will split in spectral components each one with its very particular polarization depending on the difference $M - M'$ between the atomic sublevels involved in the transition. This polarization signature in the split line is of particular relevance for solar observations because at the magnetic field strengths typical of solar dynamics, the line splitting due to the Zeeman effect is almost certainly smaller than, or at most comparable to, the thermal broadening of the line. Hence, it is very difficult and error-prone to try to observe and measure this Zeeman splitting. A much better approach is to measure the polarization of the line so that the several components make themselves apparent through their different polarizations. Particularly useful for example is the observation of circular polarization in a Zeeman-sensitive line which often results in two lobes of different sign on the wings of the spectral line. The amplitude of these lobes can be seen to be proportional to the projection of the magnetic field onto the line of sight.

In our simplified description of the Zeeman effect and its observational signatures in terms of polarization of spectral lines we jumped over many details and aspects of the correct treatment of the problem. We did so in the sake of brevity and clarity. But it should not escape the reader the fact that the exact computation of the atomic state and the radiative transitions are critical to correctly measure the magnetic field. When computing the exact atomic state of the atom, once the atomic level $|N, L>$ has split under the presence of a magnetic field, we need to figure out how the electrons originally populating that level have spread through the split Zeeman sublevels M. Of course we understand *population* in a statistical sense after considering a big set of identical atoms. Have the electrons populated the M sublevels evenly? Did they have a preference over one or several of the sublevels? To answer these questions we should remember that in the quantum world an atomic level has not a definite energy, but rather a well centered distribution of energies called the natural width of the level. Our previous questions should be recast into two different cases: the easy case, when the energy splitting due to the Zeeman effect is much larger than the natural width of the sublevels and they can be considered independent entities. The often (but not always) found answer in this case is that the electrons have spread evenly among the M sublevels. The second possible scenario has more interesting answers. If the energy splitting due to the magnetic field is zero, smaller than or comparable to the natural width, quantum coherences appear among the sublevels. In a graphical sense, the sublevels are identical or not distinct enough for the electron to be clearly in one or the other, so that it sits in all of them in a state of quantum superposition. Magnetic fields play in this case another role by subtly modifying those quantum coherences. This is the Hanle effect. The emitted radiation, its polarization, is also modified by these coherences. The observation of the polarization of lines emitted by atoms whose levels carry these coherences gives us information on those coherences and on the eventual modification induced by magnetic fields.

Our two previous descriptions, though extremely simplified, can be used as a guideline: the Hanle effect appears when the magnetic splitting is smaller than or comparable to the natural width of the line, the Zeeman effect appears when the magnetic splitting is much larger than the natural width of the line. We will have to figure out the natural widths of the atomic levels of the spectral lines we observe in prominences and whether these are smaller or comparable to the splitting induced by the magnetic fields we expect to measure. Depending on the answer, we will have to depend on the Hanle effect or on the Zeeman effect to measure the fields. And the answer will be that, in prominences, the Hanle effect is our main tool, although the Zeeman effect is also visible when strong fields make their appearance in active region filaments.

Historically, however, the first attempts to measure any magnetic field in prominences were made assuming the presence of the (easy to interpret) Zeeman effect. This had disastrous results. Curiously, the average fields found were not too erroneous (Brown et al. 2003) thanks to a lucky coincidence: in the observed lines of Helium, the Hanle and Zeeman effects have similar amplitudes as long as the field strength is comparable, even if the orientation of the field is completely unrelated in one case or the other. The first correct measurements of magnetic fields in prominence had therefore to wait for the introduction of Hanle effect by Leroy and his colleagues in the 1980s. The successful measurements made by those authors provided the confirmation that magnetic fields are essentially horizontal and twisted with field strengths of 10–20 G in average. In spite of that success, those measurements were limited by the assumption that no Zeeman effect was present in the polarized profiles. This assumption set an upper limit to the field strengths they could measure, as stronger fields would produce a level splitting larger than the natural width of the sublevels. For the observed lines, this limit was around 40 G. By the 1990s it was clear that both Hanle and Zeeman effect had to be considered simultaneously to interpret the observations. Immediately, stronger fields were measured, as the Zeeman effect took off where the Hanle effect saturated. Fields stronger than 40 G were reported by Paletou et al. (2001) and Casini et al. (2003). When active region filaments started to be observed it was seen that the Zeeman effect actually dominates the signal, while the Hanle effect is just a perturbation. Fields of up to 800 G have by now been observed in these solar features.

Before we move into a more detailed description of the Hanle and Zeeman effects and the tools used to diagnose those effects in the observed polarized spectra, it is worth ending this section with a word on the spectral lines sensitive to the Hanle and Zeeman effects. In a perfect world one would like to choose the atomic lines that are both very sensitive to those effects and at the same time easy to compute. In particular with the Hanle effect, the computation of the quantum coherences makes it desirable a simple atom. Such a perfect world almost comes true in the case of the solar photosphere. The large amount of spectral lines from atoms like Fe or Ni allows the observer to pick the ones with the larger Landé factors (or at the appropriate wavelengths) but with a small dependence on thermodynamics or plasma velocities and flows. Prominences present almost the opposite scenario. From the relatively few lines emitted by prominences, only a handful are strong enough to be worth consideration for polarimetry (typical polarization levels are at 0.1 % of the intensity). In the absence of space instruments carrying polarimeters for UV lines, the constraint to the visible part of the spectrum leaves us with just four strong spectral lines and a few other less intense. The four spectral lines strong enough to top our list are Hα, $H\beta$, and the HeI lines at 1,083 nm and D_3. Other not so intense but still available lines are the CaII IR triplet at 854 nm, the Na D_1 and D_2 lines and the CaII H and K lines. This short list has been made out of the sole consideration of emission intensity, but are they sensitive to the Hanle effect? The H lines are actually too sensitive to the Hanle effect and to everything else. Because of this they are often avoided for its interpretation is difficult. For this reason, prominence magnetometry has focused from the very first works on the two He lines D_3 and its partner at 1,083 nm. The He atom is easy enough to model, sensitive to the Hanle and Zeeman effects, yet not very sensitive to many other perturbations that may hide the magnetic information carried by its polarized spectra. These are the lines of choice, on which the rest of this chapter will focus.

8.2 The Forward Problem

Neither the Hanle or the Zeeman effects offer a measurement over the polarized spectra that can be directly relied to the magnetic field. Looking into the examples of Fig. 8.1 one can at most relate the amplitude of the Stokes V profile of the Zeeman case to the longitudinal field. And this is about all that can be said. In the absence of useful proxies directly related to the magnetic field, its measurement becomes an inference process made of two steps: a forward problem in which we compute what polarized spectra will be emitted by a prominence with known magnetic field, temperature and other relevant physical parameters, and an inverse problem in which the observed polarized spectra is compared to the computed ones. A magnetic field is inferred from the observations when the comparison of the observed and computed polarized spectra results in a satisfactory match.

We address in this section the forward problem: the computation of polarized spectra emitted by prominence plasma in the presence of known magnetic field, temperature, pressure and other physically relevant parameters. The goal here is not to give all the details of this computation, but to illustrate the physical phenomena important to the emission of polarization in prominences and their relation with the magnetic fields. A comprehensive description of the state of the art in this forward problem is given by Landi Degl'Innocenti and Landolfi (2004).

We will illustrate the computation of polarization profiles with the case of the He atom and, in particular, the D_3 line. The starting point is necessarily a model for this atom. The usually computed model is schematically shown in Fig. 8.2.

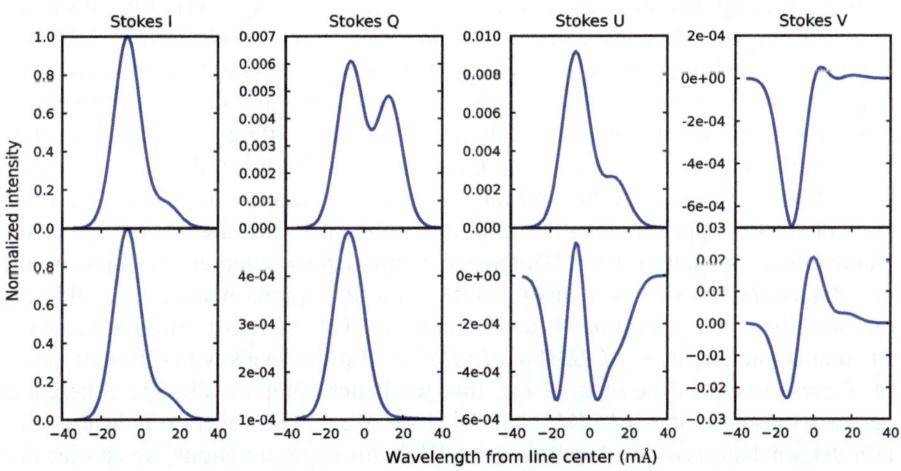

Fig. 8.1 Two examples of polarized profiles for the He D_3 line: on *top* for a field of 2 G we recover typical Hanle profiles, with strong linear polarization and a Stokes V profile small and looking like the linear polarization profiles; on *bottom* for a field of 500 G we are in the Zeeman regime, with a characteristic antisymmetric and large Stokes V profile and small and symmetric linear polarization profiles. The intensity profile is identical in both cases

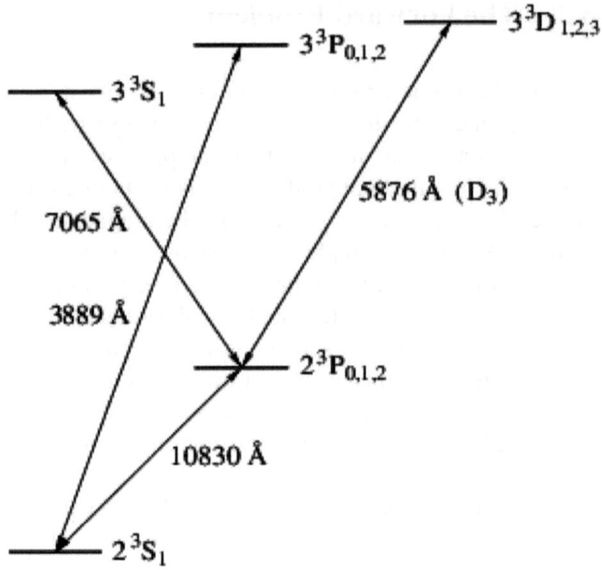

Fig. 8.2 Model for the triplet system of He, or ortho-helium. Levels and radiative transitions including the observed D_3 and 10,830 lines are indicated. Each level is labelled in the usual spectrographic convention specifying the multiplicity, the angular momentum L, and the total angular moemntum J

Each one of the levels in the atomic model is labelled by the usual spectroscopic notation, that we can easily translate into the quantum notation. Thus, of the three ^3D levels intervening in the emission of the D_3 line, the one with $J = 1$ will be written as $\|n = 3, L = 2, S = 1, J = 1, M >$, each Zeeman sublevel having a different value of M. In the absence of polarization, we only need to compute the populations of these levels. If we use the density matrix formalism, the populations are the diagonal entries of the density matrix ρ: $< nLSJM\|\rho\|nLSJM >$; and this for each one of the levels indicated in Fig. 8.2. The Zeeman effect only requires these populations, just specified for each one of the Zeeman sublevels. In the Zeeman regime, the magnetic sublevels have very different energies and there are no quantum coherences between them. As we said above, this is not the case in the Hanle regime. Here the splitting between the Zeeman sublevels is so small that coherences appear among the sublevels, coherences that are responsible of the Hanle effect we want to study. We have to compute those coherences which are the non-diagonal entries of the density matrix. At a first approximation we could only consider the coherences among the Zeeman sublevels with just different values of M, that is, the entries $< nLSJM\|\rho\|nLSJM' >$. But the levels with different values of J are so near in the case of He, that we better compute also the coherences between those levels: $< nLSJM\|\rho\|nLSJ'M' >$. We will however exclude any other non-diagonal terms of the density matrix. With this approximations, we can see that there are 423 density matrix terms to be computed to determine the atomic state of He before computing the radiative emissions.

To compute such a density matrix we will assume that it is in statistical equilibrium and that all transfers of population and coherences are due to radiative transitions between the levels due to absorption of photospheric radiation and emission (stimulated or not), and the effect of the magnetic field that will tend to diminish atomic coherences and to change the energy of the Zeeman sublevels, thus changing the conditions for the radiative transitions. The explicit form of the statistical equilibrium equation for the density matrix can be found in all detail in Landi Degl'Innocenti and Landolfi (2004); Landi Degl'Innocenti (1982); Bommier and Sahal-Brechot (1978). We exclude from our problem all kinds of collissional excitations (Bommier 1980). In consequence, to emit photons the He atom must first absorb photospheric radiation. We will also assume that the incoming photospheric light has a flat spectrum. Due to this approximation, the emitted photon has no correlation with the incoming photon: it is completely redistributed in frequency, what simplifies enormously the calculations as compared to coherent scattering or partial redistribution, in which the correlations between absorbed and emitted photons have also to be taken into account.

The basic process considered is scattering. All scattering processes result in polarization if the incoming scatteres are not isotropically distributed. This is the case of atoms in prominences illuminated by the photosphere a few thousand kilometers below. The higher the prominence, the more anisotropic is the photospheric radiation field, and the more polarized is the emitted light. This can be seen in Fig. 8.3 that shows the amount of emitted polarization in the two He lines of interest for different heights above the prominence. The actual amount of polarization depends on the atomic coherences between the different atomic levels, and thus the amount of emitted polarization is different for the two lines. This is linear polarization, in a plane which is perpendicular to the scattering plane. In the conditions of observation of prominences, this linear polarization is parallel to the nearest solar limb. It is common to redefine the Stokes parameter Q so that it is positive when parallel to the limb and therefore, at zero magnetic field, the scattering

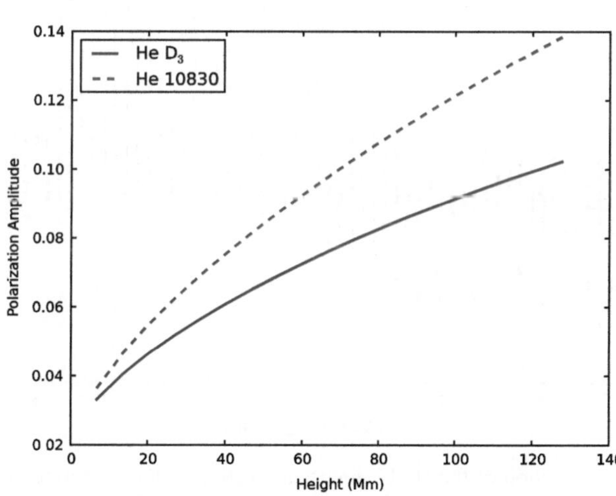

Fig. 8.3 Polarization at zero field and 90° scattering of the two He lines D_3 and 10,830 as a function of height above the photosphere

of photospheric light by the He atoms in the prominence results in a positive Q signal, as in Fig. 8.1.

The amount of Q polarization will also depend on the scattering angle. The continuous line of Fig. 8.4 shows the Q polarization expected from a prominence at 6 Mm above the photosphere with zero magnetic field at different scattering angles. A null scattering angle corresponds to forward scattering at disk center and rather than a prominence we are observing a filament. Usually beyond 60° of scattering angle we are already observing the prominence above the solar limb. The amount of polarization increases to a maximum at 90° scattering, when the prominence is observed exactly in the plane of the sky.

If there was no magnetic field in the prominence, this continuous line would be the only polarization we would observe. Magnetic fields change the atomic coherences of the atom and in consequence the amount of emitted polarization. The scattered dots in Fig. 8.4 show cases with random magnetic fields at different scattering angles. At 90° scattering, the effect of magnetic fields is to decrease coherences and diminish the amount of emitted polarization in Q, all the way to zero, while at the same time the signal in Stokes U, the orthogonal linear polarization, increases. This is the classical description of the Hanle effect: a diminution plus a rotation of the amount of linear polarization. However this is something that, for He, only happens at 90° scattering. At other angles Fig. 8.4 shows that in the presence of magnetic fields the atomic coherences can change in such ways that the

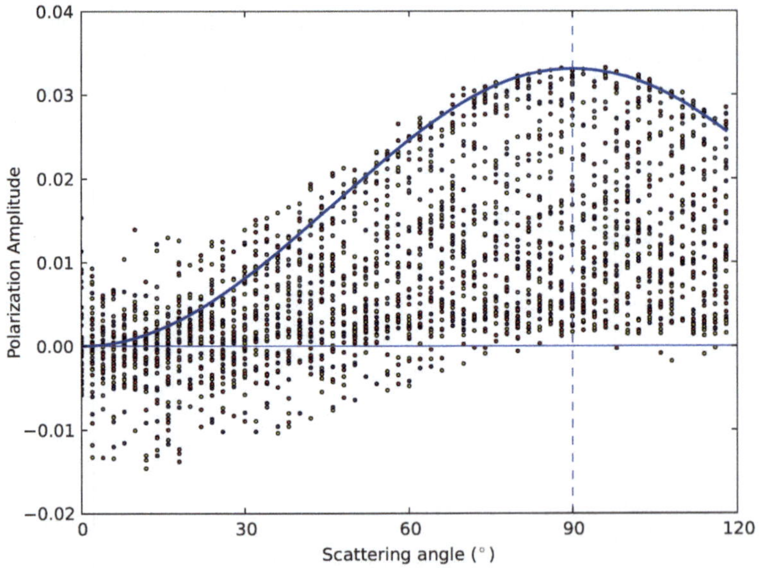

Fig. 8.4 Linear polarization of the He D_3 line as a function of the scattering angle for different magnetic fields in a prominence 6 Mm above the photosphere. A filament at disk center has zero scattering angle, while a prominence right over the limb is at 90° scattering. The continuous line is the expected polarization at zero magnetic field

amount of polarization is actually larger than the polarization at zero field, or on the contrary becomes even negative. The most extreme example is forward scattering in which there is no linear polarization at zero field, and a magnetic field introduces positive or negative polarizations. These are all quantum effects due to the changes and transfers in the atomic coherences that illustrate the importance of handling the problem of polarized line formation from the quantum point of view and avoid misleading classical approximations.

Figure 8.5 shows the expected polarization in Stokes Q and U as a function of the azimuth of the field for different field strengths (discontinuous lines) and as a function of field strength for different azimuths (continuous lines) at a given height and for fixed field inclination. They constitute the historic diagnostic tool for Hanle effect and they are a perfect illustration of the complex interplays between the atomic coherences which result in polarization signals depending on field strength and azimuth in complex manners, even with ambiguous points where the lines converge and cross. The diagnostic of magnetic fields with the Hanle effect requires inversion codes that can handle all this multi-dimensional quantum information appropriately.

Our last illustration concerns the circular polarization. Two different physical phenomena in the He atom are able to produce circular polarization. One is of course the Zeeman effect. The second one is, once again, the result of the intricated transfer of atomic coherences between the atomic levels. This will give us the excuse to introduce two important concepts in the computation of atomic coherences: alignment and orientation. Coherences were written at the beginning of this section as the non-diagonal entries of the density matrix, e.g. $\rho_{MM'} =< nLSJM\|\rho\|nLSJM' >$. We can examine the case of a level with $J = 1$ and $M = \pm 1, 0$ for illustration: the coherences inside that level that we need to compute are three: $\rho_{10} =< nLS11\|\rho\|nLS10 >$, $\rho_{-10} =< nLS1-1\|\rho\|nLS10 >$ and $\rho_{1-1} =< nLS11\|\rho\|nLS1-1 >$. The computation of these three coherences will depend on the direction and intensity of the field strength, and on the direction

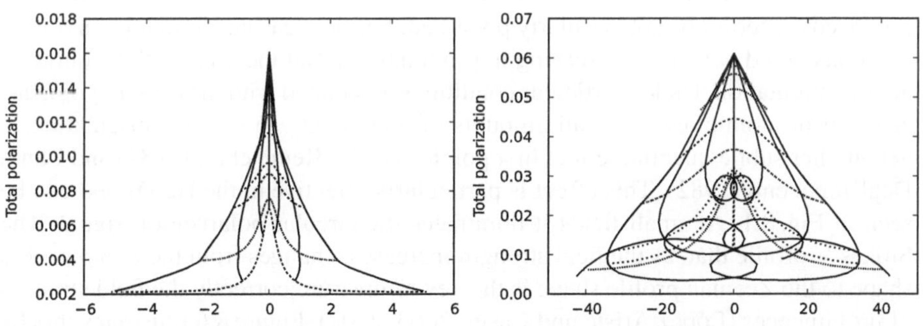

Fig. 8.5 Hanle diagram of the two components of the He D_3 line, the bigger blue component (*left*) and the smaller red component (*right*) that can be easily identified in the intensity spectra of Fig. 8.1. The diagrams represent total linear polarization vs. polarization plane angle for a series of varying azimuths at constant field strength (*dashed line*) or varying field strength at constant azimuth (*continuous lines*)

of the incoming photospheric light. The presence of two relevant directions in our problem[1] is clear indication that rotations are an important aspect of the technical details of that computation. To facilitate those rotations, it is important that the density matrix is written in a manner which is as rotationally symmetric as possible. The technical answer is to use the irreducible spherical representation of the density matrix. For any atomic level, we will introduce a tensor with indexes $K = 0, 1, 2, \ldots$ and $Q = -K \ldots 0 \ldots + K$. The index $K = 0$ has $Q = 0$ and it is a scalar: the population of the level. The index $K = 1$ has $Q = -1, 0, +1$ and is a vector called the orientation, while the index $K = 2$ has $Q = -2, -1, 0, +1, +2$ and is a second rank tensor called the alignment. The interest of this representation is that unlike the $\rho_{MM'}$ illustrated above, the new ρ^K_Q are zero most of the time, thus simplifying the description and the calculation of the problem. The $Q = 0$ alignment term is given in the particular case above by

$$\rho^2_0 = \sqrt{\frac{1}{6}}(\rho_{11} - 2\rho_{00} + \rho_{-1-1}).$$

It represents a population imbalance between the $M = \pm 1$ and the $M = 0$. It is directly connected to the emission of linear polarization and to the presence of anisotropy in the radiation field. Most of the changes in Q and U of Figs. 8.3, 8.4 and 8.5 are due to changes in the ρ^2_0 of the different atomic levels. On the other hand the $Q = 0$ orientation term is

$$\rho^1_0 = \sqrt{\frac{1}{2}}(\rho_{11} - \rho_{-1-1}).$$

This is an imbalance of population between the $M = 1$ and the $M = -1$ sublevels. The presence of atomic orientation results in the emission of circularly polarized light, even in the absence of Zeeman effect. But atomic orientation is almost permanently zero in He. The photospheric radiation can induce atomic orientation only if it is circularly polarized. But the continuum photospheric spectrum is not polarized or, certainly, not circularly polarized. For this reason, circular polarization was considered not worth observing in prominences. But there is another source of atomic orientation. Under particular conditions associated with the crossing or near-crossing of atomic levels, small amounts of alignment can become orientation in yet another subtle quantum effect first pointed out by Kemp et al. (1984) and Landi Degl'Innocenti (1982). This effect is particularly effective in the He D_3 and can be seen in Fig. 8.1. At small fields it dominates the circular polarization signal. The Stokes V profile that, as the field strength increases, changes from the *Kemp* profile shape to the Zeeman profile shape is the key to measure correctly the field strength of prominences (López Ariste and Casini 2003, 2002). Figure 8.6 illustrates this by

[1]Three, if we remember that at the end we are only interested in the polarized photons emitted in the direction of the line of sight.

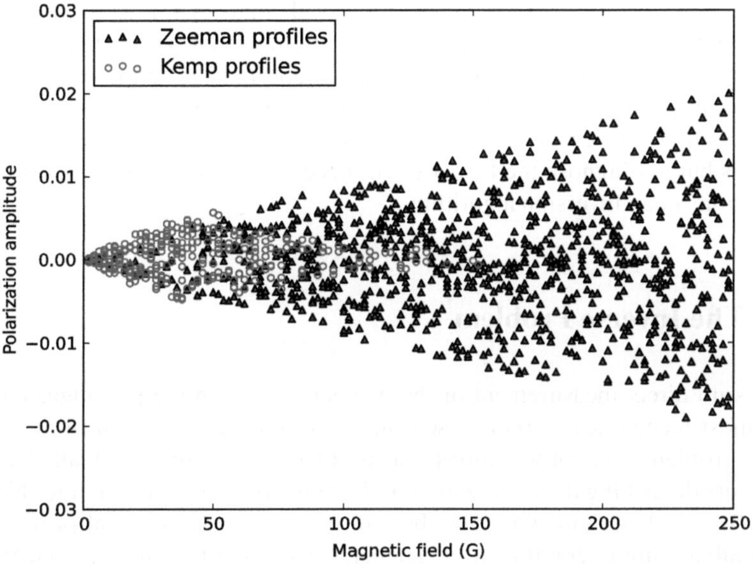

Fig. 8.6 Amplitude of the Stokes V profile of the He D_3 line in a prominence scenario for increasing field strength and random orientations. *Circles* are used when the profile is dominated by the Kemp mechanism of transfer of alignment into orientation, like those on *top* of Fig. 8.1; *triangles* mark profiles dominated by the antisymmetric Zeeman profile, as those on the *bottom* example of Fig. 8.1

plotting the amplitude of the Stokes V signal for random geometries and increasing field strength. Profiles dominated by the *Kemp* profile are marked with circles and dominate the weak fields. As the field increases, the Zeeman profiles take over and grow in amplitude while the Kemp profiles diminish their amplitude and disappear altogether.

The computation of the polarized spectra of the He lines in the presence of a known magnetic field requires the solution of a statistical equilibrium equation for the atomic density matrix. This has to be done using the quantum formalism, since quantum coherences and, in particular, alignment and orientation, are critical components of the emitted polarized spectra. The pure scattering process of photospheric light, highly anisotropic, on He atoms sitting a few Mm above the photosphere produces linearly polarized light whose amplitude depends on the height of the prominence and the scattering angle. Magnetic fields alter this picture by reducing and transferring atomic coherences among levels. The general effect is a reduction of the emitted linear polarization and a rotation of the polarization plane, but in reality everything can happen, from higher polarization amplitudes to negative polarizations in the presence of magnetic fields. Also due to these quantum subtleties we can even observe circular polarization not due to the Zeeman effect. While the proportion of the linear polarizations Q and U is often enough to determine the geometry of the field, the field strength depends mostly on the shape and amplitude of the circular polarization spectra: Kemp profiles for weak fields, Zeeman profiles for strong fields, and all kind of mixtures of Kemp and Zeeman in the middle. The richness of the scattering physics is such that is difficult to give rules of thumb or proxies of the magnetic field in terms of the observed polarization spectra. We have illustrated the many possibilities of just two He lines. We can compute those profiles at will by solving the quantum statistical equilibrium equations but to diagnose the field we will need to compare profile by profile with observations in an inversion algorithm

8.3 The Inverse Problem

There is no direct measurement of the magnetic fields in the prominence plasma. As in most techniques of remote sensing we rely on the solution of an ill-posed inverse problem. Our observations consist of spectral profiles of all the Stokes parameters describing the polarization of the emitted light in atomic lines like those of He D_3 and 1,083 nm. We know how to compute those spectral profiles under known albeit simple conditions of magnetic field, thermodynamics and radiative transfer. The previous section of this chapter is an introduction to this forward problem. But we do not know how to solve the inverse problem and deduce the magnetic, thermodynamic and radiative transfer conditions from those spectral profiles. This inverse problem is actually ill-posed. A well-posed problem, as defined by Hadamard, has a unique solution, and it changes continuously as we change the initial conditions. In our case we can be sure that there is a magnetic field in prominences and that He atoms emit the observed radiation in the presence of such magnetic field: a solution therefore exists. But it is certainly not unique. Even in the most simple model of a single atom in a single vector field, we know that the polarization profiles observed when the field is strictly vertical are exactly the same as those emitted in the zero-field case. Many other examples may be brought up. Some of them are referred to as ambiguities, and we shall come back to them later on. But even worse, one can imagine sets of atoms emitting inside the spatial and temporal resolution of our instruments but feeling a variety of spatial and temporal distributions of magnetic fields. The possibilities are infinity and cannot be distinguished: the solution is not unique in many cases. The problem of inferring the magnetic field in a prominence is a truly ill-posed problem.

Finding solutions for ill-posed problems requires the implicit or explicit assumption of regularization conditions. In a strawman parlance, what other information, beyond the observed spectral profiles, can we add that helps minimizing the number of solutions. Some of those regularizing conditions are trivial: we expect the spectral profiles to be continuous in wavelength, and the intensity is always a positive quantity. Some others are not trivial but can be accepted as true in all cases: we expect that the He atoms can be described in LS-coupling and that the Hanle and Zeeman effects are the important phenomena describing the sensitivity of these atoms and the emitted radiation to the magnetic field. We expect Hanle effect to dominate in most cases, and the atoms are polarized predominantly by anisotropic photospheric radiation, therefore the main polarization expected is linear polarization in a direction perpendicular to the local vertical (that joins the prominence to the photosphere below) and, when observed off the solar limb, parallel to the visible limb. Finally some regularization conditions are simplifying assumptions: we assume that inside our resolution element, both spatial and temporal, there is a single magnetic field vector, we neglect any radiative transfer that may change the anisotropy of the radiation field. These last set of conditions are the ones that future developments on this field should try to overcome.

In the previous section we clarified under what conditions we could solve the forward problem. Under those conditions we can compute, given the magnetic field and the thermodynamic conditions in a prominence, a set of spectral profiles for all the Stokes polarization parameters: the intensity I, the linear polarizations Q and U, and the circular polarization V. Our path to a solution will be now to compare this computed profile with the observed one. Before going into the technical meaning of *compare*, we should ensure that the computed profiles look like those provided by our instruments. The first part of this task is easy: our spectrographs have a predefined spectral resolution and spectral sampling, also a point spread function. We have to compute our spectra with the same spectral sampling and convolve them with the point spread function to obtain comparable profiles. The definitions of the directions of linear polarization and sign of circular polarization should be the same or we should know how to transform (rotate) one definition into the other. There is however one crucial aspect of the comparison that is not trivial: how many photons are coming from the prominence and how many photons do our code predict. It would appear that one needs an absolute photometric calibration of the instrument on the observational side and to solve the full statistical equilibrium of the atom (including in our case, populations of all He states, ionization and recombination rates) on the computational side to answer those questions. This is very unfortunate because, on the observational side, we cannot afford spatial, temporal and spectral resolutions simultaneously with photometric accuracies of better than 0.1 % while the expected polarization signals are rather on the level of 0.01 % of the emitted light. On the computational side, the solution to the full radiative transfer is feasible but takes the forward problem to new heights of complexity well beyond our description of the previous section. It would seem that the problem of measuring magnetic fields in prominences is doomed before the start.

The coincidence of two unrelated effects solves this conundrum. On the observational side, it is true that one cannot afford to measure absolute polarization to better than 0.1 % in acceptable observational conditions for prominences, but one can measure relative polarimetry down to 0.001 %. Thus, we may not know the actual amount of photons detected, but we know the proportion of those photons that are polarized with great accuracy, up to an unknown zero offset. The zero offset of polarization can be easily solved observationally by forcing the light observed outside of the He lines (or whatever other line emitted at the prominence) to have zero polarization. The light seen off the line may be due to the solar continuum if we are observing inside the solar disk, or just scattered light from this very same solar continuum when observing off disk. Relative polarimetry is what observations will provide with high precision.

On the side of the computation of the spectral profiles, we discover that all the quantities are proportional to $\mathcal{N}$ the total population in the atomic system of interest, for example in the triplet system of He of Fig. 8.2. Computing the value of this total population would require solving the complete statistical equilibrium of the atom. But we can also put $\mathcal{N} = 1$ and consider that all our computations are in terms of relative populations respect to the unspecified total population. And this has no impact whatsoever in the sensitivity of the emitted polarization to the magnetic field. It appears therefore that both, the observational and the computational constraints coincide in that it is relative polarimetry, as opposed to absolute, that contains the important information on the magnetic field of prominences. The first important step in our comparison of observed and computed profiles will be to scale all the profiles, both computed and observed, so that the maximum of the observed and computed intensities is 1.

8.3.1 Fitting Spectral Profiles: Dimensionality, Noise and Redundancy

The preparation of the computed profiles in terms of spectral resolution, sampling and Stokes parameter definitions, and the scaling of the computed and observed profiles to set the maximum of the observed intensity to 1, sets the path to the comparison. The purpose of this comparison is obviously not to say whether they are alike or not, but to propose the magnetic field that makes them equal. The discussion here does not differ from other inversion problems, and the questions are similar.

How many free parameters are in our model? By model we mean the ensemble of conditions under which we solve the forward problem. As presented in the previous section it is clear that the mininum, the simplest, scenario considers the following as free parameters: the magnetic field vector (three parameters), the temperature that broadens the profiles (one parameter), the height of the emitting atom above the photosphere to compute the radiation anisotropy (one parameter) and the scattering angle of the photon (one parameter). To these we can easily add a Doppler-shifting velocity (one parameter) to a total of seven free parameters in our simplest model.

How many observables are in our spectral profiles? It is important to realise that there should be at least as many observables as parameters if we want to ensure either the existence or the unicity of the solution from the present observables. A common joke among inversion experts says that often in a problem *the solution does not exist and, furthermore, it is not unique.* Using a model with more free parameters than observables may result in overfitting. The classical example is the fit of three points in the plane: a polynomial of order 3 is the minimum requirement, but any other polynomial of higher order will also fit the points. Unless other information is available, one usually concludes that the higher order polynomials have too many free parameters. So, how many observables there are in our data? More or less than the seven free parameters of our model? We are going to illustrate this question with the particular case of the He D_3 line. This line is made of four transitions, three of which fall roughly at the same wavelength while the fourth one falls 150 m Å away at larger wavelengths. In the absence of any radiative transfer, an explicit regularization constraint of our model, the ratio of intensities of the combined three transitions in the blue respect to the isolated fourth in the red is exactly eight. Strictly speaking this ratio may change slightly with the magnetic field, but it is a second order effect that we neglect at this point (though it is included in the formulae of the previous section). The intensity of D_3 contains therefore only two observables: the position of the line (the Doppler shift) and the width of the profile (its broadening that can be interpreted as temperature). Independently of how many wavelength samplings we took with our instrument over the line, there are only two observables in the intensity. In linear polarization, we also have the position and the width of the profiles, but to this we have to add the signed amplitude of the blue component and the ratio of the blue to the red component, that now depends on the magnetic field and the scattering conditions. These are two more observables for Stokes Q, and two more for Stokes U. We are at six. For many years these were the only observables measured (Leroy 1977, 1978; Leroy et al. 1984, 1977; Leroy 1981; Bommier et al. 1994). Two of them (the broadening and the wavelength position) are almost directly related to two free parameters (velocity and temperature) of the model. We are left with four observables for five free parameters. This was one of those dangerous situations that can be compared to trying to fit four points in a plane with a polynomial of fifth degree. The observation of Stokes V, the circular polarization, becomes a *sine qua non* requirement. We saw in Sect. 2.4 that our forward model suggests two sources of circular polarization: the Zeeman effect, and the transfer of atomic alignment into orientation. Therefore two more observables appear: The amplitude of the Zeeman effect and the ratio between the two mechanisms. These adds up to eight potential observables in the He D_3 line. Sometimes some of the signals may be zero or below the noise level, reducing the number of observables. But in general the He D_3 line contains eight observables to constraint the seven free parameters of our model. This favourable situation may be compared to that of the other observed line of the He triplet system, the 1,083 nm line. This line has three transitions also arranged in a blue and a red components with fixed intensity ratios in the absence of radiative transfer. Contrary to the case of the He D_3 line, the isolated transition (which lies in the blue side of the line this time) has no linear polarization in the absence of optical opacity, and at the magnetic fields of interest there is no transfer of atomic alignment into orientation. This reduces the number of observables from eight to just five, even when Stokes V is observed, clearly insufficient to constraint the seven free parameters of our simple model, unless other information is added (measured height over the limb, for example).

The accounting of free parameters and observables has been presented in a straightforward and simplistic manner. From the description of the previous paragraph one could conclude that it is sufficient to measure the several amplitudes involved and not the full spectral profiles. This is however not the case because of the presence of noise. Noise is a necessary part of any measurement. It sets the sensitivity levels of a measurement, the minimum amplitude of the errors and signals the presence of biases or parasitic signals. Observing noise is as important as the signal itself. Coming to our present problem, the measurement of a Stokes V profile made out of noise, may seem like a waste of time and photons, but it actually provides almost as much information as if an actual signal had been measured: the solution model must provide a computed Stokes V profile with amplitude smaller than the noise level. This is actually a strong constraint that may eliminate many other solutions compatible with the measured linear polarization. But noise also has an obvious downside: it obliterates the information: the amplitude of the linear polarization will only be known with a certain precision, with a standard deviation at least comparable to the noise level, unless it is repeated several times, unless there is some degree of redundancy in the data. This is a good argument for measuring the full spectral profiles of the He lines emitted in prominences, rather than just the amplitudes: to provide enough redundancy to conquer noise. If, let us say, ten points form the observed profile of the blue component of the He D_3 line and we know that the linear polarization of this blue component is a gaussian of position and width known but whose amplitude we want to measure, the ten points provide constraints to this amplitude, and not just the point in the top of the profile. Usual instruments present noises with gaussian probability distributions. When measuring spectral profiles of atomic lines, observers will try to be limited by the photon noise. This is a multiplicative noise that, for few photons, follows a Poisson probability distribution, but for large number of photons converges to a Gaussian probability distribution with standard deviation $\sqrt{N_{ph}}$ the square root of the number of photons N_{ph} measured. This will set our precision when measuring the amplitude of linear polarization in our example. Unless the ten points accross the spectral line are used what increases the number of measured photons. If the line had a square profile the improvement in the precisionof our measurement would be a straight $\sqrt{10}$ factor, but since commonly the spectral profile of a line is a gaussian, the improvement is of just a $\sqrt{2}$ factor, over 40 %.

Redundancy, in terms of the measurement of full spectral profiles in our problem, is a requirement to downplay the nasty influence of noise. Noise that nevertheless is an important part of our measurements, as a constraint to signal amplitudes or as a determination of sensitivity levels. Noise has to be measured and it should be accompanied by a certain degree of redundancy in our measurements.

8.3.2 *Inversion Techniques: Pattern Recognition and Least Squares Fitting*

Our observations have a sufficiently large amount of observables, and a degree of redundancy to compensate for the noise. Our model, with at least seven free parameters, produces spectral profiles of polarization emitted in an appropriate atomic line like those of the He triplet system. The polarization has been scaled to the intensity both in the computation and in the observations, and the spectral resolution and sampling have been made to coincide between both sets of profiles. We are ready to perform the comparison.

By *comparison* in inverse problems we should understand the choice of a merit function and the computation of that merit function with the observed and computed profiles. The more similar the two profiles, the nearer the merit function is to one of its extremes, usually a minimum of the function, often a minimum with value zero. There are many such merit functions in the literature of inverse problems with well-known relations among them and a list of advantages and disadvantages, usually related to the regularization constraints imposed to the problem. But in the actual problem of prominence magnetometry through spectropolarimetry, just two algorithms have been used. So we limit ourselves to describe those two.

The first inversion method used in prominence magnetometry after the use of the Hanle diagrams was Principal Component Analysis (PCA). The PCA method starts by computing a basis, in the algebraic sense, of the expected profiles. This basis is usually computed by determining the eigenvectors of the correlation matrix of a large set of spectral profiles computed under random parameters. The set is supposed to contain information on the main spectral features, their range of variation and their frequency of appearance. The three types of information will be included in the eigenvectors of the correlation matrix. This eigenvectors look like generalized profiles and are usually referred to as *eigenprofiles*. They are also optimal in the sense that any observed or computed profile can be written as a linear combination of just a few of these eigenprofiles. Usually the number of eigenprofiles needed is just a mere factor 2 or 3 the number of observables found in the spectral profiles. In the example above, where we computed the number of observables in the Stokes profiles of the He D_3 line, we find that 4 to 6 eigenprofiles suffice to reproduce all the observed profiles for every polarization. The coefficients of the linear combination contain all the information, the observables, in a compressed form. Through projection onto the eigenprofiles, PCA takes full advantage of the redundancy in the observations to acomplish an effective denoising and compression of the observables.

This feat is then used to generate a database of different cases computed under known conditions of magnetic field and rest of model parameters. In our illustrations each entry of the database will contain a set of seven numbers for the seven parameters of the model and a set of 4×4 coefficients of the linear combination of the computed profile in the four sets of eigenprofiles, one set per Stokes parameter. PCA also provides us with a metric for the algebraic space spanned by the eigenprofiles: this space is Euclidean. Our merit function is therefore the Euclidean distance in terms of the coordinates given by the PCA coefficients between the observed profile and each one of the entries of the database. Since the dimension of the space is just 4×4, this distance is easily computed. The case in the database which lies nearest to the observed profile as measured by this PCA distance is the solution to the problem. Ideally this distance would have to be zero in the absence of noise. One actually requires that it is of the order of the measured noise to be accepted as solution. In practice one just picks the nearest one and keeps the PCA distance as a measurement of the validity of the solution. If the database is comprehensive and all possible cases allowed by the model are present, the observations are either compatible with one case already present in the database or they cannot be explained by the model. Thus the PCA distance informs us also of the ability of our model to explain the observations and in what cases our model is not enough and must be made more complex.

But, how do we know that our database is comprehensive? A too large database may make the comparison intractable, while a too small one may leave out valid models. A solution is to fill the database through Montecarlo techniques for a given noise level typical of the observations. Model parameters are selected randomly and a profile is computed then projected onto the eigenprofiles and its PCA distance to all models in the database is measured. If this distance is smaller than the noise level, the model is considered to be already represented in the database. In the other case the model is included in the database and the process restarted with a new random model. This process may be slow, but it has to be done only once, since the database can then be used for multiple observations. In practice it is convenient to fill the database with a noise smaller than the expected one, so several models are found in the database at PCA distances smaller than the expected noise level. This opens the possibility to retrieve a list with all the models compatible up to noise, and not just the nearest one. With such a list one can then easily compute error bars for the solution model parameters from the distribution of parameters among the models in the list.

The PCA code was the first inversion algorithm used in prominence magnetometry and allowed the construction of the first maps of magnetic field vectors in prominences (Casini et al. 2003). However it has a serious drawback: if a solution is not in the database, it will not be retrieved by the inversion. The number of solutions is finite and discrete.

A solution to this problem was given by the second kind of inversion codes used in prominence magnetometry: the more common Least-Squares fitting codes. In these codes, the actual observed profiles, wavelength sample by wavelength sample, are used to compute the merite function just as the quadratic difference between the observed and the computed profile, weighted by the noise variance. This square norm, so similar to the PCA Euclidean distance, is a traditional merit function for most regression and inversion problems, and requires not much more description. These codes then use this quadratic distance to propose a change in the model parameters from which to compute a new profile to be compared. If the proposed model change goes in the right direction the iterative process converges rapidly into a model that minimizes the quadratic distance with the observations.

The connection between the quadratic distance and the modification of the model is often made through a Marquardt-Levemberg algorithm, where the derivative of the profiles respect to a change in every one of the model parameters is computed numerically, given the complexity of the forward problem. The Marquardt-Levemberg algorithm allows for a continuous change of the model parameters as long as they improve the quadratic difference between the observed and the computed profiles. A much better fit, and better precision in the final model is expected from this kind of codes than from the PCA algorithm described above. The trade-off's are first speed (including the numerical computation of the derivatives, the forward problem has to be solved at least 20 times per iteration, and convergence is attained after tens of iterations in average) and second the ever-present risk of falling into local minima that look like a solution to which the algorithm converges. Of course many recipes are known to minimize these drawbacks, but yet they can be seen as the disadvantages respect to the PCA algorithm. Presently, the most important least-squares algorithm for the inversion of polarization profiles of the He lines in prominences is Hazel (Asensio Ramos et al. 2008). One should also mention the code Helix (Lagg et al. 2004) that has been used in active region filaments and that only computes the Zeeman-effect and thus is limited to the relatively strong fields one finds in active region filaments.

8.4 The Measurement of Prominence Magnetic Fields at Present

The historic measurements of magnetic fields inprominences made use of most of the elements described in the second section of this chapter concerning the theory of atomic coherences and the Hanle effect. They missed however the interest of observing the circular polarization signal with both the Zeeman and the Kemp signatures. They also lacked the resource of inversion codes and had therefore to rely on Hanle diagrams. Despite of those drawbacks they established the main basic facts of magnetic fields in prominences: first, the field is mostly horizontal, second its field strength varies in average from a few Gauss to about 30 G, third the azimuth of the field makes a small angle with the main axis of the filament (Bommier et al. 1994). Present observations of polarization in prominences always include the circular polarization. As illustrated in Fig. 8.6, circular polarization allows to break the upper bound in field strengths that the signals of Q and U have, in what is called *Hanle saturation* and that can be seen both in the Hanle diagrams of Fig. 8.5 and in Fig. 8.4 as a concentration of lines or dots at particular places for high fields. By using the information from circular polarization Paletou et al. (2001) and Casini et al. (2003) were able to show the presence of strong fields, above 40 G, and up to 80 or 90 G in quiescent prominences. The work of Kuckein et al. (2009, 2012) and Sasso et al. (2014, 2011) shows that in active region filaments fields are even stronger, reaching hundreds of G. The second advantage of today's measurements is that they indefectively rely on inversion codes. Other than making the process of measurement automatic and fast, what has resulted in maps of prominence magnetic fields rather than just isolated points (Fig. 8.7), the use of inversion codes allows a careful analysis of errors and biases in the measurements. Figure 8.8 illustrates the importance of this on the critical subject of the inclination of the magnetic fields in prominences. The two lines (dotted and continuous) represent distributions of

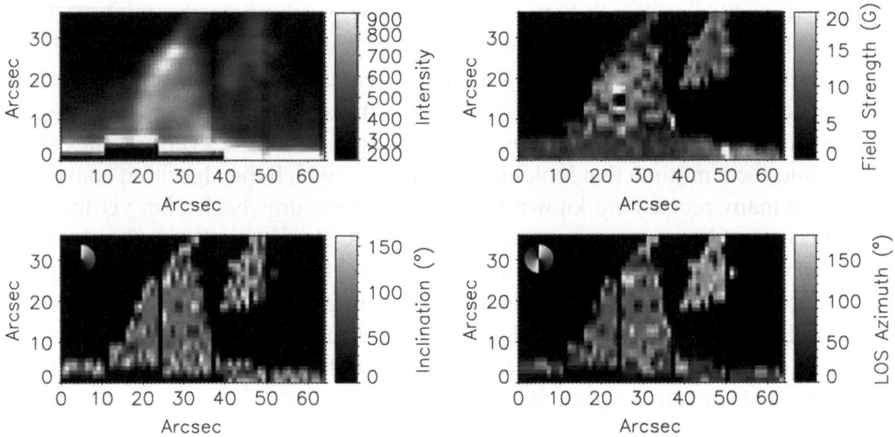

Fig. 8.7 Example of measurement of the magnetic field vector over a prominence from spectropolarimetric data of the He D_3 line, measured at THEMIS in May 2014, and inverted using a PCA code

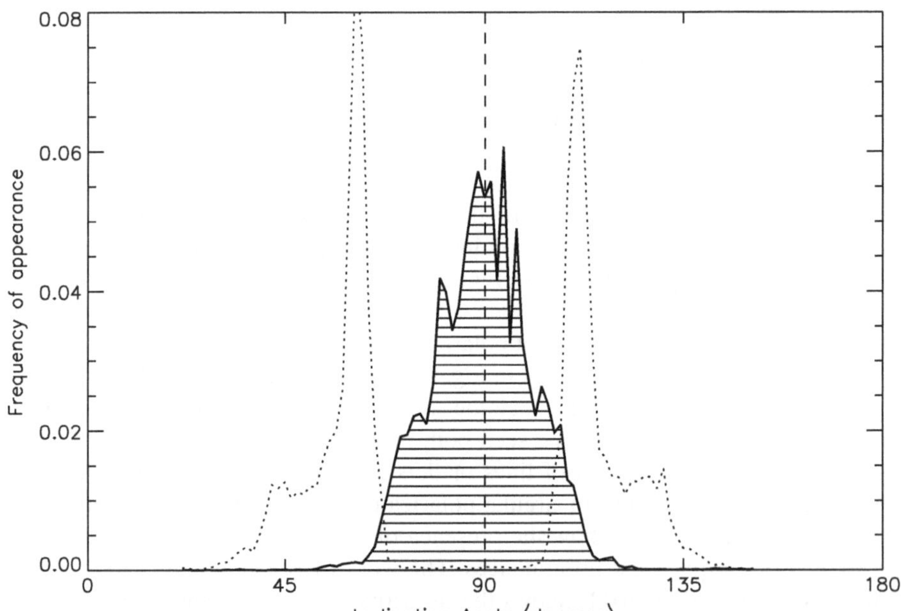

Fig. 8.8 Distribution of magnetic field inclinations in over 3,000 measurements of 58 prominences observed with THEMIS in June 2014. The shaded region, centered around an inclination of 90° or horizontal fields, corresponds to well inverted profiles, with errors below 10°. The *dotted line* corresponds to the measurements with errors larger than 30°

magnetic field inclinations measured over 58 prominences in May and June 2014 by the telescope THEMIS using the He D_3 line. Taken together one sees three characteristic peaks at roughly 60°, 90° and 120° respect to the local vertical. Horizontal fields cluster around the 90° value, but we also see two important peaks at non-horizontal fields. Without any further information one would conclude that altogether with the horizontal field, inclined fields are present and must be important in the dynamics of prominences. However there is an important difference between both distributions, one that only appears when error bars, provided by the inversion code, are examined. The distribution plotted in a continuous line corresponds to measurements with a standard deviation smaller than 10°. López Ariste and Casini (2002) and Casini et al. (2009) determined that 10° is the inherent error in the inclination of the magnetic field given the usual signal-to-noise ratios of present solar spectropolarimeters. The continuous line distribution is made of those measurements that the inversion code considers reliable measurements. It shows in average horizontal fields.

The dotted line, on the other hand, corresponds to those measurements with error bars larger than 30°. Those are not reliable measurements: a measurement from that set of data with for example 60° plus or minus 30° can be anything from vertical to horizontal. The inversion code determines that those are not reliable measurements. Having interpreted them as an indication of the presence of non-horizontal fields in prominences would have been a great mistake. Recent work (Schmieder et al. 2014) appears to suggest that many of those may actually correspond to unresolved fields made of a macroscopic horizontal field plus an unresolved turbulent field. The addition of both components could explain the observed profiles although it is to be demonstrated that such model is robust for inversion.

8.5 The Future: Ambiguities, Sources of Error and Incertitudes

This chapter was intended partly as an introduction to the methods and techniques of measuring the magnetic field in prominences through polarimetry. The two previous sections focus on those methods. But for many potential readers there may be not so much interest in knowing the details but on assessing the confidence of a measurement. In this last section we address this point: given a measurement of the magnetic field vector in a prominence, are there other solutions possible due to ambiguities? what are the error bars? what improvements can be expected in the future?

Since the relations of the observed polarization signals with the magnetic field are far from linear it is difficult to summarize errors with just a few numbers. At low fields, the Stokes Q linear polarization has a maximum amplitude and can be measured with higher precision than average fields. But if the fields grow strong, as in active region filaments, then Stokes V will show a strong Zeeman signature which will allow a precise measurement of the longitudinal field. Thus the error does not just depend on the noise, as it often happens in the measurement of photospheric magnetic fields, but also on the actual magnetic configuration that is being measured.

In spite of this some characteristic numbers have been provided by López Ariste and Casini (2002) and Casini et al. (2009) as a means to assess the confidence on the codes. In average, with typical signal to noise ratios of present instruments of 10^{-4} to 10^{-3} the magnetic field strength can be measured to better than 10 G almost always and the inclination and azimuth angles to better than 10°.

Although any improvement in the signal-to-noise ratios provided by instruments will always be welcomed, it is a common conclusion that those errors are mostly due to the simplicity of our models, and not to the noise in the observed spectra. However, before considering what improvements can be foreseen in those models it is necessary to list the ambiguities in the solutions.

The shared reason for these ambiguities is the convoluted relation between the polarization profiles and the magnetic field. It is not uncommon that two completely different magnetic configurations result in exactly the same polarization profiles and cannot be told apart. There are various situations in which this can happen but we will describe shortly the two more frequent and better known, both related to the geometry of the magnetic field: the 180° and the 90° ambiguities.

The best known of these two ambiguities is the 180° or azimuth ambiguity, intrinsic to any polarization measurement. Since we measure intensities, a quadratic quantity of the oscillating electric field of the light beam, we cannot tell apart situations which rotate this electric field by 180°. In the case of our measurements of prominence magnetic fields, as in the case of the photospheric measurements, it is the azimuth of the magnetic field in the plane of the sky that rotates the plane of the electric field of the radiation and hence we cannot distinguish two magnetic fields whose azimuth in the plane of the sky is 180° apart, all other parameters being identical. This azimuth ambiguity can be solved by considering other information, as measurements at other times under other lines of sight or from considerations on the shape and form of the prominence plasma, or the filament channel. Whatever the method, it is stranger to the magnetometry techniques and it is therefore a safe attitude to remember that all azimuths can change direction and result in the same observations.

The second ambiguity is the 90° ambiguity. It is strictly related to the scattering process: measurements using only the Zeeman effect are not affected by it. Its origin can be traced to a factor $3\cos^2\alpha - 1$ appearing in the expression for the alignment ρ_0^2 of Sect. 8.2. When $\alpha = 54.7°$ that tensor component is zero and there is no induced alignment in the atom by the radiation field. This particular angle is called the Van Vleck angle and results in a zero of polarization independent of the magnetic field or other parameters. On either side of the Van Vleck angle the signals tend to be symmetric what results in ambiguous situations.

When this Van Vleck angle is written in terms of the magnetic field geometry it is possible to see that, given certain conditions, changing the inclination of the magnetic field by 90° changes the value of α to a symmetric place beyond the Van Vleck angle and the polarization signals are identical, resulting in a second ambiguous solution. The ambiguity is not general. Not in all cases a change of 90° in the inclination results in an ambiguous solution. Landi Degl'Innocenti (1982) shows a table of cases as a function of the other angles in the model and López Ariste and Casini (2005) studied the space of parameters for regions when this ambiguity appeared or not. If it was a general ambiguity it would have been difficult to distinguish horizontal from vertical fields in prominences. In the face of important questions as whether the field is horizontal or vertical in prominences, ensuring that the 90° ambiguity is not at work may be fundamental. In this sense the He D_3 line, with its two polarizable components is less sensitive to this ambiguity than the He 1,083 nm line (Casini et al. 2009). Observing both of these lines reduces even further the risk of soluctions affected by this ambiguity. In general nevertheless all measurements of magnetic fields in prominences have to be accompanied by an assessment of the impact of this ambiguity in the results.

As said above, both these ambiguities are intrinsic to the process of generation of polarization in the emitted light. There is nothing that can be done with our methods to solve them. More information is needed: either from other lines with a different sensitivity to the magnetic fields, either from observations at different times with different perspective angles, either from considerations of the most probable magnetic configuration given our knowledge of the prominence topology.

Future improvements in the measurement of prominence magnetic fields will not come from the solution of the ambiguities but from the ability to drop some of the approximations taken at this point and described in Sect. 8.2. The first step is the inclusion of radiative transfer in the models. Some recent codes as Hazel (Asensio Ramos et al. 2008) include already the absorption of light along the line-of-sight. The atomic state along the ray path is not yet modified by this, but still defined by illumination exclusively from the photosphere. Yet one can compute both an emissivity, used in our model, and an absoprtion term. This last one results in a wavelength-dependent and polarization-dependent opacity that can be integrated along the line of sight. One immediate result of this is that the ratio of intensities between the blue and red components of either one the He D_3 or the 1,083 nm lines is not any longer 8, the ratio that results from the line strengths computations, but smaller. The larger component (the blue in the case of D_3, the red in the case of the 1,083 nm line) has a larger opacity and is farther absorbed along the line of sight, resulting in a smaller amplitude. This comforts the observations that see a distribution of ratios that, in the case of the D_3 line, peaks around 6.

But inclusion of radiative transfer should not stop at taking this extra opacity into account. As the radiation field in the prominence is not exclusively determined by the photosphere but includes a contribution from the light emitted in the prominence itself we should expect the anisotropy of the radiation field to change. As it dictates the amount of alignment in the atom, the alignment and the emitted polarization will also change. Let us suppose for a while that prominences are spherical blows of plasma. The contribution to the radiation field from the light emitted by such a prominence would be an isotropic term. It would not contribute to the alignment but only to the total intensity. A smaller radiation anisotropy translates into a smaller atomic alignment and in a smaller polarization amplitude in the emitted light. Taking into consideration the radiation emitted by the prominence itself will result in a depolarization of the spectral lines. A similar case has been modelled and studied in the case of resonance polarization in the atmosphere of the planet Mercury (López Ariste et al. 2012) but its impact on the measurement of prominence magnetic fields is yet to be determined.

Of course prominences are not spherical. Eventually a correct consideration of the light emitted by the prominence itself and its impact in the emitted polarization and in the determination of the magnetic field will have to consider the actual shape of each prominence: is it made of optically-thin threads of plasma? Are we seeing those threads along their length or accross but with many piling up along the line of sight? Our models for the measurement of the magnetic field will have to include these considerations in the near future.

References

Asensio Ramos, A., Trujillo Bueno, J., & Landi Degl'Innocenti, E. (2008, August). Advanced forward modeling and inversion of stokes profiles resulting from the joint action of the hanle and zeeman effects. *The Astrophysical Journal, 683*, 542–565.

Bommier, V. (1980, July). Quantum theory of the Hanle effect. II - Effect of level-crossings and anti-level-crossings on the polarization of the D3 helium line of solar prominences. *Astronomy and Astrophysics, 87*, 109–120.

Bommier, V., Landi Degl'Innocenti, E., Leroy, J.-L., & Sahal-Brechot, S. (1994, October). Complete determination of the magnetic field vector and of the electron density in 14 prominences from linear polarization measurements in the HeI D3 and H-alpha lines. *Solar Physics, 154*, 231–260.

Bommier, V., & Sahal-Brechot, S. (1978, September). Quantum theory of the Hanle effect - Calculations of the Stokes parameters of the D3 helium line for quiescent prominences. *Astronomy and Astrophysics, 69*, 57–64.

Brown, A., López Ariste, A., & Casini, R. (2003, August). The effect of atomic polarization on early estimates of prominence magnetic field strengths. *Solar Physics, 215*, 295–305.

Casini, R., López Ariste, A., Paletou, F., & Léger, L. (2009, September). Multi-Line Stokes Inversion for Prominence Magnetic-Field Diagnostics. *The Astrophysical Journal, 703*, 114–120.

Casini, R., López Ariste, A., Tomczyk, S., & Lites, B. W. (2003, November). Magnetic Maps of Prominences from Full Stokes Analysis of the He I D3 Line. *The Astrophysical Journal Letters, 598*, L67–L70.

Kemp, J. C., Macek, J. H., and Nehring, F. W. (1984, March). Induced atomic orientation, an efficient mechanism for magnetic circular polarization. *The Astrophysical Journal, 278*, 863–873.

Kuckein, C., Centeno, R., Martínez Pillet, V., Casini, R., Manso Sainz, R., & Shimizu, T. (2009, July). Magnetic field strength of active region filaments. *Astronomy and Astrophysics, 501*, 1113–1121.

Kuckein, C., Martínez Pillet, V., & Centeno, R. (2012, March). An active region filament studied simultaneously in the chromosphere and photosphere. I. Magnetic structure. *Astronomy and Astrophysics, 539*, 131.

Lagg, A., Woch, J., Krupp, N., & Solanki, S. K. (2004, February). Retrieval of the full magnetic vector with the He I multiplet at 1083 nm. Maps of an emerging flux region. *Astronomy and Astrophysics, 414*, 1109–1120.

Landi Degl'Innocenti, E. (1982, August). The determination of vector magnetic fields in prominences from the observations of the Stokes profiles in the D3 line of helium. *Solar Physics, 79*, 291–322.

Landi Degl'Innocenti, E., & Landolfi, M. (2004, August). In *Polarization in spectral lines* (vol. 307). Netherlands: Springer.

Leroy, J. L. (1977, August). On the intensity of magnetic field in quiescent prominences. *Astronomy and Astrophysics, 60*, 79–84.

Leroy, J. L. (1978, March). On the orientation of magnetic fields in quiescent prominences. *Astronomy and Astrophysics 64* (Mar. 1978), 247–252.

Leroy, J.-L. (1981, June). Simultaneous measurement of the polarization in H-alpha and D3 prominence emissions. *Solar Physics, 71*, 285–297.

Leroy, J. L., Bommier, V., & Sahal-Brechot, S. (1984, February). New data on the magnetic structure of quiescent prominences. *Astronomy and Astrophysics, 131*, 33–44.

Leroy, J. L., Ratier, G., & Bommier, V. (1977, February). The polarization of the D3 emission line in prominences. *Astronomy and Astrophysics, 54*, 811–816.

López Ariste, A., & Casini, R. (2002, August). Magnetic fields in prominences: Inversion techniques for spectropolarimetric data of the He I D3 line. *The Astrophysical Journal, 575*, 529–541.

López Ariste, A., & Casini, R. (2003, January). Improved estimate of the magnetic field in a prominence. *The Astrophysical Journal Letters, 582*, L51–L54.

López Ariste, A., & Casini, R. (2005, June). Inference of the magnetic field in spicules from spectropolarimetry of He I D3. *Astronomy and Astrophysics, 436*, 325–331.

López Ariste, A., Leblanc, F., Casini, R., Manso Sainz, R., Gelly, B., & Le Men, C. (2012, August). Resonance scattering polarization in the magnetosphere of Mercury. *Icarus, 220*, 1104–1111.

Paletou, F., López Ariste, A., Bommier, V., & Semel, M. (2001, August). Full-Stokes spectropolarimetry of solar prominences. *Astronomy and Astrophysics, 375*, L39–L42.

Sasso, C., Lagg, A., & Solanki, S. K. (2011, February). Multicomponent He I 10 830 Å profiles in an active filament. *Astronomy and Astrophysics, 526*, 42.

Sasso, C., Lagg, A., & Solanki, S. K. (2014, January). Magnetic structure of an activated filament in a flaring active region. *Astronomy and Astrophysics, 561*, 98.

Schmieder, B., Tian, H., Kucera, T., Lopez Ariste, A., Mein, N., Mein, P., Dalmasse, K., & Golub, L. (2014). Open questions on prominences from coordinated observations by IRIS, Hinode, SDO/AIA, THEMIS and the Meudon/MSDP. *Astronomy and Astrophysics, 569*, 85.

Chapter 9
The Magnetic Field Structure of Prominences from Direct and Indirect Observations

Sara F. Martin

Abstract Solar prominences are fascinating and unique magnetic structures in our solar system. From all observational evidence to date, their intricate structure, dynamics and plasma parameters are apparently all derived from observable source magnetic fields on the Sun. While the processes that convert these source fields to prominence magnetic fields are not fully understood, there is a trail of observational information that gives many clues about how prominence magnetic fields are derived from or related to these source fields and maintained for various lengths of time through multiple processes, over a wide range of spatial scales. This chapter highlights that trail of observational information for two primary magnetic types with very different origins: channel prominences and coronal cloud prominences.

9.1 Introduction

9.1.1 *Distinguishing Between Channel and Coronal Prominences*

The importance of detecting, deducing and modeling the magnetic field structure of channel prominences is increasingly being recognized due to their frequent eruption along with the occurrence of flares and CMEs. Considered together, these three dynamic solar features are called eruptive solar events in recognition that no one of these phenomena causes the other. Instead they have a common cause. There is a building body of evidence that channel prominences are the practical precursors to major eruptive solar events. Channel prominences are also commonly referred to as "filaments", because that is what they look like when observed against the solar disk. This linguistic distinction between filaments and prominences in general is useful to solar observers because they know prominences observed against the disk are only the more dense, hence more visible varieties or parts of prominences. The term filament is also an immediate clue that any information presented will partially

S.F. Martin (✉)
Helio Research, La Crescenta, CA, USA
e-mail: Sara@HelioResearch.org

J.-C. Vial, O. Engvold (eds.), *Solar Prominences*, Astrophysics and Space Science Library 415, DOI 10.1007/978-3-319-10416-4_9

involve a top-down view instead of only the side view of similar prominences observable on the limb.

Coronal cloud prominences, on the other hand, have only been observed above the limb because they are less dense, and they have rarely been seen to erupt. They mysteriously appear as aggregates of mass, *clouds* in the corona. They include continuously growing or shrinking clouds, flows of mass from the cloud down around or through the coronal magnetic features to the chromosphere, and coronal rain without an apparent cloud.

9.1.2 Evidence for Two Magnetically Different Types of Prominences

Evidence is presented in Table 9.1 that most prominences belong to one of these two different classes prominences. These are difficult to extract from the historical classifications of prominences reviewed by Engvold (2014). This simplification (channel prominences and coronal cloud prominences, Table 9.1) to two fundamental categories depended on the assimilation of new information mostly from ground-based imaging of high contrast and spatial resolution of less than 1 arc sec.

Channel prominences are those that are have spines and in most circumstances also develop barbs as part of their basic prominence structure. Channel prominences (channel filaments) include whole gamut of active region, intermediate and quiescent prominences as well as mini-filaments. These prominences owe their existence to the development and maintenance of a three-dimensional magnetic field environment known as a *filament channel* or prominence channel [review by Gaizauskas (1998)].

Table 9.1 Comparison of channel prominences with coronal cloud prominences

Channel prominences with spine and barbs	Coronal cloud prominences often above invisible arcs
Low < 50,000 km	High (up to ∼200,000 km)
Bright (1,011 particles/cc)	Faint ($\sim 10^{10}$ particles/cc)
Seen against disk in Hα	Rarely seen against disk in Hα
Reveal counterstreaming in spine and barbs	Only down flows from unknown sources
Located in filament channels	Not in filament channels
Mass input from injection sites where magnetic fields are cancelling	Mass input from previously ejected filament mass or CME's—two of several hypotheses
Lie above polarity reversal boundaries in filament channels	Might lie within coronal loop systems above separatrix surfaces
Have chirality (handedness)	No known chirality (handedness)
Often end lifetime by erupting with a CME	Rarely erupt; large ones form after a CME with an erupting filament

Examples of the other major category, coronal cloud prominences, appear among the earliest observations of prominences. On average, however, coronal cloud prominences are less dense and fainter than channel prominences, and to date have only been recognized above the limb. Figure 16 Engvold (2014) shows three examples among the tallest coronal cloud prominences recorded to date. Further details about the magnetic fields associated with these two types of prominences are herein discussed separately in Sects. 9.2 and 9.3 respectively.

9.2 Magnetic Field Information Related to Channel Prominences (Filaments)

9.2.1 The Overall Constraining Environment of Filaments

Channel filaments occur in a unique three-dimensional magnetic environment in the photosphere, chromosphere and low corona. Figure 9.1 schematically represents a channel and filament which are beneath an over-arching set of coronal loops. All of the structure to the sides and up to the top of the red filament is broadly defined as the filament channel. Between the filament channel and the overlying system of loops is the filament cavity, a much larger volume than depicted here. The stronger magnetic flux of the coronal loop system initially confines the lesser magnetic flux of a building filament channel, filament, and cavity below.

Filament channels are the birth sites of filaments with spine and barb structure depicted in Fig. 9.1. The channel, filament, and overlying loops are connected to the photosphere represented by a horizontal line in the end view and the white plane in the perspective view. To understand how the channel structure and filament are

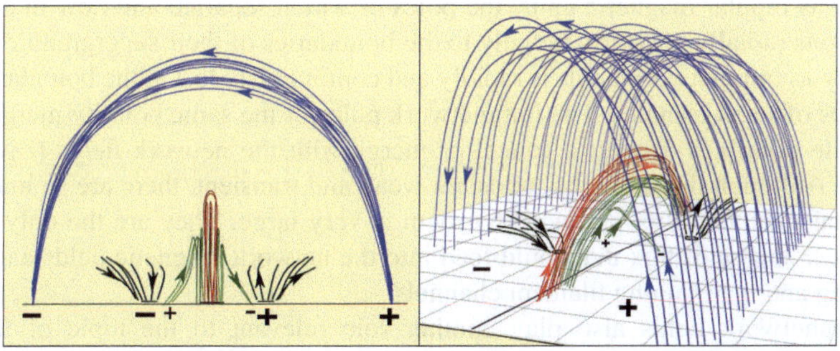

Fig. 9.1 Schematic diagram of a filament channel and its overlying cavity and loop system. This diagram is a qualitative observational representation derived from typical solar images and data on the relationship of filaments to their environment. A filament spine is represented in *red* and its barb structure in *green* with *green arrows*. The photosphere and chromospheric structure are represented by *dark green* to *black lines* with *black arrows*. Additional low coronal loops and coronal cells and their plumes to each side of the spine are not depicted for initial simplicity. Illustration credit: O. Panasenco in Martin et al. (2012)

formed, it is important first to have a basic picture of the photospheric magnetic fields in which the channel structure and the filament are rooted.

9.2.2 Magnetic Field Sources on the Sun Essential to Channel Filaments

The source magnetic fields fundamental to the existence of channel prominences (filaments) are: (1) the strong magnetic fields of active regions of all sizes and (2) the very much weaker intranetwork magnetic fields that originate in the photospheric convection cells known as supergranules (Lin 1995), and (3) related to some small filaments only, the moving magnetic features that appear around sunspots in active regions.

Filament channel and filament formation evolve most frequently and rapidly at the outer boundaries of the expanding bipolar fields of active and ephemeral active regions. These strong fields, on all scales, gradually evolve into a background network of magnetic fields due to the action of super granulation. There is no intermediate component; the strong and weak components have distinctly different origins (Lin 1995) as implied above. The strong component of solar magnetic fields includes the smallest active regions, called ephemeral active regions (Harvey 1993). Analogous to large filaments, many of the smallest filaments (often called mini-filaments) are associated with the evolution of ephemeral active regions. These are so numerous that they are possible catalysts to temporary activations that occur in filaments.

The weak component consists of intranetwork magnetic fields that appear in the centers of solar convection cells known as supergranules (Leighton 1959; Leighton et al. 1962; Leighton and Simon 1964; Janssens 1970). To a first approximation, they appear as bipolar magnetic units, the poles of which separate and flow in opposite directions radially or nearly radially to the boundaries of their supergranules. Their identity as intranetwork fields is rapidly and continuously lost at the boundaries and vertices of supergranules. The intranetwork poles of the same polarity merge while opposite polarities appear to cancel or merge with the network fields (Livi et al. 1985). Although these bipolar fields are weak and transient, there are so many that the total magnetic flux produced by them is very large. They are the only known *sufficient* source of flux that could feed into the network magnetic fields and aid in creating and maintaining filament channels.

Intranetwork fields also play another role relevant to the topic of filament magnetic fields. It is now well established that the endpoints of barbs are related to small pockets of magnetic field opposite in polarity to the network fields on each side of a filament spine (Martin 1998b; Aulanier et al. 2000; Chae et al. 2005). The only sources of magnetic flux for these pockets are ephemeral active regions and the intranetwork magnetic fields. The ubiquity of intranetwork magnetic flux points to it as the major source.

9.2.3 Photospheric Magnetic Fields in Filament Channels

The first association of filaments with photospheric magnetic fields was by Babcock and Babcock (1955) upon completion of their magnetograph at Mt Wilson. A large polar crown filament was seen to be related to a boundary between opposite polarity magnetic fields. Their subsequent work found this same relationship of filaments to sites where the magnetic field reversed from positive to negative polarity. Their work was further confirmed by Smith et al. (1965) showing that there were no exceptions to this relationship when the magnetograms were reduced to isogauss contours and compared with photographs of filaments made to the same spatial scale.

Previously polarity boundaries have been called polarity inversions, or polarity inversion lines, or neutral lines. In this chapter, the term *boundary* is more suitable than *line* because filament channels occupy a volume of space from the photosphere through the low corona.

At the polarity boundary, the magnetic field direction changes direction in space by making what is called a *rotational magnetic field configuration* (Martin et al 1994). The horizontal part of the rotational configuration, where filaments reside, is like a narrow planar boundary extending upward through the chromosphere into the corona as high as the filament spine. Consistent with this configuration, there is an absence of evidence that the magnetic field inverts in the sense of making a direct hairpin-like turn across the polarity boundary in filament channels.

Polarity reversal boundaries are characterized by at least five properties when observed in photospheric magnetograms:

a. A sharp reversal in the direction of the magnetic field of abutted strong field components of photospheric magnetic flux in the sense described above. These boundaries may be bordered by areas of relatively dense magnetic flux as in parts of active regions; more commonly they are a linear series of boundaries between separated concentrations of strong, abutted fields of both polarities.
b. Convergence of magnetic flux toward opposite polarity magnetic flux (Rompolt and Bogdan 1986; Rompolt 1990) that specifically initiates and sustains the cancellation of magnetic flux (Martin et al. 1985; Livi et al 1985; Martin 1990; Mackay et al. 2008, 2014).
c. Continuous cancelling of the abutted opposite-polarity, strong fields. Cancellation is defined as the disappearance of equal and opposite amounts of magnetic flux of opposite sign at a common interface as seen in line-of-sight magnetograms (Martin et al. 1985; Livi et al. 1985; Martin 1990, and numerous subsequent references). There are multiple interpretations for cancellation (Zwaan 1987).
d. Little intermingling of opposite polarities once a boundary is established (Martin and Panasenco 2010). The plasma flows within granules, meso-granules, and super-granules constrain the network fields to motions along the boundaries of these convection cells, tending to prevent migration of one polarity around the other. This limits the opposite polarity concentrations of flux to moving either together or apart along cell boundaries. Primarily they move together although exceptions can occasionally be found.

e. Alignment of the local magnetic field along the polarity reversal boundary. This precise alignment was initially deduced from the field-aligned property of the chromospheric fibrils as implied in Figs. 9.2 and 9.3 (Foukal 1971a, b; Rompolt and Bogdan 1986; Rompolt 1990; Martin et al. 1994). Subsequently, vector magnetograms confirmed the strong field component along the polarity boundary (Grigoryev and Ermakova 1999; Metcalf et al. 2006; Okamoto et al. 2008; Lites et al. 2010; Kuckein et al. 2012).

Figure 9.2 below depicts the above five properties of polarity reversal boundaries. In Fig. 9.2 the first image in the upper left is a mosaic of 12 Hα images from the Dutch Open Telescope (Hammerschlag et al. 2013). The next image to the right is the same field-of-view in the white square cut from the full disk SDO/HMI

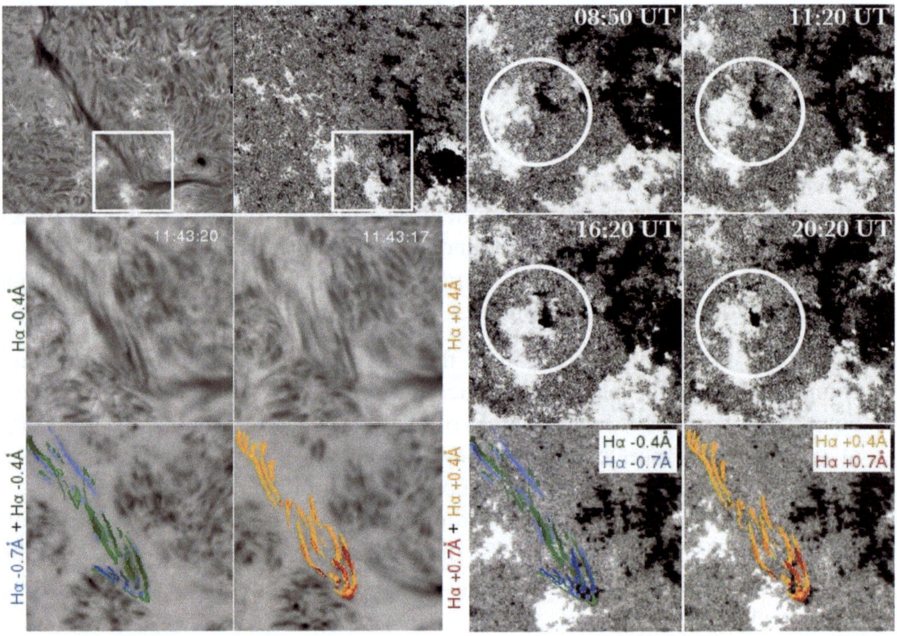

Fig. 9.2 The *white boxes* in the *upper* two images from DOT and SDO/HMI show the site of a relatively isolated cancelling feature in Hα and magnetograms respectively. Within the *white circles* in the four enlarged SDO/HMI magnetograms in the *upper right*, convergence and cancellation of flux are seen. After cancellation began at this site, the positive polarity magnetic flux (*white*) engulfs the negative polarity (*black*) as both continue to decrease in flux. The *middle pair of images* on the *left side* is filtergrams as labeled in the wings of the Hα line. The lower two images are tracings of the mass flows from the above pair and another pair of images at Hα + and −0.7 Å color coded as in the labels at the outer edges of the images. The last pair of images in the *lower right* is a composite showing the mass flows superposed on the magnetogram. These data sets confirm the close spatial association of source and sink sites of counterstreaming in the threads of filament with cancelling magnetic fields. Less clear are sites of cancelling fields close to a sunspot to the *lower right* due to new flux associated with moving magnetic features around sunpots. The filament spine lies over these cancelling fields

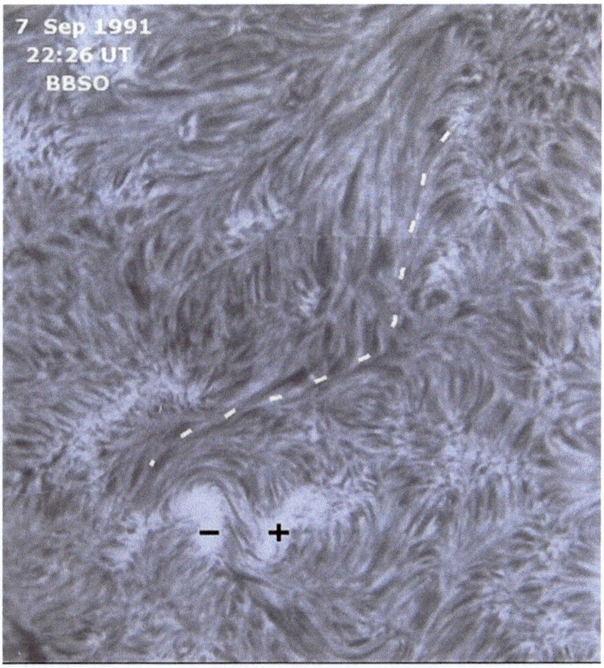

Fig. 9.3 The *dashed line* shows the *center* of a filament channel without a filament. At its *lower left end*, the characteristic path of aligned fibrils, lies on the *upper boundary* of a small decaying bipolar active region. The areas of brightest plage associated with the two halves of the bipolar active region are connected by fibrils that slant relative to the shortest distance between the plages. This slanting pattern of fibrils is a first indication that another filament channel is beginning to form between the two bipolar components of the small active region. Mosaic of two BBSO images

image closest in time to the Hα image. Within the white square is a prominent cancelling magnetic feature beneath the filament. It is shown enlarged in the four magnetograms in the upper right which begin before and continue after the DOT observations for several hours.

The *convergence* of the positive polarity toward the negative polarity and the subsequent *cancellation* flux can be seen along with the establishment of the *high magnetic flux gradient* at the cancellation sites. The cancelling feature might be an unusual example; the larger positive magnetic flux surrounds the more isolated negative polarity flux as they are cancelling. It demonstrates why network magnetic flux of opposite polarity does not intermingle very much. It cannot intermingle for a very long interval before the smaller clumps of flux are engulfed and cancelled by whichever clumps have more flux. Not obvious here is the role that supergranules play in determining which clumps of positive polarity and negative polarity flux will come into apparent contact and become the next cancelling features.

In the lower left, four images illustrate, at just one time, the flows that were continuously present in this filament and associated specifically with the cancelling magnetic feature. The traces of the threads are color coded green and blue respectively to represent threads at Hα − 0.4 Å and Hα − 0.7 Å in the blue wing. Orange and red represent threads at Hα + 0.4 Å and Hα + 0.7 Å respectively. These represent the counterstreaming flows observed to and from the cancelling site as shown in the pair of images in the lower right. Counterstreaming flows are one of the confirmations that the threads of filaments are field-aligned. This example is one of many from sets of 5–7 wavelengths recorded at 30 s to 1 min cadence at the DOT and available in the DOT archives.

Vector magnetograms are highly important for revealing the direction and magnitude of the horizontal components of the magnetic fields within the extended photospheric areas defined as filament channels. However, vector magnetograms are not as readily available as line-of-sight magnetograms, and vector measurements have a much lower temporal resolution. Unless these instrumental factors are convincingly taken into account, interpretations of reported small deviations with respect to a polarity reversal boundary remain in question. Okamoto et al. (2008), Lites et al. (2010), Kuckein et al. (2012) have all reported, from magnetograms recorded with SOTon board the Hinode satellite, that the magnetic field along the polarity reversal boundary evolves beyond complete alignment with the polarity boundary such that it has an inverse component. This is one example of where the vector magnetograms are consistent with the deductions about the magnetic field from Hα but where the higher resolution of the Hα images shows additional significant details.

In Hα, the inverse component is well seen and has been associated both with the chromospheric fibrils in filament channels and the barbs of filaments. However, there is yet no discussion of whether a photospheric inverse component could be related to pockets of flux of minority polarity which become related to the endpoints of barbs. The barbs have inverse polarity, as discussed by Martin et al. (1994). In addition, the chromospheric fibrils (Foukal 1971a, b; Martin et al. 1994) and coronal cells (Sheeley et al. 2013) close to filaments have also been observed and in some studies filament channels, to develop an inverse component over time (Su et al. 2010). It is noteworthy that this inverse component does not interfere with the maintenance of the polarity reversal boundary in filament channels in the chromosphere and low corona. The Hα and coronal observations show that the structures associated with the inverse component evolve next to the polarity boundary, not coincident with it. Because these chromospheric fields are rooted in the photosphere, it seems there should be a good agreement with vector magnetograms, although the fibrils can bend significantly in the chromospheric part of a filament channel as shown in the lower left image of Fig. 9 in Engvold (2014). Understanding the meaning of such details in vector magnetograms necessarily depends on the development of instruments with better resolution.

The five observed properties of the polarity reversal boundaries, above, are closely linked. However, cancelling magnetic fields often get the limelight while the large role played by convection in creating and sustaining cancelling boundaries is neglected. Convection enables the existence of large scale filament channels that host filaments precisely above cancelling boundaries. Realistic modeling of the formation of filament channels and filaments needs to begin with modeling the development of filament channels from known sources of magnetic flux and the convective processes that act on the source fields. Mackay et al. (2014) reviews modeling of global fields with input from real data and how this has been done to date.

The tight juxtaposition of strong fields of opposite polarities is what distinguishes the *polarity reversal boundary* within a filament channel from most other locations.

Elsewhere the change in direction of the strong magnetic field is much more gradual and spatially resolvable. The tight juxtaposition is evidence of a force or forces driving or pulling the magnetic flux of opposite polarity together. Relevant modeling has been done using the observed distributions of active regions, and iteratively improving the modeling until it fits the whole of the observational picture (Mackay et al. 2014).

Apart from modeling on the global scale, interpreting cancelling magnetic fields as magnetic reconnection at or near the photosphere has led to noteworthy progress. Detailed physical interpretation of cancelling magnetic fields is of first importance in the formation of filament channels Litvinenko (2010) and then filament formation (Litvinenko 1999; Litvinenko et al. 2007).

Convection on all scales of granulation, mesogranulation and supergranulation has the observed and proven ability to force the network field of decaying active regions to the boundaries of the supergranules and granules on smaller scales, to force opposite polarity fields and same polarity fields into contact, and to contribute to keeping them in close contact. There is no reason to believe that opposite polarity fields on the Sun are driven together initially except by chance. However, once a cancellation boundary is established, it is observed to have a strong tendency to continue cancelling. Are the cancelling features pushed together, pulled together, or both?

The highest rates of cancellation occur in the largest concentrations of magnetic flux in or around active regions. These are the sites where new filament channels and filaments form most rapidly (Mackay et al. 2008, 2014). This provides evidence that cancellation rates depend on the forces that bring opposite polarity fields together. The rate at which flux moves together and the magnitude of the flux are important inputs into the modeling details of filament formation (Litvinenko 1999). The process of channel formation followed by filament formation can be either relatively rapid in or near active regions or very slow on the quiet Sun (Martin et al. 1985; Livi et al. 1985; Litvinenko and Martin 1999; Wang and Muglach 2007).

9.2.4 Filament Channels and Filaments in Areas of High Magnetic Flux Density

The most useful feature of filament channels is that their organized fibrils are everywhere field-aligned. If the fibril structure in Hα and other lines were not seen, the role of the magnetic fields in filament channels would not be as easily recognized. While Hα and other chromospheric images should be complemented by vector magnetograms, the Hα images provide much of the needed information about the magnetic fields of channels and filaments (Smith 1968; Foukal 1971a, b; Gaizauskas 1998; Martin 1998a,b).

Specific fibril patterns in the chromosphere are the primary signature of filament channels in the chromosphere that yield information about the local magnetic fields, because the fibrils are field-aligned (Pikel'ner 1971; Foukal 1971a, b; Zirin 1972).

An example of a small channel is seen in Fig. 9.2. The most conspicuous feature in Fig. 9.2 is a small decaying active region at the bottom of the frame recognized by the two brightest patches of plage, marked with their magnetic polarity. They are connected by a few slanted fibrils. These slanted fibrils are a sign that a filament channel has begun to form at that location. To the upper left of this decaying bipolar region is a conspicuous path of fibrils aligned nearly parallel with one another. This is the left end of a filament channel that has no visible filament. Along the white dashed line in Fig. 9.3, the fibrils appear to sweep upward and to the right away from the bright plagettes. However, on the right side of the white dashed line is a similar pattern of fibrils that appear to sweep to the left and downward parallel with the white dashed line but only very close to the dashed line. This opposing directionality of the fibrils is the primary signature of filament channels in the chromosphere.

This pattern was first pointed out by Foukal (1971a, b) in his studies of the morphology of chromospheric structure. His diagram depicting the chromospheric pattern of fibrils of filament channels is shown in Fig. 9.4. As indicated in the diagram, the fibrils can be considered like small arrows on the positive polarity side of a filament channel. Their tapered ends point in the direction of the local magnetic field from which we can learn useful information about both the horizontal and vertical components at the sites of the fibrils. For the fibrils on the negative side, the direction of the field is towards the base rather than toward the tapered end of each fibril. Viewed from above, the fibrils reveal the direction but not the magnitude of the tilt. The positive field tilts upward at an angle relative to the horizontal component of the fibril on the positive side of the polarity reversal boundary while the negative field of the fibril tilts downward on the other side of its channel.

A filament can only form along the narrow boundary between the fibrils whose vertical components are tilted in different directions; into the Sun on the negative side and out of the Sun on the positive side. The long threads of the filament spine can only survive if they are in the narrow horizontal zone between the fibrils with opposite vertical components. Anywhere else, the magnetic field of the filament spine would reconnect with the channel field.

Figure 9.4 shows several examples of active region filament channels with filaments or developing filaments. The patterns are illustrated in Foukal's diagram, also included in Fig. 9.4 below the image of a forming filament channel with approximately the same configuration as in the diagram. A channel was expected to form between the two small active regions in Fig. 9.4; the curvatures of the fibrils indicate the first stage of formation has begun.

Figure 9.5 illustrates two clear examples of filaments in established filament channels. The fibrils at the sides of the filament shaped like a question mark have the same pattern as the diagram in Fig. 9.4 while the fibrils next to the smaller ear-shaped filament, with its curvature concave toward the question mark, have the opposite pattern.

A significant feature of filament spines is that they are narrow whether their channels are wide or narrow. This would be more readily appreciated if, like birds, one could fly across and around the Sun looking at filaments from different perspectives. In most images, one is viewing it from one side or the other and cannot

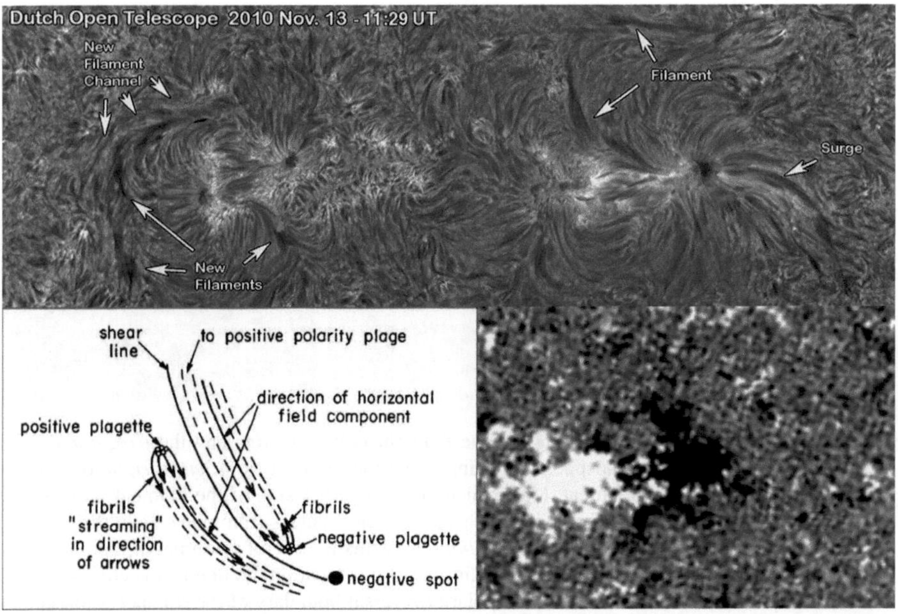

Fig. 9.4 A variety of filaments and filament channels are shown in various stages of development in this mosaic of multiple images. A new filament channel is developing along the *left* border of the small active region on the *left*. Its fibril pattern is similar to the one in the diagram *below* from Foukal (1971b) but the polarities are reversed. A filament channel is expected to form between the two small active regions with the same pattern as in Foukal's diagram, except with the positive polarity on the right and the negative polarity on the *left*. The magnetogram in the lower right is from SDO/HMI corresponding to the above H α mosaic from the DOT

see the true narrowness of a spine. Even the stem of the filament shaped like a question mark. In Fig. 9.4, is a side view of the spine but the curvatures along the two filaments show that the spine is at least as narrow as many of the fibrils. In this case, as in most examples in which the filament resides between closely-spaced areas of plage of opposite polarity, the filament channel is very narrow. Channels can be so narrow that there is no space for chromospheric fibrils between a filament and its neighboring plage. Also, a channel of fibrils may be much wider on one side of a filament than on the other side. The variety of filament channels and filaments shown in Fig. 9.5 are in relatively early stages of development; the active regions are only a few days old.

The unusually long and taller filament in Fig. 9.5 shows the thinness of the spine very clearly. In this example, both sides of the filament can be seen because it follows a curved path on the solar sphere. In the upper left, the sharp top edge of the ribbon-shaped filament is toward the top of the image; after the bend in the filament the sharp upper edge is toward the side of the image. Seeing the apparent bend makes it especially clear that the spine is like a ribbon standing on edge at the chromosphere or photosphere. The height from the chromosphere to the top of such

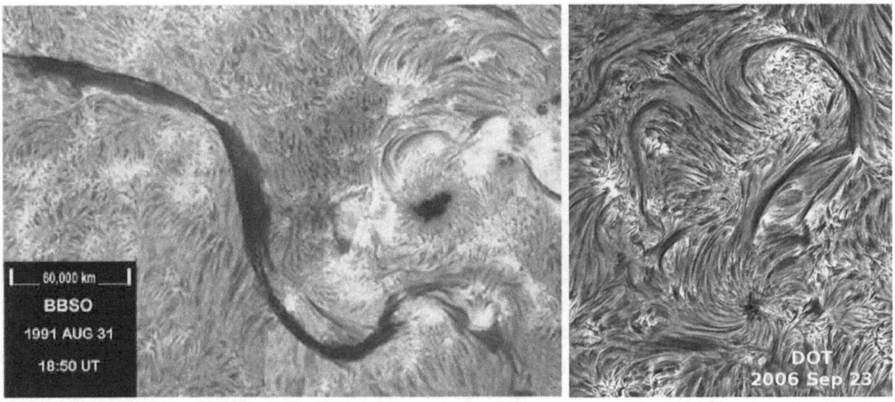

Fig. 9.5 *Left image*: The *sharper edge* of the filament is its *top edge* and the *irregular edge* is its intersection with the chromosphere. Viewing along the filament from *left* to *right*, one comes to an apparent bend in the filament, after which bend the other side of the filament is seen. The narrowness of spine is revealed at the bend where the view changes from one side to the other. At the strong curvature in the filament near the *bottom* of this mosaic of two images, both edges of the filament are sharp and cross over distinct fibrils. At this location the filament is elevated above the chromospheric part of the channel; the polarity reversal boundary of the channel is marked by low filament threads on the concave side of the much higher spine. At this location, in effect the filament is split into two levels. *Right*: The filament shaped like a *question mark* faces a smaller, *ear-shaped* filament that is almost like the *question mark* laterally reversed. The two ends of the filaments that curve away from each other in the *upper middle* of the image are a sign that these two filaments will never merge into a single filament

a filament can be calculated at any point along the spine provided its coordinates on the solar disk are known, along with assuming it is vertical.

Other information about the height of the filament in Fig. 9.5 is gleaned along the more pronounced bend in the ribbon nearest the lower edge of the image. Both edges are sharp and the fibrils below are at a large angle to the spine. At that location, all the threads of the filament are high above the chromosphere. Again, from knowing its coordinates, and deducing which side of the narrow ribbon is being observed, it is possible to calculate the height of both the bottom and top edges.

In Fig. 9.5 attention is again called to the strongly curved section near the bottom of the frame. Where the filament crosses over fibrils in the chromosphere, it is typically high. Beneath and to the right, a small filament is seen; it is in the filament channel of the higher section of the filament. This is noteworthy because a few filaments have been observed to split like this into two sections, one above the other within a few days prior to their eruption.

The long filament in Fig. 9.5 has thus served to show that it is possible to find the 3-dimensional structure of specific filaments using the aid of solar rotation to give different perspectives. Using this method, Martin and Echols (1994) created a three dimensional scale model of the filament spine of an observed filament before it was feasible to mathematically model a filament in three dimensions. By examining the barb threads in Fig. 9.6, it should be clear why only a few representative barb threads

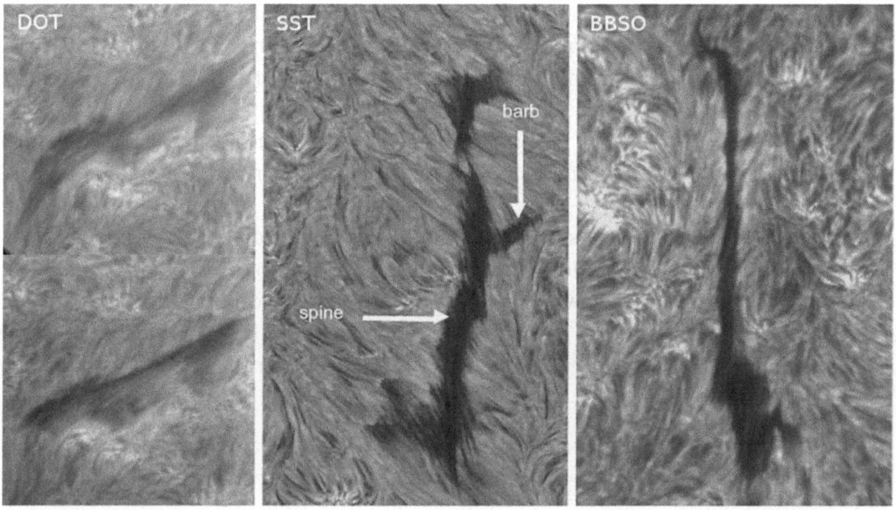

Fig. 9.6 *Left image pair*: A short filament with barbs is located in a filament channel of intermediate magnetic flux density. *Scattered bright* plagettes, corresponding to network magnetic fields in the photosphere, are the signature of intermediate magnetic flux density. This side view shows two barbs, each with many threads. Threads terminate at the borders of plagettes, not in the plagettes. *Right image pair*: The labeled image is in an area of lesser plage density than the one on the *right*. The larger barbs extend further from the spine with decreasing magnetic flux density in the environment. However many barb threads close to the spine are also roughly parallel with the surrounding chromospheric fibrils. The spine that separates the barbs on its *left* and *right* sides is not seen because it is thin and superposed on many other threads but a typical spine is clearly seen in the *lower left image* along with barbs close to the sides of the spine

were included in their scale model. Figure 9.5 can serve as a guide in re-examining the filaments in Figs. 9.4 and 9.6 for similar evidence of their width and relative height, and closeness to the chromosphere. In these Figures, one can also see that a few barbs form close to the spine on some active region filaments.

The active region filament channels and filaments shown in Figs. 9.2, 9.3, 9.4, and 9.5, illustrate the following information observed or deduced about their magnetic fields:

The field-aligned chromospheric fibrils, associated with the positive polarity network, taper to thin points at their upper end, and serve as arrowheads pointing in the horizontal direction of the local magnetic field. The fibrils on the negative polarity side serve as "inverse arrowheads" in which the magnetic field points to their base rather than to their upper tapered ends.

Summarizing the properties of filament channels and filaments in active regions:

1. Filament threads close to the chromosphere run parallel or nearly parallel with the field-aligned closely neighboring fibrils.
2. When the channels are broad enough to resolve single fibrils, the fibril pattern can be used to identify the polarity reversal boundary in the channel even when

it is devoid of a filament mass. The characteristic pattern has two components: (a) a horizontal component in common with the polarity boundary, and (b) a substantial vertical component that is visibly bent away from vertical in opposite directions on the two sides of the polarity boundary.

3. Filament spines form only along the boundary where the local magnetic field is very close to horizontal as deduced from the fibril pattern of filament channels (2 above).
4. An active region filament spine is an ultra-narrow vertical stack of largely horizontal threads remaining along the boundary between opposite polarity magnetic fields but its ends curve downward to connect to the chromosphere/photosphere.
5. The magnetic field along the mid section of a spine is usually in the same direction as the horizontal component of the adjacent fibrils, unless it is unusually high.

It is the intention that the list above is purely observational without interpretation. Some interpretive conclusions drawn from the above observations were:

a. (1) above was evidence that the threads of filaments are field-aligned (Foukal 1971a), in agreement with mass motions always being observed along the threads; and as well as various oscillations observed in threads.
b. The direction of the magnetic field along a channel can be deduced from the fibrils with only minimal knowledge of the polarity of the magnetic field on either side of a filament channel or filament (Foukal 1971a; Zirin 1972).
c. The bending of the vertical component of fibrils away from each other on the opposite sides of the filament channel implies the presence of a force strong enough to bend the tops of the fibrils downward (illustration in Foukal 1971a, b; Gaizauskas 1998).

9.2.5 Filament Channels and Filaments in Areas of Intermediate Magnetic Flux Density

Intermediate filament channels and filaments are found in areas where active regions are in a relatively recent stage of decay. To simplify this discussion, a decaying region will be considered one in which the appearance of new strong flux has ceased. A recent stage of decay is hours for ephemeral regions, a day to a few days for small active regions and many days to weeks for large active regions or complexes of active regions.

As soon as new flux has ceased, the process of decay becomes clear. Mesogranules and/or supergranules begin to appear amidst active regions and begin the process of dispersal of their magnetic flux. When the dispersed magnetic flux of active regions has been spread to the boundaries of supergranules, the dispersed magnetic flux is loosely defined as the *network or network magnetic flux.*

Intermediate filament channels share all of the same properties as the filament channels of active regions described above but the channels are typically wider, have more dispersed magnetic flux, and change more slowly. These channels typically vary in width from about one to several supergranule diameters ($\sim$30 mm) in association with the decay of active regions or complexes of active regions.

The spine threads of intermediate filaments also have all the same properties as the spines of active region filaments, including a similarly narrow spine. However, they also develop a lot of barbs threads along the sides of the spine. Some barbs are composed of clusters of threads that extend further from the spine. These barbs are known to have the approximate spacing of supergranules. The side view of a filament in two images about two hours apart is shown in the smaller left panels in Fig. 9.6. The left-most barb appears to terminate in the chromosphere, noticeably outlining a plagette, not connecting to it but rather its edge. Such plagettes correspond to network magnetic fields in the photosphere. The endpoints of barbs do not connect to the network fields on the each side of a filament but instead occur near the boundaries between network and intranetwork magnetic fields or, alternatively, close to whichever ephemeral region pole is opposite in polarity with the network. The intranetwork fields are so weak that it is not easy to illustrate this point with available magnetograms. However, it is known that intranetwork fields are continuously interacting with network magnetic fields everywhere on the Sun (Wang et al. 2000). Therefore, the example in the left side of Fig. 9.6, showing that the threads stop at the periphery of the plagettes, is consistent with the termination of the threads at the interface between intranetwork field and opposite polarity network magnetic field at the edges of the plagettes.

The changing intensities of the threads in the spine and barbs are related to counterstreaming mass motions along the threads. Counterstreaming is readily seen in the time lapse movies in the center line and wing movies of the data sets represented in Fig. 9.6, and most other high resolution data sets. Counterstreaming motions along the spine (Fig. 9.2), are well seen in the side view of intermediate filaments, as in the left-most images in Fig. 9.6. The density of the multiple threads in the line of sight obscures the threads along the spine in the two filaments in the middle and right panels of Fig. 9.6. However, counterstreaming in the barbs is often seen effectively from the top view of intermediate filaments.

Comparison of the side view of the filament in the first two panels of Fig. 9.6 with the top down view in the other two examples, show why the barb threads appear slanted along each side of the spine in the top down view. The spine is so narrow that it is actually not seen clearly against the numerous threads of the barbs. Where the filament is dark, numerous threads are stacked in the line of sight. The threads slant down in the barbs of the filament in the left panels. Similar slanting threads are seen from the top-down view on each side of the spine. Although they are not observed to cross through the spine during the normal quiescent state of a filament, they are approximately parallel on the two sides. When the spine threads are not seen, the mind tends to see the barbs on each side as connected across the spine, counter to our evidence from other perspectives. Therefore, the study of the geometry of filament threads needs to be done as much as feasible with images from different perspectives along with time-lapse movies. Doppler images are needed to enable one to trace the motions of threads at multiple wavelengths around the Hα line center. This is a fundamental point which warrants close attention. It is often overlooked because in most data sets, only one view of a filament is seen at any given time.

These two examples in the middle and right in Fig. 9.6 were also chosen for comparison of the channel structure and the filament structure. The filament on the left has a barb with many threads near its lower left end. The barb threads are nearly aligned with the chromospheric threads as expected. The fine threads give the true direction of the magnetic field in the barb.

Also, the barb threads are much darker near their endpoints. It is suggested comparing this barb with the one that is labeled "barb" on the right side of the spine. The barb threads in the labeled barb also consist of a series of ends of threads with high density near their footpoints. The upper parts of these threads become invisible between the dense feet and the spine. The thread direction, also confirmed from mass motions along the threads, is shown in the schematic diagram in Fig. 9.7, a similar case is also shown in Fig. 9.5, Martin et al. (2008).

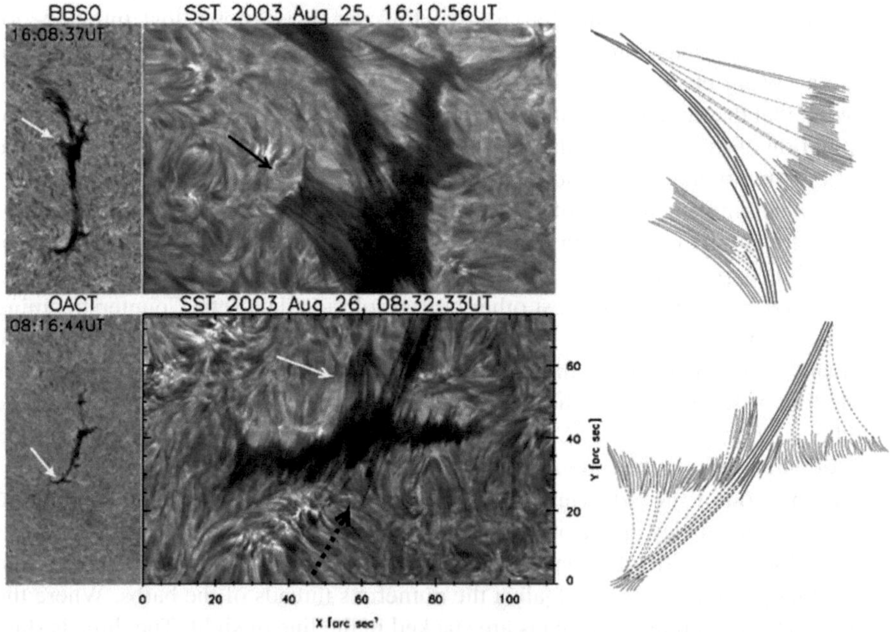

Fig. 9.7 The *white arrows* on the BBSO images at the left show the sections of the filament observed at the SST. The schematic drawing on the right represents the observed parts of barb threads as *solid lines* and the inferred invisible parts of the threads as *dashed line*. The orientations of the invisible threads are represented from flows observed in time-lapse sequences of images. When the threads are extended as *dashed lines* in the direction of the observed flows, the normal pattern of a spine with barbs is more clearly identified. Illustration by Yong Lin

The threads at the visible ends of the filaments in Fig. 9.6 are aligned quite well with fibrils.

Note further that for the low sections of filaments, the alignment of the filament threads with the fibrils leaves no space for any other magnetic field beneath these parts of the filaments.

These three examples of intermediate filaments in Fig. 9.6 are relatively short but have the same basic spine and barb structure possessed by longer and taller intermediate filaments. In these cases and all other examples, the barbs are some of the same threads as in the spine but they bend and extend sideways from the spine to the chromosphere having found appropriate footpoints. However, not all of the barb threads extend far from the spine; many remain close to the spine. The filament on the right has barbs but they are not large and do not extend far from the spine. It is common that when many barb threads remain close to the spine, it creates an illusion that they cross the spine. Special care needs to be taken in interpretation.

Adding to the summary at the end of the previous section are these additional observational points illustrated here or found in the literature:

6. The spines of intermediate filaments are narrow, like the spines of active region filaments.
7. Intermediate filaments have conspicuous barbs that are seen as continuations of spine threads that deviate gradually from the spine to connect with the chromosphere/photosphere to the sides of filaments; or stated in reverse, barb threads are observed or originate in the photosphere and move upward from the chromosphere and asymptotically merge with threads of the spine.
8. Barbs do not connect to network magnetic fields.
9. The endpoints of barbs are related to the fields opposite in polarity to the network on each side of a filament or to cancelling boundaries at the interface of these fields.

Interpretative conclusions about the magnetic fields of barbs are:

d. The horizontal component of barbs has a component in common with the spine but the barb threads turns inversely away from the spine meaning the barb component orthogonal to the spine is opposite to the orthogonal component of the coronal loops above the spine.
e. Barb threads could be a consequence of a complex series of magnetic reconnections of spine threads with the magnetic fields of newly appearing intranetwork magnetic fields or the magnetic fields of ephemeral active regions. However, the latter are more amenable to study because the fields are more readily detected.

9.2.6 Filament Channels and Filaments in Areas of Low Magnetic Flux Density

When the magnetic flux density is low, the plagettes in Hα are more sparse and do not stand out as bright background features. Filaments that form in such areas or evolve into areas where the magnetic field has low magnetic flux density are called *quiescent* filaments. This is a traditional name for large filaments on the quiet Sun. They tend to be long-lived and taller than intermediate prominences and change slowly from day to day but are nevertheless continuously changing.

Because individual quiescent prominences appear to have very different structure from active region prominences, one of the most frequently posed academic questions has been: are quiescent prominences and active region prominences different in their basic physics? Figure 9.8 addresses that question with a selection of images from the Big Bear Solar Observatory and SOHO EIT images.

The upper image shows a single filament which is an active region filament at its west end, an intermediate filament in the neighboring area of semi-bright plage and is a quiescent filament at its east end. While there appears to be some small spaces between the segments of the filament, it is now well known that a spine almost always connects all segments that are as close together and end-to-end as in this image. The evidence comes from movies in Hα that show the moving threads over long distances along the thin spine. However, these spine threads are not always seen in single images. Spine threads of quiescent prominences are much better seen in 304Å images from SOHO/EIT and SDO/AIA.

The row of three images in the middle of Fig. 9.8 continues through the changes in structure of quiescent filaments that evolve further but have no special name. In areas of yet lower magnetic flux density, additional changes in their appearance are seen in Hα images. The images in the middle row of Fig. 9.8 show an unusually rapid evolution in a single quiescent filament but it is representative of the evolution of filaments in areas of ultra low magnetic flux density. In the first image in the middle row, the filament is continuous and therefore has a spine and barbs. The second image two days later shows little evidence of the spine and the third image, two more days later, shows only barbs. Continuing in the bottom row, the filament is seen with its barbs against the disk in Hα and some of the high parts of the barbs above the limb but no spine. The image brightness has been increased to show some of the faint upper part of the barbs again the sky. Further increasing the brightness might have made the higher parts more visible, but not as clearly, as typically seen in the next image the following day in a 304Å image from SOHO/EIT. In this image, the prominence spine is complete as it has been all along, just not seen in Hα. In the final 304Å image the prominence is erupting.

Most single filaments evolve through only part of the whole process of starting as an active region filament and evolving to an extremely quiescent filament without a visible spine in Hα. Filaments can form at any of these evolutionary stages and either progress or not progress to a next stage before eruption. Their formation and their structure are dependent on the magnetic flux density of areas of opposite polarity

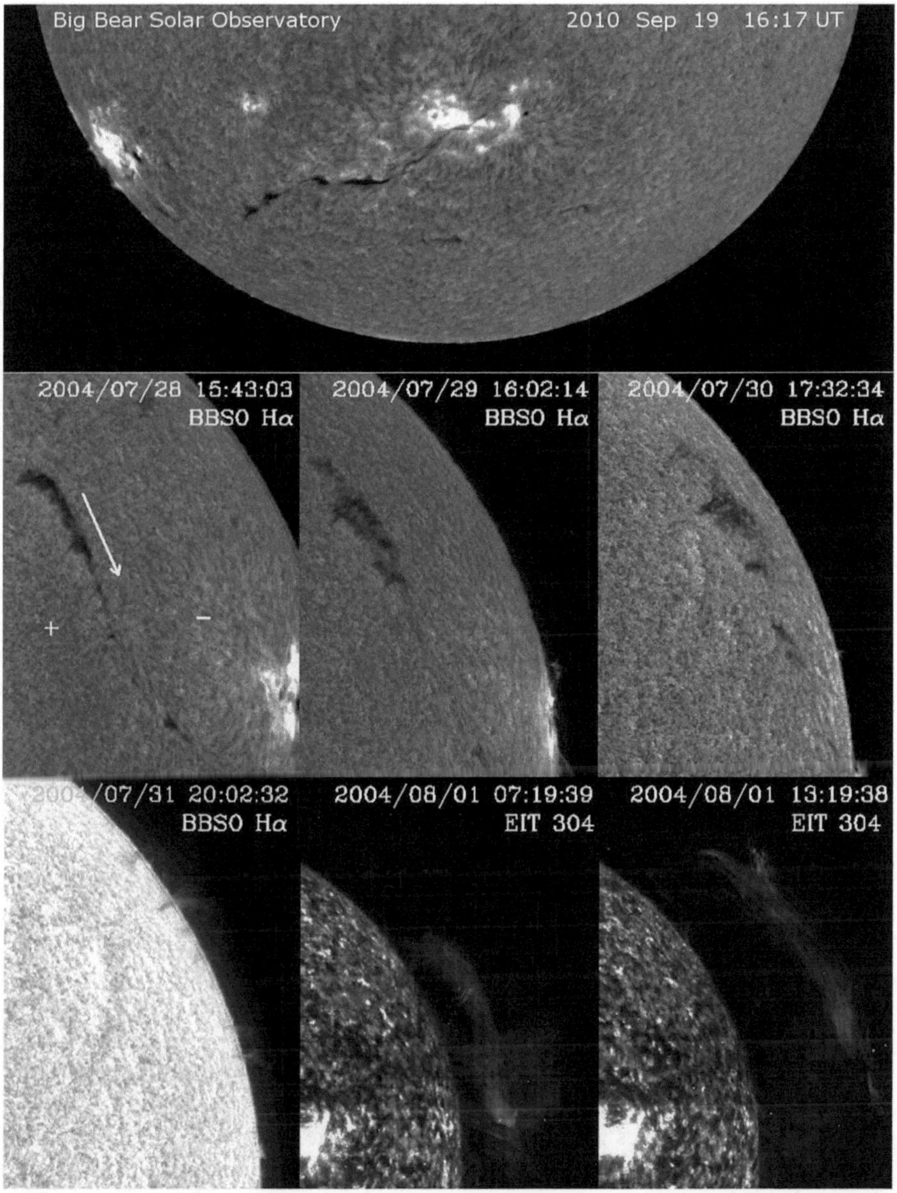

Fig. 9.8 The first image shows how the structure of a filament changes from one end in an active region to the other end on the quiet Sun. The differences are attributed to the filament responding to an environment with lower and lower magnetic flux density with increasing distance from the active region. In the *middle*, a filament with spine and barbs loses its spine as seen in Hα and evolves to showing only its large barbs without a clear spine. The spine however was visible in SOHO EIT 304Å images at the limb for one day before the filament erupted and clearly revealed its whole spine as it began to erupt

that migrate together and establish a boundary of cancelling fields. Additionally, multiple eruptions are common; filaments erupt and reform in the same channels so often that a single filament rarely survives a single solar rotation without erupting. Therefore it is the evolution of their filament channels that gives continuity in the appearance of the evolution of filaments from "active region" to "quiescent".

The point here is that there is continuity in filament structure, which parallels the evolution of active regions, from their birth, through the time when they exist between the remnants of many previous active regions. Three parameters that can be used to study and show this continuity are: (1) the mean magnetic flux density of the line-of-sight component of photospheric magnetic fields within filament channels, (2) the degree of organization of fibril structure of filament channels and (3) chirality, the "handedness" property of filaments and their surrounding environment (discussed in Sect. 9.2.8). When the mean magnetic field density is low as in Figs. 9.8 and 9.9, the filament channel is difficult to discern. It can only be found in images of higher resolution close to the places where the spine has threads intersecting the chromosphere.

For very tall filaments, as in Fig. 9.9, often the spine and the places where the spine connects to the chromosphere are no long clearly seen in Hα. However, from 304Å images as seen in Fig. 9.8, it is certain that the spine still exists. On the other hand, the barbs in Hα that connect the spine to the chromosphere are clearly seen. As the filament in Fig. 9.8 comes across the east limb, it is seen to have at least four extremely wide and tall barbs which can be identified from day to day as our perspective of it changes due to solar rotation. In the last three images, our view is changed from seeing the right side of the filament to viewing its left side. The arrow in the last image points to the perspective from which it would be most desirable to view a large arc-shaped structure such as the large arcs seen in Hinode images (Berger et al. 2008).

The last two images in Fig. 9.9 show an SST image and an image from SOT on Hinode. The images are all arranged so that the arrows on the image represent the direction from which the filaments should be viewed to mentally compare them with the large arc seen in the Hinode/SOT image. The dashed arrow on the Hinode/SOT image shows a thin linear absorption feature perpendicular to the solar limb. This is likely threads of the spine that are seen high in the image and above the large structure suggested to be a large giant barb, as in the Hα images in the middle row. The SST image is included to show that possible arc structures might also be seen against the disk in images with spatial resolution of the order of 0.3–5 arc sec. Time-lapse sequences of images are necessary for definitive identifications of arc structure against the disk.

SST images have revealed the counterstreaming threads in filaments more clearly than any other data sets to date (Lin et al. 2003, 2005; Lin 2004). They show many intricate details about the motions and oscillations along threads as well as bodily oscillations and swaying of threads, also reviewed in Kucera (2014). Because threads represent the magnetic structure of filaments, another detail deserving attention here is the evidence that the visible threads of barbs taper into invisible magnetic threads. This is seen in the spines of filaments (Fig. 9.8) and in their barbs (Fig. 9.7).

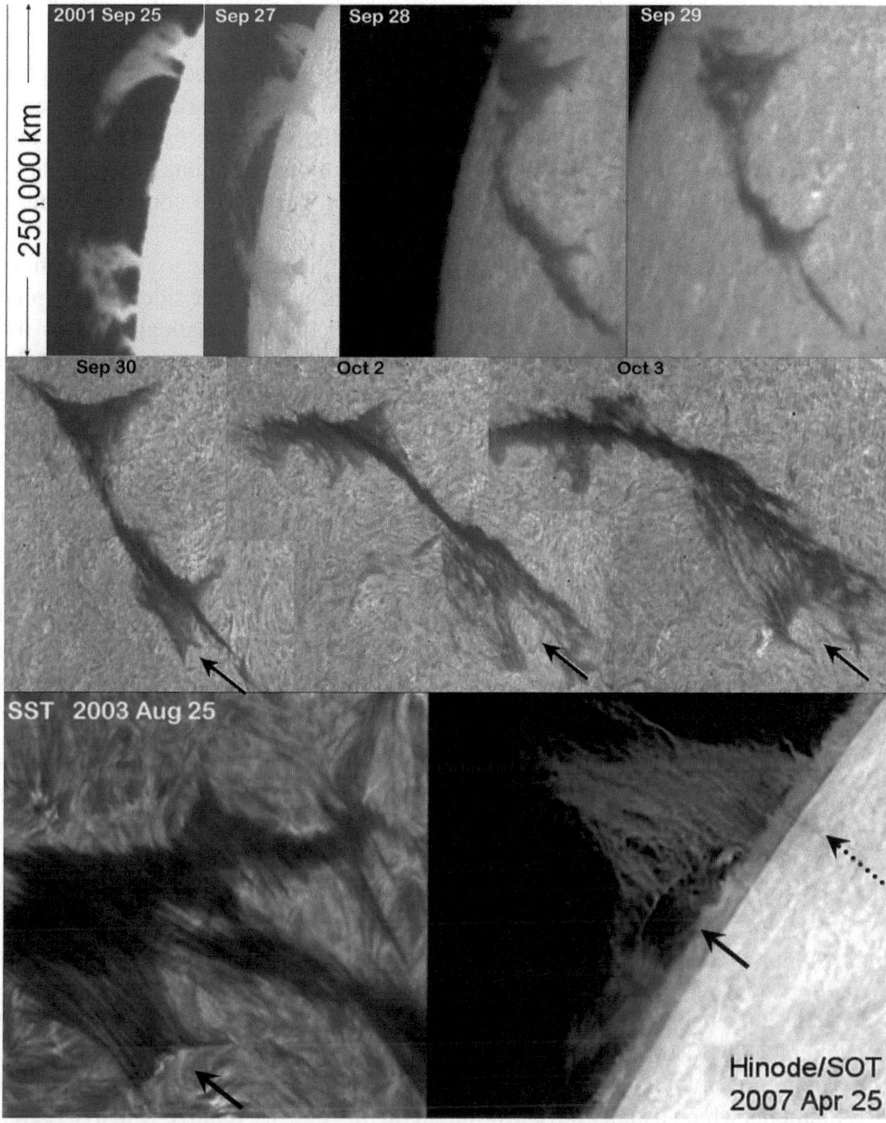

Fig. 9.9 Images from Helio research and the SST show structures that resemble the rising arches seen in Hinode/SOT images. It is thought that the arcs represent the presence of magnetic bubbles coming up through the chromosphere. These images show the potential for new research on this topic if they can be identified in time-lapse series of images against the disk. First two rows of images recorded at Hello Research in southern California, USA

In general, there is more mass near the footpoints of barbs than in the higher parts just outside of the spine (Lin 2004). This property is seen in Hα in the lower image of Fig. 9.12 in the short barb to the left of the spine near the top of the image.

In the 10 arc sec of the barb threads above their endpoint, the threads are dark and appear to merge. The barb on the right side of the spine is an extreme example. The schematic diagram to the right of the image is given to aid in recognizing that the two barbs to the right of the spine in both images have many dark parallel threads at the chromospheric ends of the barbs. Closer to the spine the threads taper and vanish. Time-lapse images verify that the mass motions are along the threads as shown in the schematic. These are therefore normal barbs with counterstreaming mass between the spine and the footpoints.

Once the direction of threads is recognized, it is then possible to interpret more extreme examples such as the structure of the barb on the left side of the spine in the lower images in Fig. 9.7. The dashed black line schematically shows the direction of the spine threads and the barb threads that invisibly extend out of the frame of the image. This evidence of invisible threads is not often seen among active region filaments or intermediate filaments but examples are known where segments of intermediate filaments are relatively high above the chromosphere. This is an essential observation to understanding the magnetic structure of filaments. This knowledge of high threads, which trail into invisibility, does not change any of the observational information or interpretations cited here. To date, it only helps to comprehend structure that initially seemed confusing until tracing extensions of their threads and seeing that they conform to the ordinary pattern of the spine and barb structure of filaments.

In summary, the key observations for quiescent filament channels and filaments are:

11. Taking into account the evidence for invisible threads in the spines and parts of the middle sections of barbs, quiescent filaments are seen to have the same basic spine and barb structure as intermediate filaments but the spines and barbs are taller and barbs extend further from the spine.
12. The spines are as narrow, as for active region and intermediate filaments, and also have threads that are invisible in Hα that can be seen in 304Å images.
13. The spines of prominences in 304Å are well seen in images above the limb where they are bright against the black sky background, but not nearly as well seen as absorption features against the disk and against the dark filament channels in 304Å.
14. The filament channel in the chromosphere is much less organized in quiescent filaments which usually have low magnetic flux density relative to active region and intermediate channels.
15. Quiescent prominences have revealed dynamic rising arch-shaped structures at the base of prominences seen above the limb (Berger et al. 2008). They have also been detected in archives of Hα movies above the limb (de Toma et al. 2009). Amateur astronomers have recently been successfully at observing them. It might be possible to also detect them from mass motions observable in projection against the disk in the barbs of large filaments in Hα or in other lines.

Interpretive conclusions are:

f. The threads of quiescent prominences are field-aligned.
g. The Raleigh–Taylor instability (Berger et al. 2010) has been suggested as the explanation of many of the observed dynamics in filament structure around the rising bubbles. The effect has implications for the filament cavity (Gibson 2014).

9.2.7 *Filament Channels and Filaments in the Low Corona*

A pattern has been found in coronal cells in filament channels that is similar to the pattern of chromospheric fibrils. The pattern is described by Sheeley et al. (2013) for coronal cells in EUV images from SDO/AIA and an example is shown in Fig. 18 in Engvold (2014). Coronal cells recently identified and described by Sheeley and Warren (2012) are most clearly seen at 193 Å. An example is shown in Fig. 9.10 reproduced from Fig. 2 in Sheeley et al (2013). A filament channel is the dark corridor in the middle of the 193 Å image and the dark area to the extreme left is a coronal hole. Coronal cells are seen on both sides of the dark, elongated filament channel in the middle of the image. As described by Sheeley and Warren (2012), the cells are centered on clumps of network magnetic field and the cells are capped with streamers emanating into the corona from each cell.

The cells and their streamers have the shape of tadpoles. Filament channels are recognizable from the opposite directions of the plumes of the cells on the two sides of the filament channel. In Fig. 9.10, the positive polarity is on the right side of the channel. The tails of the tadpoles over the positive network clumps of field bend

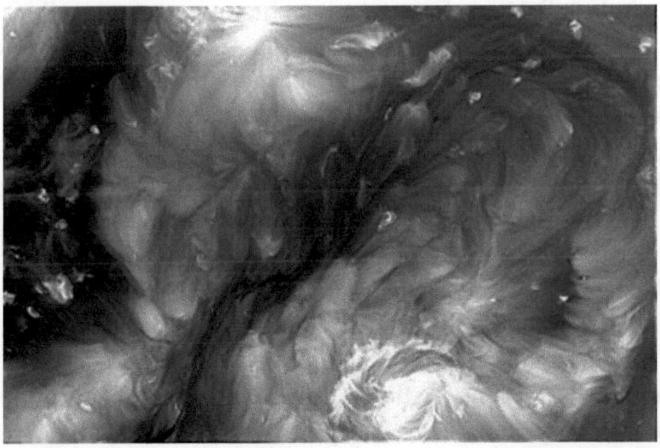

Fig. 9.10 The *dark lane* in the middle of the SDO/AIA image at 193 Å, is a filament channel. Coronal cells are the *tadpole-shaped* features on both sides of the channel. The negative polarity side of the channel is on the *left*. Therefore the field points southward along the filament and has a southward component throughout the channel. The *dark* filament is difficult to see because it is viewed against the *dark* channel in 304 Å. From Fig. 9.2 Sheeley et al. (2013)

into the channel and southward while on the negative polarity side, the tails point northward. Therefore the field along the channel in Fig. 9.10 is south and eastward along it.

The advantage of the 193 Å images is that the coronal cells are bright and large in comparison with Hα fibrils. Therefore they enable (1) finding filament channels of quiescent prominences where the previous illustrations have shown that the channel is difficult to recognize from Hα fibril structure, (2) ascertaining the direction of the magnetic field along the polarity reversal boundary, and hence, (3) along the spine of any filament along the boundary no matter how high.

9.2.8 Chirality and Chiral Systems

By definition, all features with chirality are either left-handed or right handed. Filament channels and filaments are called either sinistral or dextral according to the pattern of their barbs or the corresponding pattern of their filaments channels. Examples of a sinistral and dextral filament are shown respectively in the middle and right images in Fig. 9.6. Details for recognizing chirality in a wide range of filaments are illustrated in Martin et al. (2008).

The chiralities of solar features fit together one-for-one into two larger groupings called *chiral systems* (Martin et al. 2012), designated as the N and S Systems according to their strong trend to be dominant in the northern and southern solar hemispheres respectively (Martin 1998b). The chiralities of the features related most closely to filaments within these two systems are outlined in Table 9.2. Thus if the chirality of one feature is definitively determined, one knows the chirality of all other features within the same system by referring to Table 9.2 and/or the graphical representation in Fig. 9.11, or from previous literature referenced in Table 9.2 with the addition of the plumes of coronal cells in Fig. 9.10 (Fig. 1 in Sheeley et al. 2013).

The above two paragraphs give the reader more than enough information to find the chirality of all of the filament channels and filaments in Figs. 9.1, 9.2, 9.3, 9.4, 9.5, 9.6, 9.7, 9.8, 9.9, 9.10, and 9.11 herein. Figure 9.8 is the most challenging because the resolution of the images is too low to see the threads in the filaments or the fibrils in the chromosphere. Only subtle indications allow seeing that the long filament in the BBSO image is sinistral from two of its clearest barbs that appear like left-hand exits from a high speed roadway. In the lower images in Fig. 9.8, three of its filament barbs in the first two days slant from the spine downward to the right, the side view signature of a dextral filament.

As seen in Table 9.2, identifications of the chirality of most of the solar features that belong to chiral systems were being found in the early 1990s (review Martin 1998b; Martin et al. 2012). A significant addition is the chiral patterns of coronal cells that define filament channels in the corona (Sect. 9.2.7). Early interpretations were expressed by Rust (1994) and Low (1994) that eruptive solar events enabled the Sun to shed magnetic energy in the form of helicity.

Table 9.2 Chiral forms in solar features and groupings within chiral systems

Solar feature	N system	S system	First references
Polarity reversal boundaries or Filament spine	0 0	0 0	Represent 0 chirality
(1) Filament in channels in chromosphere	Dextral	Sinistral	Martin et al. (1994)
(2) Coronal cell plumes Filament channels—low corona	Dextral	Sinistral	Sheeley et al. (2013)
(3) Filaments barbs	Dextral	Sinistral	Martin et al. (1994)
(4) Cavities represented by sigmoids	Reverse S-writhe	S-writhe	Pevtsov et al. (1996)
(5) Coronal loop systems	Left-skewed	Right-skewed	Martin and McAllister (1996)

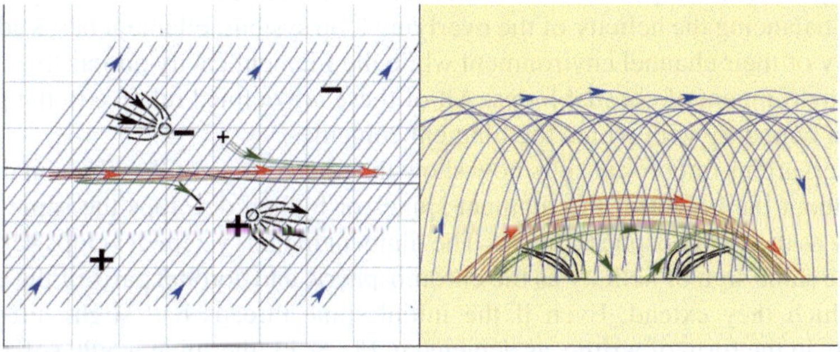

Fig. 9.11 Components of a N chiral system represented area: (1) left-skewed coronal loop system (*blue*), (2) dextral filament channel (*green*), (3) dextral filament barbs (*green*), (4) purely filament spine threads (*red*). The filament spine can be regarded as having no chirality or very slight left-hand writhe. The filament cavity is implied by the space above the channel component and the filament. Illustration credit: O. Panasenco in Martin et al. (2012)

Separate from the above interpretations, filaments are now recognized to lie at the heart of the large-scale magnetic *chiral systems* that develop within filament channels beneath coronal loop systems. Chiral systems can be regarded as the magnetic containers of the buildup of the magnetic energy that is released in solar eruptive events (Martin et al. 2012). During a solar eruptive event, the channel base of the container is forced to remain at the Sun while the contents undergo explosive or enduring reconnection and take the magnetic lid of loops with it into interplanetary space as the magnetic reconnection converts the loop system and its writhing contents into twist. There are and will be many more theoretical ideas developed on the details of how this happens (Gibson 2014). More observational details about chiral systems are important as guides for future models or tests for existing ones.

Some of the key features of an N-chiral system are depicted in Fig. 9.11 from a top and side view. The directions are reversed 180° for S-Systems (not illustrated).

Because filament channels develop in the core of decaying active regions (after flux emergence has ceased), on the boundaries of active regions, and along other polarity reversal boundaries on the quiet Sun, it appears that the Sun is continuously converting the magnetic fields of all active regions, and their dispersed remnants, into chiral systems. Apparently, the Sun does not expel simple bipolar regions into the corona; it first reorganizes their magnetic fields via interplays of various regimes of magnetic reconnections and converts them into chiral systems which serve to both store and release accumulating energy and helicity.

A significant property of individual N or S chiral systems is that each develops both left-handed and right-handed components. This is consistent with the hypothesis that each chiral system is an entity that conforms to the conservation of helicity (Martin et al. 2012). If so, a system is expected to have equal and opposite helicity in different parts of the total system which is approximately what the observations of a chiral systems show. In the schematic of a chiral system in Fig. 9.10, the coronal loop system depicted in blue lines, is left-skewed. However, a right-handed filament channel, without exception, develops beneath a left-handed coronal loop system as if it were balancing the helicity of the overlying loop system. Filament barbs adopt the chirality of their channel environment while the filament cavity adopts the chirality of its environment of coronal loops. All of these are defined relative to the polarity reversal boundary with which the filament coincides.

Observations of coronal cells now show that the volume of the filament channel, up to about the height of the filament spines or higher, develops the same sign of helicity as the chromospheric fibrils. The filament in this system develops barbs only with the same sign of helicity as the chromospheric and coronal parts of the channel into which they extend. Even if the initial spine threads had slight left-handed chirality in the form of writhe, as depicted in Fig. 9.11, the spine would still develop right-handed barb threads because only right handed barbs could survive in a right handed channel.

At the top of the spine, the sign of chirality changes to be consistent with the sign of the overlying loop system. This is verified by the occasional formation of an X-ray or EUV sigmoid in the cavity above the filament spine. The reverse S-shaped writhe in Table 9.2 is left-handed, the same chirality as the overlying coronal loop system. For the simplest helicity-balanced field system in a least energy state, it is reasonable that everything below the filament spine would have one sign of helicity and everything above and far out around the filament spine would have the opposite sign of helicity. If the magnitudes of the left-handed and right-handed parts of the chiral system are equal, helicity is conserved.

As shown herein, the filament spine serves as a visible magnetic signature of largely horizontal magnetic threads appearing concurrently with the disappearance of line-of-sight magnetic fields, the observed cancelling magnetic fields at its base. The consistent interpretation analytically developed by Litvinenko (1999) is that magnetic reconnection at or near the photosphere converts line-of-sight magnetic flux into horizontal flux that can rise into the corona. Once in the corona, counterstreaming reveals filament threads both gaining and losing their mass. Considered in the light of the theory of Bellan (2003) of why flux tubes are thin, Martin et al. (2012) suggested that the role of filaments in chiral systems is shedding the mass of their field-aligned threads and passing these invisible magnetic threads continuously into the cavity where they are significant contributors to energy storage in their chiral system. This reasoning is presented as just one context in which chiral systems are relevant to further research and modeling to understand how and why solar eruptive events occur.

9.3 Coronal Cloud Prominences

Coronal cloud prominences differ in major ways from channel prominences (see Table 9.1, Sect. 9.1). One such difference is that they do not have chirality, and in this sense are peripheral to chiral systems. A major observational difference is that cloud prominences are generally much fainter. This is well illustrated in two comparison photographs of a cloud prominence taken with short and long exposures in Fig. 9.12 by Leroy (1969). In exposures suitable for the disk and low prominences, the only sign of the coronal cloud prominence was a short streak of coronal rain descending from a higher and more cloud-like part of the prominence. Without the longer exposure, the streak would not have been recognized as part of a large cloud prominence. Consequently, relatively few examples of coronal cloud prominences (also called suspended prominences in the Menzel–Evans classification) have been recorded and studied.

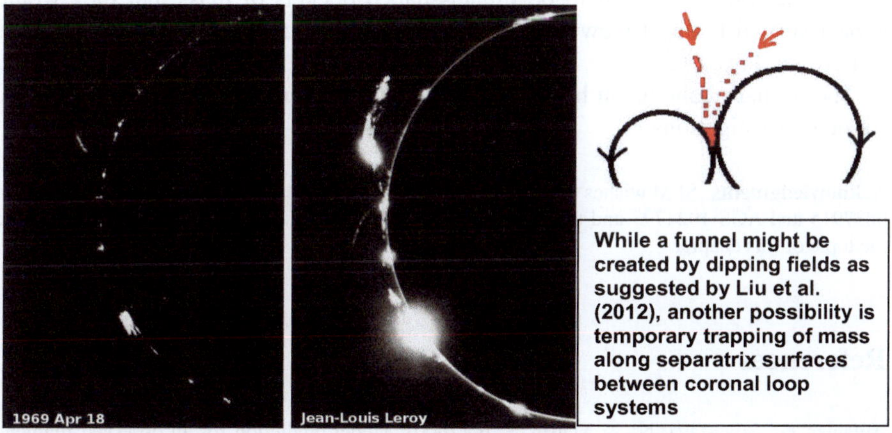

Fig. 9.12 A tall coronal cloud prominence at the northeast limb and coronal rain at the east limb is only seen well in the second long-exposure photograph through an H α filter. However, the longer exposure results in over-exposing all of the channel prominences which have much lower heights. Credit: Jean-Louis Leroy (1969) for left and middle images

Many possibilities for finding new information about coronal cloud prominences are effectively demonstrated in (Liu et al 2014), and in recent observations during solar eclipses (Habbal et al. 2011). The latter reveal many previously undetected features in the solar corona which could temporarily halt, slow, or divert mass falling inward to the Sun from various sources.

One of the basic questions that have begun to be addressed is, the source of the mass for coronal cloud prominences. Lin et al (2006) observed and studied four tall coronal cloud prominences. They noted that these prominences appeared within about a day following a CME in the same solar quadrant and suggested that the source of their mass comes from mass falling back to the Sun after eruptive solar events. Previously, Wang et al. (1999) analyzed inflows to the Sun, and also found a tendency for them to occur about a day after a CME. If the Lin et al. hypothesis proves to be true, at least some coronal cloud prominences are a by-product of chiral systems rather than being an integral part of the buildup of chiral systems. In any case, it is important to remain open to other possible sources of mass for coronal cloud prominences (Liu et al. 2014).

A second major question is what type of magnetic field configuration supports these prominences. Using SDO/AIA images, Liu et al (2012) show excellent imaged evidence in one example that the falling mass encounters a coronal magnetic field that responds by sagging and appearing like a V-shaped dip that is smooth rather than sharply angled. However, in other cases their shapes occasionally give the impression that they might be mass trapped at separatrix surfaces between adjoining coronal loop systems as depicted in Fig. 9.12. This would be consistent with their mass sometimes being seen falling to large arcs and then flowing along the arcs to the chromosphere. Viewing perspective is all important. In the diagram in Fig. 9.12, viewed from the perspective of the face of a system of coronal loops, coronal cloud prominences might appear to be funnel-shaped pockets of mass, but they would appear very different if viewed at 90° to the plane depicted, more like the example of Leroy in Fig. 9.12.

The primary conclusion here is that coronal cloud prominences are a ripe topic for new investigations.

Acknowledgments SFM wishes to thank all colleagues who shared the work on NSF grants AGS-0837915 and AGS-1024793 and to express appreciation for many valuable discussions related to the topic of this chapter.

References

Aulanier, G., Srivastava, N., & Martin, S. F. (2000). Model prediction for an observed filament. *The Astrophysical Journal, 543*(1), 447–456.

Babcock, H. D., & Babcock, H. W. (1955). The Sun's magnetic field, 1952–1954. *The Astrophysical Journal, 121*, 349.

Bellan, P. M. (2003). Why current-carrying magnetic flux tubes gobble up plasma and become thin as a result. *Physics of Plasmas, 10*, 1999–2008.

Berger, T. E., Shine, R. A., Slater, G. L., et al. (2008). Hinode SOT observations of solar quiescent prominence dynamics. *The Astrophysical Journal, 676*, L89–L92.

Berger, T. E., Slater, G., Hurlburt, N., et al. (2010). Quiescent prominence dynamics observed with the Hinode solar optical telescope. I. Turbulent upflow plumes. *The Astrophysical Journal, 716*, 1288–1307.

Chae, J., Moon, Y.-J., & Park, Y.-D. (2005). The magnetic structure of filament barbs. *The Astrophysical Journal, 626*(1), 574–578.

de Toma, G., Casini, R., Berger, T. E., Low, B. C., de Wijn, A. G., Low, B. C., Burkepile, J. T., & Balasubramaniam, K. S. (2009). Observations of large-scale dynamic bubbles in prominences. In B. Lites, M. Cheung, T. Magara, J. Mariska, & K. Reeves (Eds.), *ASP conference* (Vol. 415, p. 163).

Engvold, O. (2014). Description and classification of Prominences. In J.-C. Vial & O. Engvold (Eds.), *Solar prominences, ASSL* (Vol. 415, pp. 31–60). New York: Springer.

Foukal, P. (1971a). Morphological relationships in the chromospheric Hα fine structure. *Solar Physics, 19*, 59.

Foukal, P. (1971b). Hα fine structure and the chromospheric field. *Solar Physics, 20*, 298–309.

Gaizauskas, V. (1998). Filament channels: Essential ingredients for filament formation (Review). In *ASP conference series*. (Vol. 150, pp. 257–264).

Gibson, S. (2014). Coronal cavities: Observations and implications for the magnetic environment of prominences. In J.-C. Vial & O. Engvold (Eds.), *Solar prominences, ASSL* (Vol. 415, pp. 321–351). New York: Springer.

Grigoryev, V. M., & Ermakova, L. V. (1999). On the origin of solar filament magnetic fields. *Astronomical and Astrophysical Transactions, 17*, 355–365.

Habbal, S. R., Druckmüller, M., Morgan, H., Ding, A., Johnson, J., Druckmüllerová, H., Daw, A., Arndt, M. B., Dietzel, M., & Saken, J. (2011). Thermodynamics of the solar corona and evolution of the solar magnetic field as inferred from the total solar eclipse observations of 2010 July 11. *The Astrophysical Journal, 734*(2). article id. 120, 18 pp.

Hammerschlag, R. H., Sliepen, G., Bettonvil, F. C. M., Jägers, A. P. L., Sütterlin, P., Lin, Y., Martin, S. F., Panasenco, O., & Romashets, E. P. (2013). *Optical Engineering, 52*(8), 081603.

Harvey, K. L. (1993). Doctoral Thesis, University of Utrecht.

Janssens, T. (1970). Long term observations of the Hα chromospheric network. *Solar Physics, 11*, 222–242.

Kucera, T. (2014). Derivations and observations of prominences bulk motions and mass. In J.-C. Vial & O. Engvold (Eds.), *Solar prominences, ASSL* (Vol. 415, pp. 77–99). New York: Springer.

Kuckein, C., Martinez Pillet, V., Centeno, R. (2012). An active region filament studied simultaneously in the chromosphere and photosphere. *I. Magnetic structure, Astronomy & Astrophysics, 539*. article id. A131, 15 pp.

Leighton, R. B. (1959). Observations of solar magnetic fields in plage regions. *The Astrophysical Journal, 130*, 366.

Leighton, R. B., & Simon, G. W. (1964). Velocity fields in the solar atmosphere. III. Large-scale motions, the chromospheric network, and magnetic fields. *The Astrophysical Journal, 140*, 1120.

Leighton, R. B., Noyes, R. W., & Simon, G. W. (1962). Velocity fields in the solar atmosphere. I. Preliminary report. *The Astrophysical Journal, 135*, 474.

Leroy, J.-L. (1969). Photométries comparées des émissions de l'hydrogène et de l'hélium dans les régions externes des protubérances. *Solar Physics, 7*, 221–237.

Lin, H. (1995). On the distribution of the solar magnetic fields. *The Astrophysical Journal, 446*, 421.

Lin, Y. (2004). *Magnetic field topology inferred from studies of fine threads in solar filaments*. PhD Thesis, Institute of Theoretical Astrophysics, University of Oslo.

Lin, Y., Engvold, O., & Wiik, J. E. (2003). Counterstreaming in a large polar crown filament. *Solar Physics, 216*, 109–120.

Lin, Y., Engvold, O., Rouppe van der Voort, L., Wiik, J., & Berger, T. (2005). Thin threads of solar filaments. *Solar Physics, 226*, 239–254.

Lin, Y., Martin, S. F., & Engvold, O. (2006). Coronal cloud prominences and their association with coronal mass ejections (abstract). *Bulletin of the American Astronomical Society, 38*, 219.

Lites, B. W., Kubo, M., Berger, T., Frank, Z., Shine, R., Tarbell, T., Title, A., Okamoto, T. J., & Otsuji, K. (2010). Emergence of helical flux and the formation of an active region filament channel. *The Astrophysical Journal, 718*, 474–487.

Litvinenko, Y. E. (1999). Photospheric magnetic reconnection and canceling magnetic features on the sun. *The Astrophysical Journal, 515*, 435–440.

Litvinenko, Y. E. (2010). Evolution of the axial magnetic field in solar filament channels. *The Astrophysical Journal, 720*, 948–952.

Litvinenko, Y. E., & Martin, S. F. (1999). Magnetic reconnection as the cause of a photospheric canceling feature and mass flows in a filament. *Solar Physics, 190*, 45.

Litvinenko, Y., Chae, J., & Park, S.-Y. (2007). Flux pile-up magnetic reconnection in the solar photosphere. *The Astrophysical Journal, 662*, 1302–1308.

Liu, W., Berger, T. E., & Low, B. C. (2012). First SDO/AIA observation of solar prominence formation following an eruption: Magnetic dips and sustained condensation and drainage. *The Astrophysical Journal Letters, 745*, L21. 8 pp.

Liu, W., Berger, T. E., & Low, B. C. (2014). Coronal condensation in funnel prominences as return flows of the chromosphere–corona mass cycle. In *IAU symposium* (Vol. 300, pp. 441–442).

Livi, S. H. B., Wang, J., & Martin, S. F. (1985). The cancellation of magnetic flux I – On the quiet Sun. *Australian Journal of Physics, 38*, 855.

Low, B. C. (1994). Coronal mass ejections and magnetic. *Phenomena and solar wind consequences, ESA 373* (p. 123).

Mackay, D. H., Gaizauskas, V., & Yeates, A. R. (2008). Where do solar filaments form?: Consequences for theoretical models. *Solar Physics, 248*, 51–65.

Mackay, D. H., Gaizauskas, V., & Yeates, A. R. (2014). Where do solar filaments form? Nature of prominences and their role in space weather. In B. Schmieder, J.-M. Malherbe, & S. T. Wu (Eds.), *IAU symposium* (Vol. 300, pp. 445–446).

Martin, S. F. (1990). Conditions for the formation of prominences as inferred from optical observations. In *Dynamics of solar prominences, IAU Colloquium at Hvar, Yugoslavia* (Lecture notes in physics, p. 1). Berlin: Springer.

Martin, S. F. (1998a). In D. Webb, D.M. Rust, & G. Schmieder (Eds.), *IAU Colloquium 167, ASP Conf. Series* (Vol. 150, p. 419)

Martin, S. F. (1998b). Conditions for the formation and maintenance of filaments (invited review). *Solar Physics, 182*, 107–137.

Martin, S. F., & Echols, C. R. (1994). An observational and conceptual model of the magnetic field of a filament. In R. J. Rutten & C. J. Schrijver (Eds.), *Solar surface magnetism* (pp. 339–346). Dordrecht: Kluwer Academic Publishers.

Martin, S. F., & McAllister, A. H. (1996). The skew of X-ray coronal loops overlying H α Filaments. In Y. Uchida, T. Kosugi & H. S. Hudson (Eds.), *Magnetodynamic phenomea in the solar atmosphere: Prototypes of stellar magnetic activity* (p. 497). Dordrecht: Kluwer Academic Publishers.

Martin, S. F., & Panasenco, O. (2010). On dynamical properties of filament channels. *Memorie della Societa Astronomica Italiana, 81*, 662.

Martin, S. F., Panasenco, O., Berger, M. A., Engvold, O., Lin, Y., Pevtsov, A. A., & Srivastava, N. (2012). The Build-Up to Eruptive Solar Events Viewed as the Development of Chiral Systems, The Second ATST-EAST Meeting: Magnetic Fields from the Photosphere to the Corona. In T. Rimmele, A. Tritschler, F. Woger, V. Collados, H. Socos-Navarro, R. Schlichenmaier, M. Carlsson, T. Berger, A. Cadavid, P. Gilbert, P. Goode & M. Knolker (Eds.), *ASP Conference Proceedings*. (Vol. 463, p. 157). San Francisco: Astronomical Society of the Pacific.

Martin, S. F., Livi, S. H. B., & Wang, J. (1985). The cancellation of magnetic flux. II – In a decaying active region. *Australian Journal of Physics, 38*, 929–959.

Martin, S. F., Bilimoria, R., & Tracadas, P. W. (1994). Magnetic field configurations basic to filament channels and filaments. In R. J. Rutten & C. J. Schrijver (Eds.), *Solar surface magnetism* (p. 303). Dordrecht: Kluwer Academic Publishers.

Martin, S. F., Lin, Y., & Engvold, O. (2008). A method of resolving the 180 degree ambiguity employing the chirality of solar features. *Solar Physics, 250*, 31–51.
Metcalf, T. R., Leka, K. D., Barnes, G., Lites, B. W., Georgoulis, M. K., Pevtsov, A. A., Balasubramaniam, K. S., Gary, G. A., Jing, J., Li, J., et al. (2006). An overview of existing algorithms for resolving the 180° ambiguity in vector magnetic fields: Quantitative tests with synthetic data. *Solar Physics, 237*, 267–296.
Okamoto, T. J., Tsuneta, S., Lites, B. W., Kubo, M., Yokoyama, T., Berger, T. E., Ichimoto, K., Katsukawa, Y., Nagata, S., Shibata, K., et al. (2008). Emergence of a helical flux rope under an active region prominence. *The Astrophysical Journal, 673*, L215–L218.
Pevtsov, A. A., Canfield, R. C., & Zirin, H. (1996). Reconnection and helicity in a solar flare. *Astrophysical Journal, 473*, 533–538.
Pikel'Ner, S. B. (1971). Nature of the fine structure of the chromosphere. *Astronomicheskii Zhurnal, 48*, 1212.
Rompolt, B., & Bogdan, T. (1986). On the formation of active region prominences (H αfilaments). In *NASA. Goddard space flight center, coronal and prominence plasmas* (pp. 81–87) (SEE N87-20871 13–92).
Rompolt, B. (1990). Small scale structure and dynamics of prominences. *Hvar Observatory Bulletin, 14(1)*, 37.
Rust, D. M. (1994). Spawning and shedding helical magnetic fields in the solar atmosphere. *Geophysical Research Letters, 21*, 241–244.
Sheeley, N. R., Jr., & Warren, H. P. (2012). Coronal cells. *The Astrophysical Journal, 749*, 40–54.
Sheeley, N. R., Jr., Martin, S. F., Panasenco, O., & Warren, H. P. (2013). Using coronal cells to infer the magnetic field structure and chirality of filament channels. *The Astrophysical Journal, 772*. article id. 88, 11 pp.
Smith, S. F. (1968) The formation, structure and changes in filaments in active regions. In *IAU symposium* (Vol. 35, p. 267).
Smith, S. F., Ramsey, H. E., & Howard, R. (1965). Some characteristic properties of solar magnetic fields. *Astronomical Journal, 70*, 330.
Su, J. T., Jing, J., Wang, H. M., Mao, X. J., Wang, X. F., Zhang, H. Q., Deng, Y. Y., Guo, J., & Wang, G. P. (2010). Observational evidence of changing photospheric vector magnetic fields associated with solar flares. *The Astrophysical Journal, 721*, 901.
Wang, Y.-M., & Muglach, K. (2007). On the formation of filament channels. *The Astrophysical Journal, 666*, 1284–1295.
Wang, Y.-M., Sheeley, N. R., Jr., Howard, R. A., Cyr, O. C. S., & Simnett, G. M. (1999). Coronagraph observations of inflows during high solar activity. *Geophysical Research Letters, 26*, 1203–1206.
Wang, J., Li, W., Denker, C., Lee, C., Wang, H., Goode, P. R., McAllister, A., & Martin, S. F. (2000). Minifilament eruption on the quiet Sun. I. Observations at Hα central line. *The Astrophysical Journal, 530*, 1071–1084.
Zirin, H. (1972). Fine structure of solar magnetic fields. *Solar Physics, 22*, 34–48.
Zwaan, C. (1987). Elements and patterns in the solar magnetic field. *Annual Review of Astronomy and Astrophysics, 25*, 83–111.

Chapter 10
Plasma Structure and Dynamics

Judith T. Karpen

Abstract Despite over a century of observations, the physical processes by which prominence plasma forms and evolves remain controversial. In this chapter we review the observational constraints on all mass formation models, review the four leading models—injection, levitation, evaporation–condensation, and magneto-thermal convection, describe the strengths and weaknesses of each model, and point out opportunities for future work. As needed, short tutorials are provided on fundamental physical mechanisms and concepts not covered in other chapters, including magnetic reconnection and energy balance in coronal loops.

10.1 Introduction

Although the magnetic field is largely responsible for the support, shape, and ultimate ejection of prominences, their defining characteristics are entirely derived from observations of the cool, dense plasma residing in this magnetic structure. As noted in Chap. 8; López Ariste (2014), magnetic field measurements in prominences are difficult and rare, although advances in instrumentation promise to increase direct access to this vital information. Hence, despite over a century of observations, the physical processes by which prominence plasma forms and evolves remain controversial.

Successful plasma formation models must satisfy the following primary observational constraints (see Chap. 2 for details; Engvold 2014):

- Prominences cover an enormous range in space and time. Those originating in active regions tend to be short ($\sim$10 Mm), short-lived (minutes to hours), and low ($<$10 Mm above the photosphere), In contrast, prominences formed outside active regions can extend up to hundreds of Mm, live several weeks to months, and reach maximum heights up to 100 Mm.
- Prominence plasma is highly dynamic, reaching speeds of order 10–70 km s^{-1} (Kubota and Uesugi 1986; Schmieder et al. 1991; Zirker et al. 1998; Kucera et al.

J.T. Karpen (✉)
NASA Goddard Space Flight Center, Code 671, Greenbelt, MD 20771, USA
e-mail: judith.t.karpen@nasa.gov

J.-C. Vial, O. Engvold (eds.), *Solar Prominences*, Astrophysics and Space Science Library 415, DOI 10.1007/978-3-319-10416-4_10

2003; Lin et al. 2003; Okamoto et al. 2007; Berger et al. 2008, 2010; Chae et al. 2008; Ahn et al. 2010; Hillier et al. 2011). Quiescent filaments can go through phases of enhanced internal motion and activity (Martin 1973; Tandberg-Hanssen 1995; Kucera and Landi 2006), often leading to eruption.
- Prominence plasma frequently appears in situ high in the hot corona (McMath and Pettit 1938; Schrijver 2001; Berger et al. 2012), but also flows up from the chromosphere (Chae et al. 2000; Liu et al. 2005; Schwartz et al. 2006).
- Prominence plasma exhibits fine structure down to the resolution limit ~100 km), in the form of knots and threads 3–20 Mm long (Lin et al. 2003, 2005; Heinzel and Anzer 2006; Gunar et al. 2007, 2008; Berger et al. 2008).
- Barbs (Chap. 2; Engvold 2014) appear to extend from the prominence spine toward small regions of parasitic polarity (Kippenhahn and Schlüter 1957; Malherbe et al. 1989; Martin 1998; Chae et al. 2005), but their termination at the chromosphere is very difficult to observe. Their heights exceed the gravitational scale height of the cool plasma, but it is unclear what prevents this material from collapsing.

This broad set of properties cannot be explained by any single model, so this chapter discusses several possibilities.

Over 40 years ago, researchers realized that the large prominence mass must come from the chromosphere, rather than the rarefied corona (Pikel'ner 1971; Saito and Tandberg-Hanssen 1973; Zirker et al. 1994). The chromosphere is two orders of magnitude denser than the corona, so it is much easier to extract sufficient mass to explain prominences from the chromosphere than from the corona. The fundamental question is, then, how can chromospheric mass be transformed into prominence mass far above the standard chromosphere? Physics dictates two pathways: either through magnetic forces, which inject or lift cool plasma directly into the corona, or through thermal forces, which evaporate heated plasma that subsequently condenses into prominence knots or threads. In this chapter we summarize the current status of these models for the prominence plasma, and describe the basic physical processes involved in these models.

10.2 Injection

Injection models are based on the premise that chromospheric plasma in filament channels is driven directly into the corona, and should be visible as upward jets or upflows. High-cadence images of the Sun commonly show a multitude of cool upflows, in the form of spicules, macrospicules, polar jets, etc., but most of these are not confined to filament channels or are unlikely sources of prominence mass for other reasons (e.g., mass is observed to drain rather than stay in corona). A few specific observations of strong flows into active-region prominences support this model (Chae 2003), but evidence is lacking for the mass of larger, higher quiescent prominences being supplied by this mechanism. Both large- and small-scale upflows have been recorded in quiet-Sun hedgerow prominences (Zirker et al. 1994; Berger

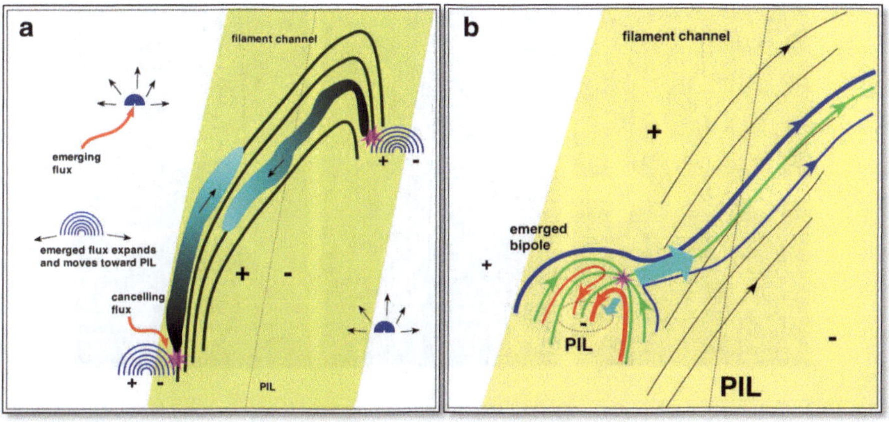

Fig. 10.1 (**a**) Illustration of injection of prominence plasma (aqua) by reconnection between emerged bipoles (*blue lines*) and preexisting filament-channel field (*black lines*). Polarity inversion lines are *dotted*; the photospheric filament channel is *yellow*. *Black arrows* indicate flow directions. (**b**) Close-up view of reconnection and associated flows. The magnetic topology is indicated by the *green field lines* of the spine and fan (separatrix) separating the intruding bipolar flux system (*red lines*) from the filament channel field (*blue lines*); thin *black field lines* in the channel are beneath the colored field lines closer to the filament-channel PIL, and unaffected by reconnection. The *thicker red* and *blue field lines* were formed by reconnection at the flux-breaking region (magenta star), driving a strong upflow and a weaker downflow (*thick aqua arrows*). The "parent" field lines (before reconnection) resembled the *thinner red* and *blue field lines* nearby, which are next in line to reconnect

et al. 2008, 2010), although it is difficult to determine whether and how these flows originate at the photosphere because of spicules and other obscuring chromospheric activity. In the magneto-thermal convection model such upflows are explained by a different process, as is discussed in Sect. 10.5.

In general, injection models (see Fig. 10.1) rely on magnetic reconnection in the low atmosphere to propel cool plasma to typical prominence heights (Wang 1999; Chae et al. 2001; Litvinenko et al. 2007). This conjecture is largely motivated by the well-observed, but poorly understood, connection between flux cancellation and filament channel formation (van Ballegooijen and Martens 1989; Martin 1998; Wang and Muglach 2007), discussed in Chap. 9; Martin 2014 and Chap. 12; Fan 2014. Magnetic reconnection is a fundamental process in magnetized plasmas in laboratory devices and throughout the universe, which is especially relevant to energetic activity throughout the heliosphere. Here we only provide a brief overview of reconnection, and refer the reader to the excellent reviews by Biskamp (2000), Priest and Forbes (2000), Kulsrud (2005), and Yamada et al. (2010) to explore this important subject thoroughly.

Magnetic reconnection is simply a mechanism that breaks and reconfigures magnetic field lines (see Fig. 10.2). In an ideal plasma—where resistivity (magnetic dissipation) is negligible by definition—the ions and electrons are forced to gyrate around field lines, and the field lines retain their original connectivity. This is

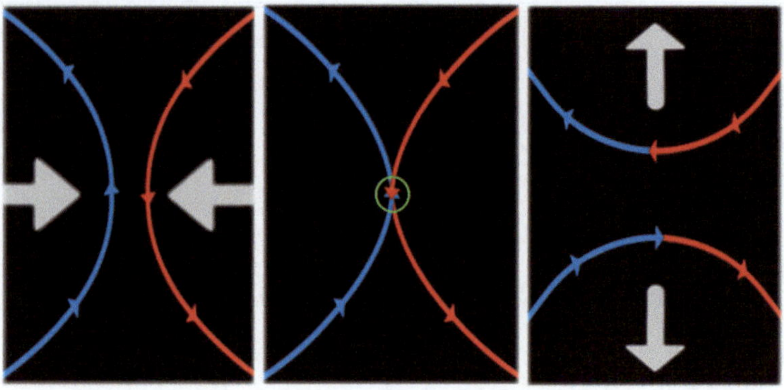

Fig. 10.2 Time sequence (*left to right*) of two reconnecting field lines (*red and blue*). *Gray arrows* indicate flow directions. The flux-breaking region is marked by a small, *green circle* in the *middle panel*

denoted the *frozen-in* condition. Conversely, high resistivity allows the plasma to slip through the field; field lines lose their identity, and the pairs of footpoints initially linked by specific field lines are no longer connected. When the *flux-breaking* region is much larger than typical kinetic scales, magnetic *diffusion* alters the connectivity of many field lines gently and smoothly. When the flux-breaking region is highly localized, however, the reconfiguration of field lines processed through that region is called *reconnection*, which can be extremely dynamic and impulsive. Note that high resistivity alone, even in a small region, generally yields diffusion rather than reconnection; the field lines can slip too quickly through the plasma, releasing energy steadily rather than explosively.

Reconnection provides one way for a stressed, magnetized plasma to lower its energy, bringing it closer to the potential (most relaxed) state. The magnetic energy released through reconnection is transferred to the plasma primarily in kinetic (so-called reconnection jets) and thermal (Ohmic heating, shocks, and thermalization of the reconnection jets) forms, as well as nonthermal particle acceleration. However, this energy-lowering process is available only where strong gradients in the magnetic field occur: at topological features called magnetic nulls and current sheets. The reason is easily revealed by examining the resistive magnetic-induction equation:

$$\frac{\partial \mathbf{B}}{\partial t} = \nabla \times (\mathbf{v} \times \mathbf{B}) + \eta \nabla^2 \mathbf{B} \tag{10.1}$$

where $\mathbf{B}$ is the vector magnetic field, $\mathbf{v}$ is the vector velocity field, and η is the resistivity. The second term on the right-hand side is the key term: if either the resistivity (η) or the magnetic-field gradient ($\nabla \mathbf{B}$) is zero, then no dissipation or reconnection can occur. In most of the solar corona the effective resistivity is very small, so reconnection is confined to special locations where either the resistivity is anomalously enhanced (e.g., by current instabilities) or where the magnetic gradients are extremely large. Current sheets and nulls meet this requirement because one or more components of the magnetic field reverse direction at those locations, and because the kinetic processes responsible for flux breaking become significant on the small spatial scales of those topological features.

On the Sun, reconnection generally can occur wherever different magnetic flux systems intersect and form current sheets. These flux systems are unlikely to be completely aligned where they intersect, so the stage is set for current sheet formation and subsequent reconnection. To drive explosive activity a source of excess magnetic energy is required, however, such as preexisting twisted flux or the buildup of twist through rotational or braiding photospheric motions. In the injection model, reconnection is likely to occur where an emerged bipole (one flux system) encounters filament-channel flux (another flux system), as sketched in Fig. 10.1a. As shown, this system already contains a null (located where the green fan and spine lines intersect), which is easily deformed into a current sheet by the slightest extra pressure or footpoint motions; similarly, current sheets form readily along the fan, providing additional sites for reconnection. Figure 10.1b shows a close-up view of the dynamic results of reconnection in this scenario.

Because injection primarily produces upflows in each flux tube, counter-streaming could arise either through reconnection occurring at the bases of different flux tubes on opposite sides of the PIL or through sequential reconnection alternating between paired footpoints. Some injection models propose that the reconnection sites are at the PIL (e.g., Wu et al. 1990; Chae 2003; Liu et al. 2005), while others suggest that the jets originate at minority–polarity intrusions offset from the PIL (e.g., Wang 1999). Regardless of location, the flow in this model is driven by the relaxation of magnetic tension on newly reconnected field lines, coupled with twist propagation if the cancelling flux systems contain different amounts of twist. Therefore one can estimate an upper limit on the energy that can be imparted to the mass—and hence how high the cool plasma can be driven, and at what speed—by estimating the conversion of magnetic energy to kinetic energy through reconnection.

It is unclear whether injection can account for all aspects of the observed dynamic evolution of quiet-Sun prominences, in particular the frequency with which cool plasma appears suddenly in the corona and the predominantly horizontal, fine-scale counter-streaming flows. Furthermore, the extent to which reconnection can drive cool filament-channel material as high as 100 Mm in the corona, without also heating this plasma, has not been demonstrated in a realistic three-dimensional (3D) geometry with applicable energy sources and sinks. In simulations of Yohkoh X-ray jets by Yokoyama and Shibata (1995), reconnection produces hot jets directly, while the associated (but not co-spatial) cool jets are produced by compression of nearby open field. A similar mechanism might explain the association between flux cancellation and injection of cool plasma, but hot jets are not typically observed in coordination with prominence mass accumulation.

We also don't know where cancellation reconnection preferentially occurs: if the interacting flux systems reconnect in the low corona instead of the chromosphere or below, then cool, dense plasma will not be lifted or injected directly. The Spitzer (collisional) resistivity is highest at the temperature minimum (Sturrock 1999), but other processes such as local turbulence or current-driven instabilities could enhance the resistivity at other locations. Although signatures of flux cancellation are observed in the photosphere or chromosphere, the associated reconnection site is not necessarily at the same level (Zwaan 1987; Harvey et al. 1999). Only by observing the vector magnetic field at multiple levels can we discern the height of the reconnection site.

In summary, the essential observable features predicted by the injection model are that photospheric or chromospheric mass is injected with substantial speed at locations of flux cancellation/reconnection, either at or away from the PIL, and that this mass rises into the corona at or near its original cool temperature.

10.3 Levitation

Levitation models propose that cool plasma is lifted by rising magnetic fields at the PIL and transported transverse to the magnetic field (see Fig. 10.3). One type of levitation model, based on the assumption that the filament channel is a highly twisted flux rope, simply carries the cool plasma along with the field as it emerges above the photosphere (Fig. 10.3a). Here the prominence plasma must reside in the upward concavities of the helical field (Rust and Kumar 1994), below the axis. However, such concave-upward formations have been observed only rarely in non-erupting prominences (see, e.g., Lites 2005; López Ariste et al. 2006; Okamoto et al. 2008; Kuckein et al. 2012; Yelles Chaouche et al. 2012), and appear confined to

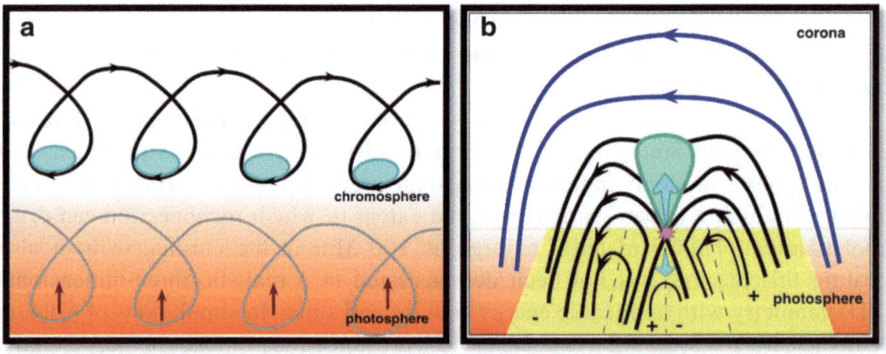

Fig. 10.3 Illustration of levitation of prominence plasma (aqua) by (**a**) flux rope emergence (*gray rope* is pre-emergent, *black* has emerged); (**b**) post-reconnection relaxation associated with flux cancellation. Here we show a 2D view of reconnecting arcades, for clarity. Overlying arcade field lines are *blue*, sheared field lines in *yellow* filament channel are *black*. Vertical *blue arrows* show bi-directional flows driven by reconnection (*magenta star*)

low-lying active region (AR) prominences. Interestingly, Kuckein et al. (2012) and Yelles Chaouche et al. (2012) found that the top of their observed AR filaments are in the chromosphere and the lower regions are trapped in the photosphere, making it unclear whether the structures have fully emerged.

The relaxation of magnetic fields during emergence of U loops (Deng et al. 2000), or after reconnection associated with flux cancellation (Ballegooijen and Martens 1989; Priest et al. 1996; Oliver et al. 1999; Litvinenko and Martin 1999; Galsgaard and Longbottom 1999; Litvinenko and Wheatland 2005; Welsch et al. 2005; Kubo and Shimizu 2007), also might levitate cool plasma (Fig. 10.3b). In this version of levitation, as in static support of prominence plasma, magnetic tension counterbalances the gravitational pull on the cool mass (see Chap. 7; Gilbert 2014); however, the levitating field deforms continuously as it rises into the corona, thus changing the force balance. If the supporting field lines become arched rather than dipped, then the magnetic force reverses sign and no longer works against gravity. As a result, the uplifted cool plasma will drain along the field to the chromosphere. This behavior is observed frequently in erupting prominences, but also in some quiescent prominences that resemble a central pillar from which cool material streams outward and downward (e.g., Okamoto et al. 2010).

Simulations of the reconnection–levitation mechanism have verified that rising field lines are indeed produced by reconnection between bipolar systems, but significant work remains to prove that this mechanism can reproduce the observed characteristics of prominence plasma. For example, the extent to which this process can lift photospheric or chromospheric material as high as 100 Mm into the corona, without significant draining, has not been demonstrated.

Detailed calculations of reconnecting bipoles (e.g., Galsgaard and Longbottom 1999; Welsch et al. 2005; von Rekowski and Hood 2008) predict significantly different plasma properties, depending on the dimensionality, the initial conditions, and the terms included in the energy equation. To date, none of the levitation simulations has allowed flux to retract beneath the photosphere during reconnection, thus preferentially favoring upward motions. Another difficulty with levitation is that most MHD simulations have shown that emerging U loops and the associated photospheric plasma do not readily rise to the coronal heights characteristic of quiet-Sun and intermediate prominences, even without the inhibiting presence of a preexisting coronal field (Fan 2001; Archontis et al. 2004, 2007; Manchester et al. 2004; Moreno-Insertis 2004; Magara 2006, 2011; Murray et al. 2006; Cheung et al. 2007; Galsgaard et al. 2007; Archontis 2008; MacTaggart and Hood 2010; Fang et al. 2010).

The partial ionization of portions of the chromosphere is likely to alter the reconnection and emergence processes, and associated magnetic and plasma structures (Leake and Arber 2006; Zweibel et al. 2011; Leake et al. 2012). Partial ionization affects magnetic reconnection primarily through enhanced cross-field diffusion (ambipolar diffusion; Braginskii 1965), which dissipates perpendicular (cross-field) currents. The parallel (field-aligned) currents will be affected as well because they are coupled to the perpendicular components. Because the reconnection rate can be determined by the Alfvén speed, which depends only on the ionized particle density, the reconnection rate can be higher for more weakly ionized plasmas. In addition, as noted previously, the resistivity reaches a maximum at the temperature minimum. On the other hand, collisions between ions and neutrals tend to synchronize the motions and temperatures of both populations, possibly counteracting some of the above effects (Zweibel 1989). Recent 2.5D emergence simulations including ion-neutral coupling show that partial ionization enhances the drainage from emerging flux and hence the slippage of the field through the chromosphere, thus allowing more flux to rise into the corona and more reconnection to occur. However, this also diminishes greatly the amount of cool mass carried up to prominence heights, as well as the amount of shear supplied to the corona (Leake and Linton 2013). Fully 3D studies are urgently needed to determine the effects of partial ionization on filament-channel formation and associated prominence mass formation.

As discussed above for the injection models, it is unclear where reconnection between cancelling flux systems would occur in the solar atmosphere; the Spitzer resistivity is highest at the temperature minimum, but many factors dictate where flux breaking is most likely to happen. For example, the reconnection in the Galsgaard and Longbottom (1999) study begins in the corona, at the apex of a separator line connecting two null points, so the bulk of the mass lifted at that location is coronal and not sufficiently cool or dense to be prominence material. An alternative mechanism proposed to lift and support prominence plasma in the corona is upward-propagating, weakly damped MHD waves (Pecseli and Engvold 2000); recent observational evidence for significant Alfvénic perturbations in chromospheric structures (De Pontieu et al. 2007) makes this an intriguing suggestion, but more quantitative work is needed to evaluate this mechanism rigorously. In general, the key features of levitation models overlap significantly with those of injection models (see above); however, the levitated mass does not travel as far or as fast as injected mass would, and typically is predicted to be located above the PIL.

10.4 Evaporation–Condensation

Evaporation–condensation models are based on two fundamental facts:

- The optically thin coronal radiative loss function peaks around 10^5 K, and falls off sharply at lower and higher temperatures (see Chap. 7; Gilbert 2014). Furthermore, the optically thin radiative losses are proportional to n^2, where n is the electron number density.
- Adding heat to a coronal loop increases the density of the corona while decreasing slightly the chromospheric mass (Fig. 10.4)—this is evaporation.

Two additional factors greatly simplify the calculation of force and energy balance in the corona. First, the low gas pressure implies that the magnetic field strongly resists cross-field plasma motions. Second, the thermal conductivity is highly anisotropic in magnetized plasmas, so conduction across the magnetic field is many orders of magnitude below that along the field. As a result, we can consider the

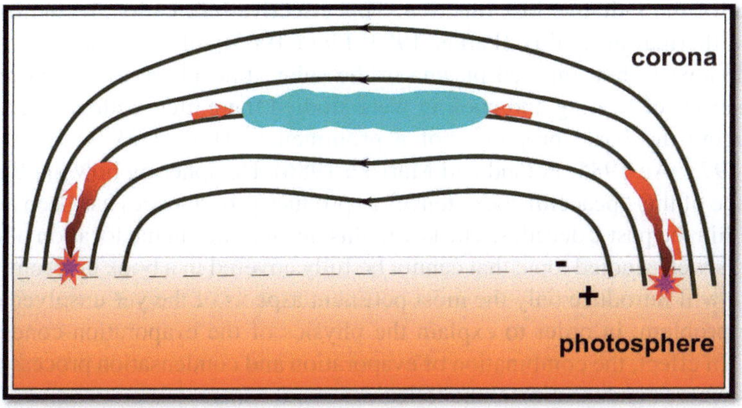

Fig. 10.4 Illustration of an evaporation-condensation model: heating localized above the footpoints (*magenta stars*) evaporate chromospheric plasma, driving hot upflows (*red*), which ultimately condense in the corona to form a cool prominence thread (aqua blob)

plasma dynamics and heat transfer to be channeled along magnetic field, allowing us to determine the plasma behavior by solving the one-dimensional hydrodynamic equations of mass, momentum, and energy conservation:

$$\frac{\partial \rho}{\partial t} + \frac{1}{A}\frac{\partial}{\partial s}(A v \rho) = 0 \tag{10.2}$$

$$\frac{\partial (v\rho)}{\partial t} + \frac{1}{A}\frac{\partial \left(A v^2 \rho\right)}{\partial s} + \frac{\partial p}{\partial s} = \rho g_{\parallel} \tag{10.3}$$

$$\frac{\partial E}{\partial t} + \frac{1}{A}\frac{\partial \left[A\left(E+p\right)v\right]}{\partial s} = \rho v g_{\parallel} + \frac{1}{A}\frac{\partial}{\partial s}\left(A\kappa_0 T^{5/2}\frac{\partial T}{\partial s}\right) - n^2\Lambda(T) + Q(s) \tag{10.4}$$

$$E = \frac{1}{2}\rho v^2 + \frac{p}{\gamma - 1}. \tag{10.5}$$

Here s represents the distance along the loop from the left base; t is time; ρ is the mass density, assuming a fully ionized hydrogen plasma where n is the electron number density; T is the temperature; v is the plasma speed; $P = 2nkT$ is the pressure; $A(s)$ is the normalized cross-sectional area; E is the total energy (kinetic plus thermal); κ_0 is the coefficient of thermal conductivity; $\gamma = 5/3$ is the ratio of the specific heats; $Q(s)$ is the volumetric heating; $\Lambda(T)$ is the optically-thin radiative loss function; and $g_{\parallel}$ is the component of gravity parallel to the loop axis.

The temperature and density dependences of coronal radiation work together to ensure that most of the mass remains at temperatures well above or well below the peak of the radiative loss function. This effective bifurcation, recognized long ago as a thermal instability (Parker 1953; Field 1965), allows cool, dense plasma to reside next to hot, rarefied plasma in the solar atmosphere. Different temporal combinations of heating and cooling were studied thereafter, without successfully reproducing the basic properties of a prominence (Hildner 1974; Engvold and Jensen 1977; An 1985; Poland and Mariska 1986). Deciphering how the Sun takes advantage of this "peaceful coexistence" to produce prominences has been achieved only within the past 2 decades. The key to this mystery lies in the location of coronal heating, a well-studied topic that cannot be fully covered in a book on prominences. Here we will introduce only the most pertinent aspects of the yet-unsolved coronal heating problem, in order to explain the physics of the evaporation-condensation models. In effect, the combination of evaporation and condensation processes led to success where condensation alone failed.

Coronal loops can remain in static equilibrium only if the energy input (heating) and output (radiation) are balanced when integrated over the entire loop. In this case

$$Q\lambda \approx n_e^2 \Lambda(T) L \tag{10.6}$$

where λ is the heating scale length and L is the loop length; note that conduction and enthalpy serve to redistribute energy within the loop, but do not add or subtract from the total. The heating scale is the characteristic distance over which most of the heat input occurs, which depends on the nature of the heat source. Most proposed coronal heating mechanisms deposit energy in very small regions with strong gradients, whether resonance absorption, waves, or reconnection. However, many such sites are needed to meet the energy requirements of observed loops, making λ the effective scale of the cumulative heat input. If the spatial distribution of the summed heating events were Gaussian, for example, then λ would equal the e-folding width of the Gaussian. If this effective heating scale equals the loop length (i.e., uniform heating), then Eq. (10.6) shows that the radiative losses simply equal the heat input (Fig. 10.5). More interesting behavior results when the heating is nonuniform, as is demonstrated below.

In order to understand how the corona remains hot and how it responds to additional energy input from flares, the detailed effects of heating in isolated loops have been explored extensively through theoretical analysis and numerical simulations (e.g., Warren et al. 2002; Klimchuk 2006; Martens 2010; Bradshaw and Cargill 2013). Early efforts determined the classic scaling laws that effectively represent the balance between energy sources and sinks in a loop (Rosner et al. 1978), starting with uniform heating and continuing on to consider localized heating at the loop apex, as occurs in flares (e.g., Antiochos and Sturrock 1978), and at the base of the corona, denoted the loop footpoints (Serio et al. 1981). Although observational evidence for the spatial distribution of coronal heating remains hotly debated, some observations have indicated that it is concentrated near the footpoints (Aschwanden et al. 2001; Winebarger et al. 2002; Antolin et al. 2010). Early theoretical studies indicated that concentrating coronal heating near the footpoints of a loop should produce a cool condensation at or near the apex (Mok et al. 1990;

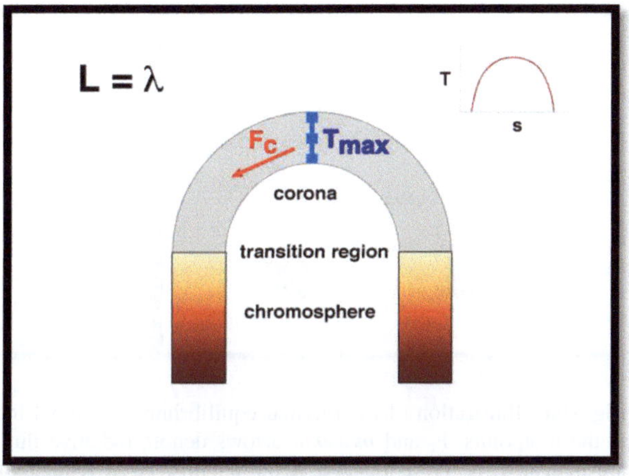

Fig. 10.5 Illustration of the thermal equilibrium of coronal loop plasma with uniform heating. Here F_c denotes thermal conductive flux, T_{max} (*blue bar*) marks the location where the temperature is highest, L is the coronal loop length, and λ is the heating scale length. The *inset* in the *upper right* shows temperature T as a function of distance s along the loop, for the L − λ relationship shown in the *upper left*

Antiochos and Klimchuk 1991; Dahlburg et al. 1998). However, confirmation of the basic principle behind this evaporation–condensation model for prominence plasma formation, denoted *thermal nonequilibrium*, became possible only with the use of adaptive-mesh numerical simulations, to handle the rapid birth and subsequent evolution of a new, thin transition region at each interface between the loop and the cool condensation (Antiochos et al. 1999).

The central concept of the thermal nonequilibrium model is that, if the heating scale is small compared to the length of a coronal loop and the deposition region is localized near the chromospheric footpoints, then the plasma in the midsection of the tube, where the heating is negligible, must undergo a radiatively driven thermal collapse to low temperatures. The energy balance and its dependence on loop length are illustrated in Fig. 10.6. As the density in the loop increases through evaporation, the radiative losses everywhere increase quadratically. Heating at both footpoints causes evaporation from both footpoints, yielding a density buildup at the site where the upflows meet (at the midpoint only for uniform cross-section and equal heating rates). The plasma responds locally by radiating more. If the enhanced radiative losses cannot be replenished locally, for reasons described below, the plasma will cool. A runaway situation develops in the coronal plasma due to the first fundamental fact noted at the start of this section: once the local plasma has cooled to the critical temperature of 10^5 K, it must cool all the way to chromospheric temperatures to regain equilibrium.

The ratio of the heating scale to the loop length is a crucial factor because the total radiative losses from the loop increase linearly with length, but thermal conduction and other energy transport or loss terms either decrease or remain constant with loop length. For shorter loops (Fig. 10.6, left panel), the conductive and enthalpy fluxes can offset the radiative losses near the midpoint, leading to a temperature dip but preventing thermal collapse. However, the radiative losses will dominate for lengths above a threshold value that is approximately an order of magnitude greater than the heating scale (Fig. 10.6, right panel). One category of magnetic-

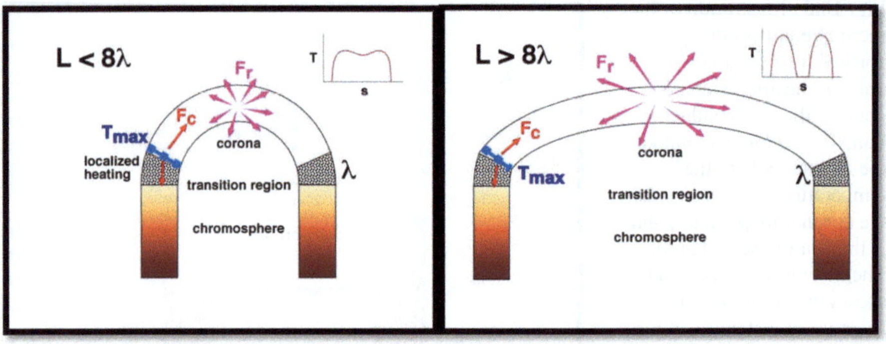

Fig. 10.6 Illustration of the thermal equilibrium of coronal loop plasma with heating localized at the footpoints. F_r and *magenta* arrows denote radiative flux. Relative strengths of F_c and F_r are indicated by the associated arrow lengths. Apex temperature drops for shorter loops (*left*); condensations form in longer loops (*right*)

field models (Mackay et al. 2010; Chap. 14; Mackay 2014) finds that the structure containing quiet-Sun prominence material is most likely to be a sheared arcade (Martin and Echols 1994; DeVore and Antiochos 2000; DeVore et al. 2005) or weakly twisted flux rope (Martens and Zwaan 2001; Bobra et al. 2008; see also Chap. 13, Fig. 9; Gibson 2014), in which many of the loops nearly aligned with the PIL are much longer than typical coronal loops. Therefore, for a given heating scale, condensations are more likely to form in these elongated loops than in shorter loops rooted within or outside the filament channel. Thermal non-equilibrium also occurs in sufficiently long, highly arched coronal loops outside filament channels, but the resulting condensations appear as small, short-lived "coronal rain" (Müller et al. 2003, 2005; Antolin et al. 2010).

A series of computational investigations of this evaporation–condensation process, denoted *thermal non-equilibrium*, has systematically explored the dynamics and energetics of the plasma within individual elongated flux tubes heated near the footpoints (Antiochos et al. 1999, 2000; Karpen et al. 2001, 2003, 2005, 2006; Karpen and Antiochos 2008; Luna et al. 2012). While the nature of the footpoint heating is not well understood, this work has established constraints on the conditions favorable to condensation formation, the magnetic structure of prominences, and the nature of the associated coronal heating. The key factors in determining the likelihood and behavior of condensations are the flux tube geometry and the localized heating properties. In long, low-lying flux tubes with shallow arches or dips, unequal quasi-steady heating yields a repetitive cycle of condensation formation, motion along the tube, and destruction by falling onto the nearest chromosphere. It is important to note that this process does not require the presence of dipped flux tubes, in contrast to the assumptions of many magnetic-structure models (e.g., Anzer and Heinzel 2003), although dips certainly facilitate the collection and retention of cool plasma. On the other hand, if the host flux tube has a deep dip, as would occur in the outer portions of a highly twisted flux rope, thermal non-equilibrium produces condensations that rapidly fall to the lowest part of the dip and remain there, stationary but growing as long as the heating remains quasi-steady. Condensations also form when the energy input is impulsive in nature and randomly distributed in time, as in nanoflare models of coronal heating, so long as the average interval between energy bursts is shorter than the radiative loss time in the ambient corona. The calculated condensation speeds, counter-streaming, lifetimes, and sizes are consistent with observations of many quiescent prominences.

However, this model does not provide a satisfactory explanation of those active-region prominences that are too short (<80 Mm) to support the thermal non-equilibrium process with typical values of the heating scale (~10 Mm; Aschwanden et al. 2001), or of the vertical structure and dynamics of hedgerow prominences (see Sect. 10.5). Barbs could be consistent with thermal non-equilibrium if they are composed of vertically aligned dips in otherwise horizontal flux tubes (Heinzel and Anzer 2001; Schmieder et al. 2013), as long as the dips are deep enough to trap the condensed matter: for example, in photospheric "bald patches" near parasitic polarity sites (Aulanier et al. 1998; Aulanier and Schmieder 2002; van Ballegooijen 2004; López Ariste et al. 2006).

Thermal non-equilibrium could occur in any favorable magnetic structure, but thus far has been considered systematically only within the context of the sheared arcade model for filament channels (Chap. 14; Mackay 2014) and in idealized arcades. Using a collection of 1D simulations, Luna et al. (2012) investigated the time-dependent plasma behavior along 125 flux tubes rooted in and outside a sheared-arcade filament channel. Steady footpoint heating with randomized asymmetry (i.e., one footpoint heated more strongly than the other) was imposed on all flux tubes. This approach has the advantage of much higher spatial resolution and faster turn-around than can be reached with current MHD models and computational resources. The results, represented by Fig. 10.7, show that thermal non-equilibrium in a sheared arcade yields both small dynamic condensations and long, thin threads suspended in the corona. Despite its simplifications and assumptions, this work constitutes an important advance toward a credible representation of a prominence and the associated cavity that can achieve closure with observations. The most recent confirmation of this process has been made by Xia et al. (2012, 2014), who performed pioneering 2D and 3D MHD simulations of the plasma structure formed within an idealized coronal arcade with localized heating, radiation, and thermal

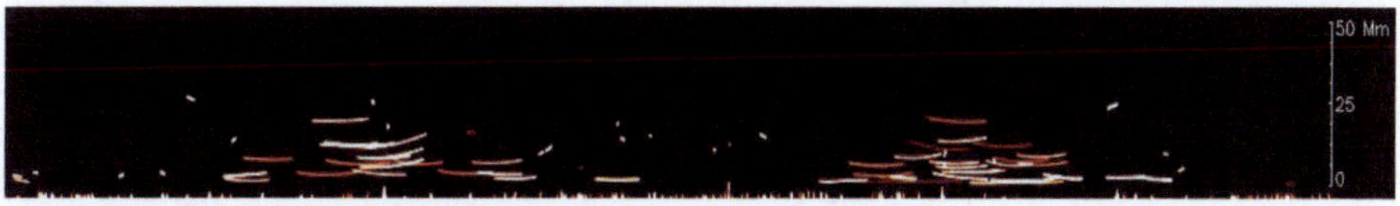

Fig. 10.7 Side-view snapshot of predicted Hα distribution from a multi-threaded simulation of prominence mass in a sheared-arcade quiescent filament channel (from Luna et al. 2012). Whiter features are brighter. The small features between the two pillars are dynamic, falling from the corona to the chromosphere. Note coronal rain high above the *left* pillar, in the surrounding coronal cavity (Chap. 13; Gibson 2014)

conduction. (For a discussion of energy balance considerations in prominences see Chap. 7; Gilbert 2014). A similar effort is urgently needed to explore thermal non-equilibrium in other filament-channel magnetic configurations, particularly flux rope models.

Evaporation–condensation models, as represented by thermal non-equilibrium, uniquely predict both stationary and highly dynamic prominence threads that condense in situ in the corona and trace the supporting flux tubes. An increase in coronal density precedes each condensation episode, while the collapse of the condensations reduces the ambient coronal density and generates waves and shocks. Many features predicted by the thermal non-equilibrium model have been observed, in particular the bright EUV emission at the ends of the cool threads (e.g., Luna et al. 2014) and the appearance of cool condensations high in the corona (Berger et al. 2012; Liu et al. 2012), but more multi-temperature, high-resolution observations and more realistic simulations are required for further validation. The long-period oscillations seen in active regions and prominences, which cannot be attributed to MHD waves, are consistent with the cycle times predicted by thermal non-equilibrium with asymmetric heating (e.g., Bocchialini et al. 2011; Auchere et al. 2014). In addition, there are many crucial unknowns requiring further study: for example, the mechanism or mechanisms responsible for coronal heating, the magnetic structure of filament channels and its origin, and the response of these channels to the ceaseless magnetic rearrangement occurring in the photosphere.

10.5 Magneto-Thermal Convection

Hedgerow and polar crown prominences (see Chap. 2; Engvold 2014) present a unique challenge to prominence formation models. Unlike other prominences, their observed filamentary structure and motions are predominantly vertical. Most models struggle to explain the simple fact that the cool plasma is extended over many gravitational scale heights without collapsing. Moreover, high-resolution observations from SOT/Hinode and EUV imagers have revealed a bewildering variety of flows and thermodynamic features, including swirling motions, hot bubbles, and hot plumes (Berger et al. 2008, 2010, 2011).

A new paradigm was proposed recently that builds upon these dynamic characteristics to formulate a comprehensive view of prominence formation and evolution: magneto-thermal convection (Berger et al. 2011). This theory was motivated by certain key observed features: the hot plasma filling the rising bubbles and plumes, the sharp boundary between the bubbles and the overlying prominence, and the penetration of filamentary, falling prominence plasma into the bubbles. Although the physical basis for this model is still under development, the fundamental concept is summarized as follows. Twisted magnetic flux emerges from the solar interior into the chromosphere beneath prominences, forming magnetic bubbles. The plasma contained in these bubbles is heated by internal reconnection or Alfvén wave dissipation (similar to standard coronal-heating models; see Sect. 10.4) while buoyancy carries them to typical prominence heights in the corona. This sets up ideal conditions for Rayleigh–Taylor instability (RTI), as the prominence mass is much denser than the underlying bubble plasma. The onset of the RTI initiates a turbulent transport mechanism wherein hot plasma and magnetic flux are carried into the corona via the observed plumes while the condensed prominence plasma shreds and drains in vertical "spikes" or streams back to the chromosphere to complete the cycle. The Kelvin–Helmholtz instability (KHI), which occurs at strong gradients (shear) in flow speed, could facilitate additional mixing of cool and hot plasma at the interface between the falling and rising material (Berger et al. 2010; Hillier et al. 2012b).

Observational support for this model thus far comes from an increasing number of intriguing prominence-formation events. Thanks to the continuous coverage provided by the Solar Dynamics Observatory (SDO), we can now observe the dynamic prominence mass from its initial appearance onward, with simultaneous images from the Atmospheric Imaging Array instrument (AIA) showing emission and absorption from plasma at $T \sim 0.1$–12 MK. The first observed example consisted of a polar-crown prominence that formed several hours after a nearby confined eruption. SDO/AIA images revealed that a funnel-shaped concentration of hot plasma appeared in the depleted corona, indicating the presence of magnetic dips, then cooled to form a dynamic quiescent prominence with strong vertical structure and flows; the estimated mass drainage and input rates were roughly comparable (Liu et al. 2012). At least 13 more funnel prominences have been identified and analyzed (Liu et al. 2013), to determine the frequency, mass input and loss rates, and other physical characteristics of a larger sample of these features as they are born and evolve. In the second case, a quiescent polar-crown prominence disappeared over several hours without erupting, then reformed in situ over a similar interval. Hinode/SOT and SDO/AIA images showed that a concentration of hot plasma appeared high in the cavity, then cooled while descending to form another hedgerow prominence of similar but not identical size and shape (Berger et al. 2012).

This model treats prominence formation and evolution as an integral part of coronal cavity formation and evolution, including eruptions. Specific links in this chain have been explored in greater detail, although the full cycle has not been reproduced as yet. Recent theoretical analyses suggest that the mass condenses and collects where current sheets spontaneously form and dissipate in the magnetic configuration yielding intermittent localized episodes of thermal collapse, magnetic reconnection, and cross-field slippage and downflows of cool plasma (Low et al. 2012a, b). 3D MHD simulations of an idealized slab threaded by dipped field lines have found that RTI can cause hot plumes to rise in polar-crown prominences (Hillier et al. 2012a), while the nonlinear phase of the RTI leads to reconnection and supersonic downflows, as well as KHI eddies at the boundary of the falling condensations (Hillier et al. 2012b). The magneto-thermal convection hypothesis is global in scale, encompassing and connecting the entire height range from chromosphere to coronal cavity, but relies on kinetic-scale processes such as reconnection and cross-field diffusion. Therefore comprehensive simulation of this cycle poses a difficult challenge to present-day computational resources, which must be overcome in order to validate its far-reaching implications for mass formation and magnetic stability of solar prominences.

Acknowledgments JK would like to thank S. Antiochos, T. Berger, C.R. DeVore, S. Guidoni, J. Leake, W. Liu, M. Luna Benassar, and E. Zweibel for their helpful suggestions and comments.

References

Ahn, K., Chae, J., Cao, W., & Goode, P. R. (2010). Patterns of flows in an intermediate prominence observed by Hinode. *The Astrophysical Journal, 721*, 74.

An, C.-H. (1985). Formation of prominences by condensation modes in magnetized cylindrical plasmas. *The Astrophysical Journal, 298*, 409.

Antiochos, S. K., & Klimchuk, J. A. (1991). A model for the formation of solar prominences. *The Astrophysical Journal, 378*, 372.

Antiochos, S. K., & Sturrock, P. A. (1978). Evaporative cooling of flare plasma. *The Astrophysical Journal, 220*, 1137.

Antiochos, S. K., MacNeice, P. J., Spicer, D. S., & Klimchuk, J. A. (1999). The dynamic formation of prominence condensations. *The Astrophysical Journal, 512*, 985.

Antiochos, S. K., MacNeice, P. J., & Spicer, D. S. (2000). The thermal nonequilibrium of prominences. *The Astrophysical Journal, 536*, 494.

Antolin, P., Shibata, K., & Vissers, G. (2010). Coronal rain as a marker for coronal heating mechanisms. *The Astrophysical Journal, 716*, 154.

Anzer, U., & Heinzel, P. (2003). On the nature of extended EUV filaments. *Astronomy and Astrophysics, 404*, 1139.

Archontis, V. (2008). Magnetic flux emergence in the sun. *Journal of Geophysical Research, 113*, 12422.

Archontis, V., Moreno-Insertis, F., Galsgaard, K., Hood, A., & O'Shea, E. (2004). Emergence of magnetic flux from the convection zone into the corona. *Astronomy and Astrophysics, 426*, 1047.

Archontis, V., Hood, A. W., & Brady, C. (2007). Emergence and interaction of twisted flux tubes in the sun. *Astronomy and Astrophysics, 466*, 367.

Aschwanden, M., Schrijver, C. J., & Alexander, D. (2001). An assessment of coronal heating models based on Yohkoh, SoHO, and TRACE observations. *The Astrophysical Journal, 550*, 1036.

Auchere, F., Bocchialini, K., Solomon, J., & Tison, E. (2014). Long-period intensity pulsations in the solar corona during activity cycle 23. *Astronomy and Astrophysics, 563*, A8.

Aulanier, G., & Schmieder, B. (2002). The magnetic nature of wide EUV filament channels and their role in the mass loading of CMEs. *Astronomy and Astrophysics, 386*, 1106.

Aulanier, G., Démoulin, P., van Driel-Gesztelyi, L., Mein, P., & Deforest, C. (1998). 3-D magnetic configurations supporting prominences. II. The lateral feet as a perturbation of a twisted flux-tube. *Astronomy and Astrophysics, 335*, 309.

Berger, T. E., Shine, R. A., Slater, G. L., Tarbell, T. D., Title, A. M., Okamoto, T. J., Ichimoto, K., Katsukawa, Y., Suematsu, Y., Tsuneta, S., Lites, B. W., & Shimizu, T. (2008). Hinode SOT observations of solar quiescent prominence dynamics. *The Astrophysical Journal, 676*, L89.

Berger, T. E., Slater, G., Hurlburt, N., et al. (2010). Quiescent prominence dynamics observed with the Hinode solar optical telescope. I. Turbulent upflow plumes. *The Astrophysical Journal, 716*, 1288.

Berger, T., Testa, P., Hillier, A., et al. (2011). Magneto-thermal convection in solar prominences. *Nature, 472*, 197.

Berger, T. E., Liu, W., & Low, B. C. (2012). SDO/AIA detection of solar prominence formation within a coronal cavity. *The Astrophysical Journal, 758*, L37.

Biskamp, D. (2000). *Magnetic reconnection in plasmas* (p. 199). Cambridge: Cambridge University Press.

Bobra, M. G., van Ballegooijen, A. A., & DeLuca, E. E. (2008). Modeling nonpotential magnetic fields in solar active regions. *The Astrophysical Journal, 672*, 1209.

Bocchialini, K., Baudin, F., Koutchmy, S., Pouget, G., & Solomon, J. (2011). Oscillatory motions observed in eruptive filaments. *Astronomy and Astrophysics, 533*, A96.

Bradshaw, S. J., & Cargill, P. J. (2013). The influence of numerical resolution on coronal density in hydrodynamic models of impulsive heating. *The Astrophysical Journal, 770*, 12.

Braginskii, S. I. (1965). *Reviews of Plasma Physics, 1*, 205.

Chae, J. (2003). The formation of a prominence in NOAA active region 8668. II. Trace observations of jets and eruptions associated with canceling magnetic features. *The Astrophysical Journal, 584*, 1084.

Chae, J., Denker, C., Spirock, T. J., Wang, H., & Goode, P. R. (2000). High-resolution Hα observations of proper motion in NOAA 8668: Evidence for filament mass injection by chromospheric reconnection. *Solar Physics, 195*, 333.

Chae, J., Martin, S. F., Yun, H. S., Kim, J., Lee, S., Goode, P. R., Spirock, T., & Wang, H. (2001). Small magnetic bipoles emerging in a filament channel. *The Astrophysical Journal, 548*, 497.

Chae, J., Moon, Y.-J., & Park, Y.-D. (2005). The magnetic structure of filament barbs. *The Astrophysical Journal, 626*, 574.

Chae, J., Ahn, K., Lim, E.-K., Choe, G. S., & Sakurai, T. (2008). Persistent horizontal flows and magnetic support of vertical threads in a quiescent prominence. *The Astrophysical Journal, 689*, L73.

Cheung, M. C. M., Schüssler, M., & Moreno-Insertis, F. (2007). Magnetic flux emergence in granular convection: Radiative MHD simulations and observational signatures. *Astronomy and Astrophysics, 467*, 703.

Dahlburg, R. B., Antiochos, S. K., & Klimchuk, J. A. (1998). Prominence formation by localized heating. *The Astrophysical Journal, 495*, 485.

De Pontieu, B., McIntosh, S. W., Carlsson, M., Hansteen, V. H., Tarbell, T. D., et al. (2007). Chromospheric Alfvénic waves strong enough to power the solar wind. *Science, 318*, 1574.

Deng, Y., Schmieder, B., Engvold, O., DeLuca, E., & Golub, L. (2000). Emergence of sheared magnetic flux tubes in an active region observed with the SVST and TRACE. *Solar Physics, 195*, 347.

DeVore, C. R., & Antiochos, S. K. (2000). Dynamical formation and stability of helical prominence magnetic fields. *The Astrophysical Journal, 539*, 954.

DeVore, C. R., Antiochos, S. K., & Aulanier, G. (2005). *Solar Prominence Interactions*, Astrophysical Journal, *629*, 1122.

Engvold, O. (2014). Description and classification of prominences. In J.-C. Vial & O. Engvold (Eds.), *Solar prominences, ASSL* (Vol. 415, pp. 31–60). New York: Springer.

Engvold, O., & Jensen, E. (1977). On Pikel'ner's theory of prominences. *Solar Physics, 52*, 37.

Fan, Y. H. (2001). The emergence of a twisted omega-tube into the solar atmosphere. *The Astrophysical Journal, 554*, L111.

Fan, Y. H. (2014). MHD equilibira and triggers for prominence eruption. In J.-C. Vial & O. Engvold (Eds.), *Solar prominences, ASSL* (Vol. 415, pp. 295–320). New York: Springer.

Fang, F., Manchester, W., Abbett, W., & van der Holst, B. (2010). Simulation of flux emergence from the convection zone to the corona. *The Astrophysical Journal, 714*, 1649.

Field, G. B. (1965). Thermal instability. *The Astrophysical Journal, 142*, 531.

Galsgaard, K., & Longbottom, A. W. (1999). Formation of solar prominences by flux convergence. *The Astrophysical Journal, 510*, 444.

Galsgaard, K., Archontis, V., Moreno-Insertis, F., & Hood, A. W. (2007). The effect of the relative orientation between the coronal field and new emerging flux. I. Global properties. *The Astrophysical Journal, 666*, 516.

Gibson, S. (2014). Coronal cavities: observations and implications for the magnetic environment of prominences. In J.-C. Vial & O. Engvold (Eds.), *Solar prominences, ASSL* (Vol. 415, pp. 321–351). New York: Springer.

Gilbert, H. (2014). Energy balance. In J.-C. Vial & O. Engvold (Eds.), *Solar prominences*. New York: Springer.

Gunar, S., Heinzel, P., Schmieder, B., Schwartz, P., & Anzer, U. (2007). Properties of prominence fine-structure threads derived from SOHO/SUMER hydrogen Lyman lines. *Astronomy and Astrophysics, 472*, 929.

Gunar, S., Heinzel, P., Anzer, U., & Schmieder, B. (2008). On Lyman-line asymmetries in quiescent prominences. *Astronomy and Astrophysics, 490*, 307.

Harvey, K. L., Jones, H. P., Schrijver, C. J., & Penn, M. (1999). Does magnetic flux submerge at flux cancelation sites? *Solar Physics, 190*, 35.

Heinzel, P., & Anzer, U. (2001). Prominence fine structures in a magnetic equilibrium: Two-dimensional models with multilevel radiative transfer. *Astronomy and Astrophysics, 375*, 1082.

Heinzel, P., & Anzer, U. (2006). On the fine structure of solar filaments. *The Astrophysical Journal, 643*, L65.

Hildner, E. (1974). The formation of solar quiescent prominences by condensation. *Solar Physics, 35*, 123.

Hillier, A., Isobe, H., & Watanabe, H. (2011). Observations of plasma blob ejection from a quiescent prominence by Hinode solar optical telescope. *Publications of the Astronomical Society of Japan, 63*, L19.

Hillier, A., Berger, T., Isobe, H., & Shibata, K. (2012a). Numerical simulations of the magnetic Rayleigh–Taylor instability in the Kippenhahn–Schlüter prominence model. I. Formation of upflows. *The Astrophysical Journal, 746*, 120.

Hillier, A., Isobe, H., Shibata, K., & Berger, T. (2012b). Numerical simulations of the magnetic Rayleigh–Taylor instability in the Kippenhahn–Schlüter prominence model. II. Reconnection-triggered downflows. *The Astrophysical Journal, 756*, 110.

Karpen, J. T., & Antiochos, S. K. (2008). Condensation formation by impulsive heating in prominences. *The Astrophysical Journal, 676*, 658.

Karpen, J. T., Antiochos, S. K., Hohensee, M., Klimchuk, J. A., & MacNeice, P. J. (2001). Are magnetic dips necessary for prominence formation? *The Astrophysical Journal, 553*, L85.

Karpen, J. T., Antiochos, S. K., Klimchuk, J. A., & MacNeice, P. J. (2003). Constraints on the magnetic field geometry in prominences. *The Astrophysical Journal, 593*, 1187.

Karpen, J. T., Antiochos, S. K., Tanner, S. E., & DeVore, C. R. (2005). Prominence formation by thermal nonequilibrium in the sheared-arcade model. *The Astrophysical Journal, 635*, 1319.

Karpen, J. T., Antiochos, S. K., & Klimchuk, J. A. (2006). The origin of high-speed motions and threads in prominences. *The Astrophysical Journal, 637*, 531.

Kippenhahn, R., & Schlüter, A. (1957). Eine theorie der solaren filamente. Mit 7 Textabbildungen. *Zeitschrift für Astrophysik, 43*, 36.

Klimchuk, J. A. (2006). On solving the coronal heating problem. *Solar Physics, 234*, 41.

Kubo, M., & Shimizu, T. (2007). Magnetic field properties of flux cancellation sites. *The Astrophysical Journal, 671*, 990.

Kubota, J., & Uesugi, A. (1986). The vertical motion of matter in a prominence observed on May 7, 1984. *Publications of the Astronomical Society of Japan, 38*, 903.

Kucera, T. A., & Landi, E. (2006). Ultraviolet observations of prominence activation and cool loop dynamics. *The Astrophysical Journal, 645*, 1525.

Kucera, T. A., Tovar, M., & De Pontieu, B. (2003). Prominence motions observed at high cadences in temperatures from 10 000 to 250 000 K. *Solar Physics, 212*, 81.

Kuckein, C., Martinez Pillet, V., & Centeno, R. (2012). An active region filament studied simultaneously in the chromosphere and photosphere. I. Magnetic structure. *Astronomy and Astrophysics, 539*, A131.

Kulsrud, R. M. (2005). *Plasma physics for astrophysics*. Princeton: Princeton University Press.

Leake, J. E., & Arber, T. D. (2006). The emergence of magnetic flux through a partially ionised solar atmosphere. *Astronomy and Astrophysics, 450*, 2.

Leake, J. E., & Linton, M. G. (2013). Effect of ion-neutral collisions in simulations of emerging active regions. *The Astrophysical Journal, 764*, 54.

Leake, J. E., Lukin, V. S., Linton, M. G., & Meier, E. T. (2012). Multi-fluid simulations of chromospheric magnetic reconnection in a weakly ionized reacting plasma. *The Astrophysical Journal, 760*, 109.

Lin, Y., Engvold, O., & Wiik, J. E. (2003). Counterstreaming in a large polar crown filament. *Solar Physics, 216*, 109.

Lin, Y., Engvold, O., Rouppe van der Voort, L., Wiik, J. E., & Berger, T. E. (2005). Thin threads of solar filaments. *Solar Physics, 226*, 239.

Lites, B. W. (2005). Magnetic flux ropes in the solar photosphere: The vector magnetic field under active region filaments. *The Astrophysical Journal, 622*, 1275.

Litvinenko, Y. E., & Martin, S. F. (1999). Magnetic reconnection as the cause of a photospheric canceling feature and mass flows in a filament. *Solar Physics, 190*, 45.

Litvinenko, Y. E., & Wheatland, M. S. (2005). A simple dynamical model for filament formation in the solar corona. *The Astrophysical Journal, 630*, 587.

Litvinenko, Y. E., Chae, J., & Park, S.-Y. (2007). Flux pile-up magnetic reconnection in the solar photosphere. *The Astrophysical Journal, 662*, 1302.

Liu, Y., Kurokawa, H., & Shibata, K. (2005). Production of filaments by surges. *The Astrophysical Journal, 631*, L97.

Liu, W., Berger, T. E., & Low, B. C. (2012). First SDO/AIA observation of solar prominence formation following an eruption: Magnetic dips and sustained condensation and drainage. *The Astrophysical Journal, 745*, L21.

Liu, W., Berger, T. E., & Low, B. C. (2013). Nature of prominences and their role in space weather. In B. Schmieder, J.-M. Malherbe, & S.-T. Wu (Eds.), *Proceedings of IAU symposium* (Vol. 300, p. 441).

López Ariste, A., Aulanier, G., Schmieder, B., & Sainz Dalda, A. (2006). First observation of bald patches in a filament channel and at a barb endpoint. *Astronomy and Astrophysics, 456*, 725.

Low, B. C., Berger, T., Casini, R., & Liu, W. (2012a). The hydromagnetic interior of a solar quiescent prominence. I. Coupling between force balance and steady energy transport. *The Astrophysical Journal, 755*, 34.

Low, B. C., Liu, W., Berger, T., & Casini, R. (2012b). The hydromagnetic interior of a solar quiescent prominence. II. Magnetic discontinuities and cross-field mass transport. *The Astrophysical Journal, 757*, 21.

Luna, M., Karpen, J. T., & DeVore, C. R. (2012). Formation and evolution of a multi-threaded solar prominence. *The Astrophysical Journal, 746*, 30.

Luna, M., Knizhnik, K., Muglach, K., Karpen, J., Gilbert, H., Kucera, T., et al. (2014). Observations and implications of large-amplitude longitudinal oscillations in a solar filament. *The Astrophysical Journal, 785*, 79.

Mackay, D. H. (2014). Formation and large-scale patterns of filament channels and filaments. In J.-C. Vial & O. Engvold (Eds.), *Solar prominences, ASSL* (Vol. 415, pp. 353–378). New York: Springer.

Mackay, D. H., Karpen, J. T., Ballester, J. L., Schmieder, B., & Aulanier, G. (2010). Physics of solar prominences: II—Magnetic structure and dynamics. *Space Science Review, 151*, 333.

MacTaggart, D., & Hood, A. W. (2010). Simulating the "sliding doors" effect through magnetic flux emergence. *The Astrophysical Journal, 716*, L219.

Magara, T. (2006). Dynamic and topological features of photospheric and coronal activities produced by flux emergence in the sun. *The Astrophysical Journal, 653*, 1499.

Magara, T. (2011). A possible mechanism of flux cancellation via U-loop emergence on the sun. *Publications of the Astronomical Society of Japan, 63*, 417.

Malherbe, J.-M. (1989). The formation of solar prominences. In E.R. Priest (Eds.), *Dynamics and structure of quiescent solar prominences* (p. 115) Dordrecht: Kluwer.

Manchester, W., IV, Gombosi, T., DeZeeuw, D., & Fan, Y. (2004). Eruption of a buoyantly emerging magnetic flux rope. *The Astrophysical Journal, 610*, 588.

Martens, P. C. H. (2010). Scaling laws and temperature profiles for solar and stellar coronal loops with non-uniform heating. *The Astrophysical Journal, 714*, 1290.
Martens, P. C., & Zwaan, C. (2001). Origin and evolution of filament-prominence systems. *The Astrophysical Journal, 538*, 872.
Martin, S. F. (1973). The evolution of prominences and their relationship to active centers (a review). *Solar Physics, 31*, 3.
Martin, S. F. (1998). Conditions for the formation and maintenance of filaments (invited review). *Solar Physics, 182*, 107.
Martin, S. F. (2014). The magnetic field structure of prominences from direct and indirect observations. In J.-C. Vial & O. Engvold (Eds.), *Solar prominences*. New York: Springer.
Martin, S. F., & Echols, C. R. (1994). An observational and conceptual model of the magnetic field of a filament. In R. J. Rutten, & C. J. Schrijver (Eds.), *Solar surface magnetism* (p. 339) Dordrecht: Kluwer.
McMath, R. R., & Pettit, E. (1938). Prominence studies. *The Astrophysical Journal, 88*, 244.
Mok, Y., Drake, J. F., Schnack, D. D., & Van Hoven, G. (1990). Prominence formation in a coronal loop. *The Astrophysical Journal, 359*, 228.
Moreno-Insertis, F. (2004). The emergence of magnetic field into stellar atmospheres. *Astrophysics and Space Science, 292*, 587.
Müller, D. A. N., Hansteen, V. H., & Peter, H. (2003). Dynamics of solar coronal loops. I. Condensation in cool loops and its effect on transition region lines. *Astronomy and Astrophysics, 411*, 605.
Müller, D. A. N., De Groof, A., Hansteen, V. H., & Peter, H. (2005). High-speed coronal rain. *Astronomy and Astrophysics, 436*, 1067.
Murray, M. J., Hood, A. W., Moreno-Insertis, F., Galsgaard, K., & Archontis, V. (2006). 3D simulations identifying the effects of varying the twist and field strength of an emerging flux tube. *Astronomy and Astrophysics, 460*, 909.
Okamoto, T. J., Tsuneta, S., Berger, T. E., Ichimoto, K., Katsukawa, Y., Lites, B. W., Nagata, S., et al. (2007). Chromospheric anemone jets as evidence of ubiquitous reconnection. *Science, 318*, 1577.
Okamoto, T. J., Tsuneta, S., Lites, B. W., Kubo, M., Yokoyama, T., Berger, T. E., et al. (2008). Emergence of a helical flux rope under an active region prominence. *The Astrophysical Journal, 673*, L215.
Okamoto, T. J., Tsuneta, S., & Berger, T. E. (2010). A rising cool column as a signature of helical flux emergence and formation of prominence and coronal cavity. *The Astrophysical Journal, 719*, 583.
Oliver, R., Cadez, V. M., Carbonell, M., & Ballester, J. L. (1999). Coronal potential magnetic fields from photospheric sources with finite width. *Astronomy and Astrophysics, 351*, 733.
Parker, E. N. (1953). Instability of thermal fields. *The Astrophysical Journal, 117*, 431.
Pecseli, H., & Engvold, O. (2000). Modeling of prominence threads in magnetic fields: Levitation by incompressible MHD waves. *Solar Physics, 194*, 73.
Pikel'ner, S. B. (1971). Origin of quiescent prominences. *Solar Physics, 17*, 44.
Poland, A. I., & Mariska, J. T. (1986). A siphon mechanism for supplying prominence mass. *Solar Physics, 104*, 303.
Priest, E. R., & Forbes, T. E. (2000). *Magnetic reconnection*. Cambridge: Cambridge University Press.
Priest, E. R., van Ballegooijen, A. A., & Mackay, D. H. (1996). A model for dextral and sinistral prominences. *The Astrophysical Journal, 460*, 530.
Rosner, R., Tucker, W. H., & Vaiana, G. S. (1978). Dynamics of the quiescent solar corona. *The Astrophysical Journal, 220*, 643.
Rust, D. M., & Kumar, A. (1994). Helical magnetic fields in filaments. *Solar Physics, 155*, 69.
Saito, K., & Tandberg-Hanssen, E. (1973). The arch systems, cavities, and prominences in the helmet streamer observed at the solar eclipse, November 12, 1966. *Solar Physics, 31*, 105.
Schmieder, B., Raadu, M. A., & Wiik, J. E. (1991). Fine structure of solar filaments. II – Dynamics of threads and footpoints. *Astronomy and Astrophysics, 252*, 353.

Schmieder, B., Kucera, T. A., Knizhnik, K., Luna, M., López Ariste, A., & Toot, D. (2013). Propagating waves transverse to the magnetic field in a solar prominence. *The Astrophysical Journal, 777*, 108.

Schrijver, C. J. (2001). Catastrophic cooling and high-speed downflow in quiescent solar coronal loops observed with TRACE. *Solar Physics, 198*, 325.

Schwartz, P., Heinzel, P., & Schmieder, B. (2006). Study of an extended EUV filament using SoHO/SUMER observations of the hydrogen Lyman lines. *Astronomy and Astrophysics, 459*, 651.

Serio, S., Peres, G., Vaiana, G. S., Golub, L., & Rosner, R. (1981). Closed coronal structures. II – Generalized hydrostatic model. *The Astrophysical Journal, 243*, 288.

Sturrock, P. A. (1999). Chromospheric magnetic reconnection and its possible relationship to coronal heating. *The Astrophysical Journal, 521*, 451.

Tandberg-Hanssen, E. (1995). *The nature of solar prominences* (p. 358). Dordrecht: Kluwer.

van Ballegooijen, A. A. (2004). Observations and modeling of a filament on the sun. *The Astrophysical Journal, 612*, 519.

van Ballegooijen, A. A., & Martens, P. C. H. (1989). Formation and eruption of solar prominences. *The Astrophysical Journal, 343*, 971.

Von Rekowski, B., & Hood, A. (2008). Photospheric cancelling magnetic features and associated phenomena in a stratified solar atmosphere. *Monthly Notices of the Royal Astronomical Society, 385*, 1792.

Wang, Y.-M. (1999). The jetlike nature of He II lambda304 prominences. *The Astrophysical Journal, 520*, L71.

Wang, Y.-M., & Muglach, K. (2007). On the formation of filament channels. *The Astrophysical Journal, 666*, 1284.

Warren, H. P., Winebarger, A. R., & Hamilton, P. S. (2002). Hydrodynamic modeling of active region loops. *The Astrophysical Journal, 579*, L41.

Welsch, B. T., DeVore, C. R., & Antiochos, S. K. (2005). Magnetic reconnection models of prominence formation. *The Astrophysical Journal, 634*, 1395.

Winebarger, A. R., Warren, H., van Ballegooijen, A., DeLuca, E. E., & Golub, L. (2002). Steady flows detected in extreme-ultraviolet loops. *The Astrophysical Journal, 567*, L89.

Wu, S. T., Bao, J. J., An, C. H., & Tandberg-Hanssen, E. (1990). The role of condensation and heat conduction in the formation of prominences – An MHD simulation. *Solar Physics, 125*, 277.

Xia, C., Chen, P. F., & Keppens, R. (2012). Simulations of prominence formation in the magnetized solar corona by chromospheric heating. *The Astrophysical Journal, 748*, L26.

Xia, C., Keppens, R., & Guo, Y. (2014). Three-dimensional prominence-hosting magnetic configurations: Creating a helical magnetic flux rope. *The Astrophysical Journal, 780*, 130.

Yamada, M., Kulsrud, R., & Ji, H. (2010). Magnetic reconnection. *Review of Modern Physics, 82*, 603.

Yelles Chaouche, L., Kuckein, C., Martinez Pillet, V., & Moreno-Insertis, F. (2012). The three-dimensional structure of an active region filament as extrapolated from photospheric and chromospheric observations. *The Astrophysical Journal, 748*, 23.

Yokoyama, T., & Shibata, K. (1995). Magnetic reconnection as the origin of X-ray jets and Halpha surges on the sun. *Nature, 375*, 42.

Zirker, J. B., Engvold, O., & Yi, Z. (1994). Flows in quiescent prominences. *Solar Physics, 150*, 81.

Zirker, J. B., Engvold, O., & Martin, S. F. (1998). Counter-streaming gas flows in solar prominences as evidence for vertical magnetic fields. *Nature, 396*, 440.

Zwaan, C. (1987). Elements and patterns in the solar magnetic field. *Annual Review of Astronomy and Astrophysics, 25*, 83.

Zweibel, E. (1989). Magnetic reconnection in partially ionized gases. *The Astrophysical Journal, 340*, 550.

Zweibel, E., Lawrence, E., Yoo, J., Ji, H., Yamada, M., & Malyshkin, L. (2011). Magnetic reconnection in partially ionized plasmas. *Physics of Plasmas, 18*, 111211.

Chapter 11
Magnetism and Dynamics of Prominences: MHD Waves

José Luis Ballester

Abstract Quiescent solar prominences are highly dynamic structures which, among other features, display oscillatory motions. The presence of these oscillations has been confirmed by means of ground- and space-based observations, and they have been classified in small and large amplitude oscillations. Since prominences are magnetized structures, the theoretical interpretation of their oscillations has been mostly done in terms of magnetohydrodynamic (MHD) waves. This interpretation has allowed the development of prominence seismology as a tool to determine prominence physical parameters (magnetic field, Alfvén speed, inhomogeneity scale, etc.) which are difficult to measure by direct means.

11.1 Introduction

Reports about the presence of oscillatory motions in prominences go back to the 1950s and 1960s. Early studies on this subject were related with large amplitude oscillations induced by disturbances coming from a nearby flare (Hyder 1966). Later on, Harvey (1969) reported that quiescent prominences and filaments also displayed small amplitude oscillations, and Tandberg-Hanssen (1974, 1995) reviewed known evidences about prominence oscillations. Nowadays, the presence of oscillatory motions in prominences and filaments is well established from ground- and space-based observations thanks to improved observational capabilities and tools for data analysis.

Based on the velocity amplitudes, prominence oscillations have been broadly classified in two categories: large and small amplitude oscillations (Oliver and Ballester 2002). In the case of large amplitude oscillations, the prominence displays large displacements, up to 4×10^4 km, with respect to its equilibrium position and, in general, it oscillates as a whole with a velocity amplitude greater than 20 km/s. Excitation of large amplitude oscillations in filaments was attributed to travelling perturbations, caused by flares, which disturb the filaments and induce damped

J.L. Ballester (✉)
Departament de Física, Universitat de les Illes Balears, E-07122 Palma de Mallorca, Spain
e-mail: joseluis.ballester@uib.es

J.-C. Vial, O. Engvold (eds.), *Solar Prominences*, Astrophysics and Space Science Library 415, DOI 10.1007/978-3-319-10416-4_11

oscillations (Moreton and Ramsey 1960). In some cases, during the course of the oscillations, the filament becomes visible in the Hα image when the prominence is at rest, but when its line-of-sight velocity is sufficiently large, the emission from the material falls outside the bandpass of the filter and the prominence becomes invisible in Hα. This process is repeated periodically and for this reason this type of event was called a "winking filament" (Ramsey and Smith 1966; Hyder 1966). Detailed observations about large amplitude oscillations in filaments and prominences have been obtained in Hα, EUV, microwaves and He10830 Å (Tripathi et al. 2009), and the exciters seem to be Moreton, EIT or EUV waves (Eto et al. 2002; Okamoto et al. 2004; Gilbert et al. 2008; Asai et al. 2012), nearby jets and subflares (Jing et al. 2003, 2006; Vršnak et al. 2007; Li and Zhang 2012; Luna et al. 2014) and the pre-eruptive phase of a filament (Isobe and Tripathi 2006; Isobe et al 2007; Chen et al. 2008). Taking into accout the polarization of the observed motions, large amplitude oscillations can be classified in transversal, longitudinal and mixed. Transversal oscillations consist of periodic motions transverse to the main filament axis (Isobe and Tripathi 2006; Isobe et al 2007; Hershaw et al. 2011), longitudinal oscillations consist of periodic motions along the axis of a filament (Jing et al. 2003, 2006; Vršnak et al. 2007; Li and Zhang 2012; Luna et al. 2014), while mixed oscillations (Gilbert et al. 2008) show a complicated mixture of transverse and perpendicular motions with respect to the filament spine (see reviews by Oliver and Ballester 2002; Tripathi et al. 2009; Arregui et al. 2012; Ballester 2014).

For small amplitude oscillations, the velocity amplitude ranges from the noise level up to 2–3 km/s and, contrary to large amplitude oscillations, the prominence is only locally disturbed. The triggering mechanisms of these oscillations remain unknown, although these periodic perturbations could be produced by a continuous agent, as may be the case with the 5-min photospheric and 3-min chromospheric oscillations, or by an external impulsive agent (reconnection event, weakly energetic flare disturbance) that excites different eigenmodes of the structure. The investigation of small amplitude prominence oscillations has mostly been done using spectroscopical means, and most of the spectroscopic reports are based on the analysis of the Doppler velocity although some other spectral indicators such as line intensity and line width have also been used. However, only in rare occasions have the oscillations been detected in several of these spectral indicators at the same time and with the same period (Landman et al. 1977; Yi et al. 1991; Suematsu et al. 1990). To understand this issue, a clear identification of the relationship between spectral indicators, such as Doppler shift, line intensity and line half-width, with perturbed physical variables, such as density, pressure, temperature and magnetic field strength, is required. Another feature of interest in these oscillations is their spatial distribution. Two-dimensional, high-resolution observations of a limb-prominence (Molowny-Horas et al. 1999; Terradas et al. 2002) allowed to construct Doppler, period, damping time and wavevector maps. For instance, these authors reported the existence of large regions with periodic Doppler velocity oscillations having similar periods and damping times, as well as the presence, along two selected paths in the prominence region, of a plane propagating wave and a standing wave which seemed to originate in the same prominence location. On the other hand, filaments are composed by a myriad of fine structures which seem to be field aligned, outlining magnetic flux tubes (Engvold 1998, 2008; Lin 2004; Lin et al. 2005, 2007, 2008). These magnetic flux tubes are fully or partially filled with cold plasma condensations called threads (Lin 2004; Okamoto et al. 2007). Observations with high spatial resolution have shown that individual threads or small groups of threads may oscillate independently from the rest of the prominence with their own periods (Thompson and Schmieder 1991; Yi et al. 1991; Lin 2004), and have also provided with detailed information about wave features such as the period, wavelength and phase speed. Lin et al. (2007) reported the presence of travelling waves in some threads, and after averaging the Doppler signal over a small area, containing several individual threads, they found a significant periodicity at 3.6 min. Therefore, it seems that neighbouring threads tended to oscillate coherently in the considered area. Regarding oscillations lifetime, Molowny-Horas et al. (1997, 1998) used wavelet analysis to show that, in general, the periodic perturbations are not present during the whole duration of the observational data and are of limited time duration.

These small amplitude oscillations have been mostly interpreted in terms of magnetohydrodynamic (MHD) waves in simple prominence equilibria and the reported temporal damping (Landman et al. 1977; Tsubaki and Takeuchi 1986; Wiehr et al. 1989; Molowny-Horas et al. 1999; Terradas et al. 2002; Lin 2004; Berger et al. 2008; Ning et al. 2009a,b) has been interpreted as a sign of wave damping (see reviews by Oliver and Ballester 2002; Arregui and Ballester 2011; Arregui et al. 2012; Ballester 2014). Seismology refers to the process of estimating the physical conditions of a medium by analysing properties of the oscillations or the waves travelling through the medium. MHD seismology is a method of remote diagnostics of magnetized plasma structures combining observations of oscillatory motions with an interpretation in terms of standing or propagating MHD waves, and it involves the solution of the forward and inverse problems. In the forward problem, a theoretical model is built and used to predict the oscillations of many different modes. If the predictions do not agree with observations, then, we modify the model somehow and start the comparison again. On the contrary, in the inverse problem, instead of computing frequencies from a theoretical model, we construct the model from the observed frequencies. Solar Atmospheric Seismology was proposed by Rosenberg (1970), Uchida (1970), and Roberts et al. (1984) and since in prominences the magnetic field structure as well as physical plasma properties are hard to infer directly, Roberts and Joarder (1994) and Vial (1998) suggested that prominence seismology could be a useful tool to obtain a local diagnostics (local seismology) of prominence internal structure.

This chapter is organized as follows: Sect. 11.2 is devoted to introduce single-fluid MHD equations while in Sect. 11.3 linear MHD wave equations, needed for the interpretation of small amplitude oscillations, are derived. This section also includes two examples of prominence seismology applied to prominence slabs and filament threads; next, in Sect. 11.4, two damping mechanisms, resonant absorption and ion-neutral collisions, for linear MHD waves in prominence fine structures are briefly discussed and, considering the damping by resonant absorption, one example of prominence thread seismology is included; in Sect. 11.5, a seismological analysis of flowing and oscillating prominence threads is presented and, finally, in Sect. 11.6, a few examples of prominence seismology using large amplitude oscillations are also shown.

11.2 Magnetohydrodynamic Equations

The macroscopic behaviour of a plasma in the presence of a magnetic field can be described using the magnetohydrodynamic (MHD) theory. Magnetohydrodynamic equations can be derived following different approaches. The first one starts directly from Boltzmann kinetic theory and combines it with Maxwell's equations of electromagnetism, taking into account certain approximations and properties of the plasma state (Goossens 2003; Goedbloed and Poedts 2004). The second formalism introduces MHD from fundamental equations of fluid dynamics together with Maxwell's equations, considering that a plasma is a fluid made of charged particles permeated by electric and magnetic fields (Priest 2014). The last approach starts from the fundamental equations of fluid dynamics and electromagnetism, together with the general description for a partially ionized plasma composed by different species (Braginskii 1965). Later, the equations of particular species are combined (Goossens 2003; Zaqarashvili et al. 2011) and, here, we briefly summarize this last approach. The use of the single-fluid approximation is justified in the solar atmosphere when the collision times between the different species in the plasma are much shorter than the time scales of the phenomena under study. The single-fluid MHD equations for a partially ionized plasma are relevant for the study of prominence's core, while in the prominence-corona transition region (PCTR) ideal MHD equations for a fully ionized plasma could be applied.

In the single-fluid approach and assuming a hydrogen plasma composed by ions (protons), electrons, and neutrals, the center-of-mass plasma velocity, $\mathbf{v}$, is defined as

$$\mathbf{v} = \xi_{\mathrm{e}}\mathbf{v}_{\mathrm{e}} + \xi_{\mathrm{i}}\mathbf{v}_{\mathrm{i}} + \xi_{\mathrm{n}}\mathbf{v}_{\mathrm{n}}, \tag{11.1}$$

with ξ_β the relative density of species β, and $\mathbf{v}_\beta$ the corresponding species velocity, and where subscripts e, i, and n explicitly denote electron, ion, and neutral species, respectively. Due to the small electron mass, the electron contribution can be neglected from Eq. (11.1), so

$$\mathbf{v} \approx \xi_{\mathrm{i}}\mathbf{v}_{\mathrm{i}} + \xi_{\mathrm{n}}\mathbf{v}_{\mathrm{n}}. \tag{11.2}$$

On the other hand, the fluid total density, ρ, and gas pressure, p, are

$$\rho = \rho_{\mathrm{e}} + \rho_{\mathrm{i}} + \rho_{\mathrm{n}} \approx \rho_{\mathrm{i}} + \rho_{\mathrm{n}}, \tag{11.3}$$

$$p = p_{\mathrm{e}} + p_{\mathrm{i}} + p_{\mathrm{n}}, \tag{11.4}$$

where $\rho_\beta = n_\beta m_\beta$ and $p_\beta = n_\beta k_\mathrm{B} T_\beta$, with n_β, m_β, T_β and k_B, the number density, mass particle, temperature of species β, and Boltzmann constant, respectively. Since $\rho_\beta = \xi_\beta \rho$, we get the relation $\xi_\mathrm{i} + \xi_\mathrm{n} \approx 1$. In addition, the macroscopic neutrality of the hydrogen plasma imposes $n_\mathrm{e} = n_\mathrm{i}$. Next, assuming a strong thermal coupling between species, it is possible to define a single temperature, T, for the whole fluid. Then, the total density, gas pressure, and temperature are related by

$$p = \rho R \frac{T}{\tilde{\mu}}, \tag{11.5}$$

which is the equation of state and where R is the ideal gas constant and $\tilde{\mu}$ the mean atomic weight is defined as

$$\tilde{\mu} = \frac{1}{1 + \xi_\mathrm{i}}, \tag{11.6}$$

For fully ionized plasmas $\xi_\mathrm{i} = 1$ and $\tilde{\mu} = 0.5$, whereas for a neutral gas $\xi_\mathrm{i} = 0$ and $\tilde{\mu} = 1$. Therefore, the quantity $\tilde{\mu}$ can be used to indicate the plasma ionization degree.

On the other hand, since plasma is treated as a continuous medium, the equation of mass conservation must be satisfied. This equation can be easily obtained by adding the mass conservation equations of each species, resulting in

$$\frac{\mathrm{D}\rho}{\mathrm{D}t} + \rho \nabla \cdot \mathbf{v} = 0, \tag{11.7}$$

where $\frac{\mathrm{D}}{\mathrm{D}t} \equiv \frac{\partial}{\partial t} + \mathbf{v} \cdot \nabla$ is the material (or total) derivative for time variations following the plasma motion. Next, in an inertial frame the momentum equation of species β takes the following form,

$$m_\beta n_\beta \frac{D\mathbf{v}_\beta}{Dt} = -\nabla p_\beta + Z_\beta n_\beta e \left(\mathbf{E} + \mathbf{v}_\beta \times \mathbf{B}\right) + m_\beta n_\beta \mathbf{g} - \nabla \cdot \Pi_\beta - \sum_{\beta' \neq \beta} \mathscr{R}_{\beta\beta'}. \tag{11.8}$$

The terms on the right-hand side of Eq. (11.8) correspond to, from left to right, the gas pressure gradient force, the Lorentz force, the gravity force, the viscous force, and a term that accounts for the transfer of momentum due to collisions between different species. In Eq. (11.8), $\mathbf{E}$ and $\mathbf{B}$ are the electric and magnetic fields, respectively, $\mathbf{g}$ is the gravitational acceleration, Π is the viscosity tensor, and $\mathscr{R}_{\beta\beta'}$ is the collision term between species β and β', namely

$$\mathscr{R}_{\beta\beta'} = \alpha_{\beta\beta'} \left(\mathbf{v}_\beta - \mathbf{v}_{\beta'}\right), \tag{11.9}$$

with $\alpha_{\beta\beta'}$ the friction coefficient of species β and β'. The factor Z_β in Eq. (11.8) is $Z_i = 1$, $Z_e = -1$, and $Z_n = 0$ for ions, electrons, and neutrals, respectively. Next, the corresponding equations of ions, electrons, and neutrals, have been added and the collisional terms cancel each other because of the assumption of elastic collisions. In addition, we define the density current, **j**, in terms of the difference of electron and ion velocities as

$$\mathbf{j} = e\left(n_i \mathbf{v}_i - n_e \mathbf{v}_e\right) = e n_e \left(\mathbf{v}_i - \mathbf{v}_e\right). \tag{11.10}$$

Hence, the total momentum equation is

$$\rho \frac{D\mathbf{v}}{Dt} = -\nabla p + \mathbf{j} \times \mathbf{B} + \mathbf{F}_g + \mathbf{F}_\nu, \tag{11.11}$$

where $\mathbf{F}_g$ and $\mathbf{F}_\nu$ are the total gravity and viscous forces, respectively. The Lorentz force is usually rewritten as

$$\mathbf{j} \times \mathbf{B} = \frac{1}{\mu}\left[(\mathbf{B} \cdot \nabla)\mathbf{B} - \frac{1}{2}\nabla(\mathbf{B} \cdot \mathbf{B})\right]. \tag{11.12}$$

The first term on the right-hand side of Eq. (11.12) represents the magnetic tension, that appears when magnetic field lines are curved. The second term is the gradient of a scalar quantity called the magnetic pressure force, namely

$$-\frac{1}{2\mu}\nabla(\mathbf{B} \cdot \mathbf{B}) = -\nabla\left(\frac{B^2}{2\mu}\right) \equiv -\nabla p_m. \tag{11.13}$$

where μ is the magnetic permittivity. Due to the presence of the magnetic field, an element of plasma is affected by two kinds of pressure: the gas pressure, p, and the magnetic pressure, p_m. Finally, the sum of gas and magnetic pressure gives us the total pressure, p_T, namely

$$p_T = p + p_m = p + \frac{B^2}{2\mu}. \tag{11.14}$$

and the ratio of the gas pressure to the magnetic pressure is the plasma β parameter,

$$\beta = \frac{p}{p_m}, \tag{11.15}$$

which quantifies the importance of the former with respect to the latter.

11.2.1 Generalized Induction Equation for a Partially Ionized Plasma

To derive the induction equation, we must combine Maxwell's equations with an appropriate expression of Ohm's Law for a partially ionized plasma. Following Forteza et al. (2007), Pinto and Galli (2008), and Soler (2010), the general form of the induction equation for a partially ionized plasma is,

$$\begin{aligned}\frac{\partial \mathbf{B}}{\partial t} &= \nabla \times (\mathbf{v} \times \mathbf{B}) - \nabla \times \left[\frac{1}{\mu e^2 n_e^2}\left(\alpha_e - \frac{\alpha_{en}^2}{\alpha_n}\right) \nabla \times \mathbf{B}\right] \\ &- \nabla \times \left\{\frac{1}{\mu e n_e}\left[1 - 2\xi_n \frac{\alpha_{en}}{\alpha_n}\right] (\nabla \times \mathbf{B}) \times \mathbf{B}\right\} \\ &+ \nabla \times \left\{\frac{\xi_n^2}{\mu \alpha_n}\left[(\nabla \times \mathbf{B}) \times \mathbf{B}\right] \times \mathbf{B}\right\} \\ &- \nabla \times \left[\frac{\xi_n}{\alpha_n} \mathbf{G} \times \mathbf{B}\right] - \nabla \times \left[\frac{1}{e n_e}\left(\frac{\alpha_{en}}{\alpha_n} \mathbf{G} - \nabla p_e\right)\right]. \end{aligned} \tag{11.16}$$

The terms on the right-hand side of Eq. (11.16) are: the advective term, Ohm's diffusion, Hall's diffusion, the ambipolar diffusion, the diamagnetic current term, and Biermann's battery, respectively. Among these terms, Biermann's battery term can be safely neglected in solar atmospheric plasmas, and Eq. (11.16) can be written in a more compact form such as,

$$\begin{aligned}\frac{\partial \mathbf{B}}{\partial t} &= \nabla \times (\mathbf{v} \times \mathbf{B}) - \nabla \times (\eta \nabla \times \mathbf{B}) - \nabla \times \left[\eta_H (\nabla \times \mathbf{B}) \times \mathbf{B}\right] \\ &+ \nabla \times \left\{\eta_A \left[(\nabla \times \mathbf{B}) \times \mathbf{B}\right] \times \mathbf{B}\right\} - \nabla \times \left[\tilde{\Xi} \mathbf{G} \times \mathbf{B}\right], \end{aligned} \tag{11.17}$$

with

$$\eta = \frac{1}{\mu e^2 n_e^2}\left(\alpha_e - \frac{\alpha_{en}^2}{\alpha_n}\right), \tag{11.18}$$

$$\eta_H = \frac{1}{\mu e n_e}\left(1 - 2\xi_n \frac{\alpha_{en}}{\alpha_n}\right), \tag{11.19}$$

$$\eta_A = \frac{\xi_n^2}{\mu \alpha_n}, \tag{11.20}$$

$$\tilde{\Xi} = \frac{\xi_n}{\alpha_n}. \tag{11.21}$$

being Ohm's, Hall's, and ambipolar magnetic diffusivities, respectively, and $\tilde{\Xi}$ the diamagnetic current coefficient. On the other hand, $\alpha_e = \alpha_{ei} + \alpha_{en}$ is the total electron friction coefficient, with $\alpha_{\rm ei}$ and $\alpha_{\rm en}$ the electron-ion and electron-neutral friction coefficients, respectively, $\alpha_{\rm n} = \alpha_{\rm en} + \alpha_{\rm in}$ is the total neutral friction coefficient, with $\alpha_{\rm en}$ and $\alpha_{\rm in}$ the electron-neutral and ion-neutral friction coefficients, respectively, and **G**, the pressure function defined as

$$\mathbf{G} = \xi_{\rm n}\nabla\left(p_{\rm e} + p_{\rm i}\right) - \xi_{\rm i}\nabla p_{\rm n}. \tag{11.22}$$

Equation (11.17), together with $\nabla \cdot \mathbf{B} = 0$, governs the magnetic field evolution. The non-ideal terms appear due to different effects. For example, Ohm's diffusion is mainly governed by electron-ion collisions and ambipolar diffusion is mostly caused by ion-neutral collisions. On the other hand, Hall's effect is also present in the fully ionized case, but this mechanism is enhanced by ion-neutral collisions since they tend to decouple ions from the magnetic field while electrons remain able to drift with the magnetic field. Therefore, in a partially ionized plasma, the relative importance of Hall's effect grows with the density of neutrals, but in prominence conditions it can be still safely neglected (Pandey and Wardle 2008; Krishan and Varghese 2008).

On the contrary, the diamagnetic current term couples the magnetic field evolution with pressure gradients, and since **G** vanishes in both the fully ionized and fully neutral limits, the effect of the diamagnetic term is larger for intermediate values of the ionization fraction. The ambipolar diffusivity, η_A, is commonly expressed in terms of the Cowling's coefficient, η_C, as

$$\eta_{\rm A} = \frac{\eta_C - \eta}{|\mathbf{B}|^2}. \tag{11.23}$$

and it is also appropriate to define Ohm's, σ, and Cowling's, $\sigma_{\rm C}$, conductivities, as

$$\sigma = \frac{1}{\mu\eta}, \qquad \sigma_{\rm C} = \frac{1}{\mu\eta_C}. \tag{11.24}$$

Expressions for the friction coefficients between species are needed to compute Eqs. (11.18)–(11.21). Each particular friction coefficient, $\alpha_{\beta\beta'}$, is computed as

$$\alpha_{\beta\beta'} = n_\beta m_{\beta\beta'} \nu_{\beta\beta'}, \tag{11.25}$$

with $\nu_{\beta\beta'}$ the collisional frequency between species β and β', and

$$m_{\beta\beta'} = \frac{m_\beta m_{\beta'}}{m_\beta + m_{\beta'}}. \tag{11.26}$$

A property of the friction coefficient is that $\alpha_{\beta\beta'} = \alpha_{\beta'\beta}$, and expressions for the collisional frequencies are provided by Spitzer (1962), Braginskii (1965) and De Pontieu et al. (2001).

11.2.2 Non-adiabatic Energy Equation

A frequent form for the non-adiabatic energy equation is,

$$\frac{\mathrm{D}p}{\mathrm{D}t} - \frac{\gamma p}{\rho}\frac{\mathrm{D}\rho}{\mathrm{D}t} + (\gamma - 1)\,\mathscr{L} = 0. \tag{11.27}$$

where $\mathscr{L}$ represents the net effect of all the sources and sinks of energy, which in general can be written as

$$\mathscr{L} = \nabla\cdot\mathbf{q} + \rho L - \mathbf{j}\cdot\mathbf{E}^* - Q_\nu, \tag{11.28}$$

where $\mathbf{q}$ is the heat flux due to particle thermal conduction, L is the heat-loss function which balances radiative losses with an arbitrary external heating input, $\mathbf{j}\cdot\mathbf{E}^*$ is the generalized Joule heating, and Q_ν is the viscous heating. The conductive heat vector is expressed as

$$\mathbf{q} = -\boldsymbol{\kappa}\nabla T, \tag{11.29}$$

where $\boldsymbol{\kappa}$ is the thermal conductivity tensor. For convenience, the divergence of the heat flux is splitted into the components parallel ($\nabla_\parallel$) and perpendicular ($\nabla_\perp$) to the magnetic field lines as

$$-\nabla\cdot\mathbf{q} = \nabla_\parallel\cdot\left(\kappa_\parallel\nabla_\parallel T\right) + \nabla_\perp\cdot\left(\kappa_\perp\nabla_\perp T\right),$$

where $\kappa_\parallel$ and $\kappa_\perp$ are the scalar components of the thermal conductivity tensor parallel and perpendicular to the magnetic field, respectively. In a fully ionized medium, $\kappa_\parallel$ is governed by electrons, whereas $\kappa_\perp$ is caused mainly by the ions. In the partially ionized case, one has to add the contribution of neutrals, κ_n, to both scalar conductivities, since the thermal conduction by neutrals is isotropic. Thus,

$$\kappa_\parallel = \kappa_{\parallel\mathrm{e}} + \kappa_\mathrm{n}, \qquad \kappa_\perp = \kappa_{\perp\mathrm{i}} + \kappa_\mathrm{n}. \tag{11.30}$$

and expressions for these thermal conductivities are given by Spitzer (1962), Braginskii (1965), and Parker (1953). The difference between the heat input, $H(T,\rho)$, and radiative losses, $C(T,\rho)$, is often evaluated through the heat-loss function as

$$L(T,\rho) = C(T,\rho) - H(T,\rho), \tag{11.31}$$

which depends on the local plasma parameters. Assuming an optically thin plasma, Hildner (1974) performed a piecewise fit for the radiative losses as a function of temperature. The functional expression for this fit in the case of a partially ionized plasma is,

$$C(T,\rho) = \rho\chi^* T^\alpha, \tag{11.32}$$

where χ^* and α are piecewise functions depending on the temperature. While the assumption of an optically thin plasma seems a reasonable approach for coronal conditions, prominence plasmas may be considered optically thick. Other functional expressions for radiative losses have been provided by Cox and Tucker (1969), Rosner et al. (1978), Milne et al. (1979), Klimchuk and Cargill (2001), Parenti and Vial (2007), Schure et al. (2009), Soler et al. (2012a).

On the other hand, the processes involved in the solar atmospheric heating are still not well-known. A common expression for the heating function is,

$$H(T, \rho) = h\rho^{a^*} T^{b^*}, \tag{11.33}$$

where the exponents a^* and b^* can be chosen according to different heating scenarios (Rosner et al. 1978; Dahlburg and Mariska 1988), and h is a constant parameter whose value is chosen to satisfy the energy balance condition. Hence, the general expression for the heat-loss function in terms of the plasma conditions is

$$L(T, \rho) = \rho\chi^* T^{\alpha} - h\rho^{a^*} T^{b^*}. \tag{11.34}$$

The expression of the generalized Joule heating in a partially ionized plasma is

$$\mathbf{j} \cdot \mathbf{E}^* \approx \frac{1}{\sigma} |\mathbf{j}_{\parallel}|^2 + \frac{1}{\sigma_{\mathrm{C}}} |\mathbf{j}_{\perp}|^2, \tag{11.35}$$

where only the Ohm's and Cowling's heating have been considered, whereas Hall's term does not contribute at all since $(\mathbf{j} \times \mathbf{B})$ is perpendicular to $\mathbf{j}$. Finally, the general expression for the viscous heating in terms of the complete viscosity tensor is (Braginskii 1965)

$$Q_{\nu} = \sum_{m,n} \Pi_{mn} \frac{\partial v_m}{\partial x_n}, \tag{11.36}$$

where v_m is the m-th component of the velocity vector, and x_n is the n-th coordinate.

11.2.3 Summary of Single-Fluid MHD Equations

We summarize here the basic single-fluid MHD equations for a partially ionized plasma, namely

$$\frac{\mathrm{D}\rho}{\mathrm{D}t} = -\rho \nabla \cdot \mathbf{v}, \tag{11.37}$$

$$\rho \frac{\mathrm{D}\mathbf{v}}{\mathrm{D}t} = -\nabla p + \frac{1}{\mu} (\nabla \times \mathbf{B}) \times \mathbf{B} - \rho \mathbf{g} - \nabla \cdot \Pi, \tag{11.38}$$

$$\frac{\partial \mathbf{B}}{\partial t} = \nabla \times (\mathbf{v} \times \mathbf{B}) - \nabla \times (\eta \nabla \times \mathbf{B}) - \nabla \times [\eta_H (\nabla \times \mathbf{B}) \times \mathbf{B}] +$$
$$+\nabla \times \left\{ \frac{\eta_C - \eta}{|\mathbf{B}|^2} [(\nabla \times \mathbf{B}) \times \mathbf{B}] \times \mathbf{B} \right\} - \nabla \times \left[\tilde{\Xi} \mathbf{G} \times \mathbf{B} \right], \quad (11.39)$$

$$\frac{\mathrm{D}p}{\mathrm{D}t} - \frac{\gamma p}{\rho} \frac{\mathrm{D}\rho}{\mathrm{D}t} = (\gamma - 1) \left[\nabla \cdot (\boldsymbol{\kappa} \nabla T) - \rho L(T, \rho) \right] +$$
$$+ (\gamma - 1) \left[\frac{1}{\sigma} |\mathbf{j}_\parallel|^2 + \frac{1}{\sigma_C} |\mathbf{j}_\perp|^2 + \sum_{m,n} \Pi_{mn} \frac{\partial v_m}{\partial x_n} \right], \quad (11.40)$$

$$p = \rho R \frac{T}{\tilde{\mu}}, \quad (11.41)$$

$$\nabla \cdot \mathbf{B} = 0 \quad (11.42)$$

11.2.4 MHD Equations for an Ideal and Fully Ionized Plasma

In the case of a fully ionized plasma, $\xi_\mathrm{n} = 0$, therefore $\eta_A = \tilde{\Xi} = 0$. Furthermore, neglecting other dissipative effects, the MHD equations for an ideal and fully ionized plasma are,

$$\frac{\mathrm{D}\rho}{\mathrm{D}t} = -\rho \nabla \cdot \mathbf{v}, \quad (11.43)$$

$$\rho \frac{\mathrm{D}\mathbf{v}}{\mathrm{D}t} = -\nabla p + \frac{1}{\mu} (\nabla \times \mathbf{B}) \times \mathbf{B} - \rho \mathbf{g}, \quad (11.44)$$

$$\frac{\partial \mathbf{B}}{\partial t} = \nabla \times (\mathbf{v} \times \mathbf{B}), \quad (11.45)$$

$$\frac{\mathrm{D}p}{\mathrm{D}t} - \frac{\gamma p}{\rho} \frac{\mathrm{D}\rho}{\mathrm{D}t} = 0, \quad (11.46)$$

$$p = \rho R \frac{T}{\tilde{\mu}}, \quad (11.47)$$

$$\nabla \cdot \mathbf{B} = 0 \quad (11.48)$$

11.3 Linear Ideal Magnetohydrodynamic Waves

11.3.1 Linearized Ideal Magnetohydrodynamic Equations

One of the typical applications of the MHD theory is the study of MHD waves in magnetized plasmas. Since Eqs. (11.37)–(11.42) form a system of coupled non-linear differential equations, their solution, even numerically, is extremely complicated. Therefore, if we only consider small amplitude oscillations in comparison to the sound and Alfvén speeds, non-linear effects are not very important and it is enough to consider the linear regime.

We start from Eqs. (11.43)–(11.48), with gravity neglected in the momentum equation, and assume that each physical quantity, f, can be written as the sum of an equilibrium value, f_0, and a perturbation, f_1. Thus, $\mathbf{B}(t,\mathbf{r}) = \mathbf{B}_0 + \mathbf{B}_1(t,\mathbf{r})$, $p(t,\mathbf{r}) = p_0 + p_1(t,\mathbf{r})$, $\rho(t,\mathbf{r}) = \rho_0 + \rho_1(t,\mathbf{r})$, $T(t,\mathbf{r}) = T_0 + T_1(t,\mathbf{r})$, and $\mathbf{v}(t,\mathbf{r}) = \mathbf{v}_1(t,\mathbf{r})$, where subscripts 0 and 1 denote equilibrium and perturbed quantities, respectively, and $\mathbf{r}$ is the position vector. Hereafter, the equilibrium values are taken homogeneous and constant in time. Next, we replace these expressions into Eqs. (11.43)–(11.47) and assume small perturbations, so we neglect all non-linear terms in the perturbed quantities. The resultant linear equations are

$$\frac{\partial \rho_1}{\partial t} = -\rho_0 \nabla \cdot \mathbf{v}_1, \tag{11.49}$$

$$\rho_0 \frac{\partial \mathbf{v}_1}{\partial t} = -\nabla p_1 + \frac{1}{\mu} (\nabla \times \mathbf{B}_1) \times \mathbf{B}_0, \tag{11.50}$$

$$\frac{\partial \mathbf{B}_1}{\partial t} = \nabla \times (\mathbf{v}_1 \times \mathbf{B}_0), \tag{11.51}$$

$$\frac{\partial p_1}{\partial t} - c_s^2 \frac{\partial \rho_1}{\partial t} = 0, \tag{11.52}$$

$$\frac{p_1}{p_0} = \frac{\rho_1}{\rho_0} + \frac{T_1}{T_0}, \tag{11.53}$$

along with $\nabla \cdot \mathbf{B}_1 = 0$, where $c_s^2 = \frac{\gamma p_0}{\rho_0}$ is the adiabatic sound speed squared.

11.3.2 MHD Waves in Unbounded Homogeneous Medium

Because of the presence of a magnetic field, waves in a plasma are also driven by the Lorentz force besides the pressure force. In the absence of gravity, two kinds of waves can occur in a magnetized, ideal plasma: Alfvén waves (Alfvén 1942) and magnetoacoustic (slow and fast) waves.

We assume a uniform and unbounded medium with density ρ_0, pressure p_0, and temperature T_0, permeated by a homogeneous magnetic field, $\mathbf{B}_0$. The propagation of small perturbations from the equilibrium state is governed by Eqs. (11.49)–(11.53), which, following (Lighthill 1960), can be combined to obtain,

$$\frac{\partial^2}{\partial t^2}\left[\frac{\partial^2}{\partial t^2} - \left(c_s^2 + v_a^2\right)\nabla^2\right]\Delta + \frac{c_s^2 v_a^2}{|\mathbf{B}_0|^2}\left(\mathbf{B}_0 \cdot \nabla\right)^2 \nabla^2 \Delta = 0, \tag{11.54}$$

$$\left[\frac{\partial^2}{\partial t^2} - \frac{v_a^2}{|\mathbf{B}_0|^2}\left(\mathbf{B}_0 \cdot \nabla\right)^2\right]\Psi = 0, \tag{11.55}$$

where $v_a^2 = \frac{|\mathbf{B}_0|^2}{\mu\rho_0}$ is the Alfvén speed squared, and Δ and Ψ are defined as,

$$\Delta = \nabla \cdot \mathbf{v}_1, \tag{11.56}$$

$$\Psi = \frac{(\nabla \times \mathbf{v}_1) \cdot \mathbf{B}_0}{|\mathbf{B}_0|}. \tag{11.57}$$

where Δ represents the compressibility of the perturbed velocity field, while Ψ is the component of the vorticity of the perturbed velocity field along the magnetic field. Equation (11.54) governs magnetoacoustic waves, whereas Eq. (11.55) stands for Alfvén waves. Now, we take the x-axis orientated along the magnetic field direction, $\mathbf{B}_0 = B_0\hat{e}_x$, and consider perturbations in the form of plane waves, so they are written proportional to $\exp(i\mathbf{k}\cdot\mathbf{r} - i\omega t)$, where ω is the frequency and $\mathbf{k}$ the wave vector. With no loss of generality, we choose the z-axis so that the wave vector is in the xz-plane, so $\mathbf{k} = k_x\hat{e}_x + k_z\hat{e}_z$. For arbitrary Δ and Ψ, Eqs. (11.54) and (11.55) provide us with the dispersion relation for magnetoacoustic waves, namely

$$\omega^4 - \left(v_a^2 + c_s^2\right)k^2\omega^2 + v_a^2 c_s^2 k^4 \cos^2\theta = 0, \tag{11.58}$$

and for Alfvén waves, namely

$$\omega^2 - k_x^2 v_a^2 = 0, \tag{11.59}$$

with $k^2 = k_x^2 + k_z^2$ and $\theta = \arctan(k_z/k_x)$. Linear Alfvén waves have no associated density, temperature, or gas pressure changes, are driven exclusively by the magnetic tension and propagate non-dispersively along the field lines with a constant phase speed, $\omega/k_x = v_a$. Regarding magnetoacoustic waves, the analytical solution of Eq. (11.58) is

$$\omega^2 = \frac{k^2}{2}\left[\left(v_a^2 + c_s^2\right) \pm \sqrt{\left(v_a^2 + c_s^2\right)^2 - 4v_a^2 c_s^2 \cos^2\theta}\right], \tag{11.60}$$

where the $-$ sign corresponds to the slow wave and the $+$ sign to the fast wave, whose properties depend on the relation between c_s and v_a, and they arise from the combined effect of the pressure gradient and the Lorentz force. For typical physical conditions in the solar corona and prominences, $c_s < v_a$, so the slow wave is essentially a modified acoustic wave guided by the magnetic field and the fast wave is a magnetic wave which is driven by magnetic pressure and weakly affected by acoustic effects. This behavior is the opposite when $c_s > v_a$ (Goossens 2003). The distinction between Alfvén, fast, and slow waves can be more ambiguous in a inhomogeneous equilibrium, since these three modes are in general coupled.

11.3.3 MHD Waves in Prominence Slabs

Disregarding their fine structure, many theoretical models consider prominences as thin sheets (slabs) of cool plasma threaded by a magnetic field, and the linear modes of oscillation of these magnetized prominence slabs embedded, or not, in the solar corona have been studied (Joarder and Roberts 1992a,b, 1993a,b; Oliver et al. 1992, 1993; Oliver and Ballester 1995, 1996). Since magnetic field lines seem to be oriented at a rather small angle (around 20°) with the long axis of the prominence sheet (Leroy 1988, 1989), Joarder and Roberts (1993a) considered a prominence slab embedded in a uniform magnetic field that makes an angle with the longitudinal axis of the slab. The influence of gravity was neglected and so the plasma variables (temperature, pressure and density) are uniform both in the prominence and in the coronal region, the k_y and k_z components of the wavenumber must be considered and coupled fast, Alfvén and slow modes are obtained. Therefore, the distinction between the three MHD modes is lost and, probably, there are no characteristic oscillatory directions associated to the various modes and the actual velocity field in prominences can be much more complex. The resulting dispersion diagram displays a very rich mode structure with plenty of mode couplings, and this theoretical model provides with analytical expressions for the frequency of six fundamental modes: the symmetric Alfvén, slow and fast kink modes and the antisymmetric Alfvén, slow and fast sausage modes, as a function of the prominence parameters. Regrettably, a complete determination of the physical properties of perturbations for the modes in the dispersion diagram is still lacking.

11.3.3.1 Seismology of Prominence Slabs

Few attempts to perform prominence seismology based on the consideration of prominences as plasma slabs have been done. For instance, Régnier et al. (2001) observed an active region filament with SUMER/SoHO detecting oscillations covering different ranges of periods: < 5 min; 6–20 min; > 40 min and, as theoretical counterpart, they considered the prominence slab model with a uniform and skewed magnetic field proposed by Joarder and Roberts (1993a). Observations provided with estimates for the width (8,000 km) and length (63,000 km) of the filament, and assumptions on other parameters, such as the temperature of the filament

(8,000 K) and of its environment (10^6 K), the density of the slab (10^{12} cm^{-3}), the magnetic field strength (20 G) and for the angle between the magnetic field and the long axis of the slab (25°), were made. Using these parameters, the dispersion relations were solved and the corresponding periods were obtained and classified. The frequency ratio of the fundamental even Alfvén mode to the fundamental odd Alfvén mode only depends on the ratio of the half-width of the slab to the half-length of the filament, which is a measurable quantity, and the same applies to the frequency ratios involving the slow kink/sausage and fast kink/sausage modes. Then, parametric calculations for the frequencies as a function of the magnetic field strength and the inclination angle, while keeping the slab density constant, were performed, and a diagnostic of the observed filament was obtained by looking for the parameters values that enable the matching of theoretical and observed frequencies. By following this method, the angle between the magnetic field and the long axis of the slab is estimated to be 18° and, using this value, an algebraic relation for the magnetic field strength as a function of the slab density was also derived.

Using the same theoretical model, Pouget et al. (2006) performed a more complete analysis of the long duration and high temporal resolution observations of three filaments made with CDS/SoHO. These observations allowed to detect and measure the entire range of periodicities theoretically expected in a filament. In particular both the short (less that 10 min) and the long ones (more than 40 min) were detected. As before, the frequency ratios between fundamental even/odd (kink/sausage) modes only depend on the ratio of the filament half-width to its half-length, Once this ratio was measured, Pouget et al. (2006) assume that their 16 h long observation had allowed them to observe the six modes of interest, since the slowest mode is expected at a period of 5 h, for standard prominence parameters. The inversion method first assigns a possible triplet of measured frequencies to the three odd fundamental frequencies (odd Alfvén, slow sausage, and fast sausage modes). The coherence of each choice is examined against two tests. The first requires to find three corresponding even frequencies, with the condition that the even/odd frequency ratios are consistent with the measured half-width to half-length ratio. The second involves the inferred values for the density, temperature, magnetic field inclination angle, and magnetic field strength to be consistent with typical values reported in the literature. For each test, if the test was negative, the full triplet was changed and the series started again. On the contrary, if the tests succeeded, they considered that the six fundamental modes were identified. The three filament observations led to coherent diagnostics and a single possible set of frequencies was found for each observation. The most important feature of this study is that it allows to simultaneously determine the values of the inclination angle, temperature, and Alfvén speed for the same prominence. For instance, for the three observed filaments, Pouget et al. (2006) obtained values of 19°, 33°, and 35° for the inclination angle; 6,200, 11,700 and 6,800 K for the temperatures, and 157, 490, 488 km/s for the Alfvén speeds.

Finally, Heinzel et al. (2014) have used the linear perturbations of a bounded, homogeneous prominence slab threaded by a transverse magnetic field, as the input variables for a one-dimensional radiative transfer code which calculates the full spectral profile of the hydrogen Hα and Hβ lines. The obtained results (Heinzel 2014) represent a first step for the understanding of the relationship between spectral indicators (Doppler shift, line intensity and line half-width) with density, pressure, temperature and magnetic field perturbations in prominences. By considering more elaborated slab prominence models together with sophisticated radiative transfer codes, a more complete understanding of the wave behaviour in realistic prominence conditions will be reached, in a similar way to what has been done in chromospheric studies (Carlsson and Stein 1997; Heggland et al. 2011).

11.3.4 MHD Waves in Filament Threads

High-resolution Hα observations (Engvold 2004; Lin 2004; Lin et al. 2005) have allowed to observe filament fine structure in great detail. The measured average width of resolved thin threads is about 0.3 arcsec (∼210 km), while their length is between 5 and 40 arcsec (∼3,500–28,000 km). The fine threads of solar filaments seem to be partially filled with cold plasma (Lin et al. 2005), typically two orders of magnitude denser and cooler than the surrounding corona, and it is generally assumed that they outline the magnetic flux tubes in which they reside (Engvold 1998, 2008; Lin 2004; Lin et al. 2005; Okamoto et al. 2007; Martin et al. 2008; Lin et al. 2008). This idea is strongly supported by observations which suggest that threads are skewed with respect to the filament long axis in a similar way to what was early found for the prominence magnetic field (Leroy 1980; Bommier et al. 1994; Bommier and Leroy 1998). On the other hand, observations have also given us detailed information about the local oscillatory behaviour of the fine structure of filaments, and two different scenarios can be considered. In the first one, propagating MHD waves along a thread have wavelengths much shorter than the thread length, then, infinitely long threads can be considered. The second case includes propagating MHD waves with wavelengths comparable to or larger than the length of the thread, and standing modes whose wavelength is of the order of the length of the supporting magnetic tube and thus much larger than the thread length, then, finite length threads (Díaz et al. 2001, 2002, 2003, 2005, 2010; Díaz and Roberts 2006) must be considered.

Here, we assume a gravity-free, straight, unlimited in the longitudinal direction, cylindrically symmetric flux tube of radius a with the internal and external magnetic field pointing in the z-direction, filled with prominence material and immersed in the coronal environment (Fig. 11.1), with

$$B_0(r), \rho_0(r), p_0(r) = \begin{cases} B_f, \rho_f, p_f, & r < a \\ B_c, \rho_c, p_c, & r > a \end{cases}$$

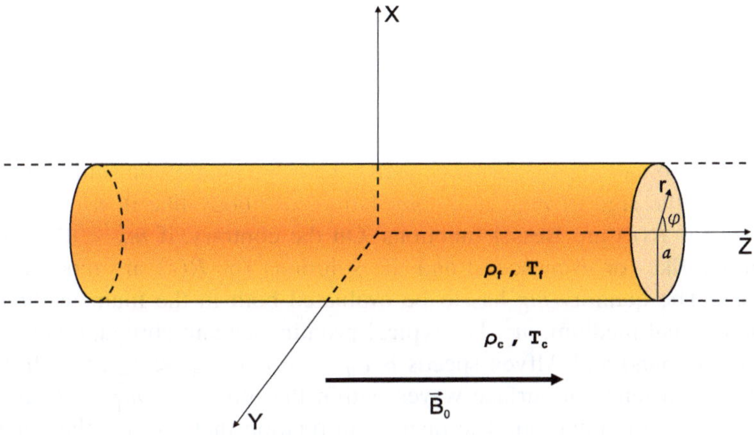

Fig. 11.1 Sketch of the homogeneous cylindrical flux tube model. From Soler (2010)

where B_f, ρ_f, p_f and B_c, ρ_c, p_c, are filament and coronal magnetic field, density and pressure, respectively. The stationary state of their oscillations is governed by the trapped normal modes (Terradas et al. 2007; Edwin and Roberts 1983) and the linear, ideal MHD magnetosonic modes supported by this model are described by Eq. (11.54) which in cylindrical coordinates (r, φ, z) becomes

$$\frac{\partial^2}{\partial t^2}\left[\frac{\partial^2}{\partial t^2} - \left(c_s^2 + v_a^2\right)\nabla^2\right]\Delta + c_s^2 v_a^2 \frac{\partial^2}{\partial z^2}\nabla^2 \Delta = 0, \tag{11.61}$$

Next, since φ and z are ignorable directions, we write

$$\Delta = R(r)\exp\left(im\varphi + ik_z z - i\omega t\right), \tag{11.62}$$

where m is an integer that plays the role of the azimuthal wavenumber, k_z is the longitudinal wavenumber, ω is the frequency, and $R(r)$ is a function accounting for the radial dependence. Now, applying this last expression to Eq. (11.61), one finds that $R(r)$ satisfies the well-known Bessel equation of order m, namely

$$\frac{d^2R(r)}{dr^2} + \frac{1}{r}\frac{dR(r)}{dr} + \left(m_0^2 - \frac{m^2}{r^2}\right)R(r) = 0, \tag{11.63}$$

with

$$m_0^2 = \frac{\left(\omega^2 - k_z^2 v_a^2\right)\left(\omega^2 - k_z^2 c_s^2\right)}{\left(v_a^2 + c_s^2\right)\left(\omega^2 - k_z^2 c_t^2\right)}, \tag{11.64}$$

and

$$c_t^2 = \frac{v_a^2 c_s^2}{v_a^2 + c_s^2}, \tag{11.65}$$

which is the so-called cusp (or tube) speed. The character of the solutions for $R(r)$ depends on the sign of m_0^2. Thus, oscillations are body-like if $m_0^2 > 0$ and the solutions for $R(r)$ are Bessel functions. On the contrary, if $m_0^2 < 0$ oscillations are surface-like (or evanescent) and the solutions for $R(r)$ are modified Bessel functions. The quantity m_0^2 has to be evaluated both in the filament thread, m_{f}^2, and the coronal medium, m_{c}^2. For typical prominence and coronal conditions, the ordering of sound and Alfvén speeds is $c_{sf} < v_{af} < c_{sc} < v_{ac}$, which does not permit the existence of surface waves within the thread, so $m_{\mathrm{f}}^2 > 0$ and in the corona $m_c^2 < 0$ is imposed. The dispersion relation that governs the behavior of wave modes is,

$$\frac{n_{\mathrm{c}}}{\rho_{\mathrm{c}}\left(\omega^2 - k_z^2 v_{ac}^2\right)} \frac{K_m'(n_{\mathrm{c}}a)}{K_m(n_{\mathrm{c}}a)} - \frac{m_{\mathrm{f}}}{\rho_{\mathrm{f}}\left(\omega^2 - k_z^2 v_{af}^2\right)} \frac{J_m'(m_{\mathrm{f}}a)}{J_m(m_{\mathrm{f}}a)} = 0, \tag{11.66}$$

where the prime denotes the derivative with respect to the argument, and $n_{\mathrm{c}}^2 = -m_{\mathrm{c}}^2$. The solutions of Eq. (11.66) can be classified according to several criteria. Considering the value of the azimuthal wavenumber, solutions with $m = 0$ are called sausage modes, solutions with $m = 1$ are kink modes, and solutions with $m \geq 2$ are fluting modes. Considering their magnetoacoustic properties, wave modes with a phase velocity, ω/k_z, in the range $c_{tf} < \omega/k_z < c_{sf}$ and whose dominant velocity component is along magnetic field lines are identified as internal slow modes. On the other hand, the modes with $v_{af} < \omega/k_z < v_{ac}$, which are mainly polarized transversely to magnetic field lines, are usually denoted as fast modes. However, Goossens et al. (2009) pointed out that the mode with $m = 1$, the so-called kink mode, has mixed alfvénic and fast properties, the magnetic tension being the dominant restoring force. For this reason, we use the more general expression transverse modes, based on their displacement polarization, to refer to solutions with $v_{af} < \omega/k_z < v_{ac}$.

An analytical approximation to Eq. (11.66) can be obtained by considering the thin tube (TT) approximation, i.e., $k_z a \ll 1$. Performing a first order, asymptotic expansion for small arguments of the Bessel functions in Eq. (11.66) for $m \neq 0$, the dispersion relation then becomes

$$\rho_{\mathrm{f}}\left(\omega^2 - k_z^2 v_{af}^2\right) + \rho_{\mathrm{c}}\left(\omega^2 - k_z^2 v_{ac}^2\right) = 0, \tag{11.67}$$

whose analytical solution is

$$\omega^2 = \frac{\rho_{\mathrm{f}} v_{af}^2 + \rho_{\mathrm{c}} v_{ac}^2}{\rho_{\mathrm{f}} + \rho_{\mathrm{c}}} k_z^2 \equiv \omega_{\mathrm{k}}^2, \tag{11.68}$$

where $\omega_{\rm k}$ is the so-called kink frequency, with v_{af} and v_{ac} the filament (f) and coronal (c) Alfvén velocities, respectively. Note that Eq. (11.68) is only valid for transverse modes with $m \neq 0$. An important remark is that the kink mode is the only one producing a significant transverse displacement of the cylinder axis and, also, producing short-period oscillations of the order of minutes. These two features are very important in order to identify the MHD mode responsible for filament fine structure oscillations (see Sects. 11.3.4.1 and 11.4.1.1).

11.3.4.1 Seismology Using the Period of Filament Thread Oscillations

Lin et al. (2009) analyzed observations of thread oscillations by combining simultaneous recordings of motions along the line-of-sight and in the plane of the sky, which provides with information about the orientation of the oscillatory velocity vector. From the measurements of swaying motions in the plane of the sky, several threads presented travelling disturbances whose main features were characterized (period, phase velocity and oscillatory amplitude). Moreover, two of these threads also showed Doppler velocity oscillations with a period similar to that of the swaying motions, so that the threads had a displacement that was neither in the plane of the sky nor along the line of sight. By combining the observed oscillations in the two orthogonal directions, these authors derived the full velocity vectors, and in the case of the two analyzed threads, oscillatory motions were reasonably close to the vertical direction. The observed events were interpreted as propagating MHD kink waves supported by the thread body. This interpretation also implies that the measured phase velocity is equal to the kink speed. Considering the theoretical model (Fig. 11.1) introduced in Sect. 11.3.4, a comparison between the observed wave properties and the theoretical prediction can be made. This enables to obtain estimates for some physical parameters of interest, namely the Alfvén speed and the magnetic field strength in the studied threads. Assuming the thin tube (TT) approximation, and that inside and outside the magnetic flux tube the density is given by,

$$\rho_0(r) = \begin{cases} \rho_f, & r \leq a \\ \rho_c, & r > a \end{cases}$$

the kink speed, c_k, is,

$$c_k = \sqrt{\frac{2B_0^2}{\mu(\rho_f + \rho_c)}}$$

and the above expression can be written as,

$$c_k = v_{Af}\sqrt{\frac{2\zeta}{\zeta + 1}}$$

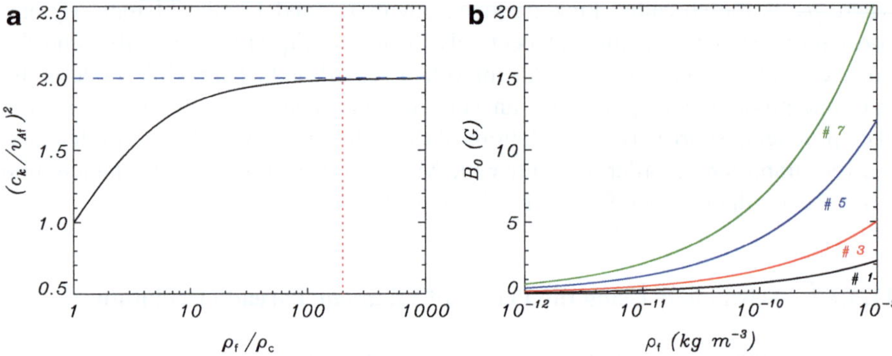

Fig. 11.2 (**a**) Ratio c_k^2/v_{Af}^2 (*solid line*) as a function of the density contrast, ρ_f/ρ_c. The *dotted line* corresponds to the value of the ratio c_k^2/v_{Af}^2 for $\rho_f/\rho_c \to \infty$. (**b**) Magnetic field strength as a function of the filament density, ρ_f, corresponding to four selected threads. From Soler (2010)

with

$$\zeta = \frac{\rho_f}{\rho_c}$$

and when this ratio becomes very large (see Fig. 11.2a),

$$c_k \sim \sqrt{2} v_{Af}$$

where v_{Af} is the thread Alfvén speed. The results for the internal Alfvén speed show a strong dispersion (see Table 2 in Lin et al. 2009), suggesting that the physical conditions in different threads are very different in spite of belonging to the same filament. This result clearly reflects the highly inhomogeneous nature of solar filaments. Once the Alfvén speed in each thread was determined, the magnetic field strength could be computed after a value for the thread density was assumed or measured (Fig. 11.2b)

11.4 Damping Mechanisms for Prominence Oscillations

Observational evidence reveals that small amplitude oscillations, once they have been excited, decay in short spatial and temporal scales. Evidences for the damping of small amplitude oscillations in prominences can be found in Landman et al. (1977), Tsubaki and Takeuchi (1986), Tsubaki et al. (1988), Wiehr et al. (1989), Molowny-Horas et al. (1999), Terradas et al. (2002), Ballai (2003), Lin (2004), Berger et al. (2008), Ning et al. (2009b), and Lin et al. (2009). Reliable values for the damping time have been derived, from different Doppler velocity time series, by Molowny-Horas et al. (1999) and Terradas et al. (2002) in prominences, and by Lin (2004) in filaments. The values thus obtained are usually between one and four times the corresponding period, and large regions of prominences/filaments display similar damping times. Several theoretical mechanisms have been proposed in order to explain the observed damping (Arregui and Ballester 2011). Linear non-adiabatic MHD waves have been studied by Carbonell et al. (2004), Terradas et al. (2001), Terradas et al. (2005), Soler et al. (2007, 2008). The overall conclusion from these studies is that thermal mechanisms can only account for the damping of slow waves, although in a not very efficient manner, while fast waves remain almost undamped, therefore, these mechanisms could be disregarded. On the other hand, since prominences can be considered as partially ionized plasmas, a possible mechanism to damp fast waves (as well as Alfvén waves) could come from ion-neutral collisions (Forteza et al. 2007; Forteza et al. 2008; Soler et al. 2009a), although the theoretical ratio of the damping time to the period, due to ion-neutral collisions, does not completely match the observations. Besides non-ideal mechanisms, another possibility to attenuate fast waves in thin filament threads comes from resonant wave damping (Goossens et al. 2010). In the following we consider the effects of resonant absorption and ion-neutral collisions as damping mechanisms for the transverse oscillations of a filament thread.

11.4.1 Damping of Kink MHD Waves in a Filament Thread

Consider now a gravity-free, straight, unlimited, in the longitudinal direction, cylindrically symmetric flux tube with a mean radius a (see Fig. 11.3) with the internal and external magnetic field pointing in the z-direction, filled with prominence material and inmersed in the coronal environment. We assume the zero-β approximation; i.e., gas pressure is neglected, so slow waves are absent and this allows us to concentrate on the oscillatory properties of fast and Alfvén MHD waves and their mutual interaction (Arregui et al. 2008; Soler et al. 2009c,d). We assume a one-dimensional non-uniform distribution of density, $\rho(r)$, across the structure. The internal prominence plasma, with uniform density, ρ_f, occupies the full length of the tube and is connected to the coronal medium, with uniform density, ρ_c, by means of a non-uniform transitional layer of thickness l. The ratio l/a provides us with a measure of the transverse inhomogeneity length-scale, that can vary in between $l/a = 0$ (homogeneous tube) and $l/a = 2$ (fully non-uniform tube). Considering perturbations with $m = 1$, which represent kink waves that produce the transverse displacement of the tube as they propagate along it, its frequency in the long wavelength or thin tube (TT) approximation ($k_z a \ll 1$) is given by Eq. (11.68). Then, by defining $\zeta = \rho_f/\rho_c$, for the density contrast, the period of kink oscillations with a wavelength $\lambda = 2\pi/k_z$ can be written as

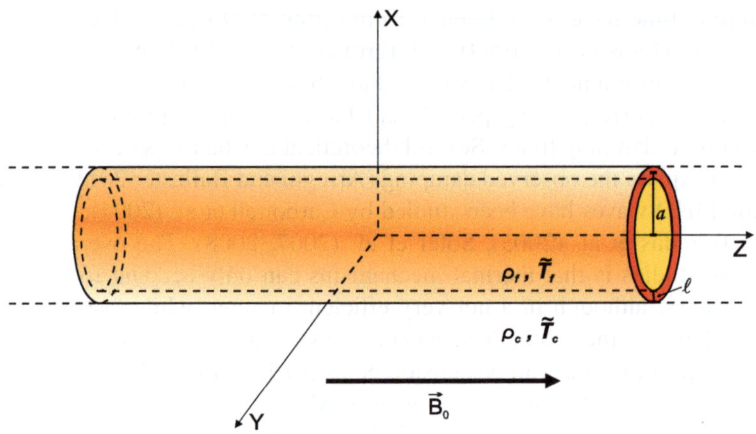

Fig. 11.3 Sketch of the inhomogeneous filament thread. From Soler (2010)

$$P = \frac{\sqrt{2}}{2}\frac{\lambda}{v_{af}}\left(\frac{1+\zeta}{\zeta}\right)^{1/2}. \tag{11.69}$$

For kink waves to be damped by resonant absorption, transverse inhomogeneity in the Alfvén velocity needs to be considered. In our uniform field model this is obtained by considering $l \neq 0$. Then, the $m = 1$ solution is resonantly coupled to local Alfvén waves. The coupling produces the temporal attenuation of transverse motions which are converted into localized azimuthal Alfvénic oscillations. Asymptotic analytical expressions for the damping time can be obtained under the assumption that the transverse inhomogeneity length-scale is small ($l/a \ll 1$). This is the so-called thin boundary approximation (TB). When the TT and TB approximations are combined, the analytical expression for the damping time (τ_D) over period (P) can be written as (Hollweg and Yang 1988; Sakurai et al. 1991a,b; Goossens et al. 1992, 1995; Ruderman and Roberts 2002)

$$\frac{\tau_D}{P} = F\ \frac{a}{l}\ \frac{\zeta+1}{\zeta-1}. \tag{11.70}$$

where F is a numerical factor that depends on the density profile in the non-uniform layer. For a linear variation, $F = \frac{4}{\pi^2}$ (Goossens et al. 1992), while for a sinusoidal variation, $F = \frac{2}{\pi}$ (Ruderman and Roberts 2002). Consider for example $\zeta = 200$ as a typical density contrast and $l/a = 0.1$. Then, Eq. (11.70) predicts a damping time of ~6 times the oscillatory period, thus producing a time-scale compatible with observations. From Eq. (11.70) we can observe that the damping ratio τ_D/P rapidly decreases when the density contrast is increased, tending to a constant value in the large density contrast regime. Also, the damping time over period is independent of the wavelength of perturbations, but rapidly decreases with increasing inhomogeneity length-scale. These results suggest that resonant absorption could be a very efficient mechanism for the attenuation of kink waves in inhomogeneous cylindrical flux tubes representing filament fine threads. The efficiency of the resonant damping beyond the TB approximation was explored by Soler et al. (2013), who showed that the error associated to Eq. (11.70) can be significant when the condition $l/a \ll 1$ is not satisfied.

Now, using the same theoretical model, consider a partially ionized inhomogeneous filament thread in which the ionization fraction $\tilde{\mu}$ denotes the plasma degree of ionization. A similar functional dependence for the density and ionization fraction in the radial direction has been chosen,

$$\tilde{\mu}(r) = \begin{cases} \tilde{\mu}_{\mathrm{f}}, & \text{if} \quad r \leq a - l/2, \\ \tilde{\mu}_{\mathrm{tr}}(r), & \text{if}\ a - l/2 < r < a + l/2, \\ \tilde{\mu}_{\mathrm{c}}, & \text{if} \quad r \geq a + l/2, \end{cases} \tag{11.71}$$

with

$$\tilde{\mu}_{\mathrm{tr}}(r) = \frac{\tilde{\mu}_{\mathrm{f}}}{2}\left\{\left(1 + \frac{\tilde{\mu}_{\mathrm{c}}}{\tilde{\mu}_{\mathrm{f}}}\right) - \left(1 - \frac{\tilde{\mu}_{\mathrm{c}}}{\mu_{\mathrm{f}}}\right)\sin\left[\frac{\pi}{l}(r-a)\right]\right\}, \tag{11.72}$$

where the filament ionization fraction, $\tilde{\mu}_{\mathrm{f}}$, is considered a free parameter and the corona is assumed to be fully ionized, so $\tilde{\mu}_{\mathrm{c}} = 0.5$. After considering linear perturbations and removing gas pressure terms by setting $\beta = 0$, the relevant equations for our investigation are the linearized momentum and induction equations which, neglecting other terms, contain only Ohm's, ambipolar, and Hall's diffusion. Since η, η_C, and η_H are functions of the plasma physical conditions (Eqs. (11.18)–(11.20)), their values in our equilibrium depend on the radial coordinate. In the TTTB approximation and considering weak damping, the final relation for the ratio of the damping time to the period is

$$\frac{\tau_D}{P} = F\left[m\left(\frac{l}{a}\right)\left(\frac{\rho_{\mathrm{f}} - \rho_{\mathrm{c}}}{\rho_{\mathrm{f}} + \rho_{\mathrm{c}}}\right) + \frac{2\left(\rho_{\mathrm{f}}\tilde{\eta}_{Cf} + \rho_{\mathrm{c}}\tilde{\eta}_{Cc}\right)k_z a}{\sqrt{2\rho_{\mathrm{f}}\left(\rho_{\mathrm{f}} + \rho_{\mathrm{c}}\right)}}\right]^{-1}, \tag{11.73}$$

where both Cowling's diffusivities are expressed in dimensionless form. To perform a simple application, we compute τ_D/P from Eq. (11.73) in the case $m = 1$, $k_z a = 10^{-2}$, and $l/a = 0.2$, resulting in $\tau_D/P \approx 3.18$ for a fully ionized thread ($\tilde{\mu}_{\mathrm{f}} = 0.5$), and $\tau_D/P \approx 3.16$ for an almost neutral thread ($\tilde{\mu}_{\mathrm{f}} = 0.95$). The obtained damping times are consistent with the observations, but the ratio τ_D/P depends only very slightly on the ionization degree suggesting that resonant absorption is the dominant damping mechanism.

11.4.1.1 Seismology Using the Period and Damping Time of Filament Thread Oscillations

Lin (2004) detected several periodicities over large areas of a filament, with maximum power at periods of 26, 42 and 78 min, and Doppler velocity oscillations with 26 min period were only observed for 2–3 periods, after which they became strongly damped. Also, Lin et al. (2009) clearly observed that the amplitudes of the waves passing through two different cuts along a thread are notably different and these apparent changes can be due to damping of the waves. The damping of prominence oscillations is a common feature in many observed events and damping time-scales provide with an additional source of information that can be used when performing parameter inference using seismology inversion techniques.

We consider here the theoretical model introduced in Sect. 11.4, with a density contrast between the fully ionized thread and the external corona given by $\zeta = \frac{\rho_f}{\rho_c}$. Since $l/a \neq 0$, the kink MHD mode is resonantly coupled to Alfvén continuum modes and is damped in time (see Sect. 11.4). For standing kink waves, and without using the thin tube and thin boundary approximation, the normal mode period and damping ratio are functions of the relevant equilibrium parameters,

$$P = P(k_z, \zeta, l/a, v_{Af}), \quad \frac{\tau_D}{P} = \frac{\tau_D}{P}(k_z, \zeta, l/a), \tag{11.74}$$

with k_z the longitudinal wavenumber. However, in the thin tube and thin boundary approximations (TTTB), the period does not depend on l/a (Eq. (11.69)) and the damping ratio (τ_D/P) is independent of the wavelength (Eq. (11.70)). Then, if the wavelength, period and damping time of the oscillations are known (Fig. 11.4), we can solve the forward problem using Eqs. (11.69) and (11.70) to obtain the thread Alfvén speed (v_{Af}) and the inhomogeneity scale (l/a). On the opposite, regarding the inverse problem, if no assumption is made on any of the physical parameters of interest, we have,

$$P_{obs} = P(k_z, \zeta, l/a, v_{Af}), \quad (\frac{\tau_D}{P})_{obs} = \frac{\tau_D}{P}(k_z, \zeta, l/a), \tag{11.75}$$

therefore, knowing wavelength, period and damping time, we are left with two equations and three unknown parameters (ζ, l/a, v_{Af}) and there are infinite different equilibrium models that can equally well explain the observations. The parameter values that define these valid equilibrium models are displayed in Fig. 11.4, where the analytical algebraic expressions in the TTTB approximation by Goossens et al. (2008) have been used to invert the problem. It can be appreciated that, even if an infinite number of solutions is obtained, they define a rather constrained range of values for the thread Alfvén speed. Because of the insensitiveness of the damping rate with the density contrast for the typically large values of this parameter in prominence plasmas, the obtained solution curve displays an asymptotic behaviour for large values of ζ. This makes possible to obtain precise estimates for the thread Alfvén speed $v_{Af} \simeq 12$ km/s, and the transverse inhomogeneity length scale,

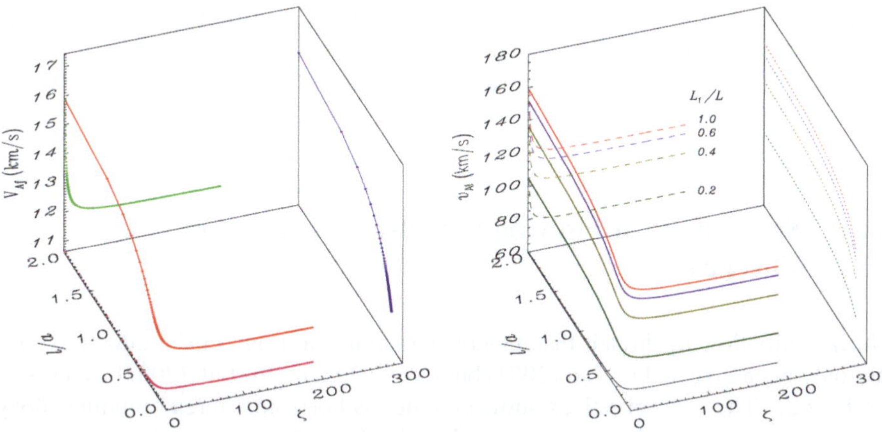

Fig. 11.4 *Left panel*: Analytic inversion of physical parameters in the $(\zeta, l/a, v_{Af})$ space for a filament thread with $P = 3\,\text{min}$, $\tau_d = 9\,\text{min}$ and a wavelength $\lambda = 3{,}000\,\text{km}$. From Arregui et al. (2012); *Right panel*: Analytic inversion of physical parameters in the $(\zeta, l/a, v_{Af})$ space for filament threads of different L_f/L, with $P = 20\,\text{min}$, $\tau_d = 60\,\text{min}$ and $L = 10^5\,\text{km}$. (*credit*: Figure 6a from Soler et al. 2010)

$l/a \simeq 0.16$. The computation of the magnetic field strength from the obtained seismological curve requires the assumption of a particular value for either the filament or the coronal density. In the inversion curve displayed in Fig. 11.4 (left panel), a change in the period produces a vertical shift of the solution curve, hence the period influences the inferred values for the Alfvén speed. The main shortcoming of this technique is the use of thread models in which the full magnetic tube is filled with cool and dense plasma. An example of the inversion of physical parameters for different values of the thread length (L_f) was presented by Soler et al. (2010). When partially filled threads, i.e., with the dense part occupying a length shorter than the total length of the tube L, are considered, one curve is obtained for each value of the length of the thread. Even if each curve gives an infinite number of solutions, again each of them defines a rather constrained range of values for the thread Alfvén speed and the ratio L_f/L is a fundamental parameter in order to perform an accurate seismology of prominence threads, since different curves produce different estimates for the prominence Alfvén speed (Fig. 11.4 right panel). Because of the insensitiveness of the damping ratio with respect to the length of the thread, all solution curves for different lengths of the threads produce the same projection onto the (ζ, l/a)-plane. Hence, the same precise estimates of the transverse inhomogeneity length scale obtained from infinitely long thread models are valid, irrespective of the length of the thread.

On the other hand, and because of the incompleteness of the observational information, to solve the inversion problem is not an easy task. Therefore, Arregui et al. (2014) have applied Bayesian formalism, widely used in other astrophysical areas, to obtain a consistent solution for the inversion problem and the correct propagation of errors from observations to inferred parameters, pointing out that this could be a very promising approach. Using the same values as in Fig. 11.4, and considering uncertainties of the order of 10 %, they obtained $v_{Af} = 11 \pm 1$ km/s; $l/a = 0.21 \pm 0.02$

11.5 Seismology of Flowing and Oscillating Prominence Threads

Horizontally flowing threads that undergo simultaneous transverse oscillations have not only been detected by Lin (2004) but also by Okamoto et al. (2007) using SOT on Hinode. These observations show continuous horizontal thread motions along an active region prominence, and the threads also suffer apparently synchronous vertical oscillatory motions, showing that waves and flows can coexist. Six threads displaying the same behaviour were studied and periods in the range 135–250 s were measured. The thread flow velocities range from 15 to 46 km/s and the vertical oscillation amplitudes range from 408 to 1,771 km. A particularly interesting feature of these oscillations is that points along each thread oscillate transversally with the same phase. To perform a seismological analysis of these oscillations, Terradas et al. (2008) first neglected the mass flows and interpreted these events in terms of the standing kink mode of a finite-length thread in a magnetic flux tube. Using previous theoretical results (Díaz et al. 2002; Dymova and Ruderman 2005), they found that a one-to-one relation between the thread Alfvén speed and the coronal Alfvén speed could be established. This relation is highlighted by means of a number of curves relating the two Alfvén speeds for different values of the length of the magnetic flux tube and the density contrast between the filament and coronal plasma. The most interesting property of these curves is that they display an asymptotic behaviour for large values of the density contrast and hence a lower limit for the thread Alfvén speed can be obtained (Fig. 11.5). Considering one of the threads as a magnetic flux tube with a length of 100 mm, a value of 120 km/s for the thread Alfvén speed is obtained. Next step was to incorporate mass flows into the analysis by considering the numerical solution of the non-linear, ideal, low-β MHD equations. The numerical results indicated that the effect of the flow on the obtained periods is weak, and because of the small value of the measured flow speeds there are no significant variations of the wave properties, and hence of the inferred Alfvén speeds. However, Soler and Goossens (2011) and Soler et al. (2012b) studied the joint effects of flows and resonant absorption on the amplitude of standing kink waves in finite and inhomogeneous filament threads, and they found that the flowing thread can affect significantly the determination of the inhomogeneity scale.

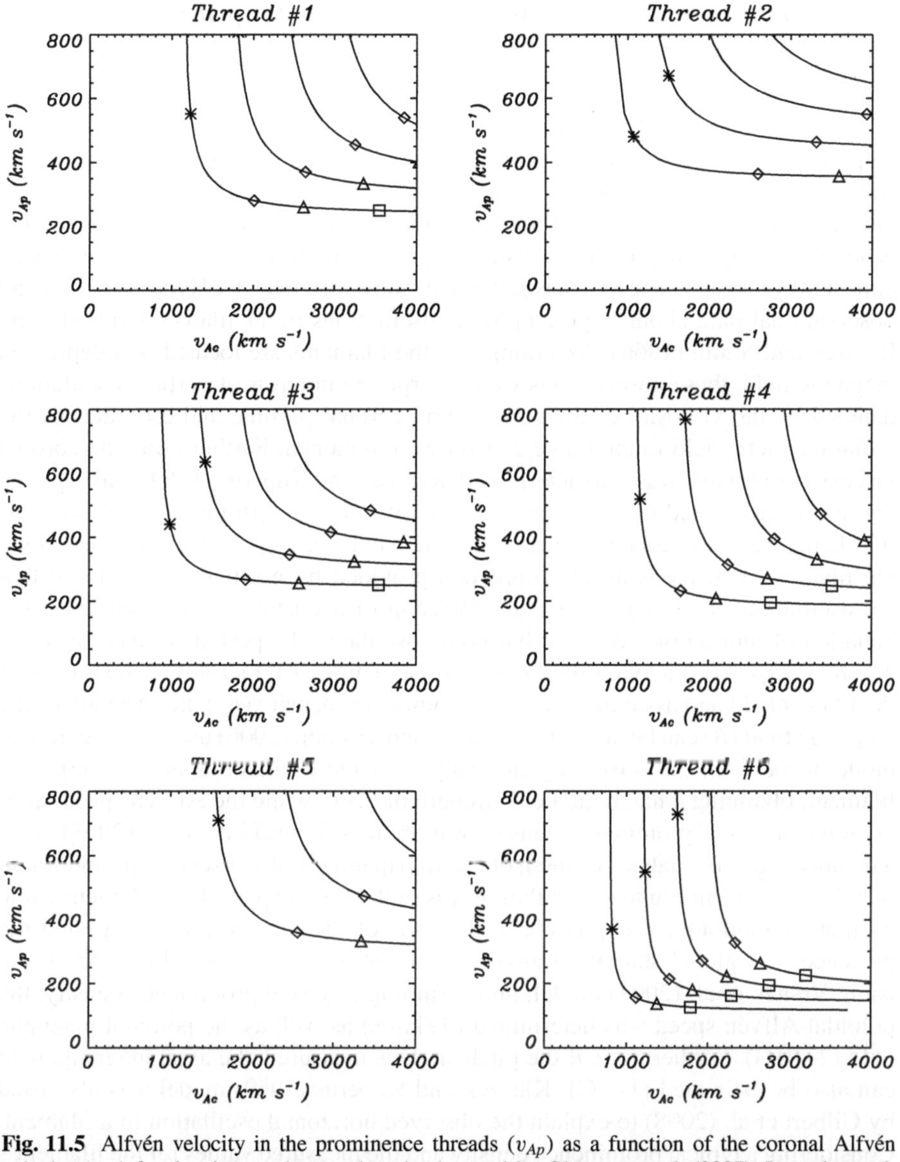

Fig. 11.5 Alfvén velocity in the prominence threads (v_{Ap}) as a function of the coronal Alfvén velocity (v_{Ac}) for the six prominence threads observed by Okamoto et al. (2007). In each panel, from bottom to top, the curves correspond to a length of magnetic field lines of 100,000, 150,000, 200,000, and 250,000 km, respectively. Asterisks, diamonds, triangles and squares correspond to density ratios of the prominence thread to the coronal gas $\zeta \simeq 5, 50, 100, 200$. (*credits*: Arregui et al. 2012 and Figure 2 from Terradas et al. 2008)

11.6 Prominence Seismology Using Large Amplitude Oscillations

Most of the theoretical models proposed to explain the different types of large amplitude oscillations, produced by a variety of exciters, do not rely on MHD waves to describe the observed oscillations. Therefore, MHD seismology has not been applied to large amplitude ocillations yet. However, attempts to derive prominence physical properties, using large amplitude oscillations, were started almost 50 years ago. The first seismological study was made by Hyder (1966) using observational data about large amplitude oscillations of 11 filaments reported by Ramsey and Smith (1966). Assuming that the filaments are located in a depressed magnetic field, these observations were interpreted in terms of vertical oscillations damped by the viscosity of the surrounding coronal plasma, and estimates of the radial magnetic field in the range 2–30 G were obtained. Furthermore, the coronal viscosity coefficient was also determined. Kleczek and Kuperus (1969) reinterpreted the above mentioned observations in terms of horizontal (transverse) oscillations of filaments. They assumed that a line-tied magnetic field was directed along the filament, that the restoring force was provided by magnetic tension and that the oscillations were damped by the emission of acoustic waves. Then, from the equation of motion of a damped harmonic oscillator, the period is given by $P = 4\pi LB^{-1}\sqrt{\pi\rho_p}$ so, knowing the period of oscillation (P), measuring the length of the filament (L) and assuming a typical prominence density (ρ_p), the strength of the magnetic field (B) can be determined. Isobe and Tripathi (2006) used this theoretical model to perform the seismological analysis of the oscillations in a pre-erupting filament, obtaining a magnetic field strength of 9.8 G while the Alfvén speed, after assuming a typical prominence density, was about 87 km/s. Pintér et al. (2008) made a detailed wavelet analysis of the Isobe and Tripathi (2006) observations and found that the largest amplitude of oscillation was in the middle part of the filament while oscillatory motions are negligible at the ends of the filament, and suggested the presence of a global standing transverse oscillation in a twisted flux rope. Then, using Vršnak et al. (2007) model, and assuming a typical prominence density, the poloidal Alfvén speed was determined (49 km/s) as well as the poloidal magnetic field (2–10 G). Furthermore, if the pitch angle is measured, the axial magnetic field can also be estimated (1–5 G). Kleczek and Kuperus (1969) model was also used by Gilbert et al. (2008) to explain the observed horizontal oscillation in a filament. Considering a typical prominence density and the measured values for the filament's length and period of oscillation, they inferred a magnetic field strength of 30 G.

In the case of longitudinal oscillations, Vršnak et al. (2007) assumed that the filament was embedded in a flux rope and suggested that the oscillations were triggered by additional poloidal flux injected at one of its legs. This new magnetic flux creates a magnetic pressure gradient along the filament, which is the restoring force. After linearising the equation of motion, the expression for the longitudinal displacement (x), in dimensionless form, is $\ddot{x} = -\frac{2v_{A\varphi}^2}{L^2}x$ which provides with an expression for the period, $P = 4.4L/v_{A\varphi}$, as a function of the poloidal Alfvén speed

($v_{A\varphi}$) and the length of the filament (L). Then, knowing the period and the length of the filament, the poloidal Alfvén speed ($v_{A\varphi} = 100\,\text{km/s}$) can be determined. Furthermore, assuming a set of typical prominence densities, the poloidal magnetic field strength is in the range 5–15 G, and measuring the pitch angle, the longitudinal magnetic field strength can be also obtained (10–30 G). Vršnak et al. (2007) and Pintér et al. (2008) assume that the filament is embedded in a twisted flux rope and that the observed oscillations correspond to oscillatory modes of this magnetic configuration. However, this should be taken with care since to perform a seismological analysis, based on this or another magnetic configuration, we should have a detailed knowledge of its oscillatory modes. For instance, the polarization of the oscillatory motions of an untwisted and twisted flux tube is completely different, and varies with the amount of twist (Terradas and Goossens 2012).

Luna and Karpen (2012), Luna et al. (2012), Zhang et al. (2013) and Luna et al. (2014) have proposed a quite different theoretical model for large amplitude longitudinal oscillations in filaments. When an energetic event happens close to a filament, the injected energy evaporates plasma at the fluxtube footpoint closest to the energetic event. Then, the flow of hot plasma pushes the cold plasma condensations (threads) located at the dips of the magnetic configuration, and the longitudinal oscillations start. After some time, they lose coherence due to period differences. The restoring force seems to be the projected solar gravity directed towards the bottom of the dip and since the magnetic tension in the dip must be larger than the weight of the threads, we have, $\frac{B^2}{R} - mng \geq 0$ where R is the dip's radius of curvature, m the particle mass, n the particle density, and g the gravitational acceleration. On the other hand, since the oscillation is gravity driven, like in a pendulum, $\omega = \sqrt{\frac{g}{R}}$ and combining the above two expressions, we obtain, $B \geq \sqrt{\frac{g^2 mn}{4\pi^2}}\,P$. Then, knowing the period (P) and assuming a typical density (n), the strength of the magnetic field (B) can be determined. Following this model, the damping of the oscillations is basically dominated by continuous mass accretion onto the threads coming from chromospheric evaporation produced by a continuous and localized heating at the footpoints. The time oscillatory behaviour of the filament can be described in terms of a Bessel function instead of a sinusoid, and applying this model to the SDO/AIA observations of longitudinal oscillations in a filament (Luna et al. 2014), a seismological analysis was performed. The best fit to the observations gives a mean period of oscillation of 0.82 h, the maximum speed amplitude is in the range 17 and 47 km/s for different positions along the filament, the minimum magnetic field strength, using a typical prominence density, is given by $B[G] \geq (17 \pm 9) P[hours]$, and the radius of curvature of the dips is between 43 and 66 mm. On the other hand, initially the damping is very strong with a damping time similar to the period of oscillation, which can be explained in terms of mass accretion, later, the damping becomes weaker with a characteristic time of hours. Finally, the magnetic structure of the observed filament is suggested to be a double sheared magnetic arcade which, as well as a flux rope, is a magnetic configuration which can support longitudinal oscillations of the cool filament plasma located in the dips.

On the other hand, taking into account the maximum velocity amplitude and displacement shown by large amplitude oscillations, one may wonder whether or not nonlinear oscillations should be considered as a possible theoretical explanation. In this regard, Vršnak et al. (2007) studied how the period of oscillation is modified when a nonlinear oscillator is considered. They found that the period decreases when the initial amplitude is increased in such a way that for large initial amplitudes the deviation of the period with respect to that of an harmonic oscillator becomes larger than 40 %. Furthermore, since the oscillation is damped in time the decrease of the amplitude would mean that the period should increase with time, and they suggested that the presence of this feature can be seen in the reported oscillations.

11.7 Final Remarks

Although prominence seismology has undergone a rapid development during last years, and a firm basis to this subject already exists, it is still a young science. In the case of small amplitude oscillations, prominence seismology is based on the interpretation of those oscillations in terms of MHD waves in highly idealized prominence theoretical models which support a variety of MHD modes with their own distinguishing periods. These periods may depend on several equilibrium parameters: prominence and/or coronal density and temperature, magnetic field strength, length of magnetic field lines, magnetic field orientation relative to the filament axis, etc., which means that by modifying some of these parameters any detected period could be reproduced.

For a further improvement of prominence seismology results, constraints on the free variables of theoretical models must be imposed by determining, if possible, the physical properties of prominences at the same time as they are observed for oscillations. Thus, information on the prominence geometry together with the spatial distribution of temperature, density, flows, etc., as well as the magnetic field structure and orientation, could help to construct more realistic theoretical models, and to determine which ones better reproduce the observed oscillatory features. In order, to provide with enough information for a theoretical interpretation, long observational times and two-dimensional high-resolution observations are needed to determine the spatial arrangement of parameters of interest. Furthermore, these observations should be analysed using techniques allowing to determine the dominant spatial and temporal structures which could help to identify propagating or standing features hidden in the observational data.

Another important step ahead for prominence seismology would be to couple radiative transfer with magnetohydrodynamic waves as a mean to establish a relationship between velocity, density, magnetic field and temperature perturbations, and the observed signatures of oscillations like spectral line Doppler shift, half-width and intensity (Heinzel 2014). Also, partial ionization is another topic of interest since, apart from influencing the behaviour of magnetohydrodynamic waves, it poses an important problem for prominence equilibrium models since cross-field diffusion of neutral atoms can give place to flows and draining of prominence material. In the case of filaments, if oscillations turn out to be confined to filament threads, cylindrical multithread models should be explored (Soler et al. 2009b), as well as the effect of the counterflows (Zirker et al. 1998; Lin 2004) on the oscillatory frequency and spatial distribution of perturbations.

On the other hand, MHD seismology has not been used to interpret large amplitude oscillations yet. Up to now, these oscillations have been explained in terms of linear oscillators whose restoring forces are magnetic tension, magnetic pressure gradient or projected gravity, and the performed seismology has been based in the analysis of these oscillations. However, taking into account the large amplitudes involved in these oscillations, some of the observed large amplitude transverse oscillations in filaments could be nonlinear, therefore, this should be taken into account when setting up theoretical models to explain these oscillations. Furthermore, since in these oscillations we also observe different polarizations of motions, three dimensional theoretical models for prominences, based on arcades or flux ropes, should be explored as well as their oscillatory properties when submitted to large disturbances. Finally, the phenomenon of "winking" filaments deserves further investigation, paying special attention to the mechanisms producing the observed damping of oscillations.

Acknowledgements This chapter is dedicated to the memory of Josip Kleczek who woke up my interest for solar prominences.

References

Alfvén, H. (1942). Existence of electromagnetic-hydrodynamic waves. *Nature, 150*, 405–406. doi:10.1038/150405d0.

Arregui, I., & Ballester, J. L. (2011). Damping mechanisms for oscillations in solar prominences. *Space Science Review, 158*, 169–204. doi:10.1007/s11214-010-9648-9, 1002.3489.

Arregui, I., Oliver, R., & Ballester, J. L. (2012). Prominence oscillations. *Living Reviews in Solar Physics, 9*, 2. doi:10.12942/lrsp-2012-2.

Arregui, I., Ramos, A. A., & Díaz, A. J. (2014). The promise of Bayesian analysis for prominence seismology. In *IAU Symposium* (Vol. 300, pp. 393–394). doi:10.1017/S1743921313011241.

Arregui, I., Terradas, J., Oliver, R., & Ballester, J. (2008). Damping of fast magnetohydrodynamic oscillations in quiescent filament threads. *Astrophysical Journal Letters, 682*, L141–L144. doi:10.1086/591081.

Asai, A., Ishii, T. T., Isobe, H., et al. (2012). First simultaneous observation of an Hα moreton wave, EUV wave, and filament/prominence oscillations. *Astrophysical Journal Letters, 745*, L18. doi:10.1088/2041-8205/745/2/L18, 1112.5915.

Ballai, I. (2003). On dissipative effects in solar prominences. *Astronomy & Astrophysics, 410*, L17–L19. doi:10.1051/0004-6361:20031401.

Ballester, J. L. (2014). Prominence seismology. In *IAU Symposium* (Vol. 300, pp. 30–39). doi:10.1017/S1743921313010703.

Berger, T., Shine, R., Slater, G., et al. (2008). Hinode SOT observations of solar quiescent prominence dynamics. *Astrophysical Journal Letters, 676*, L89–L92. doi:10.1086/587171.
Bommier, V., Landi Degl'Innocenti, E., Leroy, J. L., & Sahal-Bréchot, S. (1994). Complete determination of the magnetic field vector and of the electron density in 14 prominences from linear polarization measurements in the He I D3 and Hα lines. *Solar Physics, 154*, 231–260.
Bommier, V., & Leroy, J. (1998). Global pattern of the magnetic field vector above neutral lines from 1974 to 1982: Pic-du-Midi observations of prominences. In D. Webb, B. Schmieder, & D. Rust (Eds.), *New perspectives on solar prominences*. ASP Conference Series (Vol. 150, pp. 434–438). San Francisco: Astronomical Society of the Pacific.
Braginskii, S. I. (1965). Transport processes in a plasma. *Reviews of Plasma Physics, 1*, 205.
Carbonell, M., Oliver, R., & Ballester, J. (2004). Time damping of linear non-adiabatic magnetohydrodynamic waves in an unbounded plasma with solar coronal properties. *Astronomy & Astrophysics, 415*, 739–750. doi:10.1051/0004-6361:20034630.
Carlsson, M., & Stein, R. F. (1997). Chromospheric dynamics - what can be learnt from numerical simulations. In G. M. Simnett, C. E. Alissandrakis, & L. Vlahos (Eds.), *European meeting on solar physics*. Lecture Notes in Physics (Vol. 489, p. 159). Berlin: Springer Verlag. doi:10.1007/BFb0105675.
Chen, P., Innes, D., & Solanki, S. (2008). Soho/sumer observations of prominence oscillation before eruption. *Astronomy & Astrophysics, 484*, 487–493. doi:10.1051/0004-6361:200809544, 0802.1961.
Cox, D. P., & Tucker, W. H. (1969). Ionization equilibrium and radiative cooling of a low-density plasma. *Astrophysical Journal, 157*, 1157. doi10.1086/150144.
Dahlburg, R. B., & Mariska, J. T. (1988). Influence of heating rate on the condensational instability. *Solar Physics, 117*, 51–56. doi10.1007/BF00148571.
De Pontieu, B., Martens, P., & Hudson, H. (2001). Chromospheric damping of alfvén waves. *Astrophysical Journal, 558*, 859–871. doi10.1086/322408.
Díaz, A., Oliver, R., & Ballester, J. (2002). Fast magnetohydrodynamic oscillations in cylindrical prominence fibrils. *Astrophysical Journal, 580*, 550–565. doi:10.1086/343039.
Díaz, A., Oliver, R., & Ballester, J. (2003). Fast MHD oscillations of a 3-dimensional prominence fibril. *Astronomy & Astrophysics, 402*, 781–789. doi:10.1051/0004-6361:20030285.
Díaz, A., Oliver, R., & Ballester, J. (2005). Fast magnetohydrodynamic oscillations in a multifibril cartesian prominence model. *Astronomy & Astrophysics, 440*, 1167–1175. doi10.1051/0004-6361:20052759.
Díaz, A., Oliver, R., & Ballester, J. (2010). Prominence thread seismology using the P1/2P2 ratio. *Astrophysical Journal, 725*, 1742–1748. doi10.1088/0004-637X/725/2/1742.
Díaz, A., Oliver, R., Erdélyi, R., & Ballester, J. (2001). Fast MHD oscillations in prominence fine structures. *Astronomy & Astrophysiocs, 379*, 1083–1097. doi:10.1051/0004-6361:20011351.
Díaz, A., & Roberts, B. (2006). Fast magnetohydrodynamic oscillations in a fibril prominence model. *Solar Physics, 236*, 111–126. doi10.1007/s11207-006-0137-y.
Dymova, M., & Ruderman, M. (2005). Non-axisymmetric oscillations of thin prominence fibrils. *Solar Physics, 229*, 79–94. doi10.1007/s11207-005-5002-x.
Edwin, P., & Roberts, B.: (1983). Wave propagation in a magnetic cylinder. *Solar Physics, 88*, 179–191. DOI 10.1007/BF00196186.
Engvold, O. (1998). Observations of filament structure and dynamics. In D. Webb, B. Schmieder, & D. Rust (Eds.), *New perspectives on solar prominences*. ASP Conference Series (Vol. 150, pp. 23–31). San Francisco: Astronomical Society of the Pacific.
Engvold, O. (2004). Structures and dynamics of solar filaments - challenges in observing and modeling. In A. V. Stepanov, E. E. Benevolenskaya & A. G. Kosovichev (Eds.), *Multi-wavelength investigations of solar activity, IAU symposium* (Vol. 223, pp. 187–194). doi10.1017/S1743921304005575.
Engvold, O. (2008). Observational aspects of prominence oscillations. In R. Erdélyi & C. Mendoza-Briceño (Eds.), *Waves & oscillations in the solar atmosphere: Heating and magneto-seismology, IAU symposia* (Vol. 247, pp. 152–157). Cambridge/New York: Cambridge University Press. doi10.1017/S1743921308014816.

Eto, S., Isobe, H., Narukage, N., et al. (2002). Relation between a Moreton wave and an EIT wave observed on 1997 november 4. *Publications of the Astronomical Society of Japan, 54*, 481–491.

Forteza, P., Oliver, R., Ballester, J., & Khodachenko, M. (2007). Damping of oscillations by ion-neutral collisions in a prominence plasma. *Astronomy & Astrophysics, 461*, 731–739. doi:10.1051/0004-6361:20065900.

Forteza, P., Oliver, R., & Ballester, J. (2008). Time damping of non-adiabatic MHD waves in an unbounded partially ionised prominence plasma. *Astronomy & Astrophysics, 492*, 223–231. doi:10.1051/0004-6361:200810370.

Gilbert, H., Daou, A., Young, D., Tripathi, D., & Alexander, D. (2008). The filament-moreton wave interaction of 2006 December 6. *Astrophysical Journal, 685*, 629–645. doi10.1086/590545.

Goedbloed, J. P. H., & Poedts, S. (2004). *Principles of magnetohydrodynamics*. Cambridge: Cambridge University Press.

Goossens, M. (2003). *An introduction to plasma astrophysics and magnetohydrodynamics. Astrophysics and space science library* (Vol. 294). Dordrecht/Norwell, MA: Kluwer.

Goossens, M., Arregui, I., Ballester, J., & Wang, T. (2008). Analytic approximate seismology of transversely oscillating coronal loops. *Astronomy & Astrophysics, 484*, 851–857. doi:10.1051/0004-6361:200809728.

Goossens, M., Erdélyi, R., & Ruderman, M. (2010). Resonant MHD waves in the solar atmosphere. *Space Science Reviews*. doi10.1007/s11214-010-9702-7.

Goossens, M., Hollweg, J., & Sakurai, T. (1992). Resonant behaviour of MHD waves on magnetic flux tubes. iii. Effect of equilibrium flow. *Solar Physics, 138*, 233–255. doi10.1007/BF00151914.

Goossens, M., Ruderman, M., & Hollweg, J. (1995). Dissipative MHD solutions for resonant alfven waves in 1-dimensional magnetic flux tubes. *Solar Physics, 157*, 75–102. doi:10.1007/BF00680610.

Goossens, M., Terradas, J., Andries, J., Arregui, I., & Ballester, J. (2009). On the nature of kink MHD waves in magnetic flux tubes. *Astronom & Astrophysics, 503*, 213–223. doi:10.1051/0004-6361/200912399, 0905.0425.

Harvey, J. (1969). Magnetic fields associated with solar active-region prominences. Ph.D. thesis, University of Colorado at Boulder, Boulder, CO.

Heggland, L., Hansteen, V. H., De Pontieu, B., &Carlsson, M. (2011). Wave propagation and jet formation in the chromosphere. *Astrophysical Journal, 743*, 142. doi:10.1088/0004-637X/743/2/142, 1112.0037.

Heinzel, P. (2014). Radiative transfer in solar prominences. In J.-C. Vial & O. Engvold (Eds.), *Solar prominences, ASSL* (Vol. 415, pp. 101–128). Springer.

Heinzel, P., Zapiór, M., Oliver, R., & Ballester, J. L. (2014). Synthetic hydrogen spectra of prominence oscillations. *Astronomy & Astrophysics, 562*, A103. doi:10.1051/0004-6361/201322346, 1401.2131.

Hershaw, J., Foullon, C., Nakariakov, V. M., & Verwichte, E. (2011). Damped large amplitude transverse oscillations in an EUV solar filament, triggered by large-scale transient coronal waves. *Astronomy & Astrophysics, 531*, A53. doi10.1051/0004-6361/201116750.

Hildner, E. (1974). The formation of solar quiescent prominences by condensation. *Solar Physics, 35*, 123–136. doi:10.1007/BF00156962.

Hollweg, J. V., & Yang, G. (1988). Resonance absorption of compressible magnetohydrodynamic waves at thin 'surfaces'. *Journal of Geophysical Research, 93*, 5423–5436. doi:10.1029/JA093iA06p05423.

Hyder, C. (1966). Winking filaments and prominence and coronal magnetic fields. *Zeitschrift für Astrophysik, 63*, 78–84.

Isobe, H., & Tripathi, D. (2006). Large amplitude oscillation of a polar crown filament in the pre-eruption phase. *Astronom & Astrophysics, 449*, L17–L20. doi:10.1051/0004-6361:20064942, arXiv:astro-ph/0602432.

Isobe, H., Tripathi, D., Asai, A., & Jain, R. (2007). Large-amplitude oscillation of an erupting filament as seen in EUV, Hα, and microwave observations. *Solar Physics, 246*, 89–99. doi10.1007/s11207-007-9091-6, 0711.3952.

Jing, J., Lee, J., Spirock, T., & Wang, H. (2006). Periodic motion along solar filaments. *Solar Physics, 236*, 97–109. doi10.1007/s11207-006-0126-1.
Jing, J., Lee, J., Spirock, T., Xu, Y., Wang, H., & Choe, G. (2003). Periodic motion along a solar filament initiated by a subflare. *Astrophysical Journal Letters, 584*, L103–L106. doi10.1086/373886.
Joarder, P., & Roberts, B. (1992a). The modes of oscillation of a prominence. i. The slab with longitudinal magnetic field. *Astronomy & Astrophysics, 256*, 264–272.
Joarder, P., & Roberts, B. (1992b). The modes of oscillation of a prominence. ii - The slab with transverse magnetic field. *Astronomy & Astrophysics, 261*, 625–632.
Joarder, P., & Roberts, B. (1993a). The modes of oscillation of a prominence. iii. The slab in a skewed magnetic field. *Astronomy & Astrophysics, 277*, 225–234.
Joarder, P. S., & Roberts, B. (1993b). The modes of oscillation of a Menzel prominence. *Astronom & Astrophysics, 273*, 642–646.
Kleczek, J., & Kuperus, M. (1969). Oscillatory phenomena in quiescent prominences. *Solar Physics, 6*, 72–79. doi10.1007/BF00146797.
Klimchuk, J. A., & Cargill, P. J. (2001). Spectroscopic diagnostics of nanoflare-heated loops. *Astrophysical Journal, 553*, 440–448. doi10.1086/320666.
Krishan, V., & Varghese, B. A. (2008). Cylindrical Hall - MHD waves: A nonlinear solution. *Solar Physics, 247*, 343–349. doi10.1007/s11207-008-9117-8.
Landman, D., Edberg, S., & Laney, C. (1977). Measurements of Hβ, He D_3, and Ca^+ λ8542 line emission in quiescent prominences. *Astrophysical Journal, 218*, 888–897. doi:10.1086/155744.
Leroy, J. L. (1980). Mass balance and magnetic structure in quiescent prominences. In F. Moriyama & J. Henoux (Eds.), *Proceedings of the Japan-France seminar on solar physics* (p. 155). Tokyo, Nihon Gakujutsu Shinkokai and CNRS.
Leroy, J. L. (1988). Observations of prominence magnetic field. In J. Ballester & E. Priest (Eds.), *Dynamics and structure of solar prominences, Universitat de les Illes Balears, Palma de Mallorca, Conferències i comunicacions* (Vol. 5, pp. 33–40).
Leroy, J. L. (1989). Observation of prominence magnetic fields. In E. Priest (Ed.), *Dynamics and structure of quiescent solar prominences* (Vol. 150, pp. 77–113). Dordrecht/Boston: Kluwer/Astrophysics and Space Science Library.
Li, T., & Zhang, J. (2012). SDO/AIA observations of large-amplitude longitudinal oscillations in a solar filament. *Astrophysical Journal Letters, 760*, L10. doi:10.1088/2041-8205/760/1/L10, 1210.5110.
Lighthill, M. J. (1960). Studies on magneto-hydrodynamic waves and other anisotropic wave motions. *Royal Society of London Philosophical Transactions Series A, 252*, 397–430. doi10.1098/rsta.1960.0010.
Lin, Y. (2004). Magnetic field topology inferred from studies of fine threads in solar filaments. Ph.D. thesis, University of Oslo, Oslo.
Lin, Y., Engvold, O., Rouppe van der Voort, L., & van Noort, M. (2007). Evidence of traveling waves in filament threads. *Solar Physics, 246*, 65–72. doi10.1007/s11207-007-0402-8.
Lin, Y., Engvold, O., Rouppe van der Voort, L., Wiik, J., & Berger, T. (2005). Thin threads of solar filaments. *Solar Physics, 226*, 239–254. doi:10.1007/s11207-005-6876-3.
Lin, Y., Martin, S., & Engvold, O. (2008). Filament substructures and their interrelation. In R., Howe, R. Komm, K. Balasubramaniam, & G. Petrie (Eds.), *Subsurface and atmospheric influences on solar activity*. ASP Conference Series (Vol. 383, pp. 235–242). San Francisco: Astronomical Society of the Pacific.
Lin, Y., Soler, R., Engvold, O., Ballester, J., Langangen, Ø., Oliver, R., & Rouppe van der Voort, L. (2009). Swaying threads of a solar filament. *Astrophysical Journal, 704*, 870–876. doi10.1088/0004-637X/704/1/870.
Luna, M., Díaz, A. J., & Karpen, J. (2012). The effects of magnetic-field geometry on longitudinal oscillations of solar prominences. *Astrophysical Journal, 757*, 98. doi:10.1088/0004-637X/757/1/98, 1207.6358.
Luna, M., & Karpen, J. (2012). Large-amplitude longitudinal oscillations in a solar filament. *Astrophysical Journal Letters, 750*, L1. doi:10.1088/2041-8205/750/1/L1, 1203.5027.

Luna, M., Knizhnik, K., Muglach, K., et al. (2014). Observations and implications of large-amplitude longitudinal oscillations in a solar filament. *Astrophysical Journal, 785*, 79. doi:10.1088/0004-637X/785/1/79, 1403.0381.

Martin, S., Lin, Y., & Engvold, O. (2008). A method of resolving the 180-degree ambiguity by employing the chirality of solar features. *Solar Physics, 250*, 31–51. DOI 10.1007/s11207-008-9194-8.

Milne, A. M., Priest, E. R., & Roberts, B. (1979). A model for quiescent solar prominences. *Astrophysical Journal, 232*, 304–317. DOI 10.1086/157290.

Molowny-Horas, R., Baudin, F., Oliver, R., & Ballester, J. (1998), He I 10830 Å Doppler oscillations in a solar filament. In R. Donahue & J. Bookbinder (Eds.), *Cool stars, stellar systems, and the sun*. ASP Conference Series (Vol. 154, pp. 650–657). San Francisco: Astronomical Society of the Pacific.

Molowny-Horas, R., Oliver, R., Ballester, J., & Baudin, F. (1997). Observations of Doppler oscillations in a solar prominence. *Solar Physics, 172*, 181–188.

Molowny-Horas, R., Wiehr, E., Balthasar, H., Oliver, R., & Ballester, J. (1999). Prominence Doppler oscillations. In A. Antalová, H. Balthasar, & A. Kučera (Eds.), *JOSO annual report 1998* (pp. 126–127). Tatranská Lomnica, Slovakia: Astronomical Institute of Slovak Academy of Sciences.

Moreton, G. E., & Ramsey, H. E. (1960). Recent observations of dynamical phenomena associated with solar flares. *Publications of the Astronomical Society of the Pacific, 72*, 357. doi10.1086/127549.

Ning, Z., Cao, W., & Goode, P. (2009a). Behavior of the spines in a quiescent prominence observed by Hinode/SOT. *Astrophysical Journal, 707*, 1124–1130. doi:10.1088/0004-637X/707/2/1124.

Ning, Z., Cao, W., Okamoto, T., Ichimoto, K., & Qu, Z. (2009b). Small-scale oscillations in a quiescent prominence observed by Hinode/SOT. *Astronomy & Astrophysics, 499*, 595–600. doi10.1051/0004-6361/200810853.

Okamoto, T., Nakai, H., Keiyama, A., et al. (2004). Filament oscillations and moreton waves associated with EIT waves. *Astrophysical Journal, 608*, 1124–1132. doi10.1086/420838.

Okamoto, T., Tsuneta, S., Berger, T., et al. (2007). Coronal transverse magnetohydrodynamic waves in a solar prominence. *Science, 318*, 1577–1580. doi10.1126/science.1145447.

Oliver, R., & Ballester, J. (1995). Magnetohydrodynamic waves in a bounded inhomogeneous medium with prominence-corona properties. *Astrophysical Journal, 448*, 444–458. doi10.1086/175975.

Oliver, R., & Ballester, J. (1996). The influence of the temperature profile on the magnetohydrodynamic modes of a prominence-corona system. *Astrophysical Journal, 456*, 393–398. doi10.1086/176661.

Oliver, R., & Ballester, J. (2002). Oscillations in quiescent solar prominences observations and theory (invited review). *Solar Physics, 206*, 45–67. doi10.1023/A:1014915428440.

Oliver, R., Ballester, J., Hood, A., & Priest, E. (1992). Magnetohydrodynamic waves in a solar prominence. *Astrophysical Journal, 400*, 369–379. doi:10.1086/172003.

Oliver, R., Ballester, J., Hood, A., & Priest, E. (1993). Oscillations of a quiescent solar prominence embedded in a hot corona. *Astrophysical Journal, 409*, 809–821. doi:10.1086/172711.

Pandey, B., & Wardle, M. (2008). Hall magnetohydrodynamics of partially ionized plasmas. *Monthly Notices of the Royal Astronomiocal Society, 385*, 2269–2278. doi:10.1111/j.1365-2966.2008.12998.x, 0707.2688.

Parenti, S., & Vial, J. C. (2007.) Prominence and quiet-sun plasma parameters derived from FUV spectral emission. *Astronomy & Astrophysics, 469*, 1109–1115. doi:10.1051/0004-6361:20077196.

Parker, E. N. (1953). Instability of thermal fields. *Astrophysical Journal, 117*, 431. doi:10.1086/145707.

Pintér, B., Jain, R., Tripathi, D., & Isobe, H. (2008). Prominence seismology: Wavelet analysis of filament oscillations. *Astrophysical Journal, 680*, 1560–1568. doi:10.1086/588273.

Pinto, C., & Galli, D. (2008). Three-fluid plasmas in star formation. II. Momentum transfer rate coefficients. *Astronomy & Astrophysics, 492*, 1–1. doi:10.1051/0004-6361:20078819e.

Pouget, G., Bocchialini, K., & Solomon, J. (2006). Oscillations in a solar filament: First observation of long periods in the He I 584.33 Å line, modelling and diagnostic. *Astronomy & Astrophysics, 450*, 1189–1198. doi:10.1051/0004-6361:20053886.
Priest, E. R. (2014). Magnetohydrodynamics of the sun. Cambridge: Cambridge University Press.
Ramsey, H., & Smith, S. (1966). Flare-initiated filament oscillations. *Astronomical Journal, 71*, 197–199. doi10.1086/109903.
Régnier, S., Solomon, J., & Vial, J. (2001). Oscillations in an active region filament: Observations and comparison with MHD waves. *Astronomy & Astrophysics, 376*, 292–301. doi:10.1051/0004-6361:20010972.
Roberts, B., Edwin, P., & Benz, A. (1984). On coronal oscillations. *Astrophysical Journal, 279*, 857–865.
Roberts, B., & Joarder, P. (1994). Oscillations in quiescent prominences. In G., Belvedere, M. Rodono, & G. Simnett (Eds.), *Advances in solar physics*. Lecture Notes in Physics (Vol. 432, pp. 173–178). Berlin: Springer. DOI 10.1007/3-540-58041-7215.
Rosenberg, H. (1970). Evidence for MHD pulsations in the solar corona. *Astronomy & Astrophysics, 9*, 159.
Rosner, R., Golub, L., Coppi, B., & Vaiana, G. S. (1978). Heating of coronal plasma by anomalous current dissipation. *Astrophysical Journal, 222*, 317–332. doi:10.1086/156145.
Ruderman, M., & Roberts, B. (2002). The damping of coronal loop oscillations. *Astrophysical Journal, 577*, 475–486. doi10.1086/342130.
Sakurai, T., Goossens, M., & Hollweg, J. V. (1991a). Resonant behaviour of magnetohydrodynamic waves on magnetic flux tubes - part two. *Solar Physics, 133*, 247–262. doi10.1007/BF00149889.
Sakurai, T., Goossens, M., & Hollweg, J. V. (1991b). Resonant behaviour of MHD waves on magnetic flux tubes. I - connection formulae at the resonant surfaces. *Solar Physics, 133*, 227–245. doi:10.1007/BF00149888.
Schure, K. M., Kosenko, D., Kaastra, J. S., Keppens, R., & Vink, J. (2009). A new radiative cooling curve based on an up-to-date plasma emission code. *Astronomy & Astrophysics, 508*, 751–757. doi:10.1051/0004-6361/200912495, 0909.5204.
Soler, R. (2010). Damping of magnetohydrodynamic waves in solar prominence fine structures. Ph.D. thesis, Departament de Fisica, Universitat de les Illes Balears.
Soler, R., Arregui, I., Oliver, R., & Ballester, J. (2010). Seismology of standing kink oscillations of solar prominence fine structures. *Astrophysical Journal, 722*, 1778–1792. doi:10.1088/0004-637X/722/2/1778, 1007.1959.
Soler, R., Ballester, J. L., & Parenti, S. (2012a). Stability of thermal modes in cool prominence plasmas. *Astronomy & Astrophysics, 540*, A7. doi:10.1051/0004-6361/201118492, 1201.4668.
Soler, R., & Goossens, M. (2011). Kink oscillations of flowing threads in solar prominences. *Astronomy & Astrophysics, 531*. doi10.1051/0004-6361/201116536, 1106.3937.
Soler, R., Goossens, M., Terradas, J., & Oliver, R. (2013). The behavior of transverse waves in nonuniform solar flux tubes. I. Comparison of ideal and resistive results. *Astrophysical Journal, 777*, 158. doi10.1088/0004-637X/777/2/158, 1309.3423.
Soler, R., Oliver, R., & Ballester, J. (2007). The effect of the solar corona on the attenuation of small-amplitude prominence oscillations. I. Longitudinal magnetic field. *Astronomy & Astrophysics, 471*, 1023–1033. doi:10.1051/0004-6361:20077633, arXiv:0704.1566.
Soler, R., Oliver, R., & Ballester, J. (2008). Nonadiabatic magnetohydrodynamic waves in a cylindrical prominence thread with mass flow. *Astrophysical Journal, 684*, 725–735. doi:10.1086/590244, 0803.2600.
Soler, R., Oliver, R., & Ballester, J. (2009a). Magnetohydrodynamic waves in a partially ionized filament thread. *Astrophysical Journal, 699*, 1553–1562. doi:10.1088/0004-637X/699/2/1553, 0904.3013.
Soler, R., Oliver, R., & Ballester, J. (2009b). Propagation of nonadiabatic magnetoacoustic waves in a threaded prominence with mass flows. *Astrophysical Journal, 693*, 1601–1609. doi:10.1088/0004-637X/693/2/1601, 0809.4765.

Soler, R., Oliver, R., & Ballester, J. (2009c). Resonantly damped kink magnetohydrodynamic waves in a partially ionized filament thread. *Astrophysical Journal, 707*, 662–670. doi:10.1088/0004-637X/707/1/662, 0909.3599.

Soler, R., Oliver, R., Ballester, J., & Goossens, M. (2009d). Damping of filament thread oscillations: Effect of the slow continuum. *Astrophysical Journal Letters, 695*, L166–L170. doi:10.1088/0004-637X/695/2/L166, 0902.0572.

Soler, R., Ruderman, M. S., & Goossens, M. (2012b). Damped kink oscillations of flowing prominence threads. *Astronomy & Astrophysics, 546*, A82. doi:10.1051/0004-6361/201220111, 1209.3382.

Spitzer, L. (1962). *Physics of fully ionized gases*. New York: Interscience.

Suematsu, Y., Yoshinaga, R., Terao, N., & Tsubaki, T. (1990). Oscillatory and transient features detected simultaneously in the Ca II K and Hβ line spectra of a quiescent prominence. *Publications of the Astronomical Society of Japan, 42*, 187–203.

Tandberg-Hanssen, E. (1974). *Solar prominences* (Vol. 12). Dordrecht: D. Reidel Publishing Co.

Tandberg-Hanssen, E. (1995). *The nature of solar prominences. Astrophysics and space science library* (Vol. 199). Dordrecht/Boston: Kluwer.

Terradas, J., Andries, J., & Goossens, M. (2007). On the excitation of leaky modes in cylindrical loops. *Solar Physics, 246*, 231–242. doi:10.1007/s11207-007-9067-6.

Terradas, J., Arregui, I., Oliver, R., & Ballester, J. (2008). Transverse oscillations of flowing prominence threads observed with hinode. *Astrophysical Journal Letters, 678*, L153–L156. doi:10.1086/588728.

Terradas, J., Carbonell, M., Oliver, R., & Ballester, J. (2005). Time damping of linear non-adiabatic magnetoacoustic waves in a slab-like quiescent prominence. *Astronomy & Astrophysics, 434*, 741–749. doi:10.1051/0004-6361:20041984.

Terradas, J., & Goossens, M. (2012). Transverse kink oscillations in the presence of twist. *Astronomy & Astrophysics, 548*, A112. doi:10.1051/0004-6361/201219934, 1210.8093.

Terradas, J., Molowny-Horas, R., Wiehr, E., Balthasar, H., Oliver, R., & Ballester, J. (2002). Two-dimensional distribution of oscillations in a quiescent solar prominence. *Astronomy & Astrophysics, 393*, 637–647. doi:10.1051/0004-6361:20020967

Terradas, J., Oliver, R., & Ballester, J. (2001). Radiative damping of quiescent prominence oscillations. *Astronomy & Astrophysics, 378*, 635–652. doi:10.1051/0004-6361:20011148.

Thompson, W., & Schmieder, B. (1991). Oscillations in Hα filaments: Center-to-limb study. *Astronomy & Astrophysics, 243*, 501–511.

Tripathi, D., Isobe, H., & Jain, R. (2009). Large amplitude oscillations in prominences. *Space Science Reviews, 149*, 283–298. doi:10.1007/s11214-009-9583-9, 0910.4059.

Tsubaki, T., & Takeuchi, A. (1986). Periodic oscillations found in the velocity field of a quiescent prominence. *Solar Physics, 104*, 313–320. doi:10.1007/BF00159084.

Tsubaki, T., Toyoda, M., Suematsu, Y., & Gamboa, G. (1988). New evidence for oscillatory motions in a quiescent prominence. *Publications of the Astronomical Society of Japan, 40*, 121–126.

Uchida, Y. (1970). Diagnosis of coronal magnetic structure by flare-associated hydromagnetic disturbances. *Publications of the Astronomical Society of Japan, 22*, 341–364.

Vial, J. C. (1998). Solar prominence diagnostics (review). In D. F. Webb, B. Schmieder, & D. M. Rust (Eds.), *IAU colloq. 167: New perspectives on solar prominences*. Astronomical Society of the Pacific Conference Series (Vol. 150, p. 175). San Francisco: Astronomical Society of the Pacific.

Vršnak, B., Veronig, A., Thalmann, J., & Žic, T. (2007). Large amplitude oscillatory motion along a solar filament. *Astronomy & Astrophysics, 471*, 295–299. doi:10.1051/0004-6361:20077668, 0707.1752.

Wiehr, E., Balthasar, H., & Stellmacher, G. (1989). Doppler velocity oscillations in quiescent prominences. *Hvar Observatory Bulletin, 13*,131–135.

Yi, Z., Engvold, O., & Keil, S. (1991). Structure and oscillations in quiescent filaments from observations in He I 10830 Å. *Solar Physics, 132*, 63–80.

Zaqarashvili, T. V., Khodachenko, M. L., & Rucker, H. O. (2011). Magnetohydrodynamic waves in solar partially ionized plasmas: two-fluid approach. *Astronomy & Astrophysics, 529*, A82. doi:10.1051/0004-6361/201016326, 1101.3913.

Zhang, Q. M., Chen, P. F., Xia, C., Keppens, R., & Ji, H. S. (2013). Parametric survey of longitudinal prominence oscillation simulations. *Astronomy & Astrophysics, 554*, A124. doi:10.1051/0004-6361/201220705, 1304.3798.

Zirker, J., Engvold, O., & Martin, S. (1998). Counter-streaming gas flows in solar prominences as evidence for vertical magnetic fields. *Nature, 396*, 440–441. doi:10.1038/24798.

Chapter 12
MHD Equilibria and Triggers for Prominence Eruption

Yuhong Fan

Abstract Magneto-hydrodynamic (MHD) simulations of the emergence of twisted magnetic flux tubes from the solar interior into the corona are discussed to illustrate how twisted and sheared coronal magnetic structures (with free magnetic energy), capable of driving filament eruptions, can form in the corona in emerging active regions. Several basic mechanisms that can disrupt the quasi-equilibrium coronal structures and trigger the release of the stored free magnetic energy are discussed. These include both ideal processes such as the onset of the helical kink instability and the torus instability of a twisted coronal flux rope structure and the non-ideal process of the onset of fast magnetic reconnections in current sheets. Representative MHD simulations of the non-linear evolution involving these mechanisms are presented.

12.1 Introduction

Prominences/filaments are major precursors or source regions of coronal mass ejections (CMEs) as indicated by the observed close association between prominence/filament eruptions and CMEs (Munro et al. 1979; Webb and Hundhausen 1987; Gopalswamy et al. 2003). It is suggested that most CMEs are the result of the destabilization and eruption of a prominence and its overlying coronal structure, or of a magnetic structure capable of supporting a prominence (Webb and Hundhausen 1987). On the large scale, prominences/filaments (either in strong active region nests or in the weak fields of decaying active regions) represent stable structures that can exist over long periods of time (days), that are much longer than the Alfvén crossing time (minutes), before they suddenly erupt. Thus, on the large scale for the prominence structure as a whole, prominence magnetic fields represent magneto-static equilibria in the corona. Although on smaller scales, prominence plasma

Y. Fan (✉)
National Center for Atmospheric Research, 3080 Center Green Dr., Boulder, CO 80301, USA
e-mail: yfan@ucar.edu

J.-C. Vial, O. Engvold (eds.), *Solar Prominences*, Astrophysics and Space Science Library 415, DOI 10.1007/978-3-319-10416-4_12

exhibits continuous, on-going dynamic and turbulent behavior within the structure (e.g. Berger et al. 2011; de Toma et al. 2008; Liu et al. 2012 and see also Karpen 2014).

Given the low plasma β in the lower solar corona, where β denotes the ratio of the plasma pressure over the magnetic pressure, and if the gravitational force of the prominence mass is not significant to distort the magnetic fields on which it resides, one class of models considers prominence magnetic fields as approximately force free equilibrium structures with field aligned electric current, i.e.

$$\nabla \times \mathbf{B} = \alpha \mathbf{B} \tag{12.1}$$

where α is the torsion or twist parameter being constant along each field line but generally different for different field lines. The field aligned current, which manifests as the twist or shear of the magnetic field represents the free magnetic energy (in excess of of the potential field energy) stored in the equilibrium fields that can be released to drive the eruption.

Another class of models consider the weight of the prominence mass as playing a significant role in both the energy storage and release of the prominence magnetic structures (e.g. Kippenhahn and Schlüter 1957; Low and Smith 1993; Low and Zhang 2002; Fong et al. 2002; Low et al. 2003; Petrie and Low 2005; Zhang and Low 2004, 2005). In these models a significant local dip or distortion in the field lines can be created due to the weight of the prominence plasma, which produces significant cross-field current and hence additional free magnetic energy. A sudden removal of the prominence mass through some physical mechanisms can therefore release the stored free magnetic energy and drive eruptions.

Due to the nearly frozen-in evolution of the large scale coronal magnetic field in the highly conducting plasma of the solar atmosphere and corona, the magnetic helicity, a physical quantity that measures the topological complexity of the magnetic field (such as the linkage and/or twistness of the field) is nearly conserved (e.g. Berger and Field 1984; Berger 1984). In a volume V with all magnetic flux closed within the volume, the magnetic helicity is given by $H = \int_V \mathbf{A} \times \mathbf{B}\, dV$, where $\mathbf{A}$ is the vector potential of the magnetic field $\mathbf{B}$ in V, i.e. $\mathbf{B} = \nabla \times \mathbf{A}$, and it can be shown that H is invariant to any gauge transformation of $\mathbf{A} \rightarrow \mathbf{A} + \nabla \chi$ with χ being an arbitrary scalar function of position, and thus is a well defined quantity. For example, two linked, untwisted closed flux tubes with fluxes Φ_1 and Φ_2 respectively, have a magnetic helicity of $H = 2\Phi_1\Phi_2$, and a uniformly twisted closed magnetic torus with T winds of field line rotation about the axis over the length of the torus and with a total toroidal flux of Φ has a magnetic helicity $H = T\Phi^2$ (Berger and Field 1984). For the solar corona, we do not generally have an isolated, closed magnetic flux system and the magnetic flux is generally threading through the photosphere. Therefore a relative magnetic helicity for the magnetic field above the photospheric $z = 0$ is defined (Berger and Field 1984)

$$H_r = \int_{z>0} (\mathbf{A} + \mathbf{A}_p) \cdot (\mathbf{B} - \mathbf{P})\, dV, \tag{12.2}$$

where **B** is the magnetic field in the unbounded half space above $z = 0$, **A** is the vector potential for **B**, **P** is the reference potential field having the same normal flux distribution as **B** on the $z = 0$ boundary, and $\mathbf{A}_p$ is the vector potential for **P**. The relative magnetic helicity H_r is invariant with respect to the gauges for **A** and $\mathbf{A}_p$, and is thus a well-defined measure of the linkage or twistness of the coronal magnetic field (Berger and Field 1984; Démoulin 2007). The evolution of H_r in the corona is given by (e.g. Démoulin 2007):

$$\frac{dH_r}{dt} = -2 \int_S \mathbf{A}_p \times (\mathbf{v} \times \mathbf{B}) \cdot \hat{\mathbf{z}} \, dS + \left(\frac{dH_r}{dt} \right)_{\text{diss.}}, \tag{12.3}$$

where the first term on the right-hand-side corresponds to integration of helicity flux over the photospheric surface and the second term corresponds to dissipation of H_r in the corona. In the above $\mathbf{A}_p$ is the uniquely determined vector potential of the potential magnetic field with the gauge conditions: $\mathbf{A}_p \cdot \hat{\mathbf{z}} = 0$ on S, and $\nabla \cdot \mathbf{A}_p = 0$ in the corona above S. It is shown that the helicity dissipation (second term) is negligible for the nearly frozen-in evolution of the large scale corona even including magnetic reconnections during flares (Berger 1984). Such constraint of magnetic helicity conservation is playing an important role in the energy storage and ultimate eruption of the filament/prominence magnetic fields as described by Zhang and Low (2005). The net helicity transported into the corona through the photosphere (first term in the right-hand-side of Eq. 12.3) via flux emergence from the interior cannot be flared away and therefore the free magnetic energy cannot be completely dissipated down to the minimum energy potential field level (Woltjer 1958; Zhang and Low 2005; Démoulin 2007). The observed hemispheric pattern of the chirality of filament channels (see Sect. 3.1 in Engvold 2014) is directly related to the sign of magnetic helicity contained in the filament channel magnetic fields, where a dextral (sinistral) filament channel preferred in the northern (southern) hemisphere contains dominantly negative (positive) helicity or left-handed (right-handed) twist. The hemispheric pattern of the helicity of filament channels has its origin in the accumulation of the helicity in emerging active regions (e.g. Zhang and Low 2005; Mackay and van Ballegooijen 2005; Yeates et al. 2008 and see also Mackay 2014), which are observed to also show preferentially negative (positive) twist in the northern (southern) hemisphere, and this sign preference does not change with the solar cycles. Such accumulation of net helicity in each hemisphere is ultimately removed by the bodily ejection of the filament/prominence magnetic fields as coronal mass ejections (Zhang and Low 2005).

The main questions to be addressed in this chapter are (1) how sheared or twisted structures form in the corona as a result of magnetic flux emergence? and (2) what are the mechanisms that lead to the sudden disruption of the quasi-equilibrium coronal structures and an explosive release of the free magnetic energy? Question (1) is discussed in Sect. 12.2 with focus on understanding the formation of strongly twisted emerging active regions that develop X-ray sigmoids and/or sigmoid shaped filaments. Example MHD simulations of the emergence of a twisted flux tube from the solar interior into the atmosphere and corona are shown to demonstrate how helicity and free magnetic energy are transported into the corona. In Sect. 12.3 the basic mechanisms that can trigger the dynamic eruption of the twisted/sheared coronal magnetic structures are discussed and several MHD simulations of the non-linear evolution involving these mechanisms are presented. It is argued that current sheet formation and magnetic reconnection are playing an important role in all stages of the evolution of the prominence/filament magnetic fields, before, during and after the eruption.

12.2 Emergence of Twisted Magnetic Fields and Build Up of Free Energy and Helicity in the Corona

Observations suggest that flare productive active regions are associated with the emergence of twisted magnetic flux from the solar interior (see review by Schrijver 2009). Vector magnetic field observations of the photospheric layer of such active regions show that the transverse magnetic field at the polarity inversion lines (PILs) tend to be strongly sheared, i.e. tends to be closer to being parallel to the PILs rather than being perpendicular as expected for a potential field configuration (e.g. review by Schrijver 2009). And sometimes the transverse magnetic field shows an "inverse-polarity configuration" pointing from the negative polarity to the positive polarity, indicating a concave upturning field configuration at the PILs (e.g. Lites 2005; Canou et al. 2009; Okamoto et al. 2008). Continuous observations by SOT of Hinode studied by Okamoto et al. (2008) have found a temporal evolution of the transverse magnetic field at the PIL from a "normal-polarity" configuration (pointing from the positive to negative polarity as expected for a convex arcade loop field) to an "inverse-polarity configuration", which was interpreted as the signature of the emergence of a helical flux rope through the photosphere. Flare productive active regions also often show velocity shear at the PILs and rotating sunspots, indicative of transport of twist into the solar corona (e.g. Brown et al. 2003; Zhang et al. 2008; Jiang et al. 2012), and develop sigmoid-shaped X-ray loops and sigmoid-shaped filaments in the corona (e.g. Chae et al. 2001; Gibson et al. 2002, see also Fig. 9 in Engvold 2014).

MHD simulations have shown that the emergence of a twisted magnetic flux tube from the interior into the solar atmosphere and corona can qualitatively explain many of these commonly observed features associated with strongly flaring active regions (e.g. Magara 2004; Manchester et al. 2004; Archontis et al. 2009; Fan 2009; Fang et al. 2010, 2012a,b; Leake et al. 2013). Figure 12.1 shows the results from one example simulation (Fan 2009) of the 3D coronal magnetic field structure and the photospheric flux emergence patterns produced by a subsurface twisted flux tube whose central segment rises buoyantly to the photosphere and emerges into the atmosphere and the corona as a result of the non-linear evolution of the magnetic buoyancy instability. It is found that the twisted subsurface flux tube does not emerge as a whole into the corona. While the upper parts of the helical field

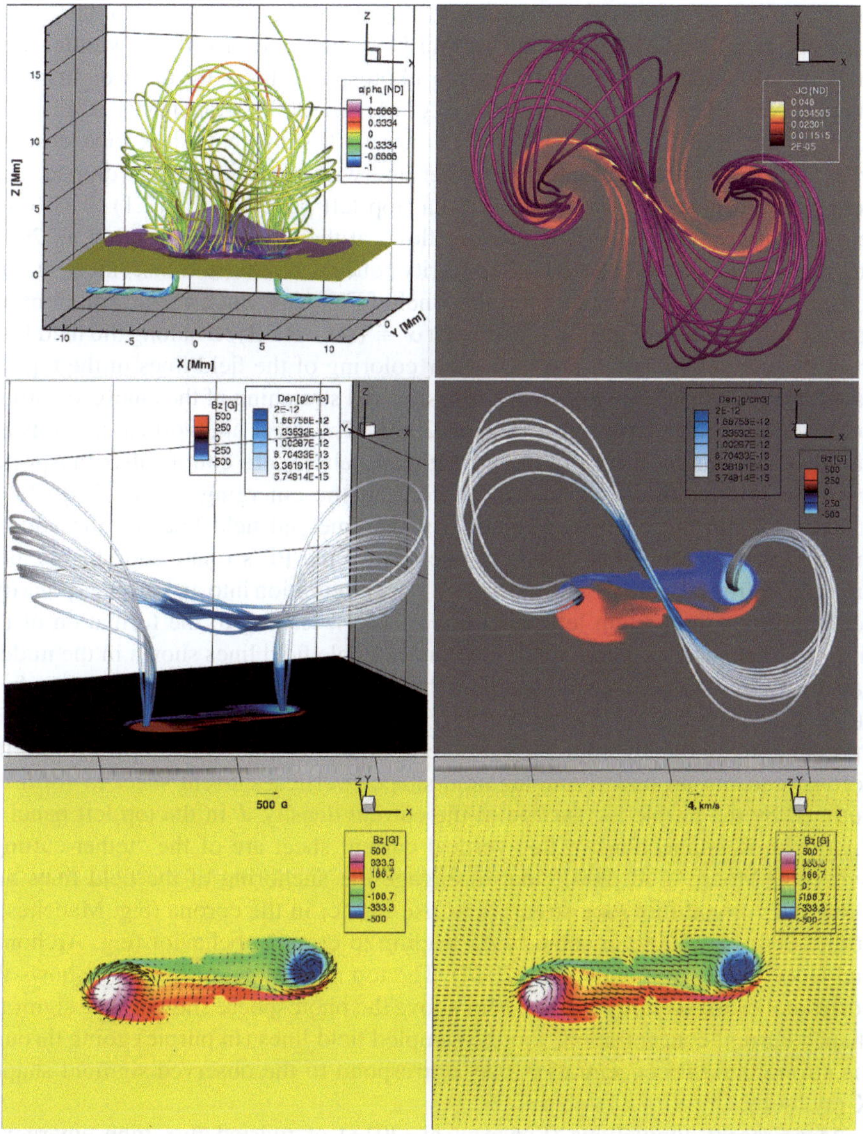

Fig. 12.1 Results from an example MHD simulation of the dynamic emergence of a twisted subsurface flux tube into the solar atmosphere and corona as described in Fan (2009). *Top-left panel* shows the 3D coronal magnetic field structure with field lines colored by the torsion parameter α. The *purple* surface is the iso-surface of electric current density J, outlining regions of high current concentration. *Top right panel* shows sample field lines (in *purple*) going through the central current sheet. The background color image shows a horizontal cross-section of the current density J (the color map is for J) at 3 mm above the photosphere. The *two middle panels* show two perspective views of a set of field lines with central dips in the corona at a height about 5 mm above the photosphere. The field lines are colored by the density, showing enhanced density at the dips. The *two bottom panels* show the vertical magnetic field pattern (*color image*) with arrows of the transverse magnetic field (*bottom left panel*) and horizontal velocity field (*bottom right panel*)

lines of the subsurface tube expand into the atmosphere due to the onset of the magnetic buoyancy instability, the bottom U-shaped portions of the winding field lines remain anchored at and below the photosphere layer by the weight of the plasma. Nevertheless, the simulations (e.g. Magara 2004; Manchester et al. 2004; Archontis et al. 2009; Fan 2009; Leake et al. 2013) show that a flux rope structure with field lines winding about each other and with sigmoid-shaped, dipped, core field lines eventually forms in the corona (top-left panel in Fig. 12.1). It is found that the Lorentz force drives both shear flows at the PIL (Manchester et al. 2004) and rotational motions in each of the polarity concentrations (reminiscent of rotating sunspots) as shown in the bottom right panel of Fig. 12.1. Such vortical motions are caused by a gradient of the torsion or twist $\alpha \equiv ((\nabla \times \mathbf{B}) \cdot \mathbf{B})/\mathbf{B}^2$ along the field lines from the interior to the corona (see the α coloring of the field lines in the top left panel of Fig. 12.1) due to the great expansion and stretching of the emerged coronal fields (Longcope and Welsch 2000; Fan 2009). The shear and vortical motions are the major means by which twist or magnetic helicity are continually transported from the interior flux rope into the corona in the emerging region. Fan (2009) found that with the continued twisting of the emerged field lines by the vortical motions at their footpoints, the field lines above the PILs rotate and change their orientation from an initial "normal-polarity" configuration into an "inverse-polarity" configuration (see bottom left panel of Fig. 12.1), leading to the formation of the sigmoid-shaped dipped core fields (see the example field lines shown in the middle panels of Fig. 12.1). This would explain the observed rotation of the transverse field at the PIL described in Okamoto et al. (2008). With continued transport of twist into the corona, the sigmoid-shaped core field also begins to rise upward into the corona, causing an underlying sigmoid-shaped vertical current sheet to form (as outlined by the purple iso-surface of the current density J in the top left panel of Fig. 12.1). Reconnections in this vertical current sheet are of the "tether-cutting" type (e.g. Moore et al. 2001) that disengage the anchoring of the field lines and allow the coronal flux rope structure to rise further in the corona (e.g. Manchester et al. 2004; Fan 2009, in some cases leading to eruptive behavior (e.g. Archontis and Hood 2012; Archontis et al. 2014). The top right panel of Fig. 12.1 shows the horizontal cross-section of J at 3 Mm above the photosphere showing the sigmoid-shaped current concentration, and the sampled field lines (in purple) going through the strong current concentration may correspond to the observed sigmoid-shaped X-ray loops.

More recent simulations (Leake et al. 2013) of twisted flux tube emergence incorporating a pre-existing dipole coronal field found clearly the formation of a stably confined coronal flux rope structure with sigmoid shaped dipped core fields and an underlying current sheet similar to that found in Fig. 12.1. They confirm the results that shear and sunspot rotation are driven by twisted flux tube emergence and they can cause the formation of stable sigmoids prior to a solar flare. Simulations that incorporate magneto-convection (Fang et al. 2012a) in the interior layer also found that shear flows at the PILs and sunspot rotation driven by the Lorentz force

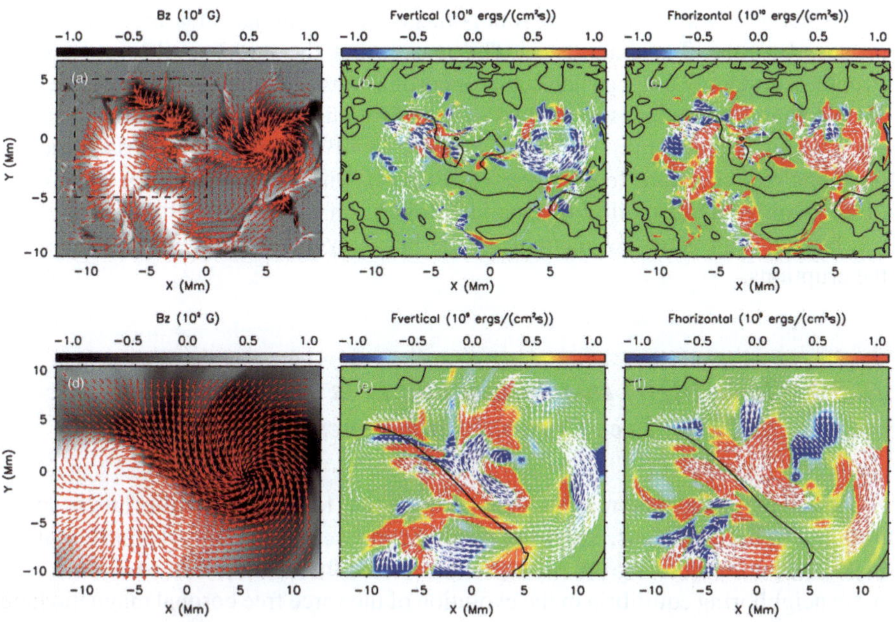

Fig. 12.2 Flux emergence pattern from a simulation of the emergence of a twisted flux tube from the interior into the atmosphere and corona including magneto-convection in the interior layer by Fang et al. (2012a): Grayscale image of B_z and *red arrows* of the transverse magnetic field (**a**), Poynting flux F_{vertical} due to direct vertical motions with *white arrows* of horizontal velocity fields (**b**), and Poynting flux $F_{\text{horizontal}}$ due to horizontal flow fields with white arrows of horizontal velocity (**c**), on the photosphere. (**d**),(**e**), and (**f**) are the same as (**a**), (**b**), and (**c**) respectively but at a height of $z = 3$ mm in the corona. PIL is shown by the *black line*. One clearly sees shear flows at the PIL, and a prominent rotation of the negative polarity spot. Transport of magnetic energy into the corona is clearly dominated by the component due to the horizontal motions (shear and sunspot rotation). Figure from Fang et al. (2012a) reproduced by permission of the AAS

are the major means twist and free-magnetic energy are transported from the interior into the corona (Fig. 12.2). It is found that sigmoid-shaped sheared fields are built up in the corona, but the formation of a coherent flux rope structure with dipped fields and inverse-polarity configuration has not been seen in these simulations.

12.3 Initiation Mechanisms for Eruption

Due to the fast Alfvén speed in the lower solar corona ($\sim 1{,}000\,\text{km}\,\text{s}^{-1}$), the process of magnetic flux emergence characterized by a photospheric flow speed of order a few km/s represents a slow driving or change of the coronal magnetic field. Thus the resulting twisted coronal structure (as discussed in the previous section) that forms is expected to evolve quasi-statically through a sequence of near force free equilibria as it is being driven slowly at the foot points by the continued shearing and twisting produced by the flux emergence. The reason that the quasi-equilibrium coronal structures suddenly erupt as flares and/or coronal mass ejections is still under debate. The mechanism that leads to the loss of a stable equilibrium and triggers the energy release and eruption may be purely ideal or involve non-ideal processes such as magnetic reconnections (e.g. Forbes et al. 2006). One likely possibility is the onset of an ideal-MHD instability or a sudden loss of an ideal-MHD equilibrium. Such mechanisms can naturally account for the fast Alfvén time scales for the onset of the eruptions.

12.3.1 Ideal MHD Instabilities and Loss of Equilibrium of Force Free Coronal Flux Ropes

A coronal magnetic field may suddenly erupt if the (force free) equilibrium becomes unstable to perturbations, i.e. if the resulting forces produced by the perturbations make the perturbations grow rather than restoring the equilibrium, or if there is no more neighboring equilibria in the evolution of the force free coronal magnetic field. Both types of theoretical analysis of (1) the ideal linear instabilities of a force free equilibrium (e.g. Hood and Priest 1981; Kliem and Török 2006; Isenberg and Forbes 2007; Démoulin and Aulanier 2010) or (2) the catastrophic loss of neighboring force free equilibrium solutions beyond a certain value of some evolutionary parameter (e.g. Forbes and Priest 1995; Lin et al. 1998; Démoulin and Aulanier 2010) have been carried out.

Two current-driven instabilities that have been extensively studied as likely triggers for flares and eruptions are the helical kink instability and the so called "torus instability" associated with a twisted flux rope. A force free cylindrically symmetric twisted flux tube of infinite length is shown to be always unstable to the helical kink instability (Anzer 1968). The kink instability can be suppressed if the ends of the cylindrical flux tube are line-tied such that within the finite length of the flux tube the total twist is not too high (Raadu 1972). Thus anchoring of the footpoints of the coronal loops by the heavy plasma of the photosphere is stabilizing for the coronal magnetic field. Again considering cylindrically symmetric force free magnetic flux tubes line-tied at both ends, Hood and Priest (1981) show that for a uniformly twisted flux tube the kink instability sets in if the angle each field line is twisted in going from one end to the other exceeds 2.49 π, or about 1.25 full rotations.

On the other hand, the torus instability is an expansion instability associated with a toroidal current ring held in equilibrium in an external potential magnetic field (Kliem and Török 2006; Démoulin and Aulanier 2010). The stability of such equilibrium configurations has been studied in Tokamaks (e.g. Bateman 1978). The torus instability for an arched coronal flux rope confined by an external potential field has been demonstrated using an analytic model (Titov and Démoulin 1999;

Isenberg and Forbes 2007) of an approximately force free coronal magnetic field as shown in Fig. 12.3, hereafter referred to as the T&D (Titov and Demoulin) flux rope model. The force free coronal magnetic field above the photosphere is constructed using three sources: a circular flux rope with a thin total current I, a pair of charges q and $-q$ below the photosphere, and a line current I_0 below the surface. The subsurface sources are just virtual sources for the analytic construction of the normal magnetic flux at the photosphere, and are not reality. With such construction, the forces at the apex of the flux rope (Titov and Démoulin 1999) can be decomposed into an outward hoop force, corresponding to the self-repulsive force of the circular current I, and an inward confining force acting on the flux rope current due to the potential field B_p produced by the charges. Note, the subsurface line current I_0 is introduced as a means of controlling the pitch of the magnetic field in the vicinity of the coronal current and it does not exert a force on the coronal current I because the field I_0 generates is parallel to I. In equilibrium the outward hoop force and the inward confining force due to the potential magnetic field should balance. Titov and Démoulin (1999) considered a sequence of equilibrium states with increasing major radius R, which could be viewed as a quasi-static emergence of the flux rope. They found that the equilibrium becomes unstable to an expansion δR, when R reaches a critical size with respect to the separation of the charges, where the decline of the potential field B_p with R becomes sufficiently steep, as measured by a decay index of $n \equiv -d \ln B_p / d \ln R > 1.5$, such that the decline of the confining force by the potential magnetic field becomes faster than the decline of the outward hoop force. Titov and Démoulin's (1999) calculation of the instability considered an azimuthally symmetric expansion δR, and therefore does not enforce line-tying by the heavy photosphere during the time scale for the onset of the instability. Later improved calculation of the torus instability for the T&D flux rope equilibrium by Isenberg and Forbes (2007) considered perturbations that truly enforce anchoring

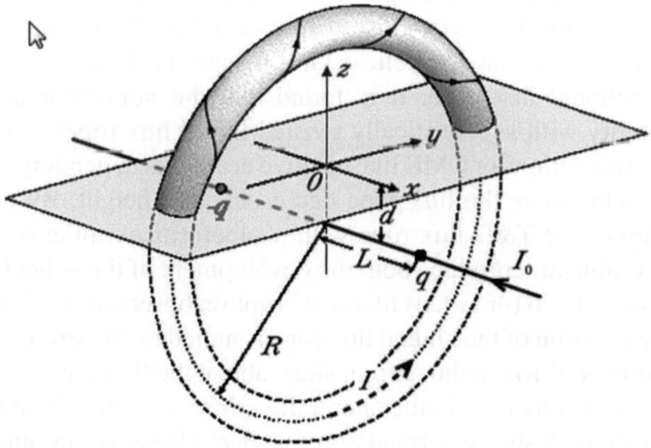

Fig. 12.3 The T&D (Titov and Demoulin) force free coronal flux rope model from Titov and Démoulin (1999). Figure reproduced with permission by Astron. & Astrophys.

of the coronal flux rope on the photosphere by using a subsurface image current to wipe out the change on the photosphere magnetic field produced by the change of the coronal current. They found similar results, that there is a critical height for the apex of the coronal flux rope, above which stable equilibrium of the flux rope confinement is not possible.

Démoulin and Aulanier (2010) examined both the loss of equilibrium and the torus instability of coronal flux ropes with concentrated thin current channels in the corona of either a straight or circular shape. They found that a global instability of the magnetic configuration is present when the current channel is located at a coronal height, h, large enough so that the decay index of the potential field, $n \equiv -d \ln B_p / d \ln h$, is larger than a critical value that is in the range of 1–1.5. They found that when a loss of equilibrium occurs the magnetic configuration is also ideally unstable to the torus instability.

12.3.2 MHD Simulations of the Eruption of Coronal Flux Ropes

The above analytical studies of the ideal linear instabilities and loss of equilibrium of coronal force free magnetic fields are necessarily limited to highly idealized field configurations with a high degree of symmetry. To study realistic three-dimensional coronal magnetic field configurations and the non-linear evolution of the instabilities and loss of equilibrium, MHD simulations are important tools to provide physical insight.

Török et al. (2004) and Török and Kliem (2005) have performed detailed studies of the helical kink instability of an arched coronal flux rope, line-tied to the photosphere, using the T&D analytical flux rope model as the initial state in three-dimensional ideal MHD simulations. They have shown that this model relaxes to a numerical equilibrium very close to the initial analytical model in the case of subcritical twist and that the helical kink instability develops for supercritical twist in the anchored flux rope. It is found that the non-linear development of the kink instability with supercritically twisted initial flux ropes can lead to either confined (failed) eruptions or CME like ejective eruptions depending on how rapidly the overlying field above the flux rope decreases with height. By using a certain parameter setup of the T&D flux rope with a supercritical initial twist, Török and Kliem (2005) is able to reproduce both the development of the helical shape and the rise profile of a confined (or failed) filament eruption observed on 27 May 2002 (see Fig. 12.4). The eruption of the kinked flux rope is halted by the strong overlying field and reconnection outflows in the current sheet above the flux rope causes expansion of the top part of the flux rope in the lateral directions as seen in both the simulation and the observation. A strong vertical current sheet also develops under the kinked flux rope between the two legs of the kinked loop (e.g. Fan and Gibson 2004; Török and Kliem 2005), which is consistent with the coronal hard X-ray emission near

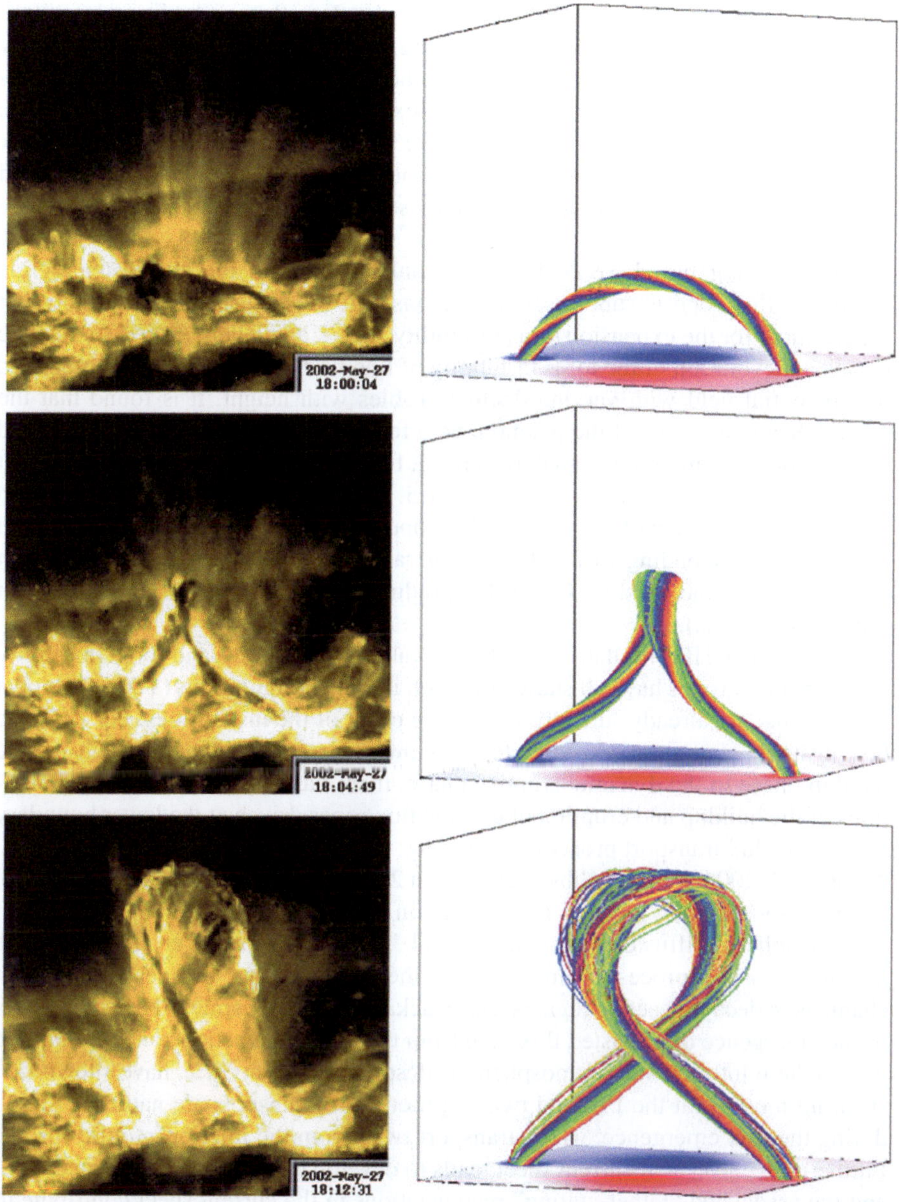

Fig. 12.4 *Left*: TRACE 195 Å images of the confined filament eruption on 2002 May 27. *Right*: Magnetic field lines outlining the core of the kink-unstable flux rope from the simulation of Török and Kliem (2005). Figure from Török and Kliem (2005) reproduced by permission of the AAS

the crossing of the loop legs observed during the flare (Liu and Alexander 2009). The similarities between the helical shape of erupting filament/prominence in many observed events and the magnetic field morphology produced by MHD simulations of the evolution of kink unstable coronal flux ropes (e.g. Török and Kliem 2005; Fan and Gibson 2004; Gibson and Fan 2006) suggest that the onset of the kink instability is a viable initiation mechanism for triggering many of such events, and that flux ropes with substantial twist can exist or form in the solar corona prior to eruption.

Further using initial T&D flux rope configurations in the parameter regime where the flux rope is subcritical for the onset of the helical kink instability but supercritical for the expansion torus instability, Török and Kliem (2007) studied the non-linear evolution of the torus instability of line-tied coronal flux rope embedded in a potential field with varying decline profiles with height. It is found that the critical decay index $n_{\rm cr}$ of the potential field for the torus instability of the line-tied 3D flux rope is similar to the analytical result for a freely expanding toroidal current ring (Kliem and Törö 2006), with $n_{\rm cr} \sim 1.5$. It is also found that the acceleration profile for the eruption of the unstable flux rope depends on the steepness of the field decrease, corresponding to fast CMEs for rapid decrease (as typical of compact active regions) and to slow CMEs for gradual decrease (as typical of quiescent filament eruptions).

The above MHD simulations of the helical kink instability and torus instability of coronal flux ropes have all started out with initial configurations (T&D analytical models) that are already supercritical for the onset of the instabilities. For studying how unstable configurations come to being and the transition from the quasi-static buildup phase to the dynamic eruptive phase, there have been many MHD simulations of the buildup and eruption of coronal flux ropes driven at the lower boundary by various flux transport processes including: a slow imposed flux emergence (e.g. Amari et al. 2004; Fan and Gibson 2007; Fan 2010, 2012; Chatterjee and Fan 2013), shearing and twisting motions in conjunction with flux cancelation at the PIL due to photospheric diffusion (e.g. Amari et al. 2003a,b; Aulanier et al. 2010). The last flux transport process is important for the the formation of quiescent filament channels in decaying active regions (see Mackay 2014). Dynamic MHD simulations of the emergence of a twisted flux tube from the solar convection zone through the photosphere into the solar atmosphere as described in Sect. 12.2, have shown that shearing motions at the PIL and twisting motions of sunspots are naturally driven during the flux emergence, which transport twist from the interior into the corona. Furthermore, a vertical current sheet tends to develop underlying the emerged field, and the associated "tether-cutting" reconnections in the current sheet contribute to the buildup of the coronal flux rope and allow it to rise into the corona (Sect. 12.2). This picture of current sheet formation and magnetic reconnections contributing to the buildup of a coronal flux rope during the quasi-static phase has also been found in coronal MHD simulations of flux rope eruption (e.g. Aulanier et al. 2010; Fan 2010, 2012). Aulanier et al. (2010) performed a coronal MHD simulation of an initially potential, asymmetric bipolar field, which evolves as it is driven at the lower boundary by sub-Alfvénic, line-tied shearing motions and a slow magnetic diffusion that causes flux cancellation at the PIL. It is found that flux cancellation at the PIL transforms sheared arcades into a slowly rising and stable flux rope. Later a quasi-separatrix layer (QSL see more discussion below) topology develops with the formation of a vertical current sheet along the QSL and "tether-cutting" reconnections allow the flux rope to continue to rise slowly. As the flux rope reaches the height at which the decay index of the potential magnetic field reaches about 1.5, the flux rope undergoes a rapid acceleration. The conclusion from this study is that the non-ideal resistive processes of photospheric flux cancellation and tether-cutting coronal reconnections are not the trigger of eruption but are the key pre-eruptive mechanisms for the build up and rise of the coronal flux rope to the critical height at which the ideal torus instability causes the eruption.

Similar evolution for the build up of torus-unstable coronal flux ropes is found in the simulations of Fan (2010, 2012). In Fan (2010) a set of simulations are carried out where the coronal evolution is driven at the lower boundary by the slow emergence of a twisted flux rope into a pre-existing coronal arcade field, with a varying amount of twist being transported into the corona for the different cases before the emergence is stopped. In all the cases the emerged flux rope settles into a quasi-equilibrium after the emergence is stopped. Subsequently, the flux rope continues to rise quasi-statically due to the "tether-cutting" reconnection in the vertical current sheet that forms underlying the flux rope (the two left panels in the top row of Fig. 12.5), even though no more Poynting flux or helicity flux is transported into the corona and the coronal magnetic energy is slowly declining (third row panel in Fig. 12.5). The flux rope is found to transition to a dynamic eruption (top right panel in Fig. 12.5) with rapid acceleration and sharp release of the magnetic energy at varying times for the different cases (the second and third-row panels of Fig. 12.5), but all corresponding to when roughly the same critical height is reached (bottom panel of Fig. 12.5) where the decay index of the potential field reaches about 1.7. This nearly uniform height dependence for the onset of eruption suggests that the trigger of the eruption is caused by the onset of the torus instability, while the tether-cutting reconnection is contributing to the build-up of the flux rope and facilitating its (quasi-static) rise to the critical height for the onset of the instability. Fan (2012) further showed that the thermal signatures of such "tether-cutting" reconnections is the formation of a hot central low-density channel containing reconnected, twisted flux threading under the flux rope axis and on top of the central vertical current sheet. When viewed in the line of sight roughly aligned with the hot channel (see Fig. 12.6), the central vertical current sheet appears as a high-density vertical column with upward extensions as a U-shaped dense shell enclosing a central hot, low-density void corresponding to the central hot channel. Such thermal features may correspond to the observed sub-structures (central hot cavity on top of prominence "horns") that have been observed within coronal prominence cavities (see Gibson 2014).

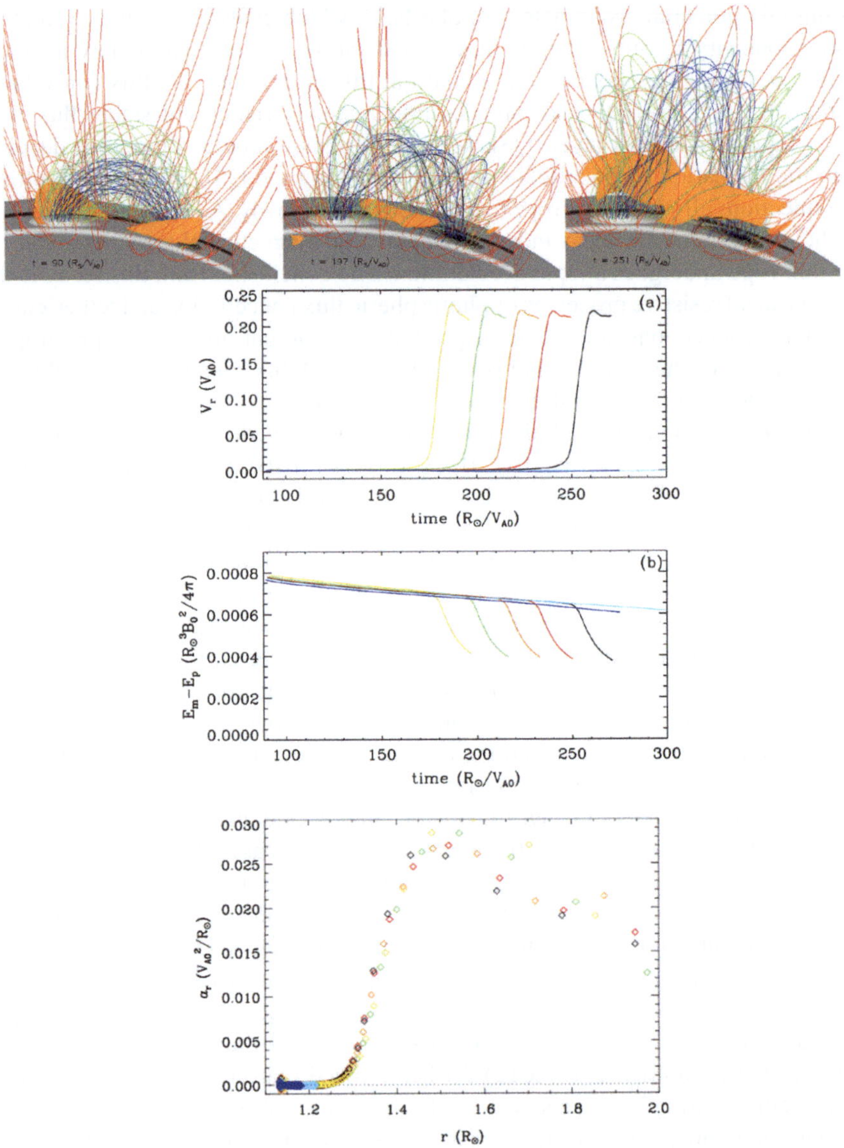

Fig. 12.5 Simulations by Fan (2010) of the buildup and eruption of torus-unstable coronal flux ropes. *Top row panels*: snapshots of 3D coronal flux rope with underlying current sheet formation (orange iso-surfaces) during the quasi-static rise phase (*two left panels*) and at the onset of dynamic eruption when the critical height for the onset of the torus instability is reached (*right panel*). The evolution corresponds to the *black curves* in the *lower panels*. *second and third-row panels* show respectively the flux rope rise speed and the free magnetic energy as a function of time after the emergence is stopped, for the different cases (*different colored curves*) where different amount of twist is transported into the corona during the imposed flux emergence. *Bottom panel*: the acceleration of the flux rope as a function of the apex height for the different cases. Rapid acceleration takes place when roughly the same critical height is reached. Figure reproduced by permission of the AAS

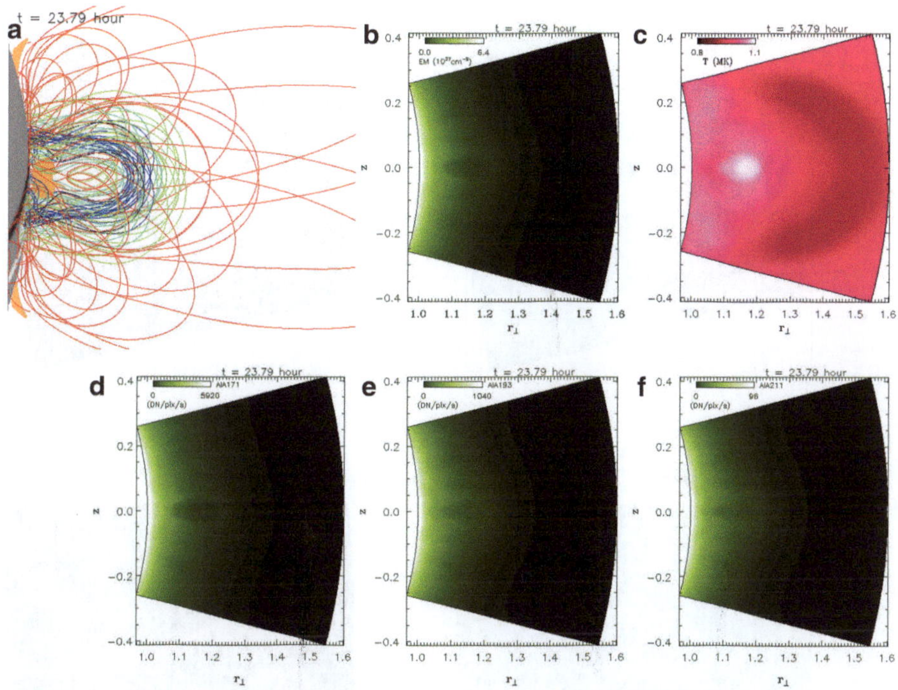

Fig. 12.6 From Fan (2012). (**a**) 3D flux rope field lines and the underlying current sheet (*orange* iso-surface of J/B) viewed above the limb along the line of sight that is roughly aligned with the hot channel. The hot channel forms due to the accumulation of the reconnected flux rising into the flux rope produced by the "tether-cutting" reconnections during the quasi-static rise phase. (**b**) and (**c**) show respectively the modeled emission measure and line-of-sight averaged temperature showing the central hot low density cavity caused by the hot channel, and (**d**), (**e**), and (**f**) show respectively the synthetic SDO/AIA intensity images at 171, 193, and 211 Å, as viewed from the same line of sight. Figure from Fan (2012) reproduced by permission of the AAS

The central vertical current sheet underlying the simulated flux rope (Fig. 12.6a) is found to have formed along topological structures identified as quasi separatrix layers (QSLs, see Fig. 12.7), which are regions of the magnetic volume where the field line connectivity to the line-tied surface experiences drastic variations (Démoulin et al. 1996a,b; Titov et al. 2002). They are identified by estimating the so-called squashing degree, Q, which measures the "squashing" of an elementary flux tube cross-section as it is mapped from one foot point to the other (Titov et al. 2002; Titov 2007; Pariat and Démoulin 2012). QSLs correspond to regions of very large Q. They are a generalization of the concept of separatrices at which the field line linkage is discontinuous. Similar to a separatrix, a QSL divides the coronal domain into quasi-connectivity domains, and due to the drastic change of the field line connectivity at the QSL, it is a site along which current sheets or magnetic tangential discontinuities tend to form and where magnetic reconnections take place (e.g. Démoulin et al. 1996b; Aulanier et al. 2005; Savcheva et al. 2012).

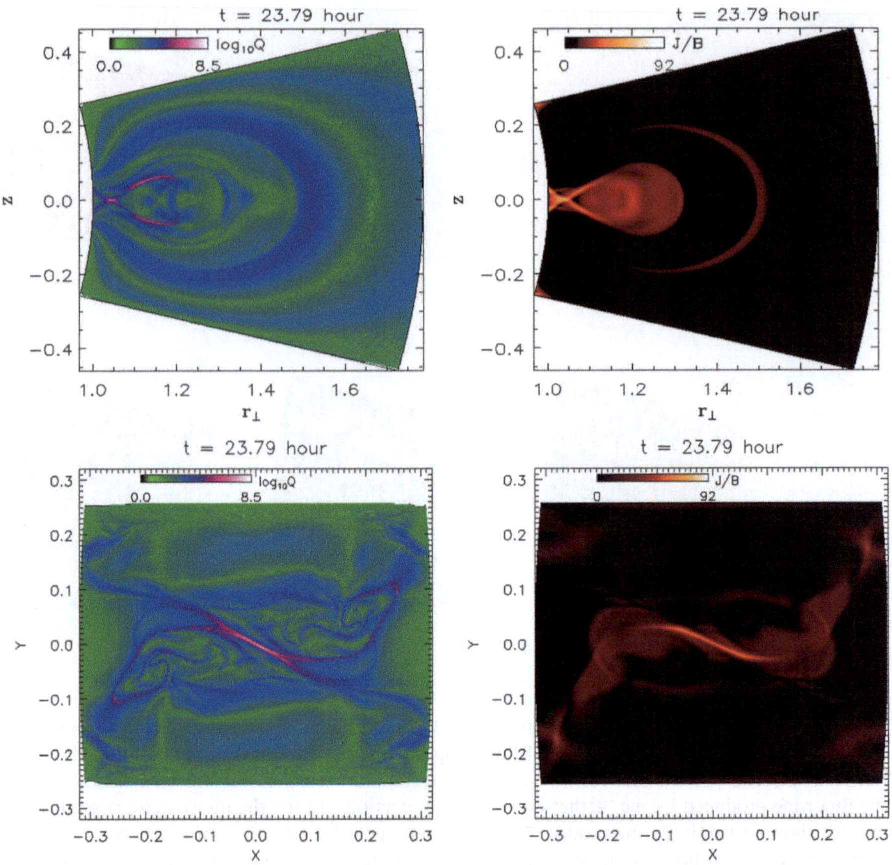

Fig. 12.7 From the simulation of Fan (2012). Squashing degree Q in respectively the central meridional cross-section of the flux rope (*upper left panel*) and a horizontal cross-section below the hot channel of the flux rope (*lower left panel*). The high Q value outlines the location of the QSLs, along which intense current sheets form as shown in the corresponding cross-sections of J/B in the right panels. Figure from Fan (2012) reproduced by permission of the AAS

The QSL with the highest Q values shown in the central meridional cross-section of the simulated flux rope (top left panel of Fig. 12.7) correspond to the mid cross-section of the so-called Hyperbolic Flux Tube (HFT), a generalization of the X-line configuration, which divides the magnetic volume into four distinct domains of magnetic field connectivity (e.g. Titov 2007; Aulanier et al. 2005; Savcheva et al. 2012). The central vertical current sheet underlying the flux rope forms along the HFT (see the top right panel of Fig. 12.7). Fan (2012) suggests that the observed density feature of a dense prominence column with upward extending "horns" (see Gibson 2014) corresponds to the current sheet that forms along the HFT.

As has been discussed above, magnetic reconnection may be playing an important role for the build up of an unstable coronal flux rope before eruption. It has also been shown that magnetic reconnection is critically important for producing an ejective eruption of the magnetic flux in a CME. The energetics for the magnetic storage models of CMEs requires that the magnetic energy of the pre-eruption field exceeds the subsequent magnetic energy during and after the eruption. If the eruption process remains ideal as the flux rope is moving out to infinity, it would have to stretch out its overlying fields to infinity. This may be energetically impossible since it has been shown that for a given normal magnetic flux distribution at the photosphere, the energy of any three-dimensional force free magnetic field with all the field lines simply connected to the photosphere is smaller than the energy of the corresponding fully open magnetic field (Aly 1984, 1991; Sturrock 1991). This is the so-called Aly-Sturrock energy constraint. Thus there is not enough energy in a pre-eruption force free field to stretch all the field lines to infinity to reach a fully open field. One way to get around this difficulty is that an ideal MHD instability or loss of equilibrium can extend the field lines to some height while driving the formation of a current sheet behind the erupting flux rope, and a fast reconnection in the driven current sheet allows the stretched-out field lines to successively reconnect and close back down behind the flux rope as the flux rope is moving out (e.g. Priest and Forbes 2002). Thus a fast magnetic reconnection is necessary to sustain the eruption and allow the flux rope to escape to infinity.

12.3.3 Sudden Onset of Fast Magnetic Reconnection as Trigger for Eruptions

Besides the ideal MHD instability and loss of equilibrium processes, another likely trigger for the onset of eruption is the sudden onset or enhancement of fast magnetic reconnection in a current sheet. Detailed theoretical and computational studies of magnetic reconnection have demonstrated that such sudden onset behavior of fast magnetic reconnection can be achieved through "collisionless" effects when the current sheet thickness falls to the order of the ion inertial length (e.g. Bhattacharjee 2004; Cassak et al. 2005), or even in the resistive MHD regime due to the onset of the plasmoid instability for extended thin current sheets of high Lundquist number (e.g. Bhattacharjee et al. 2009). In the following example MHD simulations that focus on studying the large scale dynamic evolution of eruptions triggered by the onset of fast magnetic reconnection, the reconnection processes are modeled with resistive MHD using a variable magnetic diffusion η or a scale dependent numerical diffusivity.

One of the early simulations that clearly demonstrated resistivity triggered eruption is Mikic and Linker (1994), in which an axisymmetric dipolar coronal arcade field outside the solar surface is subject to a slow shearing motion at its footpoints (see Fig. 12.8). The initial magnetic field is a Sun-centered dipole field

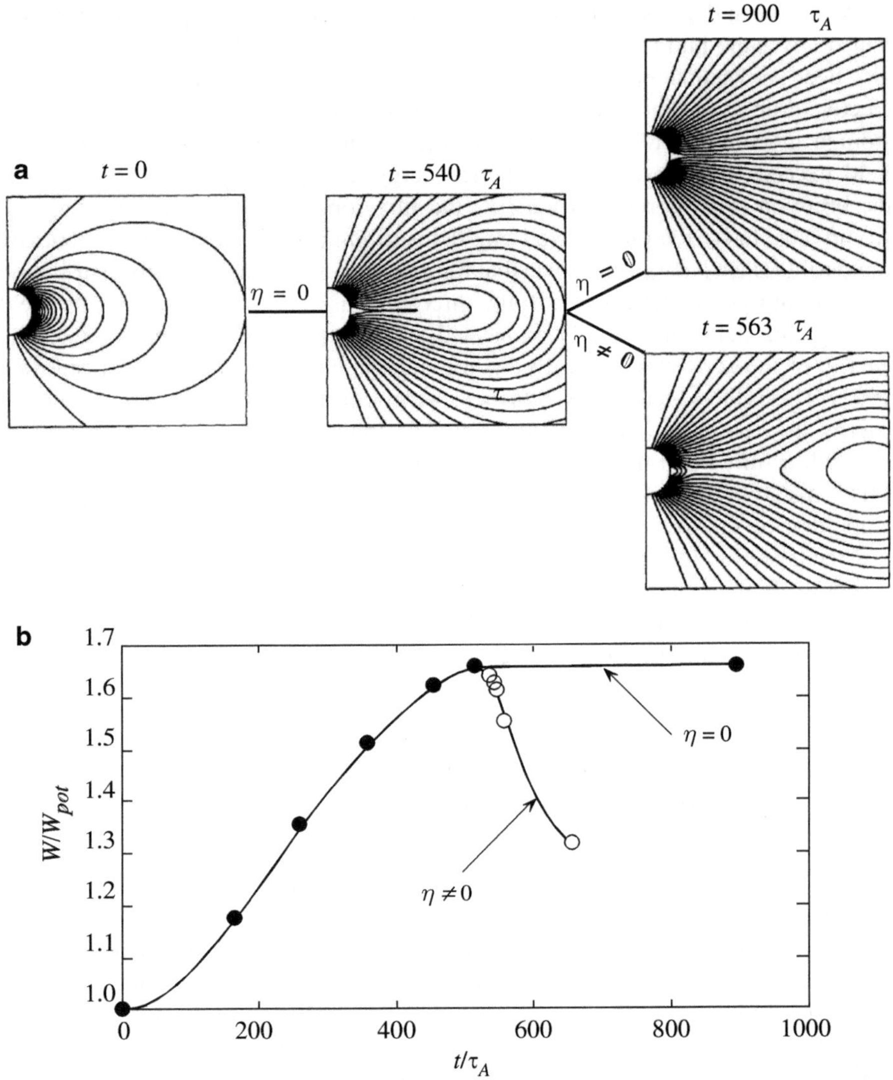

Fig. 12.8 Simulation from Mikic and Linker (1994) of an axisymmetric coronal arcade field subject to slow shearing motion at its footpoints on the solar surface with the two hemispheres moving in opposite directions. The panels in (**a**) show snapshots of the magnetic field evolution with η remaining zero .vs. being instantaneously switched on after $t = 540\tau_A$. Panel (**b**) shows the evolution of the magnetic energy for the two different cases. Figure reproduced from the review Forbes (2000) with permission

($t = 0$ panel), which inflates quasi-statically through a sequence of force free equilibria ($t = 540\tau_A$ and $t = 900\tau_A$ panels, where τ_A is the Alfvén time defined as the solar radius divided by the Alfvén speed) with increasing magnetic energy asymptotically towards a fully open field of the maximum energy (solid black dots in panel (b)) if the evolution remains ideal with the magnetic diffusivity $\eta = 0$. For such two-dimensional force free configurations with translational symmetry, shearing causes the arcade to expand quasi-statically outwards towards a fully opened state without developing an ideal loss of equilibrium. However, it is found that during the quasi-static expansion if η is instantaneously increased to a value which gives an effective magnetic Reynolds number of about 10^4, rapid reconnection takes place at the current sheet at the equator, leading to the formation of a plasmoid which is ejected dynamically (see the $t = 563\tau_A$ panel), and a sharp release of the magnetic energy (see the white circles in panel (b)).

Another well studied model for CME initiation triggered by magnetic reconnection is the "breakout" model first described by Antiochos et al. (1999). MHD simulations based on the breakout model have been carried out in both 2.5D and 3D configurations (e.g. MacNeice et al. 2004; Lynch et al. 2008), and in a 3D configuration with an ambient solar wind (e.g. van der Holst et al. 2009). Karpen et al. (2012) carried out 2.5D simulations of the breakout model with adaptive mesh refinement, achieving high resolutions in the current sheet in global scale dynamic simulations of eruption, and thus allowing identification of the various key phases of the reconnections in relation to the eruption. The basics of how the breakout model works is described as follows (see Fig. 12.9). The initial potential magnetic field outside the solar surface contains a multiflux system with three neutral lines on the surface and four distinct flux systems (Fig. 12.9a): a central arcade straddling the equator (blue field lines), two side arcades associated with the neutral lines at mid latitudes (green field lines), and a polar flux system overlying the three arcades (red lines). There are two separatrix surfaces that define the boundaries between the various flux systems, and a null point above the central arcade. In the initial quasi-static phase, the inner arcade field of the central arcade system is being sheared slowly to become the filament channel (sheared field) and remains confined by the unsheared arcade to build up free magnetic energy. As long as the reconnection between the unsheared (blue) arcade of the central system and the overlying red arcade is slow compared to the shearing, i.e. nearly ideal, energy is built up. However the current sheet formed at the null point eventually becomes

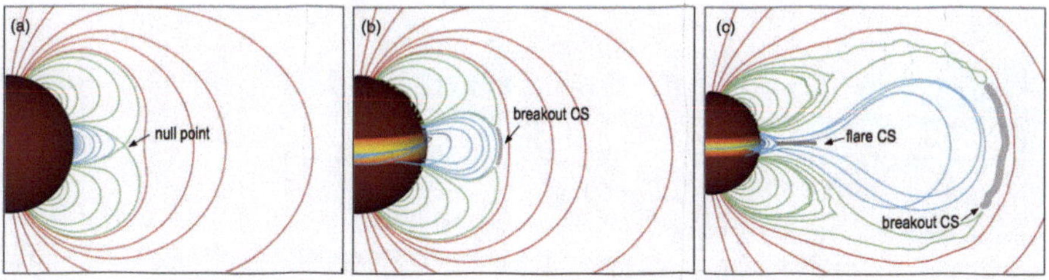

Fig. 12.9 Key steps in the simulation of Karpen et al. (2012) illustrating the basic working and structure (including current sheets CS) in the break out model. Figure reproduced from Karpen et al. (2012) with permission from the AAS

thin enough (see the breakout CS in Fig. 12.9b) and the scale-dependent numerical resistivity causes a sudden onset of the fast magnetic reconnection in the breakout CS (see Fig. 12.10b). The fast breakout reconnection rapidly transfers the confining flux in the unsheared arcade and the overlying outer arcade into the side lobes, causing an outward expansion of the filament channel field and the confinement is found to be permanently lost. Thus the onset of the breakout reconnection is the trigger of the CME and if it were to continue to act alone (Fig. 12.10c), it would have led to a complete but slow outward expansion of the filament channel field with a vertical current sheet extending downward to low heights in the corona, and thus with most of the free energy retained. But then a sudden onset of fast reconnection in the flare current sheet is found to trigger impulsive energy release with the formation of a flux rope that accelerates impulsively (Figs. 12.9c and 12.10d). Thus in this model, the resistive process of the onset of fast magnetic reconnection is the initiation mechanism (or trigger) for both the CME onset (loss of equilibrium for the filament channel field) as well as the later onset of impulsive CME acceleration. For the former it is the onset of the fast breakout reconnection and for the latter it is the onset of fast reconnection in the flare CS. The question then is whether it is possible to observationally identify these two stages of the development.

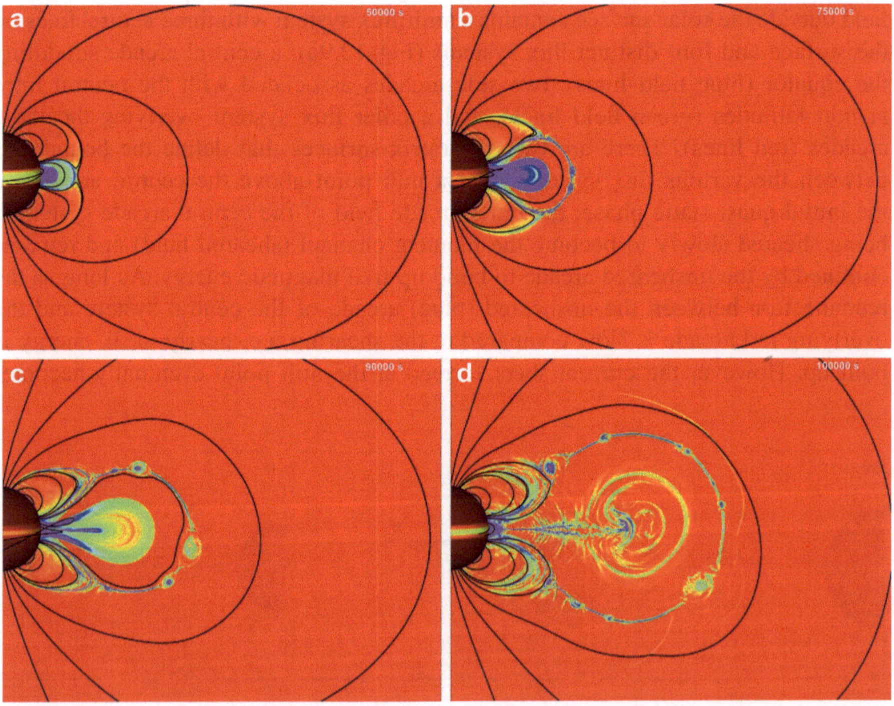

Fig. 12.10 B_ϕ on the solar surface, and normalized current density and magnetic field lines in the meridional cross-section, at selected times in the breakout simulation of Karpen et al. (2012). Figure from Karpen et al. (2012) reproduced by permission of the AAS

Concluding Remarks

The question of how twisted and sheared coronal magnetic structures capable of driving filament eruptions form is examined with focus on emerging active regions. MHD simulations of flux emergence have shown that many commonly observed features associated with flare and CME productive active regions, such as magnetic and velocity shear at the PILs, rotating sunspots, sigmoid-shaped X-ray loops and filaments, can be explained by the emergence of a twisted magnetic flux tube from the solar interior into the atmosphere. It is found that shear and twisting motions driven by the Lorentz force of the emerging tube are the major means (instead of direct vertical emergence through the photosphere) by which magnetic helicity (twist) and energy are transported into the corona. It is also found that current sheet formation and the associated tether-cutting reconnections are important for the buildup of the coronal flux rope and allow it to rise into the corona. This is found in both simulations of flux emergence through the photosphere as well as simulations of line-tied coronal flux ropes evolving quasi-statically towards the onset of eruption.

Several basic mechanisms that can trigger the sudden disruption of the quasi-equilibrium coronal magnetic structures and explosive release of the stored free magnetic energy are discussed. These include both ideal processes such as the onset of the helical kink instability and the torus instability of a coronal flux rope, and the non-ideal processes of the onset of fast magnetic reconnections in a current sheet. It is found that even when the triggering mechanism is an ideal instability or loss of equilibrium of a coronal flux rope, magnetic reconnection is playing an important role in its buildup during the quasi-static stage, as well as for sustaining the eruption and allowing for an ejection of flux rope due to the Aly-Sturrock energy constraint. MHD simulations have shown that such "hybrid" models of the quasi-static buildup of torus unstable coronal flux ropes with the development of the "HFT" topology can explain some of the thermal features observed to develop in pre-eruption coronal prominence cavities. MHD simulations of the non-linear evolution of kink-unstable coronal flux ropes have also shown magnetic field evolution that resembles remarkably well the observed morphology of some of the highly writhing filament eruptions. Fast magnetic reconnection can also trigger eruptions as described in the breakout model without the presence of a more twisted pre-eruption coronal flux rope capable of developing ideal instabilities. Simulations based on this model show that the onset of fast reconnection in the flare current sheet that rapidly creates a highly twisted flux rope is needed for the onset of the impulsive acceleration of the CMEs.

It should be noted that the discussion of this chapter has focused entirely on nearly force free coronal structures, ignoring the non force free effects of the

(continued)

prominence weight for example. There is an important body of work on the role of prominence mass in energy storage of the prominence magnetic field and CME energetics (e.g. Low and Smith 1993; Low and Zhang 2002; Fong et al. 2002; Low et al. 2003; Zhang and Low 2004, 2005), especially with regard to overcoming the Aly-Sturrock energy constraint.

Acknowledgements The author would like to thank Fang Fang, Terry Forbes, Judy Karpen, Zoran Mikic, Slava Titov, and Tibor Török for granting permission to reprint Figures from their original publications. The National Center for Atmospheric Research is sponsored by the National Science Foundation.

References

Aly, J. J. (1984). On some properties of force-free magnetic fields in infinite regions of space. *The Astrophysical Journal, 283*, 349–362. doi:10.1086/162313.

Aly, J. J. (1991). How much energy can be stored in a three-dimensional force-free magnetic field? *The Astrophysical Journal, 375*, L61–L64. doi:10.1086/186088.

Amari, T., Luciani, J. F., & Aly, J. J. (2004). Coronal magnetohydrodynamic evolution driven by subphotospheric conditions. *The Astrophysical Journal, 615*, L165–L168. doi:10.1086/426317.

Amari, T., Luciani, J. F., Aly, J. J., Mikic, Z., & Linker, J. (2003). Coronal mass ejection: Initiation, magnetic helicity, and flux ropes. I. Boundary motion-driven evolution. *The Astrophysical Journal, 585*, 1073–1086. doi:10.1086/345501.

Amari, T., Luciani, J. F., Aly, J. J., Mikic, Z., & Linker, J. (2003). Coronal mass ejection: Initiation, magnetic helicity, and flux ropes. II. Turbulent diffusion-driven evolution. *The Astrophysical Journal, 595*, 1231–1250. doi:10.1086/377444.

Antiochos, S. K., DeVore, C. R., & Klimchuk, J. A. (1999). A model for solar coronal mass ejections. *The Astrophysical Journal, 510*, 485–493. doi:10.1086/306563.

Anzer, U. (1968). The stability of force-free magnetic fields with cylindrical symmetry in the context of solar flares. *Solar Physics, 3*, 298–315. doi:10.1007/BF00155164.

Archontis, V., & Hood, A. W. (2012). Magnetic flux emergence: a precursor of solar plasma expulsion. *Astronomy & Astrophysics, 537*, A62. doi:10.1051/0004-6361/201116956.

Archontis, V., Hood, A. W., Savcheva, A., Golub, L., & Deluca, E. (2009). On the structure and evolution of complexity in sigmoids: A flux emergence model. *The Astrophysical Journal, 691*, 1276–1291. doi:10.1088/0004-637X/691/2/1276.

Archontis, V., Hood, A. W., & Tsinganos, K. (2014). Recurrent explosive eruptions and the "sigmoid-to-arcade" transformation in the sun driven by dynamical magnetic flux emergence. *The Astrophysical Journal, 786*, L21. doi:10.1088/2041-8205/786/2/L21.

Aulanier, G., Pariat, E., & Démoulin, P. (2005) Current sheet formation in quasi-separatrix layers and hyperbolic flux tubes. *Astronomy & Astrophysics, 444*, 961–976. doi:10.1051/0004-6361:20053600.

Aulanier, G., Török, T., Démoulin, P., & DeLuca, E. E. (2010). Formation of torus-unstable flux ropes and electric currents in erupting sigmoids. *The Astrophysical Journal, 708*, 314–333. doi:10.1088/0004-637X/708/1/314.

Bateman, B. (1978). *MHD Instabilities*. Cambridge, MA: MIT.

Berger, M. A. (1984). Rigorous new limits on magnetic helicity dissipation in the solar corona. *Geophysical and Astrophysical Fluid Dynamics, 30*, 79–104. doi:10.1080/03091928408210078.

Berger, M. A., & Field, G. B. (1984). The topological properties of magnetic helicity. *Journal of Fluid Mechanics, 147*, 133–148. doi:10.1017/S0022112084002019.

Berger, T., Testa, P., Hillier, A., et al. (2011). Magneto-thermal convection in solar prominences. *Nature, 472*, 197–200. doi:10.1038/nature09925.

Bhattacharjee, A. (2004). Impulsive magnetic reconnection in the Earth's magnetotail and the solar corona. *Annual Review of Astronomy and Astrophysics, 42*, 365–384. doi:10.1146/annurev.astro.42.053102.134039.

Bhattacharjee, A., Huang, Y. M., Yang, H., & Rogers, B. (2009). Fast reconnection in high-Lundquist-number plasmas due to the plasmoid instability. *Physics of Plasmas, 16*(11), 112 102. doi:10.1063/1.3264103.

Brown, D. S., Nightingale, R. W., Alexander, D., et al. (2003). Observations of rotating sunspots from TRACE. *Solar Physics, 216*, 79–108. doi:10.1023/A:1026138413791.

Canou, A., Amari, T., Bommier, V., Schmieder, B., Aulanier, G., & Li, H. (2009). Evidence for a pre-eruptive twisted flux rope using the themis vector magnetograph. *The Astrophysical Journal Letter, 693*, L27–L30. doi:10.1088/0004-637X/693/1/L27.

Cassak, P. A., Shay, M. A., & Drake, J. F. (2005). Catastrophe model for fast magnetic reconnection onset. *Physical Review Letters, 95*(23), 235002. doi:10.1103/PhysRevLett.95.235002.

Chae, J., Wang, H., Qiu, J., Goode, P. R., Strous, L., & Yun, H. S. (2001). The formation of a prominence in active region NOAA 8668. I. SOHO/MDI observations of magnetic field evolution. *The Astrophysical Journal, 560*, 476–489. doi:10.1086/322491.

Chatterjee, P., & Fan, Y. (2013). Simulation of homologous and cannibalistic coronal mass ejections produced by the emergence of a twisted flux rope into the solar corona. *The Astrophysical Journal, 778*, L8. doi:10.1088/2041-8205/778/1/L8.

de Toma, G., Casini, R., Burkepile, J. T., & Low, B. C. (2008). Rise of a dark bubble through a quiescent prominence. *The Astrophysical Journal Letter, 687*, L123–L126 . doi:10.1086/593326.

Démoulin, P. (2007). Recent theoretical and observational developments in magnetic helicity studies. *Advances in Space Research, 39*, 1674–1693. doi:10.1016/j.asr.2006.12.037.

Démoulin, P., & Aulanier, G. (2010). Criteria for flux rope eruption: Non-equilibrium versus torus instability. *The Astrophysical Journal, 718*, 1388–1399. doi:10.1088/0004-637X/718/2/1388.

Démoulin, P., Henoux, J. C., Priest, E. R., & Mandrini, C. H. (1996). Quasi-separatrix layers in solar flares. I. Method. *Astronomy & Astrophysics, 308*, 643–655.

Démoulin, P., Priest, E. R., & Lonie, D. P. (1996). Three-dimensional magnetic reconnection without null points 2. Application to twisted flux tubes. *Journal of Geophysical Research (Space Physics), 101*, 7631–7646. doi:10.1029/95JA03558.

Engvold, O. (2014). Chapter 2: Description of prominences, including classification. In J. C. Vial & O. Engvold (Eds.), *Solar prominences, ASSL* (Vol. 415, pp. 31–60). Berlin: Springer.

Fan, Y. (2009). The emergence of a twisted flux tube into the solar atmosphere: Sunspot rotations and the formation of a coronal flux rope. *The Astrophysical Journal, 697*, 1529–1542. doi:10.1088/0004-637X/697/2/1529.

Fan, Y. (2010). On the eruption of coronal flux ropes. *The Astrophysical Journal, 719*, 728–736. doi:10.1088/0004-637X/719/1/728.

Fan, Y. (2012). Thermal signatures of tether-cutting reconnections in pre-eruption coronal flux ropes: Hot central voids in coronal cavities. *The Astrophysical Journal, 758*, 60. doi:10.1088/0004-637X/758/1/60.

Fan, Y., & Gibson, S. E. (2004). Numerical simulations of three-dimensional coronal magnetic fields resulting from the emergence of twisted magnetic flux tubes. *The Astrophysical Journal, 609*, 1123–1133.

Fan, Y., & Gibson, S. E. (2007). Onset of coronal mass ejections due to loss of confinement of coronal flux ropes. *The Astrophysical Journal, 668*, 1232–1245. doi:10.1086/521335.

Fang, F., Manchester, W., Abbett, W. P., & van der Holst, B. (2010). Simulation of flux emergence from the convection zone to the corona. *The Astrophysical Journal, 714*, 1649–1657. doi:10.1088/0004-637X/714/2/1649.

Fang, F., Manchester, IV, W., Abbett, W. P., & van der Holst, B. (2012). Buildup of magnetic shear and free energy during flux emergence and cancellation. *The Astrophysical Journal, 754*, 15. doi:10.1088/0004-637X/754/1/15.

Fang, F., Manchester, IV, W., Abbett, W. P., & van der Holst, B. (2012). Dynamic coupling of convective flows and magnetic field during flux emergence. *The Astrophysical Journal, 745*, 37. doi:10.1088/0004-637X/745/1/37.

Fong, B., Low, B. C., & Fan, Y. (2002). Quiescent solar prominences and magnetic-energy storage. *The Astrophysical Journal, 571*, 987–998. doi:10.1086/340070.

Forbes, T. G. (2000). A review on the genesis of coronal mass ejections. *Journal of Geophysical Research (Space Physics), 105*, 23153–23166. doi:10.1029/2000JA000005.

Forbes, T. G., Linker, J. A., Chen, J., et al. (2006). *CME theory and models, in coronal mass ejections, space sciences series of ISSI* (Vol. 21, p. 251). Berlin: Springer. ISBN: 978-0-387-45086-5.

Forbes, T. G., & Priest, E. R. (1995). Photospheric magnetic field evolution and eruptive flares. *The Astrophysical Journal, 446*, 377. doi:10.1086/175797.

Gibson, S. E. (2014). Chapter 13: Coronal cavities: observations and implications for the magnetic environment of prominences. In J. C. Vial & O. Engvold (Eds.), *Solar prominences, ASSL* (Vol. 415, pp. 321–351). Berlin: Springer.

Gibson, S. E., & Fan, Y. (2006). Coronal prominence structure and dynamics: A magnetic flux rope interpretation. *Journal of Geophysical Research (Space Physics), 111*, A12103. doi:10.1029/2006JA011871.

Gibson, S. E., Fletcher, L., Del Zanna, G., et al. (2002). The structure and evolution of a sigmoidal active region. *The Astrophysical Journal, 574*, 1021–1038. doi:10.1086/341090.

Gopalswamy, N., Shimojo, M., Lu, W., Yashiro, S., Shibasaki, K., & Howard, R. A. (2003). Prominence eruptions and coronal mass ejection: A statistical study using microwave observations. *The Astrophysical Journal, 586*, 562–578. doi:10.1086/367614.

Hood, A. W., & Priest, E. R. (1981). Critical conditions for magnetic instabilities in force-free coronal loops. *Geophysical and Astrophysical Fluid Dynamics, 17*, 297–318. doi:10.1080/03091928108243687.

Isenberg, P. A., & Forbes, T. G. (2007).A three-dimensional line-tied magnetic field model for solar eruptions. *The Astrophysical Journal, 670*, 1453–1466. doi:10.1086/522025.

Jiang, Y., Zheng, R., Yang, J., Hong, J., Yi, B., & Yang, D. (2012). Rapid sunspot rotation associated with the X2.2 flare on 2011 February 15. *The Astrophysical Journal, 744*, 50. doi:10.1088/0004-637X/744/1/50.

Karpen, J. (2014). Chapter 10: Plasma structure and dynamics. In J. C. Vial & O. Engvold (Eds.), *Solar prominences, ASSL* (Vol. 415, pp. 235–255). Berlin: Springer.

Karpen, J. T., Antiochos, S. K., & DeVore, C. R. (2012). The mechanisms for the onset and explosive eruption of coronal mass ejections and eruptive flares. *The Astrophysical Journal, 760*, 81. doi:10.1088/0004-637X/760/1/81.

Kippenhahn, R., & Schlüter, A. (1957). Eine Theorie der solaren Filamente. Mit 7 Textabbildungen. *Zeitschrift für Astrophysik 43*, 36.

Kliem, B., & Török, T. (2006). Torus instability. *Physical Review Letters, 96* (25), 255002. doi:10.1103/PhysRevLett.96.255002.

Leake, J. E., Linton, M. G., & Török, T. (2013). Simulations of emerging magnetic flux. I. The formation of stable coronal flux ropes. *The Astrophysical Journal, 778*, 99. doi:10.1088/0004-637X/778/2/99.

Lin, J., Forbes, T. G., Isenberg, P. A., & Démoulin, P. (1998). The effect of curvature on flux-rope models of coronal mass ejections. *The Astrophysical Journal, 504*, 1006–1019. doi:10.1086/306108.

Lites, B. W. (2005). Magnetic flux ropes in the solar photosphere: The vector magnetic field under active region filaments. *The Astrophysical Journal, 622*, 1275–1291. doi:10.1086/428080.

Liu, R., & Alexander, D. (2009). Hard X-ray emission in kinking filaments. *The Astrophysical Journal, 697*, 999–1009. doi:10.1088/0004-637X/697/2/999.

Liu, W., Berger, T. E., & Low, B. C. (2012). First SDO/AIA observation of solar prominence formation following an eruption: Magnetic dips and sustained condensation and drainage. *The Astrophysical Journal Letter, 745*, L21. doi:10.1088/2041-8205/745/2/L21.

Longcope, D. W., & Welsch, B. T. (2000). A model for the emergence of a twisted magnetic flux tube. *The Astrophysical Journal, 545*, 1089–1100. doi:10.1086/317846.

Low, B. C., Fong, B., & Fan, Y. (2003). The mass of a solar quiescent prominence. *The Astrophysical Journal, 594*, 1060–1067. doi:10.1086/377042.

Low, B. C., & Smith, D. F. (1993). The free energies of partially open coronal magnetic fields. *The Astrophysical Journal, 410*, 412–425. doi:10.1086/172758.

Low, B. C., & Zhang, M. (2002). The hydromagnetic origin of the two dynamical types of solar coronal mass ejections. *The Astrophysical Journal, 564*, L53–L56 . doi:10.1086/338798.

Lynch, B. J., Antiochos, S. K., DeVore, C. R., Luhmann, J. G., & Zurbuchen, T. H. (2008). Topological evolution of a fast magnetic breakout CME in three dimensions. *The Astrophysical Journal, 683*, 1192–1206. doi:10.1086/589738.

Mackay, D. (2014). Chapter 14: Formation and Large-scale patterns of filament channels and filaments. In J. C. Vial & O. Engvold (Eds.), *Solar prominences, ASSL* (Vol. 415, pp. 353–378). Berlin: Springer.

Mackay, D. H., & van Ballegooijen, A. A. (2005). New results in modeling the hemispheric pattern of solar filaments. *The Astrophysical Journal Letter, 621*, L77–L80. doi:10.1086/428904.

MacNeice, P., Antiochos, S. K., Phillips, A., Spicer, D. S., DeVore, C. R., & Olson, K. (2004). A numerical study of the breakout model for coronal mass ejection initiation. *The Astrophysical Journal, 614*, 1028–1041. doi:10.1086/423887.

Magara, T. (2004). A model for dynamic evolution of emerging magnetic fields in the sun. *The Astrophysical Journal, 605*, 480–492.

Manchester, IV, W., Gombosi, T., DeZeeuw, D., & Fan, Y. (2004). Eruption of a buoyantly emerging magnetic flux rope. *The Astrophysical Journal, 610*, 588–596. doi:10.1086/421516.

Mikic, Z., & Linker, J. A. (1994). Disruption of coronal magnetic field arcades. *The Astrophysical Journal, 430*, 898–912. doi:10.1086/174460.

Moore, R. L., Sterling, A. C., Hudson, H. S., & Lemen, J. R. (2001). Onset of the magnetic explosion in solar flares and coronal mass ejections. *The Astrophysical Journal, 552*, 833–848. doi:10.1086/320559.

Munro, R. H., Gosling, J. T., Hildner, E., MacQueen, R. M., Poland, A. I., & Ross, C. L. (1979). The association of coronal mass ejection transients with other forms of solar activity. *Solar Physics 61*, 201–215. doi:10.1007/BF00155456.

Okamoto, T. J., Tsuneta, S., Lites, B. W., et al. (2008). Emergence of a helical flux rope under an active region prominence. *The Astrophysical Journal Letter, 673*, L215–L218. doi:10.1086/528792.

Pariat, E., & Démoulin, P. (2012). Estimation of the squashing degree within a three-dimensional domain. *Astronomy & Astrophysics, 541*, A78. doi:10.1051/0004-6361/201118515.

Petrie, G. J. D., & Low, B. C. (2005). The dynamical consequences of spontaneous current sheets in quiescent prominences. *The Astrophysical Journal Supplement, 159*, 288–313. doi:10.1086/431149.

Priest, E. R., & Forbes, T. G. (2002). The magnetic nature of solar flares. *Astronomy & Astrophsical Review, 10*, 313–377. doi:10.1007/s001590100013.

Raadu, M. A. (1972). Suppression of the kink instability for magnetic flux ropes in the chromosphere. *Solar Physics, 22*, 425–433. doi:10.1007/BF00148707.

Savcheva, A. S., van Ballegooijen, A. A., & DeLuca, E. E. (2012). Field topology analysis of a long-lasting coronal sigmoid. *The Astrophysical Journal, 744*, 78. doi:10.1088/0004-637X/744/1/78.

Schrijver, C. J. (2009). Driving major solar flares and eruptions: A review. *Advances in Space Research, 43*, 739–755. doi:10.1016/j.asr.2008.11.004.

Sturrock, P. A. (1991). Maximum energy of semi-infinite magnetic field configurations. *The Astrophysical Journal, 380*, 655–659. doi:10.1086/170620.

Titov, V. S. (2007). Generalized squashing factors for covariant description of magnetic connectivity in the solar corona. *The Astrophysical Journal, 660*, 863–873. doi:10.1086/512671.

Titov, V. S., & Démoulin, P. (1999). Basic topology of twisted magnetic configurations in solar flares. *Astronomy & Astrophysics, 351*, 707–720.

Titov, V. S., Hornig, G., & Démoulin, P. (2002). Theory of magnetic connectivity in the solar corona. *Journal of Geophysical Research (Space Physics), 107*, 1164. doi:10.1029/2001JA000278.

Török, T., & Kliem, B. (2005). Confined and ejective eruptions of kink-unstable flux ropes. *The Astrophysical Journal Letter, 630*, L97–L100. doi:10.1086/462412.

Török, T., & Kliem, B. (2007). Numerical simulations of fast and slow coronal mass ejections. *Astronomische Nachrichten, 328*, 743. doi:10.1002/asna.200710795.

Török, T., Kliem, B., & Titov, V. S. (2004). Ideal kink instability of a magnetic loop equilibrium. *Astronomy & Astrophysics, 413*, L27–L30. doi:10.1051/0004-6361:20031691.

van der Holst, B., Manchester, IV, W., Sokolov, I. V., et al. (2009). Breakout coronal mass ejection or streamer blowout: The bugle effect. *The Astrophysical Journal, 693*, 1178. doi:10.1088/0004-637X/693/2/1178.

Webb, D. F., & Hundhausen, A. J. (1987). Activity associated with the solar origin of coronal mass ejections. *Solar Physics, 108*, 383–401. doi:10.1007/BF00214170.

Woltjer, L. (1958). A theorem on force-free magnetic fields. *Proceedings of the National Academy of Science, 44*, 489–491. doi:10.1073/pnas.44.6.489.

Yeates, A. R., Mackay, D. H., & van Ballegooijen, A. A. (2008). Modelling the global solar corona II: Coronal evolution and filament chirality comparison. *Solar Physics, 247*, 103–121. doi:10.1007/s11207-007-9097-0.

Zhang, Y., Liu, J., & Zhang, H. (2008): Relationship between rotating sunspots and flares. *Solar Physics, 247*, 39–52. doi:10.1007/s11207-007-9089-0.

Zhang, M., & Low, B. C. (2004). Magnetic energy storage in the two hydromagnetic types of solar prominences. *The Astrophysical Journal, 600*, 1043–1051. doi:10.1086/379891.

Zhang, M., & Low, B. C. (2005). The hydromagnetic nature of solar coronal mass ejections. *Annual Review of Astronomy and Astrophysics, 43*, 103–137. doi:10.1146/annurev.astro.43.072103.150602.

Chapter 13
Coronal Cavities: Observations and Implications for the Magnetic Environment of Prominences

Sarah Gibson

Abstract Dark and elliptical, coronal cavities yield important clues to the magnetic structures that cradle prominences, and to the forces that ultimately lead to their eruption. We review observational analyses of cavity morphology, thermal properties (density and temperature), line-of-sight and plane-of-sky flows, substructure including hot cores and central voids, linear polarization signatures, and observational precursors and predictors of eruption. We discuss a magnetohydrodynamic interpretation of these observations which argues that the cavity is a magnetic flux rope, and pose a set of open questions for further study.

13.1 Introduction

The twisted or sheared magnetic field associated with prominences represents stored magnetic energy that may be explosively released in coronal mass ejections (CMEs). It is therefore essential to establish the full three-dimensional (3D) nature of this magnetic field, in order to be alert to topologies prone to eruption and/or approaching thresholds of instability (Fan 2014). However, the prominence itself traces only a portion of the 3D field, being generally localized to a narrow structure above the neutral line (Engvold 2014). For this reason, coronal prominence cavities, which represent a much larger volume than that of the prominence, provide a compelling avenue for exploring the larger-scale magnetic structure associated with the prominence.

13.2 Background

Coronal cavities extend, tunnel-like, above photospheric neutral lines, and are usually observed as dark ellipses or partial ellipses at the limb (Fig. 13.1). They are associated with prominences in a manner similar to how filament channels

S. Gibson (✉)
High Altitude Observatory, National Center for Atmospheric Research, P.O. Box 3000, Boulder, CO 80301, USA
e-mail: sgibson@ucar.edu

J.-C. Vial, O. Engvold (eds.), *Solar Prominences*, Astrophysics and Space Science Library 415, DOI 10.1007/978-3-319-10416-4_13

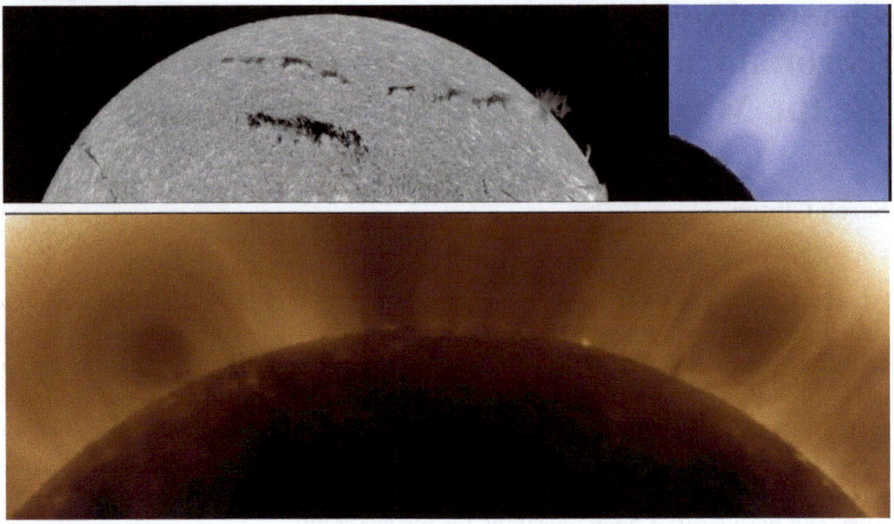

Fig. 13.1 Polar crown filament (PCF) cavities present optimal viewing conditions. *Top* July 22 2002 PCF and associated cavity, (*left*) as seen in Hα by Big Bear Solar Observatory (BBSO) and (*right*) in white light by Mauna Loa Solar Observatory Mk4 (MLSO/Mk4). *Bottom* Example of PCF cavity manifesting on both limbs, as seen by Solar Dynamics Observatory/Atmospheric Imaging Assembly (SDO/AIA) in 193 Å, processed with a radial gradient filter as described in Forland et al. (2013)

are associated with filaments (Engvold 2014): they are often seen surrounding the prominence at the limb, but clear cavities may exist even when no prominence material can be seen. Cavities may last for days or even weeks (Gibson et al. 2006), but even when not erupting, like prominences (Kucera 2014), they have flows associated with them (Berger et al. 2008; Schmit et al. 2009; Li et al. 2012). Cavities are most visible when they are lined up along the line of sight, with few neighboring bright features such as active regions to obscure them. Therefore, polar crown filament (PCF) cavities are particularly good candidates for study.

Cavities can be seen during solar eclipses, so they were initially observed in white light (Waldmeier 1970; Saito and Hyder 1968; Saito and Tandberg-Hanssen 1973). Radio observations provided further evidence that cavities were density depletions (Straka et al. 1975; Kundu et al. 1978). As space observations became possible, they were also observed in soft X-ray (SXR) (Vaiana et al. 1973; McIntosh et al. 1976; Serio et al. 1978) and extreme ultraviolet (EUV) (Schmahl 1979). Helium 10830 observations further demonstrated their links to prominences (McCabe and Mickey 1981). See Tandberg-Hanssen (1974); Engvold (1989); Tandberg-Hanssen (1995); Gibson et al. (2006) for reviews of these early observations.

In this paper we will present current understanding of coronal cavities, using multiwavelength observations and MHD theoretical interpretations of prominence/cavity systems. We will apply these data and models to address two questions in particular: What is the nature of the pre-eruption MHD equilibria represented by quiescent cavities, and what drives these structures to erupt?

13.3 Quiescent Cavities: MHD Equilibria of Energized Fields

We will begin by summarizing observations of cavities not in eruption. These show that quiescent cavities are ubiquitous, possess prolate-elliptical-arched-cylindrical (croissant-like) morphology, have low density (about a factor of two depleted relative to surrounding streamers), are multithermal, demonstrate flows spatially and temporally linking cavity to prominence, have substructure tracing nested ellipses, and possess a "lagomorphic" (rabbit-head-shaped) polarimetric signature indicating twisted or sheared magnetic field well above the height of the filament.

As we present these observations, we will discuss how they might be interpreted in the context of MHD theory. A magnetic flux rope has been proposed as a model for the cavity (Low and Hundhausen 1995), and indeed, we will demonstrate that this interpretation is basically consistent with all of the observations. Nevertheless, ambiguities and open questions remain and we will comment on these throughout this review.

13.3.1 Location and Prevalence

Coronal cavities are not hard to find, except perhaps at solar maximum. Larger cavities may extend as high as $1.6R_{\odot}$ ($0.6R_{\odot}$ above the solar surface) (Gibson et al. 2006; Fuller and Gibson 2009), and these are best observed in white light. Smaller cavities are better observed in the emission corona (e.g., EUV, SXR) as they may partly or wholly lie below the occulting disk of white-light coronagraphs.

The frequency of observed cavities was much greater in a survey of EUV cavities (Forland et al. 2013) than in a survey of white light cavities (Gibson et al. 2006): nearly 80 % of days surveyed had one or more EUV cavity, vs. 10 % of white light days surveyed. This was due in part to the greater data coverage of a space-based coronal imager (no occulting disk, 24 h coverage) compared to a ground-based coronagraph, but also to the phase of the solar cycle. The white-light survey was centered around solar maximum, when the global corona is most complex and cavities are likely to be obscured by surrounding bright features. On the other hand, the EUV study was undertaken during the ascending phase of the cycle, which is an excellent time for viewing cavities. The presence of PCFs in both hemispheres meant that cavities were frequently found in multiple solar quadrants (e.g., Fig. 13.1, bottom panel).

MHD Interpretation of Cavity Ubiquity: Minimum Energy States that Conserve Helicity As discussed in Fan (2014), prominences are generally modeled in terms of twisted or sheared magnetic fields. The non-potentiality of such fields represents stored magnetic energy, and accumulated magnetic helicity. Magnetic energy can be reduced on relatively short time scales through the spontaneous formation and dissipation of current sheets (Parker 1994; Janse et al. 2010). This mode of dissipation is subject, under condition of high electrical conductivity, to the magnetic helicity remaining approximately conserved as a global quantity (Taylor 1974; Berger 1984). It is therefore likely that, over time, excess magnetic energy that can be dissipated will be, resulting in a lower energy state that maintains an accumulated helicity.

For a given boundary condition, the minimum energy state conserving helicity is that of a constant-α force-free field (Woltjer 1958). For sufficiently large values of helicity, the constant-α force-free solution is a magnetic flux rope: a coherent structure in which magnetic field lines wind about a central axis (Low 1994). PCF evolution is likely to involve a steady accumulation of helicity over weeks and even months, resulting in the formation of stable magnetic flux ropes (Mackay 2014). Thus, a characteristic flux rope topology that creates a cavity in combination with a geometry conducive to unobstructed viewing may explain cavity clarity and prevalence in PCFs.

Open Questions

- Do all prominences have cavities extending above them, with only a subset unobscured?
- Does the presence (or lack) of a cavity depend on whether sufficient helicity can accumulate to form a stable flux rope?

13.3.2 3D Morphology

In analyzing cavities it is critical to keep in mind that they are optically thin, unlike the prominences embedded in them. It is straightforward to measure the 3D morphology of EUV cavities, which can be observed down to the limb of the Sun. In contrast, white light cavities usually have their bottom portions obscured by the occulting disk of a coronagraph. The plane-of-sky projection of EUV cavities are generally well fit by ellipses. The aspect ratio for cavity ellipses shows a strong tendency towards prolateness: 93 % of 119 EUV cavities surveyed were taller than they were wide (Forland et al. 2013).

Measuring their three-dimensional extent requires multiple viewing angles and/or an observation of the structure rotating past the plane of the sky (under the assumption that the structure does not change). The extension of the cavity above the neutral line is essentially a cylindrical structure with a Gaussian variation in height (Fig. 13.2 left) (Gibson et al. 2010). Therefore the 3D morphology of cavities may be characterized as that of a prolate-elliptical-arched cylinder, or more familiarly as the shape of a croissant pastry.

MHD Interpretation of Cavity Morphology: Expanded, but Trapped, Twisted Flux Simulations have demonstrated that a flux rope expanding outwards into closed magnetic fields may find an equilibrium configuration as the forces causing the expansion are countered by confining magnetic tension forces (Gibson and Fan 2006a). Because magnetic field strength drops off with height, there is greater lateral confinement than vertical, resulting in an equilibrium flux rope that is taller than it is wide. The equilibrium flux rope will then have an arched, tunnel-like morphology with narrow aspect ratio (Fig. 13.2 (right)).

Open Questions

- Do larger cavities observed in white light have the same 3D morphology as measured for smaller, EUV cavities?

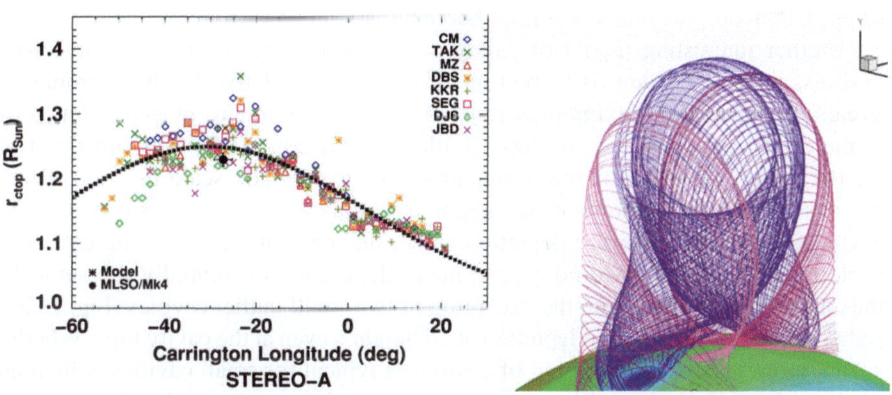

Fig. 13.2 Cavities have arched, tunnel-like morphologies with elliptical cross-sections. *Left* Multiple-observer measurements of cavity ellipse height vs. longitude/date (from Gibson et al. (2010), reproduced by permission of the American Astronomical Society (AAS)). *Right* flux surfaces of a magnetic flux rope as described in Gibson and Fan (2006a)

13.3.3 Thermal Properties

Once the three-dimensional morphology of a cavity is established, it is possible to quantify density and temperature and so gain insight into its thermal properties.

13.3.3.1 Density

Cavities observed simultaneously in multiple wavelengths provide strong evidence that they represent a depletion of coronal plasma density (Gibson et al. 2010). Perhaps the most straightforward density analysis utilizes white light data, which has no temperature sensitivity. An analysis of 24 white-light cavities (Fuller and Gibson 2009) used geometric arguments and Van de Hulst (1950) inversion to measure density within cavities and surrounding streamers. This study found that the average density depletion (relative to surrounding bright coronal streamers) as measured at the center of the cavity and just above the occulter ($1.16R_{\odot}$) was 25 %, and the maximum depletion was 60 %. An analysis using EUV line ratios (to reduce temperature dependency) in conjunction with white light data and explicitly accounting for 3D cavity morphology was able to measure density at heights above the limb as low in height as $1.08R_{\odot}$, where it was found to be 30 % depleted relative to a surrounding streamer (Schmit and Gibson 2011). These measurements are consistent with previous analyses using radio data (Straka et al. 1975; Kundu et al. 1978; Marqué 2004). Thus, cavities are by no means empty, and are significantly brighter than coronal holes at similar heights.

Another interesting aspect of cavity density structure, at least for white light cavities, is that the density depletion of the cavity relative to the surrounding streamer is generally maximum at low heights, and minimum (indeed, often zero) at the cavity top (Fuller et al. 2008; Fuller and Gibson 2009) (see Gibson et al. (2010) for a discussion of the implications for hydrostatic scale heights). When cavities are normalized to their top height, a curious self-similarity is apparent. All cavities studied have similar depletion slope, independent of size or time of solar cycle (Fig. 13.3—left). Moreover, the mean depletion is essentially the same for the majority of cavities, with the exception of two small outlier cavities (Fig. 13.3—right) which possess greater depletion at all heights, even at the cavity tops. Whether such a jump in density at the top of cavities is typical for small cavities, which are likely to be under-selected in any white-light study, remains to be demonstrated through a quantitative survey of emission cavity depletion.

MHD Interpretation of Density Depletion: Enhanced Magnetic Pressure, Field-Line-Length Dependence, and/or Stability Selection Effect? Low density within a magnetic flux rope might be expected if there is a jump in magnetic field strength across its boundary, for example due to an enhanced axial field relative to surrounding fields, and with it a current sheet or layer. Then, total pressure continuity would require a decrease in thermal pressure within the cavity to balance the increase in magnetic pressure, and if the temperature is essentially the same in

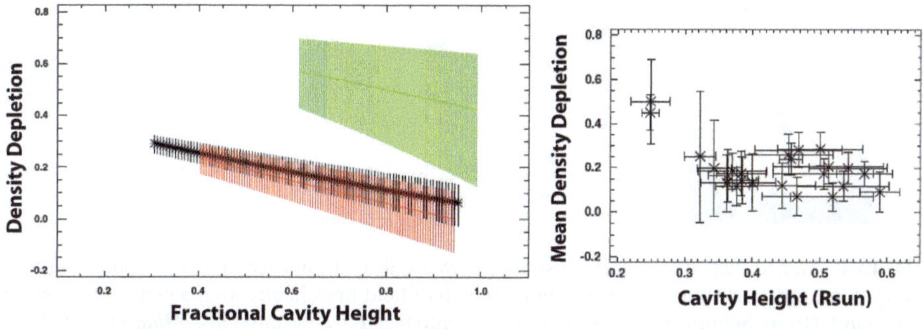

Fig. 13.3 Fractional density depletion of cavity relative to surrounding streamer, exhibiting self-similarity. *Left* normalized radial profiles of depletion for three cavities. *Right* average depletion for 24 cavities. Green profile (*left*) corresponds to one of the small outlier cavities (*right*) (from Fuller and Gibson (2009), reproduced by permission of the AAS).

the cavity and surrounding streamer (see Sect. 13.3.3.2) this would mean a decreased density.

Such continuity arguments are most relevant at the flux rope boundary, however, and depending upon the rope's twist profile there may be little (or no) jump in axial field at this boundary. Indeed, simulations of flux ropes emerging into a potential field (e.g. Gibson and Fan (2006a)) result in a near-force-free equilibrium with a smoothly varying boundary. It is true that current sheets do form at the boundary of these simulated ropes when they are dynamically perturbed, at least at lower heights, and it only takes a small jump in magnetic pressure to require a relatively large jump in thermal pressure in the magnetically-dominated (low plasma β) corona. This may explain why the elliptical boundary of the cavity is generally so sharply defined (at least at lower heights). But why is the center of the cavity persistently depleted in density?

It is clear that density will be controlled by thermodynamic variation within the rope. This variation in turn has its origins in the magnetic structure, and may be affected for example by processes that depend upon magnetic field line length. Generally speaking, field line length increases from the axis of the rope outwards (Fig. 13.4—left). Arguments have been made that line-of-sight integrated density will increase from cavity center outwards if the magnetic structure is a flux rope, due to increased number of particles moving along the progressively longer, more twisted field lines towards the exterior of the flux rope, and assuming an ongoing flow throughout the cavity (Krall and Chen 2005). In fact, a flow is not generally required for a dependence of density on field line length: solutions of hydrostatic equilibrium along field lines extracted from a simulated, stable flux rope (e.g., Fig. 13.4—middle/right) indicate that the short (and low-lying) axial field lines would be depleted about 35 % relative to the outermost rope (and surrounding arcade) field lines at the same height. This arises in part from a weak linear dependence between density n and field line length L ($n \sim Q^{(4/7)} L^{(1/7)}$) under the assumptions of constant footpoint temperature and uniform heating Q along field

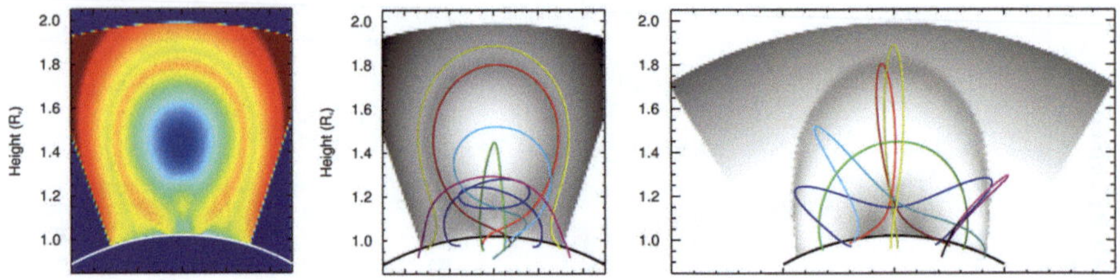

Fig. 13.4 Arched flux-rope model illustrating how axial field lines may be non-dipped and relatively short. *Left* color contours representing length of field lines intersecting a plane (*blue=short, red=long*) (from Schmit and Gibson (2014), reproduced by permission). *Middle/right* views along/normal to the LOS of sample field lines (courtesy D. Schmit). The central, axial lines are the shortest (e.g., *green line*). A slight decrease in length can be seen in the left-hand plot at the boundary between the outermost rope lines (e.g., *red line*) and surrounding arcade (e.g., *yellow line*). Note that field lines wrapping around the outer part of the rope are dipped (e.g., *dark blue*) but the central axial field lines (e.g., *green*) are not

lines. In addition, gravity requires that higher-lying field lines are supported with a higher pressure at their base, and consequently are more dense overall (Schmit and Gibson 2014). Such a hydrostatic description may only be appropriate for the short, axial field lines. Thermal nonequilibrium (TNE) (see Karpen (2014)) is likely to be significant along the longer field lines of the outer portions of the rope and the surrounding arcade (Klimchuk et al. 2010). However, this would only emphasize the contrast with the short, axial field lines at rope center, since TNE would result in further density enhancement relative to a hydrostatic state (Schmit et al. 2013).

If field line length does affect density, the degree of depletion resulting will depend upon the twist profile of a flux rope, and on its size and shape relative to overlying unsheared fields. The flux rope shown in Fig. 13.4 possesses twist between one and two full turns, and has an overall morphology (height, length, radius) similar to that measured for a PCF cavity (Gibson et al. 2010). In such a geometry, the central, axial field lines are the shortest, and indeed are significantly shorter than the outer flux rope and overlying arcade-like field lines. If the rope were less arched, the arcade field line length and height would be closer to that of the axial field line, and the depletion would be less. If the rope were more twisted, the outermost flux rope lines would be even longer than the axial field lines, and the depletion could be greater. Thus, measurements of the density depletion of cavities in combination with their 3D morphologies may constrain the possible twist profiles of the magnetic structures associated with them.

Finally, the characteristic density profiles of cavities—similar mean depletions, an upper depletion limit of about a factor of two, and decreasing depletion with height—may be a consequence of a selection effect associated with their stability. Magnetostatic solutions exist where the plasma in the flux rope has higher density than its surroundings, but these would be unstable both to magnetic curvature forces and to the Rayleigh-Taylor instability in their lower halves where heavy material lies above light (Low and Hundhausen 1995). Underdense flux ropes (thus, cavities) are possible, since magnetic curvature would be stabilizing in such cases, but if their density becomes too low they could be become unstable to buoyancy (Low et al. 1982). A much more twisted flux rope could result in longer field lines at the rope boundary, and thus potentially a more deeply-depleted rope center, but such a configuration would be prone to the kink instability. Thus, perhaps, if cavities were other than we measure them, they would not survive long enough to be measured.

Open Questions

- Is decreased density in cavities related to field line length, and if so, what are the dominant thermodynamic processes involved?
- Is the upper limit on cavity depletion (about a factor of two) due to stability selection effects, and if so what are the relevant instabilities involved?
- Could the decrease in density depletion towards the top of white light cavities be a stability selection effect where Rayleigh-Taylor-unstable boundaries are avoided?
- Could the narrow (prolate) elliptical aspect ratio seen in small EUV cavities result in a more strongly stabilizing magnetic curvature force, and thus enable a greater degree of depletion at their tops?

13.3.3.2 Temperature

Establishing the temperature of cavities requires a multiwavelength analysis, and ideally one which first takes into account both 3D morphology and density. In such a study, the average cavity temperature was found to be essentially equivalent to that of the surrounding streamer (Kucera et al. 2012). However, the cavity exhibited more thermal variability than the streamer, indicating that multiple temperatures were present for a given height in the plane of the sky (note that contributions to the line-of-sight intensity integral were established to be dominated by plasma actually lying within the extended tunnel of the cavity). Previous analyses have argued that cavities may be cooler than their surrounding streamer (Guhathakurta et al. 1992) or hotter (Fuller et al. 2008; Vasquez et al. 2009; Habbal et al. 2010). Thus there does not yet seem to be a non-ambiguous answer to the question as to whether the low-density plasma within cavities is typically hotter or cooler than that in the surrounding streamers (although the evidence for at least some hot material in the vicinity of cavities is compelling—see further discussion in Sects. 13.3.4 and 13.3.5).

Open Questions

- What is the temperature of the cavity relative to a surrounding streamer?
- Does it depend upon where in the cavity one looks?
- What is the nature of multithermal variation within the cavity?

13.3.4 Flows

Two factors may complicate cavity temperature analyses: substructure and flows. We will discuss substructure in Sect. 13.3.5, and in particular the hot cores, or "chewy nougats" apparent in SXR cavities. First, however, we will discuss the flows, both in the plane of the sky (POS) and along the line of sight (LOS), which may contribute to the thermal variability within cavities.

13.3.4.1 POS Flows

Even cavities in a quiescent, non-erupting state exhibit a range of interesting dynamics. High-time-cadence movies of EUV images indicate the presence of apparent swirling motions in plane of sky projection within cavities (Li et al. 2012) (Fig. 13.5, left). These flows have POS speeds in the range 5–10 $\mathrm{km\,s^{-1}}$, and can persist in the same sense of rotation for several days (Wang and Stenborg 2010). Another type of flow that has been associated with cavities involves plumes or bubbles (which may have sizes on the order of 10 mm) rising through the prominence and into the cavity (Fig. 13.6, right) (Berger et al. 2008; de Toma et al. 2008). The ascent speed of these bubbles is on the order of tens of $\mathrm{km\,s^{-1}}$ (Berger et al. 2010; Berger 2012), and simultaneous optical and EUV observations indicate temperatures that are 25–120 times hotter than the overlying prominence (Berger et al. 2011). Whether these represent temperatures within the bubbles and plumes themselves, or whether they are indicative of the background corona/cavity seen through gaps in the prominence opened up by the bubbles/plumes, remains a subject of debate (Dudík et al. 2012).

In addition, transient EUV brightenings within cavities occur in the form of "horns" in 171 Å which extend nonradially from the top of prominence observed in 304 Å (Fig. 13.6, left-top) (Schmit and Gibson 2013). By analyzing a database made up of 48 such horns, spatial and temporal correlations were found between the coronal flows within the cavity, and flows associated with cooler, prominence plasma: in general, the formation of horns in 171 Å preceded the formation of essentially co-spatial prominence extensions in 304 Å by about 30 min.

Fig. 13.5 Plane-of-sky flows trace out writhing field lines. *Left* SDO/AIA multi-wavelength observations of apparent swirling motion within cavities (*top* 16 June 2011, *bottom* 23 July 2012 (from Panasenco et al. (2014))). *Right* a non-flux-rope model (*top*) representation of magnetic field lines of filament spine (*green*), coronal loops (*blue*), and cavity (*black*) (from Panasenco et al. (2014)), and (*bottom*) sample field lines extracted from an MHD equilibrium flux rope model (Gibson and Fan 2006a) adapted from Schmit and Gibson (2014)

Variability in EUV, and especially at 171 Å, is much greater in cavities than in surrounding streamers. It may be that the multithermal nature of cavities arises from a dynamic prominence-cavity interface, which would project both cold and hot plasma into the main volume of cavity (Schmit and Gibson 2013).

MHD Interpretation of POS Flows: Tracing Magnetic Field Lines and Transferring Helicity and Mass The thermodynamics driving horn-like flows between prominence and cavity may arise from thermal nonequilibrium (Karpen 2014), and Fig. 13.6 (left) illustrates how a model of TNE-driven flows along a sheared-arcade magnetic configuration produces synthetic EUV images similar to observations. Likewise, a model utilizing TNE-driven flows along field lines of a flux rope found that the temporal and spatial correlations between prominence and cavity flows were well captured, and in particular that EUV brightenings (horns) would be followed by prominence formation (Schmit et al. 2013). As discussed in Sect. 13.3.3.1, TNE would be most likely to occur on longer, dipped field lines; for the flux rope these lie along the outer boundary (Fig. 13.4). Thus, the multithermal, variable, density-enhanced part of the magnetic structure would actually only fill a small fraction of a flux rope volume, but would project into some or all of the cavity at the limb.

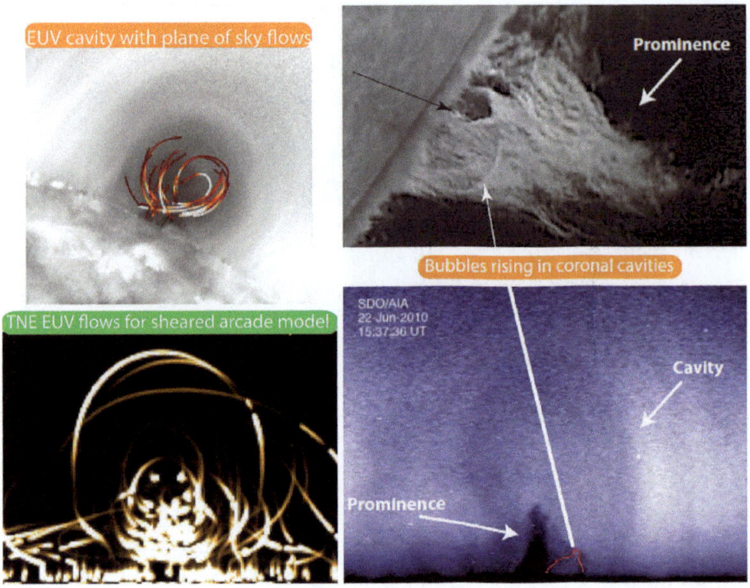

Fig. 13.6 Plane-of-sky flows connecting prominences and cavities. *Left* prominence-cavity interface flows along "horns"; (*top*) multiwavelength SDO/AIA observations, with paths of POS flows extracted (adapted from Schmit and Gibson (2013)); (*bottom*) synthesized flows driven by TNE in a sheared-arcade model (from Luna et al. (2012), reproduced by permission of the AAS). *Right* Bubble rising through prominence (Hinode/SOT) into coronal cavity (SDO/AIA) (from Berger et al. (2010), reproduced by permission of the AAS; Berger (2012); used by permission of the author)

In a low-β regime such as the corona, coronal plasma must flow along magnetic field lines (Ballester 2014). With this in mind, the swirling motions within cavities are certainly suggestive of flow along twisted magnetic field lines. One must be very cautious of projection effects, however. Indeed, analysis of these motions using multiple viewing angles implies that the motion is along strongly writhed field lines, but probably not ones that wind multiple times about an axis (Panasenco et al. 2014). The question remains, though, whether such a strongly writhed field line is part of a flux rope or not. As Fig. 13.5 (right) shows, both weakly-twisted-flux-rope and sheared-arcade models involve writhed field lines of very similar geometric appearance.

The fundamental topological difference between these models is whether the writhed field lines *wrap around an axis*. Since EUV flows light up only a portion of the magnetic field at any given moment, such observations will not generally capture both wrapping and axial lines (although see Fig. 13.7 (left) for a possible snapshot of just such a configuration). For this reason, the field-aligned flows seen in POS projection, both swirling motions and flows associated with prominence-cavity horns, are unlikely to definitively distinguish flux-rope vs. sheared-arcade models.

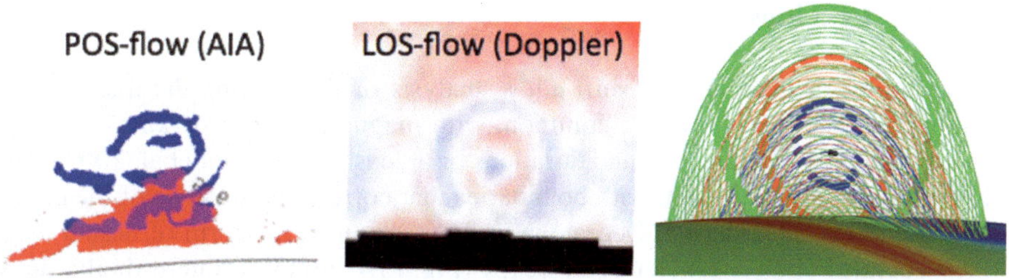

Fig. 13.7 LOS and POS flows tracing nested rings about a central axis are strong evidence for toroidal flux surfaces. *Left* Snapshot of POS flows along prominence-cavity interface horns extracted from observations of SDO/AIA analyzed in Schmit and Gibson (2013) (*blue*—171 Å, *red*—304 Å, *purple*—both) [courtesy D. Schmit]. *Middle* LOS flows measured from Doppler shift by MLSO/CoMP (adapted from Bąk-Stęślicka et al. (2013)). *Right* flux surfaces of flux-rope model described in Fan (2012) shown by colored field lines, with intersection with POS indicated by boldface colors

The rising bubbles and plumes, in conjunction with prominence condensations and other downflows within the cavity, have been proposed together as evidence of a magneto-thermal convective mass cycle in the prominence and cavity (Liu et al. 2012; Berger et al. 2012). Although previous estimates have argued that there is not enough mass in the corona to account for the mass in the prominence (Saito and Tandberg-Hanssen 1973), such arguments assume a static corona, while in fact there is likely to be a continuous input of plasma to the corona through TNE-driven flows, rising bubbles, spicules (De Pontieu et al. 2011), and reconnection-driven flows (Panasenco et al. 2014). Indeed, observations indicate that the total mass of a prominence at any given moment may be a small fraction of the mass condensing and draining through it (Liu et al. 2012).

If the bubbles seen rising through the prominence and into the cavity are associated with magnetic flux concentrations (Dudík et al. 2012), it may be that they also serve to transport small-scale magnetic twist upwards, merging with and adding to the helicity of the larger-scale fields of quiescent prominence cavities (Berger et al. 2011; Low et al. 2012a,b). Such a process could contribute to its ultimate eruption by leading up to an ideal instability or more generally to a state in which there no longer exists a stable minimum energy force-free equilibrium consistent with the increased helicity (see Fan (2014) and further discussion below in Sect. 13.4.2.2).

Open Questions

- Are rising bubbles within cavities associated with magnetic flux concentrations?

13.3.4.2 LOS Flows

Observations of Doppler line shift allow analysis of flows along the line of sight within cavities. Such flows are common, with magnitudes of 5–10 $\mathrm{km\,s^{-1}}$, length scales of tens of megameters, and durations of at least 1 h (Schmit et al. 2009). The flows are spatially coherent, with boundaries that correspond to those of the cavity or to a central substructure. Occasionally flows occur in the form of nested ring-like structures (Bąk-Stęślicka et al. 2013) (Fig. 13.7: middle). Interestingly, these nested flows may appear to be counterstreaming (alternating towards and away the viewer with radius). The Doppler measurements shown in Fig. 13.7 incorporate a subtraction of the background coronal rotation using a median filter (Tian et al. 2013), so there is some ambiguity about precisely where flows shift from towards to away in the rotating frame. Nevertheless, there are clearly strong gradients in the flow of coronal plasma within the cavity, with functional dependence on radial distance from cavity center.

MHD Interpretation of LOS Flows: Toroidal Flux Surfaces within a Flux Rope If there is a "smoking gun" observation that indicates that the cavity is a magnetic flux rope, it is probably the nested rings observed in LOS flows. When we see a coronal loop, the interpretation is that we are seeing a field line or collection of field lines along which plasma is being highlighted. In the same way, when we see nested rings of LOS flows, or other ring-like or disk-like substructure within cavities (see Sect. 13.3.5), the implication is that the underlying magnetic structure of the cavity is one in which field lines trace out concentric rings when seen in projection at the limb. Figure 13.7 shows magnetic field lines along flux surfaces, i.e., boundaries tangential to the magnetic field. It is a property *distinct to the magnetic flux rope* that its flux surfaces are nested tori, and, when centered on the POS and with axis oriented along the viewer's line of sight, the intersection of these surfaces at the limb is one of nested rings (Fig. 13.7: right).

Given a reasonably long, straight flux rope oriented along the LOS and centered on the POS, it follows that the LOS-directed component of a uniform flow will manifest as ring-like contours with a maximum at rope center, where the vector field and field-aligned flow is completely along the LOS. The fact that we do not see nested rings in every cavity must in part be due to the dependence of their observability on the orientation of the rope axis to the LOS. The cavity has a finite length, and if it is not centered on the limb the asymmetries between foreground and background can eliminate the projection of flux surfaces as rings (Gibson 2014). A large curvature of the rope axis, or angle of the axis relative to the LOS, would similarly smear the rings out.

It is less clear how the strong gradients in LOS flows arise between these rings. One possibility is that plasma-emission weighting plays a role. This could occur if a central substructure were at a temperature different enough from the rest of the cavity (see Sect. 13.3.5.1) such that it had decreased emission in the wavelengths of light where the Doppler observations were measured. Since the LOS flow measured is the integral over the line of sight, measurements at the center of the cavity would then be biased toward contributions away from the plane of the sky which, due to curvature in the rope, would sample plasma on field lines less aligned with the LOS and so contributing a reduced LOS flow. Under conditions of constant flow, the maximum LOS velocity (integrated along the line of sight) would no longer be at the center of the cavity, but would lie along a ring at some distance from that center.

However, this argument cannot explain counter-streaming flows in nested rings. For this, one requires a flow with a dependence upon radius within the rope, or equivalently that varies as a function of flux surface. If the flux rope consisted of field lines that were uninterrupted in their wrapping around nested toroidal flux surfaces (thus, ergodic), different flows for each flux surface would indeed be expected. Because the flux rope is anchored in the photosphere, however, each flux surface consists of multiple field lines (set of colored lines in Fig. 13.7). Therefore, to create nested rings with counter-flows, one would have to drive the same flow along all field lines of a particular flux surface, but a different flow along all field lines of its neighboring flux surface. If the flow is driven by differential heating at the fieldline lower boundaries, it is difficult to see why this would occur. On the other hand, it is possible that reconnection-driven flows might originate in the corona at a flux surface that represents a topological boundary (Fan 2012) (see Sect. 13.3.5.1).

Open Questions

- What drives the LOS flows seen within cavities?
- Do nested rings arise from flow excited coherently along flux surfaces, or does emission-weighting play a role?
- Are the flows truly counter-streaming (relative to the LOS) between rings?

13.3.5 Substructure

Disk-like or ring-like substructures within cavities are not limited to plasma flows. Frequent and long-lived manifestations of such substructures are also seen in EUV and SXR emission, sometimes perched atop the prominences like lollypops on their sticks (Fig. 13.8).

13.3.5.1 Chewy Nougats and Central Voids

The first substructure to be noted within cavities were the hot central cores, often referred to as "chewy nougats", that sometimes appear in SXR observations (Hudson et al. 1999) (Fig. 13.8—top). Such structures are not transient, but may

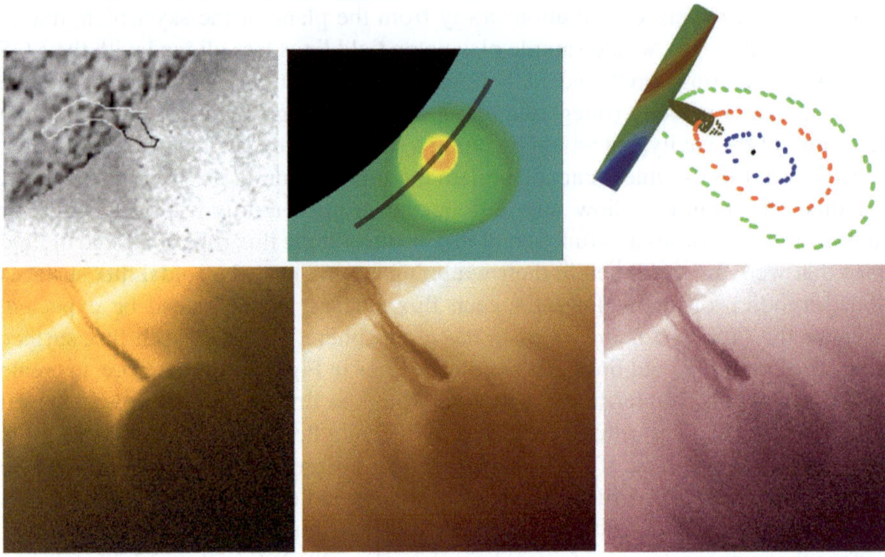

Fig. 13.8 Hot cores and central voids exist within larger cavity volume, perched atop prominences. *Top left* Yohkoh soft-Xray observation of long-lived nougat (*inverse image*); location of Hα prominence indicated by contour (from Hudson et al. (1999); reproduced by permission of the AAS). *Top middle* Temperature fit to multi-filter Hinode/XRT observations of nougat within cavity (adapted from Reeves et al. (2012)). *Bottom row* SDO/AIA images of teardrop-shaped cavity with dark disk-like void above prominence (so-called "UFO" cavity of March 11–12, 2012); *left to right*, 171, 193, and 211 Å. *Top right* intersection of flux surfaces with POS as in Fig. 13.7, but for a flux rope of greater magnetic helicity. Sheet-like localization of dipped magnetic fields (*brown*) represent likely location of prominence formation (see text)

persist for as long as the cavity is visible at the limb, and indeed reappear from rotation to rotation for months (Hudson and Schwenn 2000). Nougats may either appear in the form of a disk, or alternatively as a ring nested within the larger-scale cavity, when seen in projection at the limb. By employing the Sun's rotation to extract longitudinal information in a similar manner to that employed in determining the 3D morphology of the larger-scale cavity, the SXR nougat morphology has been modeled as a hot-walled, hollow tube lying within the cavity (Reeves et al. 2012).

There is not, generally, a corresponding signature to the SXR nougat at EUV wavelengths; that is, usually only the larger-scale cavity surrounding the nougat is apparent. However, a similar feature does sometimes occur, in the form of a central low-emission void lying above the prominence with a U-shaped lower boundary (Fig. 13.8—bottom). Unlike the nougats, these structures seem to primarily be associated with cavities that are soon to erupt (see Sect. 13.4.2.2).

MHD Interpretation of Substructure: Thermodynamic and/or Topological Interfaces As discussed in Sect. 13.3.3.1, the thermodynamic properties within a flux rope may vary significantly, depending upon field line length. An additional factor is field line curvature, which affects how dipped (or flat) a field line is. This may affect the ability of a prominence to condense and the nature of its dynamics (Karpen 2014).

Consider a magnetic flux rope with field lines that wrap between one and two times about the axis. If this rope has an arched axis, only the outermost field lines will possess a dip, while the inner, axial field lines will possess an arched geometry which will be unsuited to supporting prominence mass against gravity. This was demonstrated in Fig. 13.4, and we now consider how the location of dipped field relates to the flux rope as a whole. If we fill the dipped field lines to a prominence scale height (Aulanier et al. 2005), we obtain a sheet-like surface made up of multiple threads, lying above the neutral line (Fig. 13.9). The magnetic fields associated with these threads run essentially parallel to the underlying neutral line. This is consistent with filament and filament channel observations (Martin 2014).

There are then three distinct thermodynamic regimes likely within a flux rope:

1. The dipped field lines which lie at the outer boundary of the flux rope, but project into its POS cross section, which might be expected to be denser, dynamic, and multithermal
2. The sheet-like locus of dips of these field lines which would be the most likely place for a quiescent prominence to form (although which dips are filled at any given moment may vary, resulting in an inhomogeneous structure)
3. The central, nondipped (axial) portion of the flux rope, which, due to arched geometry and short fieldline length, would be most likely to manifest the low density of a "true" cavity

Current sheets may also form between topologically-distinct regions associated with flux ropes, for example at separatrix surfaces. One such surface (pink-purple lines of Fig. 13.9) is defined by field lines intersecting the "bald patch" of concave-up field at the photospheric polarity inversion line, and may give

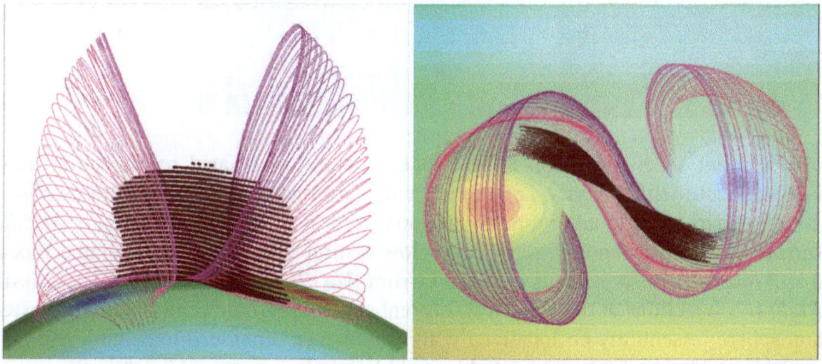

Fig. 13.9 Locus of dips in flux rope is a sheet-like structure, lying above and with fields parallel to underlying neutral line. The subset of *non-arched field lines* within the rope extend inward from the outer boundary of the rope (set of *purple/pink field lines*). *Brown dots* trace out points where the field is dipped relative to the solar radial coordinate, up to a prominence scale height (adapted from Gibson and Fan (2006a))

rise to sigmoid-shaped reconnecting field lines at its outer boundary (Titov and Demoulin 1999; Gibson et al. 2002). We can generalize to boundaries where strong gradients in field line length have functionally similar consequences as true magnetic discontinuities—so-called quasi-separatrix layers, or QSLs (Démoulin et al. 1996; Fan 2014). A QSL can exist at the center of a flux rope, as the upward stretching of the flux rope in response to ongoing helicity/flux input leads to a Hyperbolic Flux Tube (HFT) topology (Titov 2007). Reconnections occurring at the top of this HFT result in a central bundle of recently-reconnected low-density, high-temperature field lines lying above the dipped field lines of a prominence (Fig. 13.10). This has been proposed as an explanation for the chewy nougat and low density central structures within cavities, at least for pre-eruption observations (Fan 2012). We will discuss this further in Sect. 13.4.2.2.

Lollypop-like structures within cavities thus may arise as a natural consequence of these thermodynamic and topological divisions. The prominence, being associated with the sheet of dipped field, would lie below the central, arched field of the rope. This central region of the field would be thermodynamically, and potentially topologically, separated from the rest of the rope. It would thus be prone to reconnection at its outer boundary, leading to the hot ring or disk-like structure of the nougat above the prominence. When the rope axis is at its most arched, perhaps in the days/hours leading up to eruption (Sect. 13.4.2), this central portion will stretch above the prominence and may appear as a clear, central void.

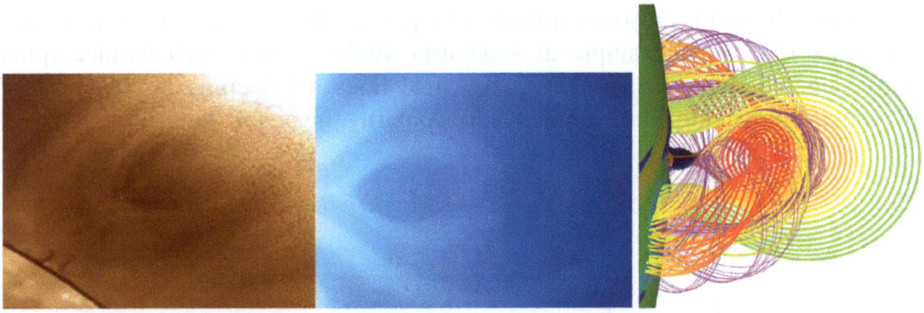

Fig. 13.10 Current sheet formation at flux rope base and associated reconnection may explain hot, low density core within cavity. *Left* central void surrounded by U-shaped horns observed atop prominence in hours prior to an eruption (case analyzed by Régnier et al. (2011)). *Middle* current sheets surrounding and extending vertically below central void within a simulated flux rope at a late (but still quasistatic) stage of its evolution (from Fan (2012); reproduced by permission of the AAS). *Right* reconnections at the top of this current sheet (*dark blue* central structure extending up from photosphere, essentially co-localized with prominence dips) lead to heating and flows along central part of rope (temperature indicated by field line color, *green-cold, red-hot*)

Open Questions

- Nougats are known to exist quiescently; is the same true for dark central voids surrounded by U-shaped horns, or do these only occur in the hours leading up to an eruption?
- Does the formation of an HFT topology inevitably lead to an eruption within a day or two?

13.3.6 Polarimetry

We have focused so far on coronal plasma observations to gain information about cavity magnetic structure. This approach is a good first step, and justified in a magnetically-dominated regime where the plasma acts as a tracer of magnetic structure. However, only part of the field is generally traced out, and the approach is limited by necessary assumptions about the nature of the sub-regions of field that are so traced (are they field lines? dipped-field? current sheets? QSLs?) and the physical mechanisms associated with their visibility (heating? cooling? flows?). Therefore, it is important to directly measure magnetic fields within the cavity. Such measurements are possible using Stokes polarimetry of forbidden infrared (IR) lines where the linear polarization is dominated by the Hanle effect in the saturated regime (see Rachmeler et al. (2013); also Lopez-Ariste (2014) for a discussion of similar techniques used in the measurement of magnetic fields of prominences).

The fraction of linearly-polarized light ($L/I = \sqrt{Q^2 + U^2}/I$, where Q and U are Stokes vectors) has turned out to be a particularly useful diagnostic for coronal cavities. It provides a measure of the direction of the magnetic field: linearly-polarized light is strongest where B lies in the plane of the sky, and goes to zero when B lies along the line of sight. In addition, L/I vanishes when the magnetic field orientation lies at the Van Vleck angle of $54.7\,^\circ$ to the solar radial. As the linear polarization vector crosses this critical angle, its orientation undergoes a $90\,^\circ$ rotation introducing an ambiguity in addition to the more standard $180\,^\circ$ one. The result is a tendency for linear polarization vectors to manifest with a radial orientation (Fig. 13.11). However, the location of nulls in L/I can be used to indicate LOS-directed field and/or Van-Vleck crossings.

The first cavity studied using IR polarimetry was from 2005, and was a large, but not PCF cavity. The linear polarization associated with the cavity was in the form of a bright ring, with a dark core and surrounding dark ring (Fig. 13.11 top). Over the past few years, synoptic observations have been available and have indicated that a much more common linear polarization signal associated with cavities, in particular PCF cavities, is a structure akin to that of a rabbit's head ("lagomorph")

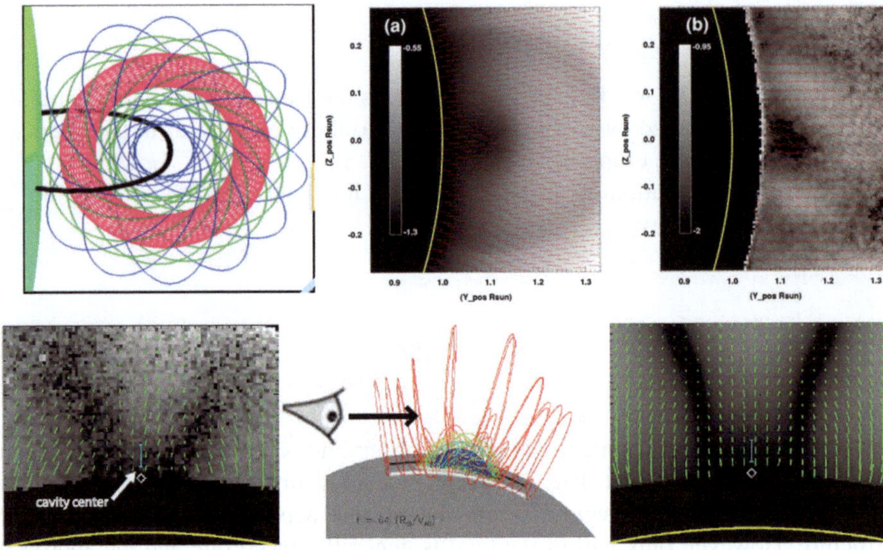

Fig. 13.11 Linear polarization measurements of cavities imply non-potential fields. *Top* toroidal (spheromak-like) flux-rope magnetic model (*left*), forward-modeled to produce synthetic L/I (*middle*) with bright-ring-like structure similar to that observed within a (non-PCF) cavity by the CoMP telescope (*right*) (adapted from Dove et al. (2011)). Red linear polarization vectors are essentially radial for both model and data. *Bottom* CoMP L/I observation of lagomorph (rabbit-head) shaped feature within a PCF cavity (*left*). Applying a flux-rope model (*middle*), the LOS-integration results in a clear L/I lagomorph (*right*). Linear polarization vectors (shown here in *green*) for model and data show similar deflection from radial at the boundary of the lagomorph (adapted from Bąk-Stęślicka et al. (2013))

(Fig. 13.11 bottom) (Bąk-Stęślicka et al. 2013). Linear-polarization lagomorphs generally scale with cavity size (Bąk-Stęślicka et al. 2014).

MHD Interpretation of Lagomorphs: Axial Field Surrounded by Poloidal Field Because of the complexities intrinsic to spectropolarimetry in combination with line-of-sight integration effects, forward modeling represents a useful means of interpreting observations. Using a quantum-electrodynamical formulation to synthesize coronal Stokes parameters (Casini and Judge 1999), the general sensitivity of L/I to the presence of coronal currents has been demonstrated (Judge et al. 2006). Moreover, forward-modeled L/I for a flux rope with a toroidal (spheromak-like) magnetic topology (Gibson and Low 1998) was shown to result in the type of nested rings observed in the 2005 cavity (Dove et al. 2011) (Fig. 13.11).

We therefore are led to consider what L/I structure would result from a simpler, cylindrical flux rope. Figure 13.11 (bottom) illustrates that, even considering LOS-integration effects, a lagomorph is precisely what we should expect to observe. A simple arcade of magnetic field lines, perpendicular to and viewed along the neutral line, will result in a V-shape because of the Van Vleck inversion where the magnetic field is oriented $54.7°$ to the local vertical (solar radial) direction.

A flux rope oriented along the LOS will possess an axial component that will distort the location of the Van Vleck nulls, and lead to vanishing L/I at its axis. LOS-integration of a curved flux rope will smear out these nulls to some extent, but if the rope is sufficiently long and straight (which is likely if it manifests as a distinct cavity), the result will be a lagomorph (Bąk-Stęślicka et al. 2013; Rachmeler et al. 2013).

The magnetic flux rope model is thus consistent with the observed L/I lagomorphs, but is it unique in this consistency? In particular, how would a sheared-arcade configuration, in which field lines exhibit writhe but do not wrap around an axis, appear? Forward-modeling (Rachmeler et al. 2013) shows that LOS-integrated L/I in the upper portions of the flux-rope and sheared-arcade models are very similar, assuming both models are surrounded by a simple arcade that produces the V-shaped rabbit's "ears". Since both models possess an axial component of field beneath this arcade, they also predict more parallel Van-Vleck inversions outlining the rabbit's "head". The flux-rope model however generally predicts a central darkening, since the axial field peaks at rope center. On the other hand, the sheared-arcade model has axial field which extends down to the photosphere. Therefore, disambiguation should be possible using observations of the polarization vector beneath the cavity center, or of the radial profile of circular polarization (Stokes V), which is proportionate to the LOS magnetic field (Rachmeler et al. 2013). However, telescopes with lower occulting disks and larger light-gathering capacity than currently available may be required for such a study.

Open Questions

- How does circular polarization (Stokes V) vary within the cavity?
- What is the orientation of linear polarization below cavity center?

13.4 Erupting Cavities and their Precursors and Predictors

Coronal mass ejections (CMEs) often exhibit a three-part structure of leading edge, following cavity, and entrained prominence core (Illing and Hundhausen 1986; Webb 2014). Such CMEs are generally pointed to as evidence for "flux rope CMEs" (Gibson and Low 1998; Dere et al. 1999; Zhang and Low 2005). Controversy remains as to whether the flux rope structure existed prior to eruption, or was formed during eruption (Fan 2014; Gopalswamy 2014). In many cases, the CME source region is low-lying, or surrounded by complex structures, making it difficult to analyze the pre-eruption configuration using observations as described in Sect. 13.3.

Cases where a clearly pre-existing cavity erupts (Fig. 13.12) are useful for studying this question in a straightforward fashion, at least as it pertains to a subset of CMEs.

13.4.1 Erupting Cavities

Figure 13.12 shows a cavity which was visible for at least 48 h before eruption and essentially unchanging for at least 24 h before signs of activation. It is an example of what we refer to as a "bodily-erupting" cavity, in which the precursor cavity morphology can be tracked throughout the eruption. Indeed, bodily-erupting cavities have been tracked from pre-eruption through the corona and out into the heliosphere (DeForest et al. 2013).

In the survey of > 100 EUV cavities (Forland et al. 2013) described in Sect. 13.3, eruptions were witnessed in approximately a third of the cavities monitored, and bodily-erupting cavities were evident for about a quarter of all cavities. The white-light survey of cavities produced similar results (Gibson et al. 2006)—roughly a third were observed to erupt within 48 h (24 out of 74 cases). Establishing which of these white-light cavities erupted bodily required monitoring the eruption onset.

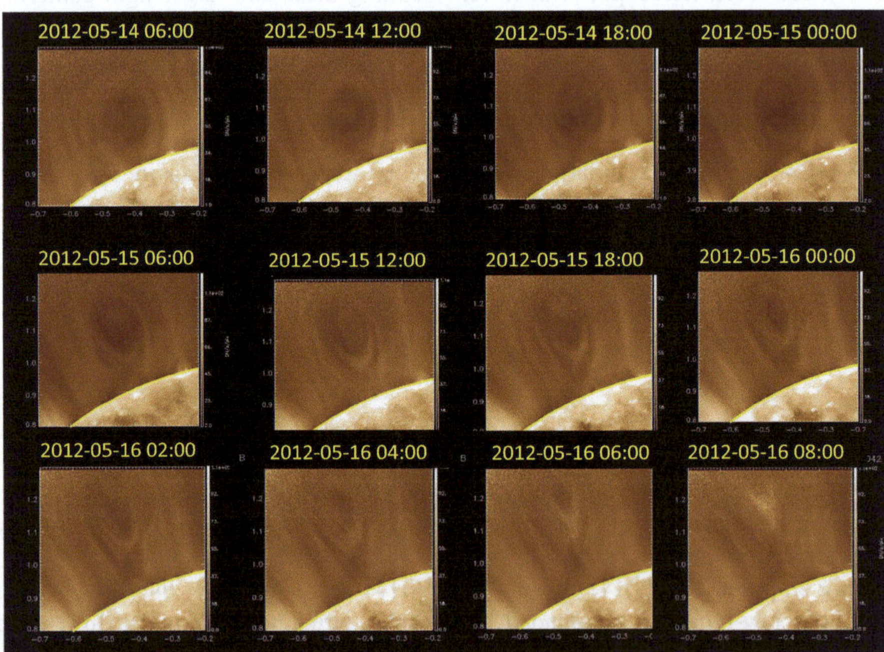

Fig. 13.12 A long-lived cavity that eventually erupts, and exhibits a slow "activation" phase in the hours leading up to eruption. SDO/AIA 193 Å observations

Due to the limited observing time window—a duty cycle that sampled about 1/4 of the hours in a given day, and was further reduced by weather—only four could be so monitored, and all of these were observed to bodily erupt. Considering the finite length of cavities and the fact that they can only be observed at the solar limbs, a rough extrapolation implies that all cavities eventually erupt, and the majority of these bodily erupt.

Further insight can be gained from a study (Maričić et al. 2009) which considered 18 3-part CMEs in which a leading edge (LE) and erupting prominence (EP) were visible from pre-eruption onwards. Note that for many of the cases (which included active region related eruptions) no pre-existing quiescent cavity could be distinguished: in such cases the appearance of a bright loop during the hours before an eruption was sufficient to establish the LE. The acceleration properties of the eruptions were analyzed, and bodily eruption quantified by tracking the kinematics of the LE vs. the EP. In the majority of the cases studied, the LE and EP acceleration began almost simultaneously (within 20 min). However, in two events the EP acceleration began considerably earlier than the LE acceleration ($>$ 40 min) and in two events the LE acceleration started $>$ 50 min before the EP acceleration.

Another study tracked the eruption of a pre-existing cavity, and found that although it clearly erupted, its associated prominence did not appear to escape with it (Liu et al. 2007). This was interpreted as a partial eruption, akin to the phenomenon where prominences seem to split in two during a CME—one part escaping, one part falling back to the Sun (Gibson and Fan 2006b).

MHD Interpretation of Erupting Cavities: Destabilization and Loss of Equilibrium As discussed in Fan (2014), CMEs may be triggered by an ideal process (e.g., the kink and torus instabilities), or by a reconnection-driven process (e.g., the breakout model). In the case of a reconnection-driven trigger, no pre-existing flux rope is required as it may form as a consequence of the eruption (although a breakout-type reconnection can also operate above a pre-existing flux rope (Török et al. 2011)). For cases in which the prominence and leading edge eruption are not synchronized, thus, no bodily eruption, it seems plausible that the rope indeed forms or at least is greatly transformed during eruption. However, for the majority of cases, a cavity or LE and following prominence exist prior to the eruption and are synchronized in their motion from its beginning. These observations provide compelling evidence for pre-existing flux ropes that are destabilized and undergo loss of equilibrium.

Open Questions

- To what extent does the behavior exhibited by bodily-erupting cavities extend to other CMEs?

13.4.2 Precursors and Predictors of Eruption

Figure 13.12 demonstrates some characteristic behavior of cavities in the hours leading up to eruption. This may be classified as "cavity activation", and is likely to be related to similar manifestations in quiescent prominences in the hours prior to their eruption (Martin 2014) as well as pre-eruption behavior of so-called "streamer blowouts" (Webb 2014). In particular, cavities appear to slowly increase in height, to narrow and become more teardrop-shaped, and to exhibit more sharply defined substructure prior to eruption. Even before activation, there is evidence that quasistatic evolution results in observational characteristics that may be used as predictors of impending eruption.

13.4.2.1 Cavity Height and Slow Rise

The hours leading up to the impulsive phase of a cavity eruption are often marked by a slow rise of the cavity and its associated prominence. An example of rise speeds for the two phases of a particular cavity eruption was found to be 0.6 $\mathrm{km\,s^{-1}}$ for the slow-rise phase and 25 $\mathrm{km\,s^{-1}}$ (average) for the subsequent impulsive phase (Régnier et al. 2011).

Before any measurable slow-rise phase, however, the height of the cavity may provide a clue to its likelihood for eruption. An absolute upper limit to cavity height appears to be approximately 1.6 $R_\odot$ (0.6 $R_\odot$ above the solar surface), beyond which no non-erupting cavities were observed in white light (Gibson et al. 2006). Due to selection effects discussed in Sect. 13.3.1, EUV cavities in general lie below this height (median value 1.2 solar radii). Nevertheless, they exhibit a marked tendency for higher cavities to be more likely to erupt: the average cavity center height was 25 % higher for cavities that erupted (measured prior to slow rise) than for a baseline set of non-eruptive cavities (Forland et al. 2013). This is no doubt related to observations indicating an upper limit for prominence height (Munro et al. 1979; Filippov and Den 2001; Liu and Schuck 2012).

13.4.2.2 Morphology and Substructure

The most obvious sign of cavity activation in Fig. 13.12 is the bright substructure that forms by 06:00, initially as U-shaped "horns" above the prominence (see Sect. 13.3.4), and progressing to a very narrow ellipse within the broader cavity envelope. The entire cavity likewise narrows during the slow-rise phase of the eruption.

This behavior is mirrored in the characteristics of pre-activation but soon-to-erupt cavities relative to the baseline, non-eruptive cavities. In Sect. 13.3.2, we discussed the tendency towards prolateness, or narrowness of the elliptical cross sections of cavities. This tendency is even more pronounced for the subset of cavities that erupt than for the baseline, non-eruptive cavities. In fact, the pre-activation cavities may be better characterized as possessing a teardrop, rather than elliptical, shape. Categorizing the EUV cavities by morphology, the likelihood of eruption of teardrop-shaped cavities was 68 %, as compared to 23 % for more elliptical cavities, and 10 % for cavities that were best described as semicircular (Forland et al. 2013). Due to the presence of an occulting disk, the full shape of white-light cavities is not generally measured, but a quality referred to as "necking" can be noted when cavities have narrower bases than tops. In the white-light cavity survey, 10/10 cases of cavities which erupted within 24 h exhibited necking, vs 25/99 of the entire sample (Gibson et al. 2006).

MHD Interpretation of Precursors and Predictors: Topological Changes Leading up to Ideal Instability The absolute upper limit on cavity height, as manifested by white light cavities, is undoubtedly related to the upper limit on the spatial scale of closed field possible in a corona before it ultimately opens into the solar wind, and may represent a limit beyond which cavities are gravitationally unstable. The smaller EUV cavities, however, lie well below this global limit, so the association of eruptions with increased cavity height may require interpretation in terms of purely magnetic forces. Indeed, the height of pre-CME cavities taken in conjunction with their narrowness and teardrop shape is consistent with the formation of a current sheet at the base of the rope (e.g., Fig. 13.10, middle), presenting an intriguing clue as to how magnetic evolution might lead up to eruption.

The cavity may be thought of potentially having three stages in its evolution. The first stage is the long-lasting period of time where a cavity exists essentially quiescently, without erupting. This stage may nevertheless represent a sequence of quasistatic equilibria, in which minimum energy states are continually updated as the lower boundary inputs magnetic helicity through flows or flux emergence over the course of days and weeks. The transition to the second stage occurs when a current sheet forms beneath the cavity and the topology changes from a flux rope grazing the photosphere (bald-patch) to one with an HFT beneath it (QSL) (see discussion in Sect. 13.3.5.1, and in Fan (2014)). At this point evolution may continue in a quasistatic fashion, but a fuse has been lit and the tether-cutting reconnections at the current sheet increase twist at the core of the flux rope (Fig. 13.10, right), creating substructure in the cavity and a slow rise of the rope axis. The third and final stage is dynamic, with eruption possibly triggered by the kink or torus instability brought on by increasing twist and/or rope height. Alternatively, the slow rise of the rope may push it into topologically distinct, overlying or adjacent magnetic fields, resulting in breakout-type reconnection. This could also occur in a manner that skips the middle stage of evolution, if, for example, the evolution of surrounding fields forces reconnections and drives a "sympathetic" eruption (Török et al. 2011).

Open Questions

- Can we establish which CMEs are predominantly ideal-instability-driven, and which are reconnection-driven, and whether certain regions (PCFs, active regions) are more likely to be one than the other?
- As we begin to measure the coronal fields themselves, is there a property: helicity, free energy, complexity, topology—that we could measure to tell us that eruption is inevitable?

Conclusions

Cavities hold unique clues to understanding the nature of pre-CME equilibria and the mechanisms that trigger their loss. They represent the bulk of the combined erupting prominence-cavity volume, so it is their magnetic structure that may map best to the magnetic cloud passing the Earth (Gopalswamy 2014; Lugaz 2014). In this paper we have presented a set of observations that offer strong physical insights into the nature of this magnetic structure.

Cavities are *ubiquitous*—at least for the longitudinally-extended filaments (e.g., PCFs) that our observations are necessarily biased towards. They have a *croissant-like morphology*, have *low density* (but not as low as a coronal hole), and are *multithermal*, at least in projection. Flows are observed that *spatially and temporally link prominence and cavity*. These flows sometimes take the form of *swirling motions* in the plane of the sky, and as *nested rings* as measured in LOS velocity. *Disk-like or ring-like sub-structure* is often seen at the center of the cavity and lying above the prominence like a lollypop on a stick. Finally, recent Stokes polarimetric observations of coronal magnetic field provide direct evidence of field oriented parallel to the underlying neutral line, *at heights well above the prominence, and corresponding to the height of the cavity center*.

Cavities erupt as CMEs, and the majority appear to exhibit *bodily eruption*. Cavities, like filaments, may be activated prior to eruption and show *slow rise, narrowing, and enhanced substructure*. Even before activation, *cavity height and teardrop morphology are good predictors of impending eruption*.

Any model of the magnetic structure of a cavity must satisfy all of these observational constraints. We have argued in this paper that a magnetic flux rope plausibly does so. In particular it is difficult to see how observations such as the nested bullseye LOS flows could be explained without invoking the toroidal flux surfaces of a flux rope. However, we have also highlighted open questions throughout our review, chief among them concerning the degree to which what we have learned about PCF cavities extends to all prominences

(continued)

and CME source regions. Mysteries remain, but cavities continue to yield intriguing glimpses into the hearts of CMEs, from pre-event out into the heliosphere.

Acknowledgements The National Center for Atmospheric Research is sponsored by the National Science Foundation. AIA data courtesy of NASA/SDO and the AIA, EVE, and HMI science teams. $H\alpha$ image courtesy of Big Bear Solar Observatory/New Jersey institute of Technology. Hinode is a Japanese mission developed and launched by ISAS/JAXA, with NAOJ as domestic partner and NASA and STFC (UK) as international partners. It is operated by these agencies in co-operation with ESA and NSC (Norway). Much of the work presented here directly relates to, or benefited greatly from research undertaken by the International Space Science Institute (ISSI) international teams on coronal cavities (2008–2010) and coronal magnetism (2013–2014). I am indebted to all of the members of both of these teams, particularly Urszula Bak-Steslicka, Terry Kucera, Laurel Rachmeler, Kathy Reeves, and Don Schmit. In addition, I thank Tom Berger, Giuliana de Toma, Yuhong Fan, Blake Forland, Jim Fuller, Judy Karpen, Jim Klimchuk, Olga Panasenco, Marco Velli, and especially B. C. Low for many helpful discussions.

References

Aulanier, G., Démoulin, P., & Grappin, R. (2005). Equilibrium and observational properties of line-tied twisted flux tubes. *Astronomy and Astrophysics*, *430*, 1067.

Ballester, J. L. (2014). Magnetism and dynamics of prominences: MHD waves. In J.-C. Vial & O. Engvold (Eds.), *Solar prominences, ASSL* (Vol. 415, pp. 257–294). New York: Springer.

Bąk-Stęślicka, U., Gibson, S. E., Fan, Y., Bethge, C., Forland, B., & Rachmeler, L. A. (2013). Twisted magnetic structure of solar prominence cavities: New observational signature revealed by coronal magnetometry. *Astrophysical Journal*, *770*, 28. Arxiv 13047388.

Bąk-Stęślicka, U., Gibson, S. E., Fan, Y., Bethge, C., Forland, B., & Rachmeler, L. A. (2014). The spatial relation between EUV cavities and linear polarization signatures. In B. Schmieder, J.-M. Malherbe & S. T. Wu (Eds.), *IAU Symposium* (Vol. 300, pp. 395–396). DOI 10.1017/S1743921313011253

Berger, M. A. (1984). Rigorous new limits on magnetic helicity dissipation in the solar corona. *Geophysical and Astrophysical Fluid Dynamics*, *30*, 79–104. DOI 10.1080/03091928408210078.

Berger, T. (2012). Quiescent prominence dynamics: An update on Hinode/SOT discoveries. In T. Sekii, T. Watanabe & T. Sakurai (Eds.) *Hinode-3: The 3rd Hinode science meeting. Astronomical Society of the Pacific conference series* (Vol. 454, p. 79).

Berger, T., Testa, P., Hillier, A., Boerner, P., Low, B. C., Shibata, K., et al. (2011). Magneto-thermal convection in solar prominences. *Nature*, *472*, 197–200. DOI 10.1038/nature09925.

Berger, T. E., Shine, R. A., Slater, G. L., Tarbell, T. D., Title, A. M., Okamoto, T. J., et al. (2008). Hinode SOT observations of solar quiescent prominence dynamics. *Astrophysical Journal Letters*, *676*, L89–L92. DOI 10.1086/587171.

Berger, T. E., Slater, G., Hurlburt, N., Shine, R., Tarbell, T., Title, A., et al. (2010). Quiescent prominence dynamics observed with the Hinode solar optical telescope. I. Turbulent upflow plumes. *Astrophysical Journal*, *716*, 1288–1307. DOI 10.1088/0004-637X/716/2/1288.

Berger, T. E., Liu, W., & Low, B. C. (2012). SDO/AIA detection of solar prominence formation within a coronal cavity. *Astrophysical Journal Letters*, *758*, L37. DOI 10.1088/2041-8205/758/2/L37, 1208.3431.

Casini, R., & Judge, P. G. (1999). Spectral lines for polarization measurements of the coronal magnetic field. II. Consistent treatment of the Stokes vector for magnetic-dipole transitions. *Astrophysical Journal, 522*, 524. DOI 10.1086/307629.

de Toma, G., Casini, R., Burkepile, J. T., & Low, B. C. (2008). Rise of a dark bubble through a quiescent prominence. *Astrophysical Journal Letters, 687*, L123–L126. DOI 10.1086/593326.

De Pontieu, B., McIntosh, S. W., Carlsson, M., Hansteen, V. H., Tarbell, T. D., Boerner, P., et al. (2011). The origins of hot plasma in the solar corona. *Science, 331*, 55. DOI 10.1126/science.1197738.

DeForest, C. E., Howard, T. A., & McComas, D. J. (2013). Tracking coronal features from the low corona to earth: A quantitative analysis of the 2008-Dec-12 CME. *Astrophysical Journal, 769*, 43. DOI 10.1088/0004-637X/769/1/43

Démoulin, P., Priest, E. R., & Lonie, D. P. (1996). Three-dimensional magnetic reconnection without null points 2. Application to twisted flux tubes. *Journal of Geophysical Research, 101*, 7631.

Dere, K. P., Brueckner, G. E., Howard, R. A., Michels, D. J., & Delaboudiniere, J. P. (1999). LASCO and EIT observations of helical structure in coronal mass ejections. *Astrophysical Journal, 492*, 804.

Dove, J., Gibson, S., Rachmeler, L. A., Tomczyk, S., & Judge, P. (2011). A ring of polarized light: Evidence for twisted coronal magnetism in cavities. *Astrophysical Journal, 731*, 1.

Dudík, J., Aulanier, G., Schmieder, B., Zapiór, M., & Heinzel, P. (2012). Magnetic topology of bubbles in quiescent prominences. *Astrophysical Journal, 761*, 9. DOI 10.1088/0004-637X/761/1/9.

Engvold, O. (1989). In E. R. Priest (Ed.), *Dynamics and structures of quiescent prominences* (p. 47). Dordrecht: D. Reidel Publishing Company.

Engvold, O. (2014). Description and classification of prominences. In J.-C. Vial & O. Engvold (Eds.), *Solar prominences, ASSL* (Vol. 415, pp. 31–60). New York: Springer.

Fan, Y. (2012). Thermal signatures of tether-cutting reconnections in pre-eruption coronal flux ropes: Hot central voids in coronal cavities. *Astrophysical Journal, 758*, 60. DOI 10.1088/0004-637X/758/1/60, 1205.1028.

Fan, Y. (2014). Magnetism and dynamics of prominences: MHD equilibria and triggers for eruption. In J.-C. Vial & O. Engvold (Eds.), *Solar prominences, ASSL* (Vol. 415, pp. 295–320). New York: Springer.

Filippov, B. P., & Den, O. G. (2001). A critical height of quiescent prominences before eruption. *Journal of Geophysical Research, 106*, 25,177–25,184. DOI 10.1029/2000JA004002.

Forland, B. C., Gibson, S. E., Dove, J. B., Rachmeler, L. A., & Fan, Y. (2013). Coronal cavity survey: Morphological clues to eruptive magnetic topologies. *Solar Physics, 288*, 603–615. DOI 10.1007/s11207-013-0361-1

Fuller, J., & Gibson, S. E. (2009). A survey of coronal cavity density profiles. *Astrophysical Journal, 700*, 1205.

Fuller, J., Gibson, S. E., de Toma, G., & Fan, Y. (2008). Observing the unobservable? Modeling coronal cavity density. *Astrophysical Journal, 678*, 515.

Gibson, S. E. (2014). Magnetism and the invisible man: The mysteries of coronal cavities. In B. Schmieder, J.-M. Malherbe, & S. T. Wu (Eds.), *IAU Symposium* (Vol. 300, pp. 139–146). DOI 10.1017/S1743921313010879

Gibson, S. E., & Fan, Y. (2006a). Coronal prominence structure and dynamics: A magnetic flux rope interpretation. *Journal of Geophysical Research, 111*. DOI 10.1029/2006JA011871.

Gibson, S. E., & Fan, Y. (2006b). The partial expulsion of a magnetic flux rope. *Astrophysical Journal Letters, 637*, 65.

Gibson, S. E., & Low, B. C. (1998). A time-dependent three-dimensional magnetohydrodynamic model of the coronal mass ejection. *Astrophysical Journal, 493*, 460.

Gibson, S. E., Fletcher, L., Del Zanna, G., Pike, C. D., Mason, H. E., Mandrini, C. H., et al. (2002). The structure and evolution of a sigmoidal active region. *Astrophysical Journal, 574*, 265.

Gibson, S. E., Foster, D., Burkepile, J., de Toma, G., & Stanger, A. (2006) The calm before the storm: The link between quiescent cavities and cmes. *Astrophysical Journal, 641*, 590.

Gibson, S. E., Kucera, T. A., Rastawicki, D., Dove, J., de Toma, G., Hao, J., et al. (2010). Three-dimensional morphology of a coronal prominence cavity. *Astrophysical Journal, 723*, 1133.
Gopalswamy, N. (2014). Eruptive prominences and their impact on the earth: The dynamic phenomenon. In J.-C. Vial & O. Engvold (Eds.), *Solar prominences, ASSL* (Vol. 415, pp. 379–408). New York: Springer.
Guhathakurta, M., Rottman, G. J., Fisher, R. R., Orrall, F. Q., & Altrock, R. C. (1992). Coronal density and temperature structure from coordinated observations associated with the total solar eclipse of 1988 March 18. *Astrophysical Journal, 388*, 633.
Habbal, S. R., Druckmueller, M., Morgan, H., Scholl, I., Rusin, V., Daw, A., et al. (2010). Total solar eclipse observations of hot prominence shrouds. *Astrophysical Journal, 719*, 1362.
Hudson, H. S., & Schwenn, R. (2000). Hot cores in coronal filament cavities. *Advances in Space Research, 25*, 1859.
Hudson, H. S., Acton, L. W., Harvey, K. A., & McKenzie, D. M. (1999). A stable filament cavity with a hot core. *Astrophysical Journal, 513*, 83.
van de Hulst, H. C. (1950). The electron density of the solar corona. *Bulletin of the Astronomical Institutes of the Netherlands, 11*, 135.
Illing, R. M., & Hundhausen, A. J. (1986). Disruption of a coronal streamer by an eruptive prominence and coronal mass ejection. *Journal of Geophysical Research, 91*, 10,951.
Janse, A. M., Low, B. C., & Parker, E. N. (2010). Topological complexity and tangential discontinuity in magnetic fields. *Physics of Plasmas, 17*(9), 092,901. DOI 10.1063/1.3474943.
Judge, P. G., Low, B. C., & Casini, R. (2006). Spectral lines for polarization measurements of the coronal magnetic field. iv. Stokes signals in current-carrying fields. *Astrophysical Journal, 651*, 1229.
Karpen, J. (2014). Plasma structure and dynamics. In J.-C. Vial & O. Engvold (Eds.), *Solar prominences, ASSL* (Vol. 415, pp. 235–255). New York: Springer.
Klimchuk, J. A., Karpen, J. T., & Antiochos, S. K. (2010). Can thermal nonequilibrium explain coronal loops? *Astrophysical Journal, 714*, 1239–1248. DOI 10.1088/0004-637X/714/2/1239, 0912.0953.
Krall, J., & Chen, J. (2005). Density structure of a preeruption coronal flux rope. *Astrophysical Journal, 628*, 1046–1055. DOI 10.1086/430810.
Kucera, T. (2014). Derivations and observations of prominence bulk motions and mass. In J.-C. Vial & O. Engvold (Eds.), *Solar prominences, ASSL* (Vol. 415, pp. 77–99). New York: Springer.
Kucera, T. A., Gibson, S. E., Schmit, D. J., Landi, E., & Tripathi, D. (2012). Temperature and EUV intensity in a coronal prominence cavity. *Astrophysical Journal, 757*, 73.
Kundu, M. R., Fuerst, E., Hirth, W., & Butz, M. (1978). Multifrequency observations of solar filaments at centimeter wavelengths. *Astronomy and Astrophysics, 62*, 431.
Li, X., Morgan, H., Leonard, D., & Jeska, L. (2012). A solar tornado observed by AIA/SDO: Rotational flow and evolution of magnetic helicity in a prominence and cavity. *Astrophysical Journal Letters, 752*, L22. DOI 10.1088/2041-8205/752/2/L22.
Liu, R., Alexander, D., & Gilbert, H. R. (2007). Kink-induced catastrophe in a coronal eruption. *Astrophysical Journal, 661*, 1260.
Liu, W., Berger, T. E., & Low, B. C. (2012). First SDO/AIA observation of solar prominence formation following an eruption: Magnetic dips and sustained condensation and drainage. *Astrophysical Journal, 745*, L21. DOI 10.1088/2041-8205/745/2/L21, 1201.0811.
Liu, Y., & Schuck, P. W. (2012). Magnetic energy and helicity in two emerging active regions in the sun. *Astrophysical Journal, 761*, 105. DOI 10.1088/0004-637X/761/2/105.
Lopez-Ariste, A. (2014). Magnetometry of prominences. In J.-C. Vial & O. Engvold (Eds.), *Solar prominences, ASSL* (Vol. 415, pp. 177–202). New York: Springer.
Low, B. C. (1994). Magnetohydrodynamic processes in the solar corona: Flares, coronal mass ejections, and magnetic helicity. *Physics of Plasmas, 1*, 1684–1690. DOI 10.1063/1.870671.
Low, B. C., & Hundhausen, J. R. (1995). Magnetostatic structures of the solar corona. ii. The magnetic topology of quiescent prominences. *Astrophysical Journal, 443*, 818.

Low, B. C., Munro, R. H., & Fisher, R. R. (1982). The initiation of a coronal transient. *Astrophysical Journal*, *254*, 335–342. DOI 10.1086/159737.
Low, B. C., Berger, T., Casini, R., & Liu, W. (2012a). The hydromagnetic interior of a solar quiescent prominence. I. Coupling between force balance and steady energy transport. *Astrophysical Journal*, *755*, 34. DOI 10.1088/0004-637X/755/1/34, 1203.1056.
Low, B. C., Liu, W., Berger, T., & Casini, R. (2012b). The hydromagnetic interior of a solar quiescent prominence. II. Magnetic discontinuities and cross-field mass transport. *Astrophysical Journal*, *757*, 21. DOI 10.1088/0004-637X/757/1/21.
Lugaz, N. (2014). Eruptive prominences and their impact on the Earth and our life. In J.-C. Vial & O. Engvold (Eds.), *Solar prominences, ASSL* (Vol. 415, pp. 431–451). New York: Springer.
Luna. M., Karpen, J. T., & DeVore, C. R. (2012). Formation and evolution of a multi-threaded solar prominence. *Astrophysical Journal*, *746*, 30. DOI 10.1088/0004-637X/746/1/30, 1201.3559.
Mackay, D. (2014). Formation and large-scale patterns of filament channels and filaments. In J.-C. Vial & O. Engvold (Eds.), *Solar prominences, ASSL* (Vol. 415, pp. 353–378). New York: Springer.
Maričić, D., Vršnak, B., & Rosa, D. (2009). Relative kinematics of the leading edge and the prominence in coronal mass ejections. *Solar Physics*, *260*, 177.
Marqué, C. (2004). Radiometric observations of quiescent filament cavities. *Astrophysical Journal*, *602*, 1037.
Martin, S. (2014). The magnetic field structure of prominences from direct and indirect observations. In J.-C. Vial & O. Engvold (Eds.), *Solar prominences, ASSL* (Vol. 415, pp. 203–233). New York: Springer.
McCabe, M. K., & Mickey, D. L. (1981). The He I 10,830 a chromosphere and filament associated structures. *Solar Physics*, *73*, 59–66. DOI 10.1007/BF00153144.
McIntosh, P. S., Krieger, A. S., Nolte, J. T., & Vaiana, G. (1976). Association of X-ray arches with chromospheric neutral lines. *Solar Physics*, *49*, 57–77. DOI 10.1007/BF00221485.
Munro, R. H., Gosling, J. T., Hildner, E., MacQueen, R. M., Poland, A.I., & Ross, C.L. (1979). The association of coronal mass ejection transients with other forms of solar activity. *Solar Phys*, *61*, 201–215. DOI 10.1007/BF00155456.
Panasenco, O., Martin, S. F., & Velli, M. (2014). Apparent solar tornado-like prominences. *Solar Phys*, *289*, 603–622. DOI 10.1007/s11207-013-0337-1, 1307.2303.
Parker, E. N. (1994). Spontaneous current sheets in magnetic fields: with applications to stellar x-rays. International Series in Astronomy and Astrophysics (Vol. 1). New York: Oxford University Press.
Rachmeler, L. A., Gibson, S. E., Dove, J. B., DeVore, C. R., & Fan, Y. (in press). Polarimetric properties of flux ropes and sheared arcades in coronal prominence cavities. *Solar Physics*. Arxiv 13047594.
Reeves, K. K., Gibson, S. E., Kucera, T. A., & Hudson, H. S. (2012). Thermal properties of coronal cavities observed with the X-ray telescope on Hinode. *Astrophysical Journal*, *746*, 146.
Régnier, S., Walsh, R. W., & Alexander, C. E. (2011). A new look at a polar crown cavity as observed by SDO/AIA. Structure and dynamics. *Astronomy and Astrophysics*, *533*, L1. DOI 10.1051/0004-6361/201117381, 1107.3451.
Saito, K., & Hyder, C. (1968). A concentric ellipse multiple-arch system in the solar corona. *Solar Physics*, *5*, 61.
Saito, K., & Tandberg-Hanssen, E. (1973). The arch systems, cavities, and prominences in the helmet streamer observed at the solar eclipse, november 12, 1966. *Solar Physics*, *31*, 105.
Schmahl, E. J. (1979). The prominence-corona interface—a review. In E. Jensen, P. Maltby & F. Q. Orrall (Eds.), *IAU Colloq. 44: Physics of solar prominences* (pp. 102–120). Oslo: Universitetet i Oslo.
Schmit, D. J., & Gibson, S. (2013). Diagnosing the prominence-cavity connection. *Astrophysical Journal*, *770*, 35. DOI 10.1088/0004-637X/770/1/35, 1304.7595.
Schmit, D. J., & Gibson, S. E. (2011). Forward modeling cavity density: A multi-instrument diagnostic. *Astrophysical Journal*, *733*, 1.

Schmit, D. J., & Gibson, S. E. (2014). The formation of a cavity in a 3d flux rope. IAU S300 proceedings. In B. Schmieder, J.-M. Malherbe & S. T. Wu (Eds.), *IAU Symposium* (Vol. 300, pp. 147–150). DOI 10.1017/S1743921313010880

Schmit, D. J., Gibson, S. E., Tomczyk, S., Reeves, K. K., Sterling, A. C., Brooks, D. H., et al. (2009). Large-scale flows in prominence cavities. *Astrophysical Journal Letters*, *700*, 96.

Schmit, D. J., Gibson, S., Luna, M., Karpen, J., & Innes, D. (2013). Prominence mass supply and the cavity. *Astrophysical Journal*, *779*,156. DOI 10.1088/0004-637X/779/2/156, 1311.2382.

Serio, S., Vaiana, G. S., Godoli, G., Motta, S., Pirronello, V., & Zappala, R. A. (1978). Configuration and gradual dynamics of prominence-related X-ray coronal cavities. *Solar Physics*, *59*, 65–86. DOI 10.1007/BF00154932.

Straka, R. M., Papagiannis, M. D., & Kogut, J. A. (1975). Study of a filament with a circularly polarized beam at 3.8 cm. *Solar Physics*, *45*, 131.

Tandberg-Hanssen, E. (1974). *Solar prominences*. Dordrecht: D. Reidel Publishing Company.

Tandberg-Hanssen, E. (1995). *The nature of solar prominences*. Dordrecht: Kluwer Academic Press.

Taylor, J. B. (1974). Relaxation of toroidal plasma and generation of reverse magnetic fields. *Physical Review Letters*, *33*, 19.

Tian, H., Tomczyk, S., McIntosh, S. W., Bethge, C., de Toma, G., & Gibson, S. (2013). Observations of coronal mass ejections with the coronal multichannel polarimeter. *Solar Physics*, *288*, 637–650. DOI 10.1007/s11207-013-0317-5, 1303.4647.

Titov, V. S. (2007). Generalized squashing factors for covariant description of magnetic connectivity in the solar corona. *Astrophysical Journal*, *660*, 863.

Titov, V. S., & Demoulin, P. (1999). Basic topology of twisted magnetic configurations in solar flares. *Astronomy and Astrophysics*, *351*, 707.

Török, T., Panasenco, O., Titov, V. S., Mikić, Z., Reeves, K. K., Velli, M., et al. (2011). A model for magnetically coupled sympathetic eruptions. *Astrophysical Journal Letters*, *739*, L63. DOI 10.1088/2041-8205/739/2/L63, 1108.2069.

Vaiana, G. S., Krieger, A. S., & Timothy, A. F. (1973). Identification and analysis of structures in the corona from X Ray photography. *Solar Physics*, *32*, 81–116. DOI 10.1007/BF00152731.

Vasquez, A. M., Frazin, R. A., & Karmalabadi, F. (2009). 3D temperatures and densities of the solar corona via multi-spacecraft EUV tomography: Analysis of prominence cavities. *Solar Physics*, *256*, 73.

Waldmeier, M. (1970). The structure of the monochromatic corona in the surroundings of prominences. *Solar Physics*, *15*, 167–175. DOI 10.1007/BF00149483.

Wang, Y. M., & Stenborg, G. (2010). Spinning motions in coronal cavities. *Astrophysical Journal Letters*, *719*, 181.

Webb, D. (2014). Eruptive prominences and their impact on the earth; the association with coronal mass ejections. In J.-C. Vial & O. Engvold (Eds.), *Solar prominences, ASSL* (Vol. 415, pp. 409–430). New York: Springer.

Woltjer, L. (1958). A theorem on force-free magnetic fields. *Proceedings of the National Academy of Science*, *44*, 489–491. DOI 10.1073/pnas.44.6.489.

Zhang, M., & Low, B. C. (2005). The hydromagnetic nature of solar coronal mass ejections. *Annual Review of Astronomy and Astrophysics*, *43*, 103.

Chapter 14
Formation and Large-Scale Patterns of Filament Channels and Filaments

Duncan H. Mackay

Abstract The properties and large-scale patterns of filament channels and filaments are considered. Initially, the global formation locations of filament channels and filaments are discussed, along with their hemispheric pattern. Next, observations of the formation of filament channels and filaments are described where two opposing views are considered. Finally, the wide range of models that have been constructed to consider the formation of filament channels and filaments over long time-scales are described, along with the origin of the hemispheric pattern of filaments.

14.1 Global Patterns of Filaments and Filament Channels

14.1.1 Global Patterns

Solar filaments (a.k.a. prominences) form over a wide range of latitudes on the Sun. These latitudes range from the active latitudes all the way to the polar crown. A typical distribution of solar filaments across the Sun during a period of high magnetic activity can be seen in Fig. 14.1a. The filaments can be seen in Hα absorption as dark features that bend and weave their way across the solar surface. The long term behavior of solar filaments over many solar cycles may be deduced from archives such as Cartes Synoptiques[1] (1919–1989 covering 7 solar cycles) or from NOAA Solar Geophysical Data[2] (1955–2009 covering 6 solar cycles). These data sets show that the number of solar filaments present on the Sun at any one time varies in a manner similar to that of the sunspot number. Over a single Carrington Rotation the total number of filaments on the Sun varies from 20 at cycle minimum

[1] ftp.ngdc.noaa.gov/STP/space-weather/solar-data/solar-features/prominences-filaments/filaments/.

[2] http://www.ngdc.noaa.gov/stp/solar/sgd.html.

D.H. Mackay (✉)
School of Mathematics and Statistics, University of St Andrews, North Haugh, St Andrews, Fife, KY16 8HB, UK
e-mail: dhm@st-and.ac.uk

J.-C. Vial, O. Engvold (eds.), *Solar Prominences*, Astrophysics and Space Science Library 415, DOI 10.1007/978-3-319-10416-4_14

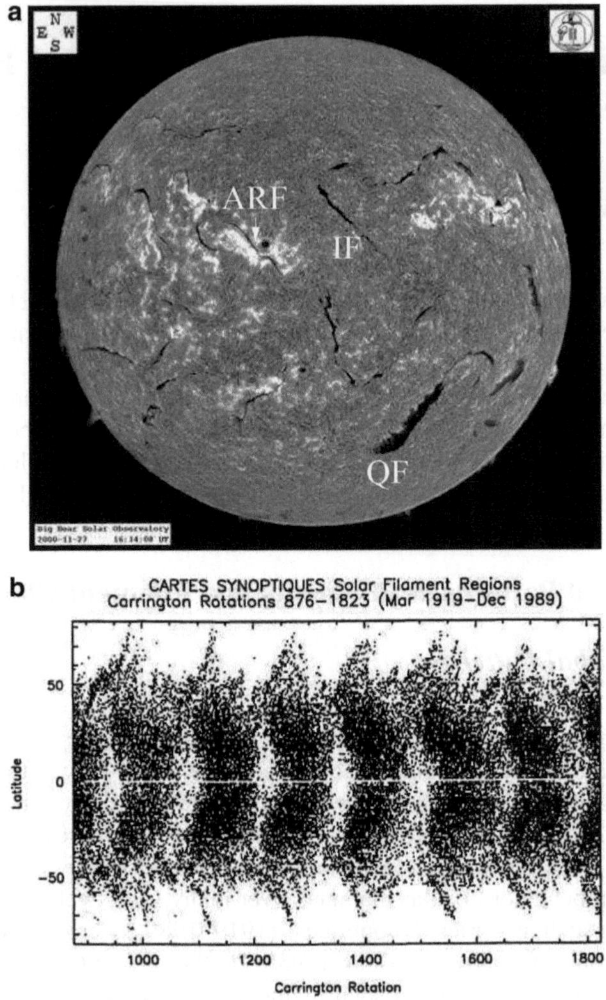

Fig. 14.1 (**a**) Full disk Hα filtergram from Big Bear Solar Observatory (BBSO) on 22nd November 2000. Examples of Quiescent (QF), Intermediate (IF) and Active Region (ARF) filaments are shown. (**b**) The filament butterfly diagram produced from Cartes Synoptiques data (from Coffey and Hanchett 1998, Fig. 2)

to over 100 at cycle maximum (Zou et al. 2014). There is also a clear latitudinal dependence to the number of filaments. At low latitudes ($\lambda < 50°$) there are more filaments and a stronger variation (10–80 from cycle min to max), while at high latitudes ($\lambda > 50°$) the variation is less strong (0–10).

As magnetic flux is transported across the surface of the Sun, solar filaments are found to migrate over a range of latitudes (McIntosh 1972; Minarovjech et al. 1998). This migration can be seen in Fig. 14.1b where the solar filament butterfly diagram is shown from 1919–1989 (Coffey and Hanchett 1998). In contrast to the sunspot butterfly diagram, the filament butterfly diagram shows a more complex structure with two branches of migration. The first branch follows a similar path to that of the sunspot butterfly diagram and extends towards lower latitudes. While it follows a similar path, solar filaments occur over a wider latitude range than sunspots (Li 2010). The second branch follows the poleward transport of magnetic flux during the rising phase of the solar cycle and is named the "rush to the poles of polar crown filaments" (Topka et al. 1982; Mouradian and Soru-Escaut 1994; Shimojo et al. 2006; Pintér et al. 2014). The disappearance of this branch signifies the reversal of the Sun's polar field around solar maximum. Due to this, filaments may be used as a tracer to infer the large-scale pattern of magnetic flux on the Sun.

Based upon the magnetic environment in which they form, filaments may be split into three broad categories (Engvold 1998, see also Chap. 2):

1. **Quiescent Filaments**: Filaments found in quiet regions of the Sun and around the polar crown (e.g. QF in Fig. 14.1a).
2. **Intermediate Filaments**: Filaments that form around the borders of active regions (IF in Fig. 14.1a).
3. **Active Region Filaments**: Filaments found within the centers of activity nests of multiple pairs of sunspots (ARF in Fig. 14.1a).

Observations tend to show that IFs and QFs are larger, much more stable structures with longer lifetimes (weeks to months) compared to ARFs, which are generally unstable with a lifetime of only a few hours to days. In later sections when discussing the possible mechanisms of filament formation, this classification scheme will prove useful in illustrating that different mechanisms may form different types of filaments.

14.1.2 Properties of Filament Channels

Although filaments may form at many locations on the Sun, they always form above Polarity Inversion Lines (PILs, Babcock and Babcock 1955), which divide regions of positive and negative flux in the photosphere. As discussed in Martin (1998) the existence of a PIL is a necessary, but not sufficient condition for a filament to form. For a filament to form, a filament channel must also exist at the height of the chromosphere (Gaizauskas 1998). Observations show that not every filament channel contains a filament. Due to this, filament channels are more fundamental than the filaments that form within them, where a single channel may survive a succession of filament formations and eruptions.

The basic magnetic structure of a filament channel is illustrated in Fig. 14.2. The properties of filament channels were first noted by Martres et al. (1966) who described them as a region in the chromosphere around a PIL where the chromospheric fibrils are aligned, (1) anti-parallel to one-another on either side and (2) nearly parallel to the path of the PIL. Foukal (1971a,b) interpreted these fibrils as

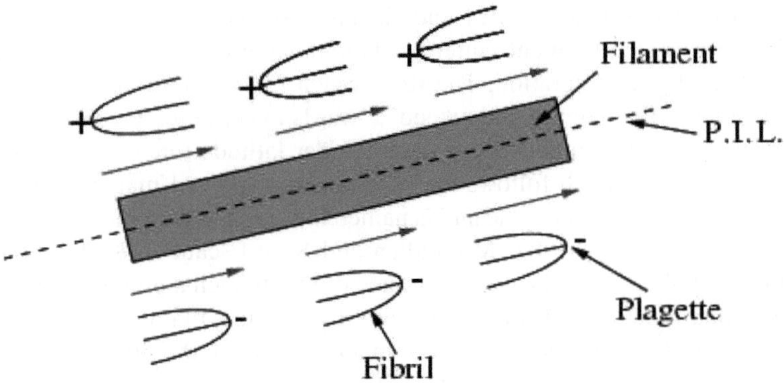

Fig. 14.2 Schematic of a filament channel with fibrils which lie (1) anti-parallel to one-another on either side of the PIL and (2) nearly parallel to the path of the PIL. The anti-parallel alignment indicates that the magnetic field (*arrows*) is dominantly horizontal and points in the same direction on either side of the channel

giving the direction of the local magnetic field in the chromosphere. Due to the anti-parallel alignment of the fibrils, Foukal deduced that the horizontal component of the magnetic field must point in the same direction on either side of the channel (Martin et al. 1992, 1994). For a more in depth discussion of the observational properties of filament channels see Chap. 2, Sect. 3.1. From this fibril alignment it can be deduced that filament channels, the birth ground of filaments, are locations of strong magnetic shear and highly non-potential magnetic fields. Solar filaments which lie in the corona are believed to be embedded in the filament channel field which extends up into the corona. Through Hanle and Zeeman effect measurements (Hyder 1965; Rust 1967; Leroy 1989; Leroy et al. 1983) the existence of a dominant horizontal field has been verified at coronal heights.

Presently it is unclear why channels and their non-potential fields build up along PILs. Karachik and Pevtsov (2014) showed that PILs with and without filaments have a similar gradient of the magnetic field across the PIL. Understanding filament channel formation is key to our understanding of the evolution of magnetic fields on the Sun and their relationship to eruptive phenomena. By observing and interpreting their formation and evolution, we may examine directly the buildup of magnetic stress and energy required for Space Weather events such as Coronal Mass Ejections. The importance of filament channels to Space Weather can be seen by the paper of Pevtsov et al. (2012) who showed that filament channels without filaments may lead to CMEs. In Sect. 14.5 a wide range of models and mechanisms for the development of the strong shear and axial fields in filament channels and filaments will be discussed. In the next section we consider the global properties and formation locations as deduced from Hα observations.

14.2 Global Formation Locations

While filaments and filament channels form at many locations on the Sun, very few studies have considered the exact nature or history of the PILs above which they form. Those studies that have considered this, are mainly restricted to studying large-scale, stable filaments and neglect smaller unstable filaments forming in the centers of activity complexes. Understanding the type of magnetic environment in which filaments and filament channels form is key to understanding the magnetic interactions that lead to the formation of strong magnetic shear.

In the past many classification schemes for filaments and prominences have been developed (d'Azambuja and d'Azambuja (1948) see Chap. 2). One classification scheme (Engvold 1998) provides a useful distinction between filaments forming inside and outside active regions. However, to understand the role that magnetic fields play in the formation, structure and evolution of filaments it is important to understand the exact type of PIL above which filaments form. One early classification scheme that aims to quantify this (Tang 1987) splits filaments into two categories. The first category is one in which the filament forms above a PIL lying within a single bipolar unit of flux, and is classified as an "Internal Bipolar Region Filament" (see Fig. 14.3a). In the second, the filament forms above a PIL which lies between two separate magnetic bipoles and is called an "External Bipolar Region Filament" (Fig. 14.3b). Observations by Tang (1987) showed that when filaments are classified into these two types, over 60 % of filaments form external to bipolar regions.

In a more recent study Mackay et al. (2008) reconsidered where large, stable solar filaments form. A key aspect of this study was that the authors followed the history and evolution of the PILs underneath filaments before each of the filaments was categorised. To distinguish the different bipole interactions that could lead to the formation of filaments, Mackay et al. (2008) introduced two additional categories in addition to those defined by Tang (1987): "Internal/External Bipolar Region Filaments" (I/EBR) and "Diffuse Bipolar Region Filaments" (DBR). The I/EBR filaments are defined as filaments that lie above both the internal PIL of a bipole and the external PIL surrounding the bipole (Fig. 14.3c), and therefore could not be classified into the scheme proposed by Tang (1987). In contrast, the DBR filaments are located in essentially a bipolar distribution of flux, but where the polarities defining the bipole did not emerge together. The formation of the bipolar distribution was the result of many flux emergences, coalescences and cancellations such that the polarities on either side of the filament could not be attributed to a single bipole emergence (Fig. 14.3d). Mackay et al. (2008) then consider the bipole interactions leading to the formation of Intermediate and Quiescent filaments over four distinct phases of the solar cycle (two before and two after cycle maximum) .

Of the 603 filaments studied by Mackay et al. (2008), 92 % formed at locations requiring multiple bipole interactions (the breakdown comprised of 62 % EBR, 17 % DBR and 13 % I/EBR). Only 7 % formed within a single bipole. These results show that large-scale filaments, namely those of the IF and QF type, preferentially form at

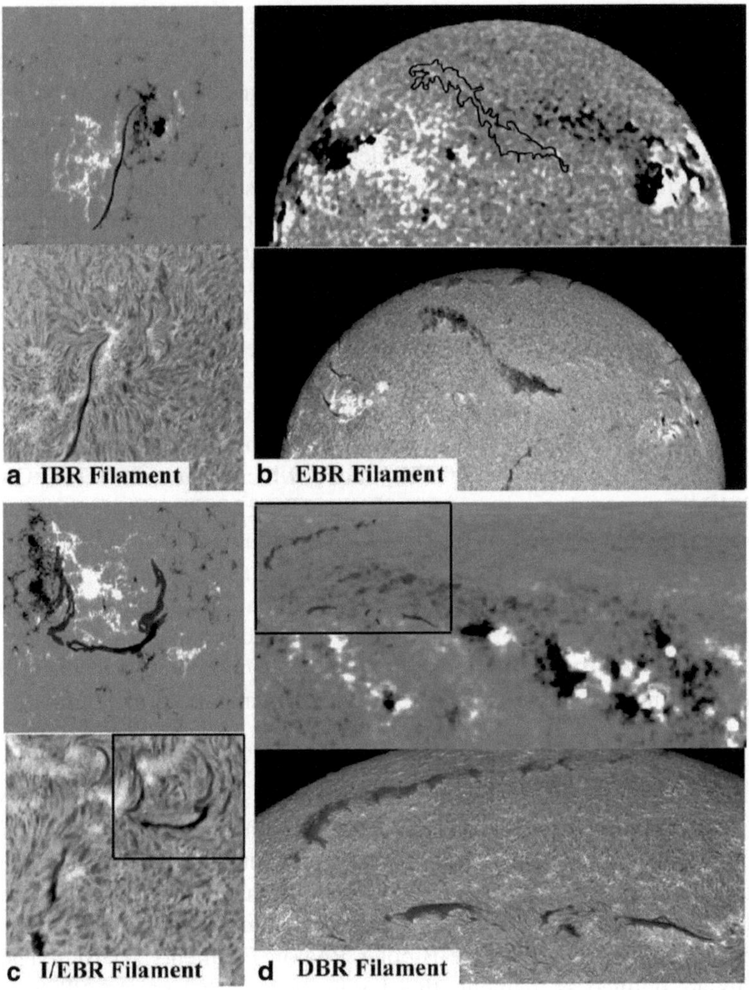

Fig. 14.3 Examples of the four categories of filaments defined by Mackay et al. (2008). In each of the panels (**a–d**), the *bottom plot* is an Hα image from the ORSO, while the *top image* shows the radial magnetic field derived from either (**a–c**) a full-disk magnetogram or (**d**) a synoptic magnetogram from Kitt Peak. Outlines of the Hα filaments are superimposed on each of the magnetograms. The dates of the observations are (**a**) 26th June 1979, (**b**) 6th May 1979, (**c**) 27th September 1979 and (**d**) 14th July 1979. For panels (**c, d**) the areas enclosed by the *boxes* denote the corresponding area of (**c**) the magnetogram and (**d**) the Hα image. In panel (**d**) (*top image*) the low latitude activity complexes which will extend poleward over time and interact to produce diffuse regions of flux at high latitudes can be clearly seen. For a schematic representation of this figure see Chap. 2

sites of multiple bipole interactions. Very few of them occur within a single bipole. In addition by considering four distinct phases of the solar cycle, Mackay et al. (2008) showed that only EBR filaments exhibit any form of solar cycle dependence, with the other three types remaining essentially constant (see Fig. 3 of Mackay et al. 2008). This dependence showed that the number of EBR filaments varied in phase with the solar cycle and matches the solar cycle variation found by Zou et al. (2014), with more filaments at cycle maximum than minimum. Such a variation indicates that the formation of EBR filaments must be strongly related to the amount of magnetic flux on the Sun.

14.3 The Hemispheric Pattern of Solar Filament Channels and Filaments

While the basic properties of solar filaments have long been known, filament channels and filaments have been classified more recently in terms of their chirality (Martin et al. 1992, 1994). This chirality may take one of two forms: dextral or sinistral. Dextral/sinistral filament channels and filaments have an axial magnetic field that points to the right/left when the main axis of the filament channel or filament is viewed from the positive polarity side of the PIL (see Fig. 14.4). In force-free field models (e.g. Aulanier and Démoulin 1998; Mackay et al. 1999; van Ballegooijen et al. 2000; Mackay and van Ballegooijen 2005) this chirality is directly related to the dominant sign of magnetic helicity that is contained within the filament channel and filament. A dextral filament or filament channel will contain dominantly negative helicity, while a sinistral one has positive helicity. The transport of filament channels and filaments across the solar surface (Sect. 14.1) is therefore an indication of the large-scale transport of magnetic helicity across the Sun (Yeates et al. 2008b), a key feature in explaining many eruptive phenomena.

A number of different techniques may be applied to determine the chirality of both filament channels and filaments depending on the resolution of the observations and strength of the underlying magnetic field. For a filament channel that does not contain a filament, high resolution Hα images are sufficient to resolve individual chromospheric fibrils and determine the chirality as long as the channel lies in a strong field region. In contrast for a channel in a weak field region, where fibril

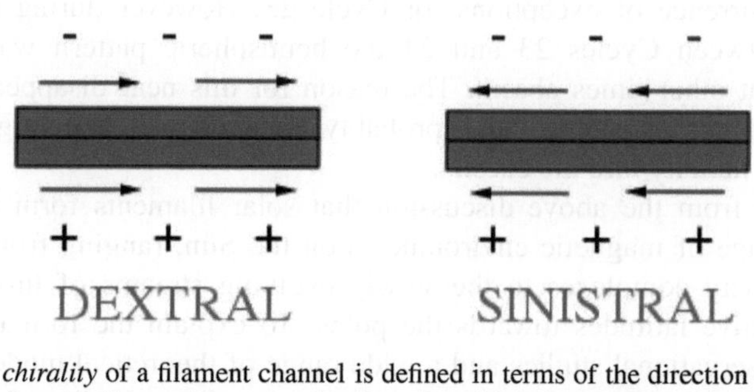

Fig. 14.4 The *chirality* of a filament channel is defined in terms of the direction of the magnetic field along the channel (denoted by *arrows*) when viewed by an observer on the positive polarity side of the channel. For a dextral (sinistral) channel, the magnetic field points to the *right* (*left*)

patterns are less distinct, magnetograms may also be used to aid the determination of chirality, by using them to determine the polarity of the magnetic elements from which the fibrils extend from or go into.

For filaments the chirality may be determined through a variety of techniques: (1) indirectly from the filament channel, as the filament is believed to lie in the extended field of the channel, (2) directly through magnetic field measurements (Leroy et al. 1983), or, (3) indirectly from the relationship of filaments to their barbs. Filaments with barbs may be classified as being either *right-bearing* or *left-bearing* depending on the direction in which the barbs extend out of the main filament axis. Martin and collaborators found a one-to-one correspondence where filaments in dextral channels have right-bearing barbs, and those in sinistral channels have left-bearing barbs. To date no studies have compared the chirality of filaments determined both directly (though magnetic field measurements) or indirectly (fibrils or barbs). Due to the lack of high resolution Hα data and direct measurements of magnetic fields within prominences, filaments are mostly classified using the relationship to barbs (Pevtsov et al. 2003; Yeates et al. 2007). Recently, a new technique for the determination of filament chirality has been described by Sheeley et al. (2013). This technique uses the plume like tails of coronal cells visible in the Fe XII 193 Å line at 1.2MK to deduce the direction of the axial field up to a height of 50 Mm, significantly higher than that of Hα. This new technique presents a powerful method due to the wide spread availability of observations in the Fe XII 193 Å line.

A surprising feature of the chirality of filament channels and filaments is that it displays an unusual large-scale hemispheric pattern: dextral/sinistral filaments dominate in the northern/southern hemispheres respectively (Martin et al. 1994; Zirker et al. 1997; Pevtsov et al. 2003; Yeates et al. 2007; Wang et al. 2013). This pattern is unusual as it is exactly opposite to that expected from differential rotation acting on a North-South coronal arcade. Although dextral/sinistral filaments dominate in the northern/southern hemisphere, observations show that exceptions to this pattern do occur. Any model which tries to explain the formation of filaments and filament channels must explain not only the origin of this hemispheric pattern but also why exceptions arise. Bernasconi et al. (2005) developed an automated detection software for filament barbs and the chirality of filaments. From applying this technique Martens et al. (2014) verified the existence of the hemispheric pattern, with the occurrence of exceptions for Cycle 23. However during the extended minimum between Cycles 23 and 24 the hemispheric pattern was sometimes present, but at other times absent. The reason for this near disappearance of the hemispheric pattern is unclear, but is probably due to the lack of emerging magnetic flux injecting helicity into the corona.

It is clear from the above discussion that solar filaments form or are found in a wide range of magnetic environments on the Sun, ranging from the rapidly evolving activity complexes to the slowly evolving streams of flux that extend out of the active latitudes towards the poles. To explain the formation of these filaments, observational studies and a wide range of theoretical models have been produced. The next section will consider observational case studies of the formation of filaments (Sect. 14.4). After discussing these, models of filaments formation will be discussed in Sect. 14.5. The models discussed in Sect. 14.5 will consider only global long-term magnetic field models for the formation of filament channels and filaments. For models that discuss the detailed thermodynamic processes and origin of the dense prominence plasma, see Chaps. 7 and 10. The observations will then be used to clarify which models of filament formation are applicable to which filament formation locations (Sect. 14.7).

14.4 Observations of Filament Channel and Filament Formation

To understand the magnetic environment and interactions leading to the formation of filament channels and filaments, it is useful to discuss test cases. To date, very few examples of filament channel formation have ever been observed, so the exact formation mechanism remains debatable. Within the published literature some papers show the formation of filament channels through surface effects that reconfigure pre-existing coronal fields, while in other papers flux emergence of horizontal flux ropes is deduced by the authors to play a critical role. Thus from interpreting the observations there appear to be two opposing views on how filament channels and filaments form. In this discussion we will consider the key observational features from case studies and determine whether the two views may be reconciled.

14.4.1 Evidence of the Reconfiguration of Pre-Existing Coronal Fields in the Formation of Filament Channels

Observations reported by Gaizauskas et al. (1997) and Gaizauskas et al. (2001) show that surface motions acting on pre-existing coronal fields play a critical role in the formation of filament channels and filaments. In the first case, an Intermediate Filament (IF) forms over a short period of a few days, while in the second a Quiescent Filament (QF) forms over a period of months. In both cases the filaments form on PILs external to any single bipole and in the classification scheme of Sect. 14.2 would be classed as External Bipolar Region Filaments. Although the two cases occur over very different time and length scales there are a number of important similarities.

Both cases begin with the emergence of a significant amount of magnetic flux in the form of an activity complex. Importantly however, no filaments form during the process of flux emergence. In fact, for the large scale QF the filament forms approximately 27 days after major flux emergence subsides. In both cases a necessary condition for the formation of the filament channels was flux convergence and cancellation at a PIL between separate bipolar regions. Such convergence and cancellation of flux was also shown to be important for filament formation in the papers by Martin (1998) and Gaizauskas (2002). Finally and most importantly, in each case a significant amount of magnetic shear was seen to build up in the activity complexes as they emerged. The redistribution of this non-potential field through surface motions towards the PIL produces a preferred direction of the coronal field above the PIL and plays a critical role in the formation of the filament channels (Mackay and Gaizauskas 2003).

Figure 14.5 illustrates the main stages in the formation of a filament channel and IF over a period of 5 days between the 20–25th July 1979 (see Figs. 2 and 4 in Gaizauskas et al. 1997). The formation of this southern-hemisphere IF involves the interaction of two distinct magnetic flux distributions, an old remnant region (McMath 16159) and a new emerging region (McMath 16166). In the Hα image of Fig. 14.5 (top left) the bright North-South plage outlines the old remnant region; new magnetic flux emerges inside the oval between the 20th and 21st July. A key feature of this image is that the chromosphere is free of any strong patterns of magnetic fields surrounding the oval. Magnetic field observations show that the new emerging region is an activity complex (Gaizauskas et al. 1983; Benevolenskaya 2005) made up of two or more sunspot pairs. Significantly, no filament forms near or around the activity complex during this period of rapid flux emergence. The key development in the formation of the filament channel occurs over a 3 h period on the 21st July

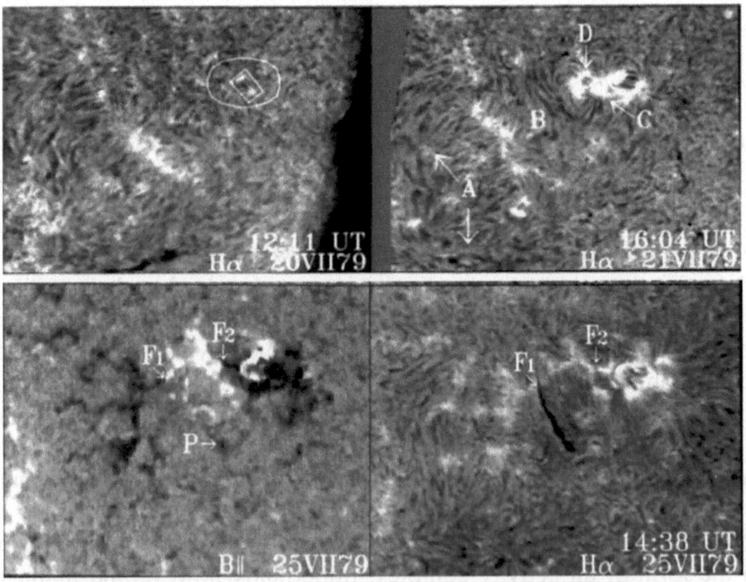

Fig. 14.5 Hα and magnetogram images from Gaizauskas et al. (1997) of the formation of an Intermediate Filament between an old remnant region and a new emerging activity complex (inside oval). The Hα images correspond to 20th (*top left*), 21st (*top right*) and 25th July 1979 where the filament channel formed on the 21st but the filament (F1) did not form until the 25th July 1979. In the magnetogram image on the 25th, *white* represents positive and *black* negative flux

(at the location denoted by B in upper right panel). Over this period a band of co-aligned fibrils form at the tail end of the new activity complex, between it and the old remnant region. These co-aligned fibrils indicate a magnetic field at this location with a dominant horizontal component, i.e. that a filament channel has formed. According to a model by Mackay et al. (1997) this pattern of co-aligned fibrils can only be explained by the extended non-potential magnetic field of the activity complex in which the field contains a large amount of positive helicity (correct sign for the southern hemisphere).

No filament forms as magnetic flux continues to emerge within the activity complex. The emergence ceases on the 23rd July after which the trailing positive polarity of the activity complex disperses or diffuses out. This dispersion causes a convergence of flux between the old and new regions. In Fig. 14.5 the distribution of magnetic flux (bottom left) and corresponding Hα image (bottom right) can be seen for the 25th July. Five days after the complex started to emerge, cancellation of flux occurs at the point F1 (bottom right). After this cancellation the filament forms and passes through the location of flux cancellation. It survived for a full solar rotation and can clearly be seen to lie on a PIL which is external to any one bipolar region. Subsequent modeling by Mackay et al. (1997) showed that the resulting magnetic structure of the filament could only be explained by the interaction of the combined fields of both the old and new magnetic distributions. Both fields were highly non-potential, again with a significant amount of positive helicity which must have originated during the creation of the new activity complex. It is clear from the observations that this filament is of EBR type and reconfiguration of the previously emerged fields played a critical role in the formation of the filament. In the recent paper of Jeong et al. (2009) the authors present observations that support the idea that the non-potential fields and helicity of IF's originate during the emergence of activity complexes.

Gaizauskas et al. (2001) described a similar process of filament channel and filament formation between two neighboring activity complexes, but this time for a QF which is nearly $1 R_{\odot}$ in length. As with the previous case, the formation of the filament channel is attributed to the extended non-potential fields of the activity complexes. The filament only forms after major flux emergence ceases and the activity complexes converge and partially cancel with one another. In contrast to the IF case, which took 5 days to form, the large-scale case of the QF takes nearly one full solar rotation (27 days) to appear.

In both cases described above no stable filaments form during the periods of the highest rates of flux emergence, and the authors concluded that surface motions acting on pre-existing coronal fields play a critical role in the formation of stable filaments through the interaction of multiple bipoles. This result is consistent with the classification of filaments given in Sect. 14.2 where the majority of filaments are found to lie in magnetic configurations that involve more than one bipole. A key role of these surface motions is to redistribute the helicity which is seen to emerge in the early stages to form the filament channel (Gaizauskas et al. 1997, 2001; Mackay and Gaizauskas 2003; Mackay and van Ballegooijen 2005, 2006). In other observations, Gaizauskas (2002) show that convergence and cancellation of flux are important for filament channel formation (Martin 1998). The study shows that, early on in the solar cycle, a unipolar region of flux has to extend 180° around the Sun to interact and cancel with an opposite polarity region before a filament can form on that PIL. For this case the redistribution of flux, after emergence, is inferred to be a key process in the formation of the filament channel (Gaizauskas 2008).

Observations by Wang and Muglach (2007) have supported the work of Gaizauskas et al. (1997, 2001). Wang and Muglach (2007) describe the formation of three filament channels and filaments (2 Intermediate or External Bipolar Regions Filaments and 1 Active Regions or I/EBR filament). The authors describe how fibrils which are initially normal to the PIL rotate to lie parallel to the PIL over a period of 1–2 days and in doing so form a filament channel. Through studying the evolution of the magnetic fields the authors deduce that flux cancellation as a result of supergranular convection plays a key role in the formation of the filament channels. They argue that this cancellation process between opposite polarity elements removes the normal component of the field but leaves the component parallel to the PIL which builds up gradually to form the axial field of the filament channel. In contrast to Gaizauskas et al. (1997, 2001) they do not observe any significant helicity resulting from the emergence of the active regions. On comparing the results of Gaizauskas et al. (1997) and Wang and Muglach (2007), while there are many similarities, there are also some differences in time scale. The clearest is the time difference required to form the filament channel. For Gaizauskas et al. (1997) the formation of the filament channel occurs over a 3 h period and is attributed to the extended non-potential field of the activity complex containing a large amount of helicity. Cancellation of flux could not produce such a strongly sheared field over such a short period of time. In contrast, Wang and Muglach (2007) do not report any strong patterns of fibrils associated with helicity emerging in the active regions but rather form the filament channel over a period of 1–2 days in a much slower process of cancellation. Therefore there appear to be two complementary methods of forming a filament channel over different time scales.

Schmieder et al. (2004) studied the formation of a filament in the complex center of a decaying active region formed out of smaller individual components. They followed the evolution of three individual filament segments denoted F1, F2 and F3 over several days, and found that F1 and F2 gently merged into a single structure, as observed by a gradual filling in Hα of the gap between them. This merging was associated with mild EUV brightenings and with small Hα Doppler shifts at the merging point. While EUV brightenings are a good indicator of magnetic reconnection (see also Wang et al. 2013), the flows revealed that the merging first took place by dynamic exchanges between the two progenitors, until they formed a single long stable filament. Two days later segments F2 and F3 came into contact and produced a confined flare, as evidenced by EUV post-flare loops (Deng et al. 2002). To determine the directions of the axial fields in the three filament segments, Schmieder et al. (2004) used the chirality rules for chromospheric fibrils and magnetic field polarity, the skew of the overlying coronal arcades, and the sense of twist in neighboring sunspots. It was then confirmed that when two filaments interact, magnetic reconnection takes place and leads to a merging when their chiralities are of the same sign, but leads to a flare when the chiralities are opposite. MHD simulations of such a process have been carried out by DeVore et al. (2005). It was also inferred that magnetic helicity must slowly accumulate prior to filament merging, as seen by the rotation of a small twisted sunspot close to the merging point.

14.4.2 Evidence of Emerging Horizontal Flux Tubes in Filament Formation

It is clear from the above observations that surface effects play a critical role in forming the studied IFs and QFs (which are long stable structures). Lites and Low (1997) describe a different process for forming short, unstable active-region filaments. In Lites and Low (1997) the emergence of a δ-spot is traced through vector magnetic field measurements using Advanced Stokes Polarimetry. Magnetic field vectors along part of the PIL within the emerging δ-spot show a concave up or dipped magnetic structure (see Fig. 1 of Lites and Low 1997; also see Lites 2005). A small active region filament forms at this location. The filament was however unstable with a lifetime of only 2 days. Lites and Low (1997) suggest photospheric material is dragged up into the corona through the levitation process, as a horizontal flux rope emerges (Rust and Kumar 1994).

A more recent example of the effect of evolving magnetic fields on the structure and stability of an active region filament is described by Okamoto et al. (2008) and Okamoto et al. (2009). In two papers, the authors present observations of a time series of vector magnetic fields taken by SOT underneath a pre-existing filament. The vector magnetic field measurements show a PIL with dominant horizontal field along it. This horizontal field probably represents that of the filament channel of the pre-existing filament. Over a period of 1.5 days the horizontal field vector changes from normal to inverse polarity and a dominant blue shift is observed. During this period the filament alters its appearance from a single structure, to a fragmented one and back again. Before returning to a single structure, brightenings are observed along the filament fragments in the Ca II H line.

From the observations the authors deduce two possible scenarios. In the first scenario they interpret the observations in terms of an emerging horizontal flux rope which fully emerges into the corona and occupies the position of the pre-existing filament. With this scenario the mass of the prominence originates from below the photosphere. In line with this scenario Lites et al. (2010) consider the formation of a filament channel within an active region and deduce that the formation is also due to the emergence of a flux rope. The second scenario interprets the brightenings in Ca II H as evidence for reconnection between the pre-existing filament and a new flux rope that emerges free of mass. The reconnection then produces a single structure along the PIL. A difficulty with both scenarios is that no simulations of magnetic flux emergence have been able to emerge a horizontal flux rope through the photosphere.

In contrast to that put forward by the authors, a third possibility also exists. As the top part of a flux rope emerges, a likely outcome is the emergence of sheared arcades. A coronal flux rope may then form out of these arcades through the process of reconnection. This reconnection may lift cool material into the corona. If the axial component of the emerging arcade lies in the same direction as that of the pre-existing filament channel, the new and old flux systems may join to produce a single structure. To consider which, if any of these three scenarios are correct, high resolution magnetic field observations at multiple levels in the solar atmosphere (e.g. photosphere, chromosphere and corona) are required. In the paper of Kuckein et al. (2012) the authors present a highly detailed study of vector magnetic fields at two heights around a compact active region filament. From the vector magnetic fields measurements they deduce that the magnetic configuration of the filament is that of a magnetic flux rope. While they show strong evidence for this, they cannot determine which scenario discussed above leads to the formation of the flux rope as the filament formed before the start of the vector magnetic field measurements. They do however find similarities in the evolution of the magnetic field to that described by Okamoto et al. (2008) and Okamoto et al. (2009).

14.4.3 Summary of Observations

The observations described in Sects. 14.4.1 and 14.4.2 provide evidence for filament formation arising from surface motions that reconfigure already existing coronal fields or, emerging flux tubes. So can the two methods be reconciled? The important distinction between these cases is the type and location of filaments formed in each case. For the cases in Sect. 14.4.1 surface motions play an important role in forming long stable Quiescent or Intermediate filaments which are External Bipolar Region Filaments, the dominant type of large-scale filament found at all latitudes on the Sun. In contrast, flux tubes emerging in a δ-spot form an Active Region or Internal Bipolar Region Filaments which are unstable, lasting merely 2 days.

While it is difficult to draw general conclusions from a few specific observations, they indicate that two different mechanisms might form filaments in different magnetic environments on the Sun. Thus large stable filaments of the IF and QF type (External or Diffuse Bipolar Region) may require surface motions to gradually reconfigure pre-existing coronal fields, while small, short-lived ARFs (Internal Bipolar Region) may form due to flux emergence. To determine whether different mechanisms do produce different types of filaments at different locations on the Sun, the formation of filaments over a wide range of latitudes needs to be considered in detail. Observational programs required to do this will be briefly discussed in Sect. 14.7.

14.5 Theoretical Models of Filament Formation

Over the years many models have been constructed to describe the formation of filaments. These models vary from descriptive papers to full numerical MHD simulations and consider two main problems. First, how to obtain the correct dipped magnetic field configuration with dominant axial magnetic field that follows the hemispheric pattern, and secondly, the origin of the dense plasma. While the second question relates more to thermodynamics (Karpen et al. 2001, see Chap. 10), this chapter is relevant to the first group of models. It is widely accepted that magnetic flux ropes are a suitable configuration to represent solar filaments; the main area of debate is how exactly these flux ropes may form. The various models which consider this may be broadly split into two distinct sub-groups: those employing surface effects to reconfigure coronal fields (Table 14.1) and those employing subsurface effects (Table 14.2). This split naturally arises from the discussion of the observations in Sect. 14.4. In these tables the surface/subsurface models have also been subdivided into those acting in single or multiple bipolar configurations in account of the observations discussed in Sect. 14.2. The list should only be regarded as representative and not exhaustive. Due to this, readers are recommended to search for other such papers in the literature. For each of the entries in Tables 14.1 and 14.2 the numbers attached correspond to the various mechanisms that the models employ, as listed in Table 14.3. From the numbers attached to each model in Table 14.1 it is clear that surface models rely on a variety of mechanisms combined together. These include: differential rotation; shear flows along a PIL (differential rotation is just a weak shear flow); and converging flows onto a PIL. For some surface models diffusion of flux towards a PIL with subsequent cancellation plays the role of the converging flow. These mechanisms generally have to occur in a specific

Table 14.1 Surface models of filament formation where the numbers attached to each paper correspond to the mechanisms given in Table 14.3

Single Bipole	Multiple Bipoles
van Ballegooijen and Martens (1989)[1,3,4,10]	Kuperus (1996)[1,3,4]
DeVore and Antiochos (2000)[1,4]	Kuijpers (1997)[3,4,8,10]
	Mackay et al. (1998)[3,4,6,8,10]
	Galsgaard and Longbottom (1999)[3,4]
	van Ballegooijen et al. (2000)[1,4,10]
	Martens and Zwaan (2001)[3,4,10]
	Lionello et al. (2002)[8,10]
	DeVore et al. (2005)[1,3,4]
	Mackay and van Ballegooijen (2005)[1,4,8,10]
	Welsch et al. (2005)[3,4,8,10]
	Litvinenko and Wheatland (2005)[3,4,8,10]
	Yeates et al. (2008a)[1,4,8,10]
	Xia et al. (2014)[3,4,8,10]

Table 14.2 Sub-surface models of filament formation where the numbers attached to each paper correspond to the mechanisms given in Table 14.3

Single Bipole	Multiple Bipoles
Low (1994)[7]	van Ballegooijen and Martens (1990)[2,3,4,7]
Rust and Kumar (1994)[7,9]	Priest et al. (1996)[2,3,4,6]
Gibson et al. (2004)[7,9]	Oliver (1999)[2,3,4,6]
Low and Hundhausen (1995)[7,9]	
Fan and Gibson (2004)[7,9]	
Fan and Gibson (2006)[7,9]	
Gibson and Fan (2006)[7,9]	
Magara (2006)[7,9]	
Fan (2009)[6,9]	

Table 14.3 Mechanisms of filament formation

Surface mechanisms	Subsurface mechanisms
(1) Differential rotation (shear flows)	(2) Subsurface motions
(3) Converging flows	
(4) Magnetic reconnection (atmosphere)	(5) Magnetic reconnection (subsurface)
(6) Flux emergence (bipoles)	(7) Flux emergence (U-loops)
(8) Magnetic helicity	(9) Magnetic helicity
(10) Flux cancellation/diffusion	

order to produce an axial magnetic field direction consistent with observations. In contrast some subsurface models apply a subsurface shear flow. In both surface and subsurface models magnetic reconnection is generally required to reconfigure the fields; the reconnection may occur either above or below the surface.

A common feature to both the surface and sub-surface models is flux emergence, but it is used in very different ways. For surface models, magnetic bipoles which emerge either untwisted or twisted are advected across the solar surface and reconfigured with other pre-existing coronal fields as discussed in the observations of Sect. 14.4.1. A key element in recent papers describing filament formation is that these bipoles are non-potential and include an initial magnetic helicity (Mackay and van Ballegooijen 2005; Yeates et al. 2008a). In contrast, flux emergence for subsurface models is presumed to occur in the form of twisted U-loops (Sect. 14.4.2).

Whilst it is impractical to describe each of the models listed in Tables 14.1 and 14.2 in detail, key elements may be considered from a few selected cases. The cases chosen are picked solely for illustrative purposes. The key feature of any sub-surface model is described in the papers by Low (1994) and Rust and Kumar (1994). For these models a filament is formed when a horizontal twisted magnetic flux tube in the convective zone emerges due to magnetic buoyancy through the photosphere into the corona, dragging cool dense material with it (Gibson and Fan 2006). In flux emergence simulations where the authors use only buoyancy and magnetic buoyancy instabilities, it is found that the axis of the flux rope does not rise through the photosphere (Archontis et al. 2004; Archontis 2008; Murray et al. 2006; Galsgaard et al. 2007). Although the axis and U-loops of the emerging tube do not rise to coronal heights, the process of flux emergence may still produce a coronal flux rope with dips. A flux rope may form through the reconnection of emerged sheared field lines that lie above the emerging tubes axis (Manchester et al. 2004; Magara 2006; Archontis and Török 2008; Fan 2009).

In contrast, one of the first surface models, by van Ballegooijen and Martens (1989), considers shearing motions acting on a coronal arcade in a bipolar configuration. The footpoints of the arcade are sheared in such a way that their separation increases and an axial field component is produced along the PIL (see Fig. 14.6a,b). In principle this shear could be a result of solar differential rotation or by other shear flows on the Sun. Next convergence, or diffusion of the flux towards the PIL, brings the foot points together where they may reconnect to produce, a long axial field line along the PIL and also a small loop which submerges through the surface (Fig. 14.6c,d). Subsequent repetition of this process creates dipped magnetic field lines consistent with the topology required for filaments (Fig. 14.6e,f). This idea was developed further by Martens and Zwaan (2001) who put forward a "head-to-tail" linkage model for the formation of filaments through the interaction of multiple bipoles. While Martens and Zwaan considered this in a conceptual model, Mackay et al. (2000) and Mackay and van Ballegooijen (2001) carried out numerical simulations of a similar process.

An alternative method of forming a strongly sheared magnetic structure was proposed by DeVore and Antiochos (2000) using a single bipolar configuration (also see Antiochos et al. 1994). In this model, a bipolar magnetic field distribution is subjected to a strong shearing motion parallel to the PIL, however, no converging flow is applied. Once the footpoints of the field lines are sheared a distance comparable to the bipole width, an untwisted dipped magnetic configuration forms. The authors show that through further shearing of the dipped field lines the initially untwisted field may form a helical structure similar to that of van Ballegooijen and Martens (1989) through a two stage reconnection process. Therefore, in contrast to van Ballegooijen and Martens (1989) and Martens and Zwaan (2001), DeVore and

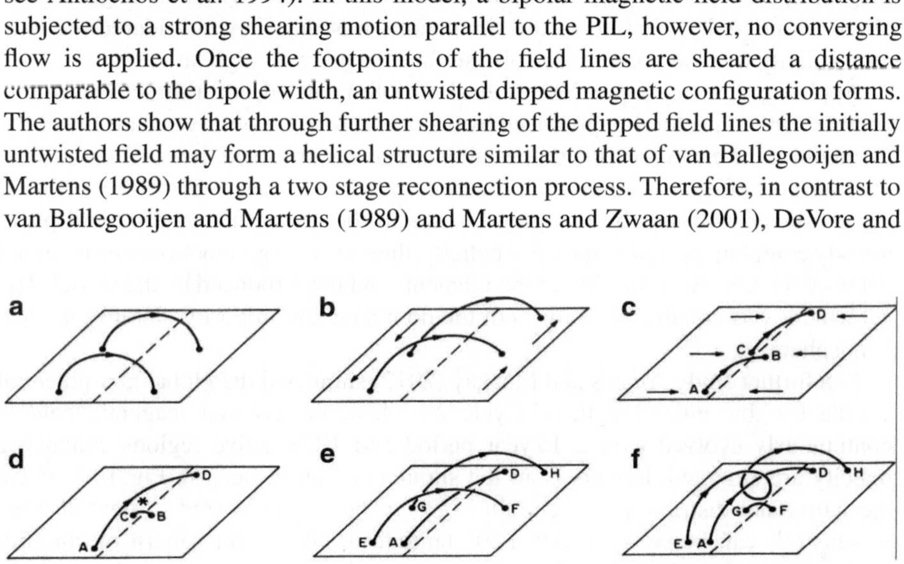

Fig. 14.6 Example of the formation of a filament's axial magnetic field through shearing motions, convergence, and reconnection as put forward by van Ballegooijen and Martens (1989)

Antiochos (2000) do not rely on convergence and cancellation of flux to produce the helical field.

From the discussion above it is clear that a wide range of theoretical models exist to explain the 3D magnetic structure of solar filaments. At the present time none of these models may be ruled out. However, by combining the observations discussed in Sects. 14.4.1 and 14.4.2 it may be argued that some models are more relevant than others for the formation of large stable filaments (Quiescent and Intermediate) compared to Active Region filaments. A full discussion along with the presented hypothesis will be carried out in Sect. 14.7.

14.6 Origin of the Hemispheric Pattern of Filaments

Any model which tries to explain the origin of a filament's magnetic field must also explain why this magnetic field exhibits a hemispheric pattern. Initial studies have considered the origin of the hemispheric pattern by modeling the evolution of either idealised magnetic distributions (Mackay and van Ballegooijen 2001, 2005) or observed distributions that can be compared directly with subsequent measurements (van Ballegooijen et al. 1998; Mackay et al. 2000). To date the most detailed investigation into the origin of the chirality of filaments has been carried out by Yeates et al. (2007) and Yeates et al. (2008a). To carry out the comparison the authors first determined the location and chirality of 109 filaments over a 6 month period. They then developed a new technique to model the continuous long term, global evolution of the Sun's magnetic field from synoptic magnetogram observations. A key feature of the simulations was that they considered the long term helicity transport across the solar surface from low to high latitudes.

Applying this technique, Yeates et al. (2008a) carried out a direct one-to-one comparison of the chirality produced by the model with the observed chirality of the filaments at the exact location that each filament was observed. It was found that, if the transport effects of differential rotation, meridional flow, and surface diffusion are combined with newly emerging bipoles in the northern/southern hemisphere already containing negative/positive helicity, then a 96 % agreement can be obtained between the observed chirality of the filaments and that produced by the model. The agreement was equally good for both the dominant and minority chirality in each hemisphere.

In a further study, Yeates and Mackay (2012) simulated the global non-potential corona for the entire length of Cycle 23 where the coronal magnetic field is continuously evolved over a 15 year period and 1838 active regions containing helicity are emerged. Results from the simulation can be seen in Fig. 14.7 where the latitudinal distribution of chirality in both hemispheres (red $\sim$ dextral, blue $\sim$ sinistral) can be seen. Below $\pm 50°$ latitude the dominant pattern of chirality predominates, i.e., dextral in the northern hemisphere and sinistral in the southern hemisphere, although the overall pattern has significant fluctuations (minority chirality exists at all latitudes). During the rush-to-the-poles between 1998 and 2001

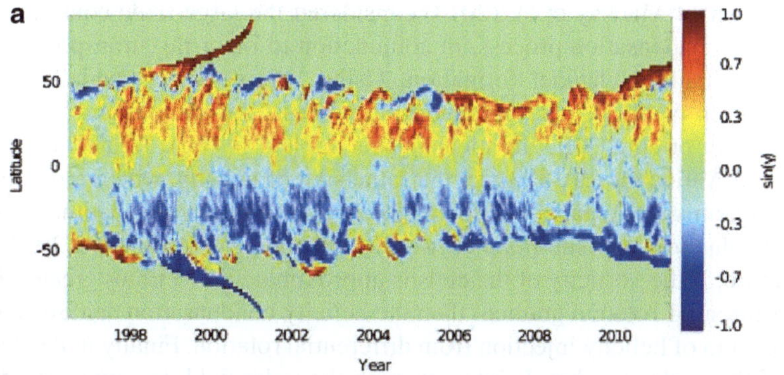

Fig. 14.7 Butterfly diagram over Cycle 23 showing the longitude-averaged skew sin γ measured at height $r = 1.033R_{\odot}$ (from Yeates and Mackay 2012). In the figure *red* represents dextral skew and *blue* sinistral skew

the polar crowns exhibit the dominant chirality pattern. In contrast during the period of low activity from 2007 to 2010, there is a more mixed chirality at lower latitudes. However from 2010 onward, the dominant chirality dominates at high latitudes once more, continuing into Cycle 24. Interestingly until 1998 and during the declining phase of cycle 23 (2001–2006), there is a tendency for minority chirality to occur on the high-latitude PILs (sinistral in the north, dextral in the south). The initial phase of minority chirality at high-latitudes in both hemispheres until 1998 is due the initial condition which is a potential field. Once this is removed by the transport of helicity poleward which occurs over a 2 year time period the correct chirality is found at high latitudes in the rising phase. In contrast in the declining phase the pattern is a true feature of the model and is not due to the initial condition. Thus far, no detailed observational studies of filament chirality have been carried out in the declining phase of the solar cycle to test these predictions.

In contrast to the simulations of Yeates and Mackay (2012) where magnetic helicity is sporadically injected into the corona through flux emergence, Antiochos (2013) proposed a new helicity-condensation model for the formation of filament channels. In this model filament channels form through a multi-stage process of helicity injection, transfer, and condensation that acts on the chromospheric and coronal magnetic fields. First helicity is injected into the overlying atmosphere by small-scale, vortical motions associated with both granular and supergranular convection. Assuming that the motions are predominantly counterclockwise in the northern hemisphere and clockwise in the southern, then the resultant magnetic twist is dextral in the north and sinistral in the south. Next reconnection within magnetically unipolar regions transfers the twist field out to the extent of the unipolar region. At the boundaries between regions of opposite magnetic polarity, the twist then accumulates at the PIL and results is an increasingly strong axial field and filament channel.

The paper of Mackay et al. (2014) considered the large-scale consequences of the helicity-condensation process, in conjunction to other flux-transport processes that affect filament-channel formation. Using a large-scale, spatially averaged representation of the helicity condensation process, the simulations show that on a north-south oriented PIL the mechanisms applied by both Yeates and Mackay (2012) and Antiochos (2013) inject the same sign of helicity, which reproduces the dominant hemispheric pattern of filaments. In contrast on a high-latitude east-west oriented polar crown or sub-polar crown PIL, the model of Antiochos (2013) adds a new feature. If the vorticity of the cells is approximately 2–3 times greater than the local differential-rotation gradient then the helicity condensation can overcome the incorrect sign of helicity injection from differential rotation. Finally in the declining phase of the cycle, as a bipole interacts with the polar field, in some cases helicity condensation can reverse the effect of differential rotation along the east-west lead arm that surround bipoles, but not in all cases. The results show that the magnetic helicity injection and condensation model of Antiochos (2013), in conjunction with the mechanisms used in Yeates et al. (2008a), is a viable explanation for the hemispheric pattern of filaments. In principle, if future observational studies of filament chirality in the declining phase disagree with the simulation results of Yeates and Mackay (2012) then the helicity condensation model may resolve any potential incompatibilities.

14.7 Discussion and Future Observations

So far we have discussed a number of properties of solar filaments and filament channels. These properties have ranged from observations and locations of their formation, to the wide variety of theoretical models used to explain them and their hemispheric pattern. We now turn our attention to tying all of these observations together, by forming a unifying hypothesis, to quantify where and at what locations the mechanisms and models discussed in Sect. 14.5 are appropriate. The aim of this hypothesis is to stimulate new observational studies to test it.

From the observations of large-scale stable filaments discussed in Sects. 14.2 and 14.4 it can be seen that IF and QF preferentially, but not exclusively, form in magnetic configurations involving multiple bipole interactions (92 %). While none of the models listed in Tables 14.1 or 14.2 can be ruled out, it is clear that those involving multiple bipole interactions are the most appropriate for these types of filament.

The question now turns to whether the IF and QF are formed due to surface motions acting on pre-existing coronal fields or whether they are due to sub-surface processes. The observations of Gaizauskas et al. (1997, 2001) and Wang and Muglach (2007) show examples of IF and QF that do not form during the emergence of flux but rather after convergence and cancellation of individual bipoles. Therefore, it may be argued that convergence leading to subsequent cancellation and reconnection (i.e., items 3, 4 and 10 in Table 14.3) are the mechanisms that result in the formation of large stable filaments found on the Sun. For these filaments flux rope emergence does not appear to play a major role, however the injection of magnetic helicity due to the process of flux emergence at an earlier time may play a key role in producing the filament channels. For those large stable filaments that do form within single bipoles (7 %) shear flows such as applied by DeVore and Antiochos (2000) may play a key role in the formation. The models in Table 14.1 which include these mechanisms appear to be the most appropriate to explain IF and QF. At the present time no further distinction can be made between these models.

In contrast the observations by Lites and Low (1997), Okamoto et al. (2009) and Lites et al. (2010) suggest that small-scale unstable active region filaments may be formed as the result of flux rope emergence dragging cool dense photospheric plasma into the corona. While this is a possibility, most numerical simulations of emerging flux ropes fail to lift the axis and cool material of the original flux tube into the corona. Therefore it remains unclear whether such a process may occur. In contrast, many authors have shown that during the process of flux emergence, after the top of the flux rope has emerged, magnetic reconnection or helicity injection (Manchester et al. 2004; Magara 2006; Archontis and Török 2008; Fan 2009) may reconfigure the emerged coronal arcade to produce a secondary coronal flux rope. During the formation of the secondary flux rope the reconnection may then lift cool dense material to coronal heights. Therefore while emerging flux appears to be important for the formation of active region filaments a key element may still be atmospheric reconnection of pre-emerged fields. To resolve this issue new observational studies similar to that of Kuckein et al. (2012), but which follow the early stages of the formation of the AR filament are required.

From the discussion above it appears that different formation mechanisms may apply to different types of filaments. Quiescent filaments and Intermediate filaments which mainly fall into the Exterior and Diffuse bipolar region types rely on surface effects acting on coronal fields. In contrast, for active region filaments a strong possibility is the emergence of flux ropes or the formation of flux ropes during emergence as a result of coronal reconnection. Therefore it is useful to distinguish between IF and QF, compared to ARF as they may have a different formation mechanism.

The formation, structure, and evolution of solar filaments is an important part of our understanding of coronal physics and the behavior of magnetic fields as they are transported across the solar surface. Present evidence suggests that a number of different mechanisms may be taking place in different magnetic environments. A better understanding of the formation of prominences requires multi-wavelength observations of prominences situated over a wide range of latitudes, from the active region belts up to the polar crowns. It is imperative to determine whether different formation mechanisms occur at different latitudes on the Sun. To distinguish this, spectral lines from $H\alpha$ to X-rays along with magnetic information are needed to provide full coverage of the wavelength ranges associated with the formation and structure of filaments. A key aspect of this study is being able to determine where and when a long-lived filament might form. Therefore, maintenance of existing synoptic data sets is a vital part of advanced studies of prominence formation.

In addition, new observational studies are required to understand the magnitude and distribution of vortical motions occurring in convective cells, as these motions may be a missing piece in our understanding of magnetic helicity generation and transport across the Sun.

Acknowledgements DHM would like to thank the members of the two ISSI teams on solar prominences led by Nicolas Labrosse for their stimulating discussions. In particular DHM would like to give special thanks to Vic Gaizauskas, Aad van Ballegooijen, Judy Karpen, Jose Luis Ballester, Brigitte Schmieder and Guillaume Aulanier who have all helped develop my understanding of solar prominences and aided my ability to write the present chapter. DHM would also like to thank STFC and the Leverhulme Trust for their financial support.

References

Antiochos, S. K. (2013). Helicity condensation as the origin of coronal and solar wind structure. *The Astrophysical Journal, 772*, 72.

Antiochos, S. K., Dahlburg, R. B., & Klimchuk, J. A. (1994). The magnetic field of solar prominences. *The Astrophysical Journal, 420*, L41.

Archontis, V. (2008). Magnetic flux emergence in the Sun. *Journal of Geophysical Research (Space Physics), 113*, 3.

Archontis, V., & Török, T. (2008). Eruption of magnetic flux ropes during flux emergence. *Astronomy and Astrophysics, 492*, L35.

Archontis, V., Moreno-Insertis, F., Galsgaard, K., Hood, A., & O'Shea, E. (2004). Emergence of magnetic flux from the convection zone into the corona. *Astronomy and Astrophysics, 426*, 1047.

Aulanier, G., & Demoulin, P. (1998). 3-D magnetic configurations supporting prominences. I. The natural presence of lateral feet. *Astronomy and Astrophysics, 329*, 1125.

Babcock, H. W., & Babcock, H. D. (1955). The sun's magnetic field, 1952–1954. *The Astrophysical Journal, 121*, 349.

Benevolenskaya, E. E. (2005). The formation and evolution of complexes of activity, activity nests and the large-scale connectivity in the solar corona. In *Large-scale Structures and their Role in Solar Activity* (Vol. 346, p. 129).

Bernasconi, P. N., Rust, D. M., & Hakim, D. (2005). Advanced automated solar filament detection and characterization code: Description, performance, and results. *Solar Physics, 228*, 97.

Coffey, H. E., & Hanchett, C. D. (1998). Digital "Cartes Synoptiques de la Chromosphere Solaire et Catalogues des Filaments et des Centres d'Activite". IAU Colloq. 167: New Perspectives on Solar Prominences, 150, 488.

d'Azambuja, L., & d'Azambuja, M. (1948). Ann. Obs. Paris-Meudon, 6, 7.

Deng, Y., Lin, Y., Schmieder, B., & Engvold, O. (2002). Filament activation and magnetic reconnection. *Solar Physics, 209*, 153.

DeVore, C. R., & Antiochos, S. K. (2000). Dynamical formation and stability of helical prominence magnetic fields. *The Astrophysical Journal*, 539, 954.

DeVore, C. R., Antiochos, S. K., & Aulanier, G. (2005). Solar prominence interactions. *The Astrophysical Journal*, 629, 1122.

Engvold, O. (1998). Observations of filament structure and dynamics (Review). IAU Colloq. 167: New Perspectives on Solar Prominences, 150, 23.

Fan, Y. (2009). The emergence of a twisted flux tube into the solar atmosphere: Sunspot rotations and the formation of a coronal flux rope. *The Astrophysical Journal, 697*, 1529.

Fan, Y., & Gibson, S. E. (2004). Numerical imulations of three-dimensional coronal magnetic fields resulting from the emergence of twisted magnetic flux tubes. *The Astrophysical Journal, 609*, 1123.

Fan, Y., & Gibson, S. E. (2006). On the nature of the X-ray bright core in a stable filament channel. *The Astrophysical Journal, 641*, L149.

Foukal, P. (1971). Morphological relationships in the chromospheric Hα fine structure. *Solar Physics, 19*, 59.

Foukal, P. (1971). Hα fine structure and the chromospheric field. *Solar Physics, 20*, 298.

Gaizauskas, V. (1998). Filament channels: Essential ingredients for filament formation (Review). IAU Colloq. 167: New Perspectives on Solar Prominences, 150, 257.

Gaizauskas, V. (2002). Formation of a switchback during the rising phase of solar cycle 21. *Solar Physics, 211*, 179.

Gaizauskas, V. (2008). Development of flux imbalances in solar activity nests and the evolution of filament channels. *The Astrophysical Journal, 686*, 1432.

Gaizauskas, V., Harvey, K. L., Harvey, J. W., & Zwaan, C. (1983). Large-scale patterns formed by solar active regions during the ascending phase of cycle 21. *The Astrophysical Journal, 265*, 1056.

Gaizauskas, V., Mackay, D. H., & Harvey, K. L. (2001). Evolution of solar filament channels observed during a major poleward surge of photospheric magnetic flux. *The Astrophysical Journal, 558*, 888.

Gaizauskas, V., Zirker, J. B., Sweetland, C., & Kovacs, A. (1997). Formation of a solar filament channel. *The Astrophysical Journal, 479*, 448.

Galsgaard, K., & Longbottom, A. W. (1999). Formation of solar prominences by flux convergence. *The Astrophysical Journal, 510*, 444.

Galsgaard, K., Archontis, V., Moreno-Insertis, F., & Hood, A. W. (2007). The effect of the relative orientation between the coronal field and new emerging flux. I. Global properties. *The Astrophysical Journal, 666*, 516.

Gibson, S. E., & Fan, Y. (2006). Coronal prominence structure and dynamics: A magnetic flux rope interpretation. *Journal of Geophysical Research (Space Physics), 111*, 12103.

Gibson, S. E., Fan, Y., Mandrini, C., Fisher, G., & Demoulin, P. (2004). Observational consequences of a magnetic flux rope emerging into the corona. *The Astrophysical Journal, 617*, 600.

Hyder, C. L. (1965). The polarization of emission lines in astronomy. II. Prominence emission-line polarization and prominence magnetic fields. *The Astrophysical Journal, 141*, 1374.

Jeong, H., Chae, J., & Moon, Y.-J. (2009). Magnetic helicity injection during the formation of an intermediate filament. *Journal of Korean Astronomical Society, 42*, 9.

Karachik, N. V., & Pevtsov, A. A. (2014). Properties of magnetic neutral line gradients and formation of filaments. *Solar Physics, 289*, 821.

Karpen, J. T., Antiochos, S. K., Hohensee, M., Klimchuk, J. A., & MacNeice, P. J. (2001). Are magnetic dips necessary for prominence formation? *The Astrophysical Journal, 553*, L85.

Kuckein, C., Martínez Pillet, V., & Centeno, R. (2012). An active region filament studied simultaneously in the chromosphere and photosphere. II. Doppler velocities. *Astronomy and Astrophysics, 539*, A131.

Kuijpers, J. (1997). A solar prominence model. *The Astrophysical Journal, 489*, L201.

Kuperus, M. (1996). The double inverse polarity paradigm—the sign of magnetic fields in quiescent prominences. *Solar Physics, 169*, 349.

Leroy, J.-L. (1989). In E. R. Priest (Ed) *Dynamics and structure of quiescent solar prominences* (p. 77). Dordrecht: Kluwer.

Leroy, J. L., Bommier, V., & Sahal-Brechot, S. (1983). The magnetic field in the prominences of the polar crown. *Solar Physics, 83*, 135.

Li, K. J. (2010). Latitude migration of solar filaments. *Monthly Notices of the Royal Astronomical Society, 405*, 1040.

Lionello, R., Mikić, Z., Linker, J. A., & Amari, T. (2002). Magnetic field topology in prominences. *The Astrophysical Journal, 581*, 718.

Lites, B. W. (2005). Magnetic flux ropes in the solar photosphere: The vector magnetic field under active region filaments. *The Astrophysical Journal, 622*, 1275.

Lites, B. W., & Low, B. C. (1997). Flux emergence and prominences: A new scenario for 3-dimensional field geometry based on observations with the advanced stokes polarimeter. *Solar Physics, 174*, 91.

Lites, B. W., Kubo, M., Berger, T., et al. (2010). Emergence of helical flux and the formation of an active region filament channel. *The Astrophysical Journal, 718*, 474.

Litvinenko, Y. E., & Wheatland, M. S. (2005). A simple dynamical model for filament formation in the solar corona. *The Astrophysical Journal, 630*, 587.

Low, B. C. (1994). Magnetohydrodynamic processes in the solar corona: Flares, coronal mass ejections, and magnetic helicity. *Physics of Plasmas, 1*, 1684.

Low, B. C., & Hundhausen, J. R. (1995). Magnetostatic structures of the solar corona. 2: The magnetic topology of quiescent prominences. *The Astrophysical Journal, 443*, 818.

Mackay, D. H., & Gaizauskas, V. (2003). Helicity as a component of filament formation. *Solar Physics, 216*, 121.

Mackay, D. H., & van Ballegooijen, A. A. (2001). A possible solar cycle dependence to the hemispheric pattern of filament magnetic fields? *The Astrophysical Journal, 560*, 445.

Mackay, D. H., & van Ballegooijen, A. A. (2005). New results in modeling the hemispheric pattern of solar filaments. *The Astrophysical Journal, 621*, L77.

Mackay, D. H., & van Ballegooijen, A. A. (2006). Models of the large-scale corona. I. Formation, evolution, and liftoff of magnetic flux ropes. *The Astrophysical Journal, 641*, 577.

Mackay, D. H., Gaizauskas, V., Rickard, G. J., & Priest, E. R. (1997). Force-free and potential models of a filament channel in which a filament forms. *The Astrophysical Journal, 486*, 534.

Mackay, D. H., Priest, E. R., Gaizauskas, V., & van Ballegooijen, A. A. (1998). Role of helicity in the formation of intermediate filaments. *Solar Physics, 180*, 299.

Mackay, D. H., Longbottom, A. W., & Priest, E. R. (1999). Dipped magnetic field configurations associated with filaments and barbs. *Solar Physics, 185*, 87.

Mackay, D. H., Gaizauskas, V., & van Ballegooijen, A. A. (2000). Comparison of theory and observations of the chirality of filaments within a dispersing activity complex. *The Astrophysical Journal, 544*, 1122.

Mackay, D. H., Gaizauskas, V., & Yeates, A. R. (2008). Where do solar filaments form?: Consequences for theoretical models. *Solar Physics, 248*, 51.

Mackay, D. H., DeVore, C. R., & Antiochos, S. K. (2014). Global-scale consequences of magnetic-helicity injection and condensation on the sun. *The Astrophysical Journal, 784*, 164.

Magara, T. (2006). Dynamic and topological features of photospheric and coronal activities produced by flux emergence in the sun. *The Astrophysical Journal, 653*, 1499.

Manchester, W., IV, Gombosi, T., DeZeeuw, D., & Fan, Y. (2004). Eruption of a buoyantly emerging magnetic flux rope. *The Astrophysical Journal, 610*, 588.

Martens, P. C., & Zwaan, C. (2001). Origin and evolution of filament-prominence systems. *The Astrophysical Journal, 558*, 872.

Martens, P. C., Yeates, A. R., & Pillai, K. G. (2014). IAU Symposium (Vol. 300, p. 135).

Martin, S. F. (1998). Conditions for the formation and maintenance of filaments (invited review). *Solar Physics, 182*, 107.

Martin, S. F., Marquette, W. H., & Bilimoria, R. (1992). The solar cycle pattern in the direction of the magnetic field along the long axes of polar filaments. *The Solar Cycle, 27*, 53.

Martin, S. F., Bilimoria, R., & Tracadas, P. W. (1994). Magnetic field configurations basic to filament channels and filaments. *Solar Surface Magnetism, 433*, 303.

Martres, M. J., Michard, R., & Soru-Iscovici (1966). Étude morphologique de la structure magnétique des régions actives en relation avec les phénomènes chromosphériques et les éruptions solaires. II. Localisation des plages brillantes, filaments et éruptions. *Annales d'Astrophysique, 29*, 249.

McIntosh, P. S. (1972). Solar magnetic fields derived from hydrogen alpha filtergrams. *Reviews of Geophysics and Space Physics, 10*, 837.

Minarovjech, M., Rybansky, M., & Rusin, V. (1998). Prominences and the green corona over the solar activity cycle. *Solar Physics, 177*, 357.
Mouradian, Z., & Soru-Escaut, I. (1994). A new analysis of the butterfly diagram for solar filaments.. *Astronomy and Astrophysics, 290*, 279.
Murray, M. J., Hood, A. W., Moreno-Insertis, F., Galsgaard, K., & Archontis, V. (2006). 3D simulations identifying the effects of varying the twist and field strength of an emerging flux tube. *Astronomy and Astrophysics, 460*, 909.
Okamoto, T. J., Tsuneta, S., Lites, B. W., et al. (2008). Emergence of a helical flux rope under an active region prominence. *The Astrophysical Journal, 673*, L215.
Okamoto, T. J., Tsuneta, S., Lites, B. W., et al. (2009). Prominence formation associated with an emerging helical flux rope. *The Astrophysical Journal, 697*, 913.
Oliver, R., Čadež, V. M., Carbonell, M., & Ballester, J. L. (1999). Coronal potential magnetic fields from photospheric sources with finite width. *Astronomy and Astrophysics, 351*, 733.
Pevtsov, A. A., Balasubramaniam, K. S., & Rogers, J. W. (2003). Chirality of chromospheric filaments. *The Astrophysical Journal, 595*, 500.
Pevtsov, A. A., Panasenco, O., & Martin, S. F. (2012). Coronal mass ejections from magnetic systems encompassing filament channels without filaments. *Solar Physics, 277*, 185.
Pintér, T., Rybanský, M., & Dorotovič, I. (2014). The polar belts of prominence occurence as an indicator of the solar magnetic field reversal. IAU Symposium (Vol. 300, p. 456).
Priest, E. R., van Ballegooijen, A. A., & Mackay, D. H. (1996). A model for dextral and sinistral prominences. *The Astrophysical Journal, 460*, 530.
Rust, D. M. (1967). Magnetic fields in quiescent solar prominences. I. Observations. *The Astrophysical Journal, 150*, 313.
Rust, D. M., & Kumar, A. (1994). Helical magnetic fields in filaments. *Solar Physics, 155*, 69.
Schmieder, B., Mein, N., Deng, Y., et al. (2004). Magnetic changes observed in the formation of two filaments in a complex active region: TRACE and MSDP observations. *Solar Physics, 223*, 119.
Sheeley, N. R., Jr., Martin, S. F., Panasenco, O., & Warren, H. P. (2013). Using Coronal Cells to Infer the Magnetic Field Structure and Chirality of Filament Channels. *The Astrophysical Journal, 772*, 88.
Shimojo, M., Yokoyama, T., Asai, A., Nakajima, H., & Shibasaki, K. (2006). One solar-cycle observations of prominence activities using the nobeyama radioheliograph 1992–2004. *Publications of the Astronomical Society of Japan, 58*, 85.
Tang, F. (1987). Quiescent prominences: Where are they formed? *Solar Physics, 107*, 233.
Topka, K., Moore, R., Labonte, B. J., & Howard, R. (1982). Evidence for a poleward meridional flow on the sun. *Solar Physics, 79*, 231.
van Ballegooijen, A. A., & Martens, P. C. H. (1989). Formation and eruption of solar prominences. *The Astrophysical Journal, 343*, 971.
van Ballegooijen, A. A., & Martens, P. C. H. (1990). Magnetic fields in quiescent prominences. *The Astrophysical Journal, 361*, 283.
van Ballegooijen, A. A., Cartledge, N. P., & Priest, E. R. (1998). Magnetic flux transport and the formation of filament channels on the sun. *The Astrophysical Journal, 501*, 866.
van Ballegooijen, A. A., Priest, E. R., & Mackay, D. H. (2000). Mean field model for the formation of filament channels on the sun. *The Astrophysical Journal, 539*, 983.
Wang, Y.-M., & Muglach, K. (2007). On the formation of filament channels. *The Astrophysical Journal, 666*, 1284.
Wang, Y.-M., Sheeley, N. R., Jr., & Stenborg, G. (2013). Fe XII stalks and the origin of the axial field in filament channels. *The Astrophysical Journal, 770*, 72.
Welsch, B. T., DeVore, C. R., & Antiochos, S. K. (2005). Magnetic reconnection models of prominence formation. *The Astrophysical Journal, 634*, 1395.
Xia, C., Keppens, R., & Guo, Y. (2014). Three-dimensional prominence-hosting magnetic configurations: Creating a helical magnetic flux rope. *The Astrophysical Journal, 780*, 130.
Yeates, A. R., & Mackay, D. H. (2012). Chirality of high-latitude filaments over solar cycle 23. *The Astrophysical Journal, 753*, L34.

Yeates, A. R., Mackay, D. H., & van Ballegooijen, A. A. (2007). Modelling the global solar corona: Filament chirality observations and surface simulations. *Solar Physics, 245*, 87.
Yeates, A. R., Mackay, D. H., & van Ballegooijen, A. A. (2008). Modelling the global solar corona II: Coronal evolution and filament chirality comparison. *Solar Physics, 247*, 103.
Yeates, A. R., Mackay, D. H., & van Ballegooijen, A. A. (2008). Evolution and distribution of current helicity in full-sun simulations. *The Astrophysical Journal, 680*, L165.
Zirker, J. B., Martin, S. F., Harvey, K., & Gaizauskas, V. (1997). Global magnetic patterns of chirality. *Solar Physics, 175*, 27.
Zou, P., Li, Q.-X., & Wu, N. (2014). Non-linear analysis of the long-term behaviour of solar filaments. *Monthly Notices of the Royal Astronomical Society, 437*, 38.

Chapter 15
The Dynamics of Eruptive Prominences

Nat Gopalswamy

Abstract This chapter discusses the dynamical properties of eruptive prominences in relation to coronal mass ejections (CMEs). The fact that eruptive prominences are a part of CMEs is emphasized in terms of their physical association and kinematics. The continued propagation of prominence material into the heliosphere is illustrated using in-situ observations. The solar-cycle variation of eruptive prominence locations is discussed with a particular emphasis on the rush-to-the-pole (RTTP) phenomenon. One of the consequences of the RTTP phenomenon is polar CMEs, which are shown to be similar to the low-latitude CMEs. This similarity is important because it provides important clues to the mechanism by which CMEs erupt. The nonradial motion of CMEs is discussed, including the deflection by coronal holes that have important space weather consequences. Finally, the implications of the presented observations for CME modeling are outlined.

15.1 Introduction

Prominence eruptions (PEs) describe the process by which a previously quasi-stationary prominence erupts and partly or wholly leaves the Sun. When the eruption happens on the disk, it is referred to as a filament eruption. The prominence visible in an instrument's field of view (FOV) in its moving phase is known as an eruptive prominence (EP). We also use the term prominence eruption (PE) as a synonym for EP akin to the usage of coronal mass ejection (CME) to denote the ejected material. The disappearance of a solar filament (DSF) from the observing pass band (usually in H-alpha) is also referred to as disparition brusque (DB). Filaments may also disappear due to local heating, but this chapter does not concern with such thermal DBs. Prominence eruptions have been known for a long time (see e.g. Kleczek 1964; Martin 1973; Engvold 1980). Kleczek (1964) published a catalog of eruptive prominences occurring between 1938 and 1961. Engvold (1980) provided a detailed discussion on the kinematics, occurrence rates, and source

N. Gopalswamy (✉)
NASA Goddard Space Flight Center, Code 671, Greenbelt, MD 20771, USA
e-mail: nat.gopalswamy@nasa.gov

J.-C. Vial, O. Engvold (eds.), *Solar Prominences*, Astrophysics and Space Science Library 415, DOI 10.1007/978-3-319-10416-4_15

regions of eruptive prominences. Prominence/filament observations exist for more than a century, so there is extensive literature covering PEs. On the other hand, complete CME observations are available only for the past few decades. Therefore we focus only on those aspects PEs that involve CMEs because we now know that PEs are an integral part of CMEs (see e.g., Hildner et al. 1975; Schmahl and Hildner 1977; Gosling et al. 1976; Hundhausen 1993; Gilbert et al. 2000; Hori and Culhane 2002; Gopalswamy et al. 2003a; Schrijver et al. 2008; Liu et al. 2012; Parenti 2014). Other chapters in this volume by Gibson (2014), Lugaz (2014), Webb (2014), and Fan (2014) provide complementary information on various aspects of eruptive prominences.

15.2 Prominence Eruptions, CMEs, and Flares

Historically, flares and PEs have been known since the nineteenth century. When CMEs were discovered, it was natural to compare PEs with flares and CMEs. In this section we would like to point out that the three processes can hardly be separated. In order to show the interconnection, we start with an example. Figure 15.1 shows a long east–west filament erupting from the northwest quadrant resulting in a two-ribbon flare and an extended post-eruption arcade (PEA). The eruption was observed by the Extreme-ultraviolet Imaging Telescope (EIT) on board the Solar

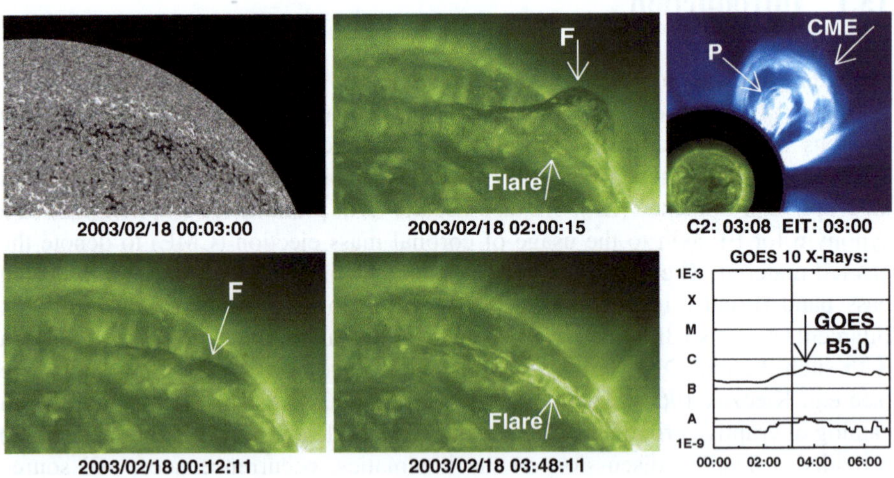

Fig. 15.1 *Left*: A SOHO/MDI magnetogram (2003 February 18 00:03:00 UT) showing the large-scale bipolar magnetic region and a EUV filament (F) overlying the polarity inversion line (00:12:11 UT). *Middle*: The filament erupts (02:00:16) accompanied by a flare arcade observed by SOHO/EIT (03:48:11 UT). *Right*: The associated CME (03:00 UT) with prominence core (P) from SOHO/LASCO and the GOES soft X-ray light curve showing a weak flare (B5.0) (Data source: SOHO/MDI, SOHO/EIT and GOES)

and Heliospheric Observatory (SOHO) mission. When the filament reached the field of view of the Large Angle and Spectrometric Coronagraph (LASCO) on board SOHO, it was found to be in the interior of a large CME. The angular width of the CME in the sky plane was ~90° and the speed was relatively high (~890 km/s). These values are above average for CMEs observed by most coronagraphs (see e.g. Gopalswamy 2004 and references therein). The GOES soft X-ray flare size was only B5.0, which means that the flare was rather weak but can be seen clearly above the background as a gradual event for ~6 h (the flare size is denoted by the letters A, B, C, M, and X in the increasing order of peak soft X-ray flux by an order of magnitude: $A1.0 = 1.0 \times 10^{-8}\ \mathrm{Wm^{-2}}$). In EUV images, the PEA remained above background for many hours. There was no active region in the vicinity of the flare, so this is purely a quiescent filament. Yet, it was associated with both a solar flare and a significant CME.

15.2.1 Statistical Associations

A high degree of association between PEs and CMEs was recognized soon after the discovery of CMEs (Munro et al. 1979; Webb and Hundhausen 1987; St. Cyr and Webb 1991). Munro et al. (1979) found that 70 % of CMEs were associated with PEs. The result was similar in studies starting with PEs and connecting them to CMEs (Gilbert et al. 2000; Hori and Culhane 2002; Gopalswamy et al. 2003b): 72 % of PEs were associated with CMEs, when all events automatically detected (Shimojo et al. 2006) from the Nobeyama Radioheliograph (NoRH) images were used. When PEs with radial trajectories were used, the association between PEs and CMEs increased to 83 %. A closer examination of the PEs without CMEs revealed that the PEs generally had transverse (parallel to the solar surface) trajectories, or they were stalled while moving in the radial direction. These PEs were the slowest and attained the lowest height (~1.2 Rs from the Sun center on the average). There were also intermediate cases in which transverse PEs attaining slightly larger heights and stalled radial eruptions resulting in detectable changes in the overlying streamers (Gopalswamy et al. 2004a). Some of these streamer-change events may indicate weakening of the pre-eruption configuration because they were followed by PEs and CMEs from the same region. The failed eruptions (Moore et al. 2001; Ji et al. 2003; Guo et al. 2010) are likely to be the "stalled radial PEs".

Small-scale energy release often takes place as a precursor to filament eruptions in the form of compact heating observed in EUV and X-rays (Gopalswamy 1999; Chifor et al. 2007; Sterling et al. 2011a) or nonthermal particles inferred from compact radio bursts (Marqué et al. 2001). These signatures indicate reconnection-favoring flux emergence and/or cancelation in the vicinity of filaments that lead to tether cutting (Feynman and Martin 1995; Wang and Sheeley 1999; Chen and Shibata 2000; Gopalswamy et al. 2006). A good example was presented in Gopalswamy et al. (2006). It must be noted that filament eruptions do occur without flux emergence, so there must be other ways in which the filament with its overlying structure gets destabilized and erupts (Schmieder et al. 2013; Aulanier 2014).

15.2.2 Solar Cycle Variations

An overall correlation between the variation of PEs and CMEs over solar cycles has been reported earlier (Webb and Howard 1994). Figure 15.2 presents a long-term comparison of the three manifestations of solar activity: PE rate, CME rate, and sunspot number (SSN). There is a clear drop in SSN between cycles 23 and 24, indicating that the latter cycle is weak. On the other hand, the PE rates are roughly the same between the two cycles, very similar to what is observed in the CME rates (see also Shimojo 2013; Gopalswamy et al. 2014). The PE rates show a clear north–south asymmetry, with the activity peaking first in the northern hemisphere and then in the south during cycle 23 with a similar trend in cycle 24. A similar asymmetry has also been reported in SSN (Svalgaard and Kamide 2013). The PE rate has a closer similarity to the CME rate than to SSN. This is consistent with the fact that PEs are the most common CME-associated phenomenon at the Sun (Munro et al. 1979). Hundhausen (1993) emphasized the tighter association between "larger-scale" activity such as filaments and helmet streamers on the one hand and CMEs on the other mainly based on similar latitudinal distribution and the long-term variation of that distribution. Hundhausen discounted the importance of "smaller-scale" phenomena such as sunspots, flares, and active regions for CMEs. However,

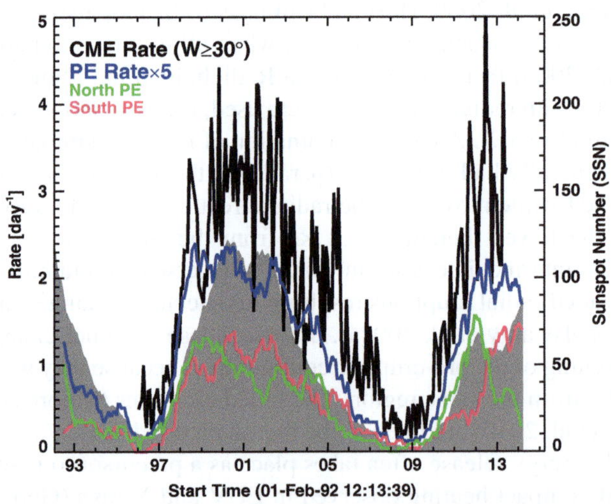

Fig. 15.2 Solar-cycle variation of CME rate (*black*), PE rate (*blue*), and sunspot number (SSN—*gray*). PE rates from the northern and southern hemispheres are distinguished by the *green* and *pink curves*, respectively. The start time coincides with the start of operations of NoRH, which is used for the automatic detection of PEs. The PE daily rate is computed as follows. The observed number of PEs in each Carrington rotation period is multiplied by a factor of three to account for NoRH duty cycle (∼8 h per day). The resulting number is divided by 27.3 days to get the daily rate. The daily rate is then multiplied by a factor of 5 to bring into the scale of the figure. The CME rate is averaged over Carrington rotation periods

it is necessary to point out that both these large- and small-scale features represent closed magnetic field regions, which can produce CMEs if free energy (the energy available for powering the eruption) can be stored in them. In fact, the most energetic CMEs originate mainly from active regions because large amounts of energy can be stored in active regions. The amount of free energy is roughly given by the potential field energy (volume $\times$ $B^2/8\pi$) (Mackay et al. 1997), which can be very large in active regions because of the high magnetic field strength (B). The special populations of CMEs that have significant space weather implications generally originate from the active region belt (see e.g. Gopalswamy et al. 2010a). Filaments are part of active regions too. Active region filaments are thin and short, but can attain much higher speeds similar to the CMEs. The famous backside solar energetic particle (SEP) event with a ground level enhancement (GLE) in cycle 23 on 2001 April 18 was produced by a fast CME ($\sim$2,500 km/s) and the prominence core had a speed of $\sim$1,650 km/s (Gopalswamy 2006a; Gopalswamy et al. 2012a). Filament eruptions have also been associated with some large SEP events, although of softer spectrum (Kahler et al. 1986). Finally, we emphasize that flares are not exclusively an active region phenomenon. Two-ribbon flares can occur from quiescent filament regions (see, e.g., Fig. 15.1).

The flares considered by Hundhausen (1993) are generally confined to the sunspot latitudes because he compiled them from the Solar Geophysical Data that lists flare locations from H-alpha observations when available. However, if we define flares by soft X-ray enhancements, every filament eruption has such an enhancement, observed as PEAs (see McAllister et al. 1996 for a good example). In fact, Gopalswamy et al. (2010a) plotted the locations of all flares from GOES Soft X-ray Imager and found flare locations extending to latitudes above 60° for the period 2004–2007 (see their Fig. 15.9). On the other hand, when the locations of GOES soft X-ray flares with size > C3.0 (i.e., 3.0×10^{-6} Wm^{-2}) alone were plotted, the locations were confined to sunspot latitudes, clearly following the sunspot butterfly diagram. Thus, there is really no clear separation between flare and filament eruption events. The average speeds of CMEs associated with the so-called filament eruption and flare events do differ (Gosling et al. 1976; Sheeley et al. 1999; Moon et al. 2002). Based on CME height-time profiles, MacQueen and Fisher (1983) had suggested that CMEs associated with prominence eruptions and flares may have different acceleration mechanisms. Another way to look at this is that the amount of free energy available in the source regions may be different, but the eruption mechanism may be the same. Fast CMEs (>1,000 km/s) associated with quiescent filament eruptions are not uncommon: Song et al. (2013) reported on 13 eruptions from cycle 23, which they referred to as “flareless CMEs”. As we noted above, such eruptions do have PEAs in soft X-ray and EUV, which become particularly clear when the intensity in a small area around the filament is monitored: the intensity gradually increases similar to other gradual flares, although the intensity is low.

15.2.3 CME and PE Kinematics

Eruptive prominences are observed as the brightest section of CMEs located in the interior of the CME structure. Soon after the discovery of white-light coronal mass ejections, it was realized that "analysis of eruptive prominence only or coronal mass ejection only would be incomplete without the other" (Schmahl and Hildner 1977). Comparing the kinematics of a CME and its prominence core, Webb and Jackson (1981) concluded that they moved out in a self-similar way. Figure 15.3 shows a recent example illustrating how the prominence core and the CME move together. The height-time plot shows the measurements close to the Sun made from NoRH images and then in the LASCO FOV. Within the NoRH FOV the prominence was still accelerating (~2.8 m/s^2) when it left the FOV and appeared as the CME core in the LASCO FOV. The CME was accelerating in the LASCO FOV (average acceleration ~14 m/s^2). The CME had an average speed of ~770 km/s in the LASCO FOV. The combined height-time plot shows that the prominence core closely followed the CME with a speed of ~660 km/s. Statistical analyses comparing the PE properties from below 2 Rs and CME properties in the coronagraphic FOV (above 2 Rs) for ~100 events yielded the following results

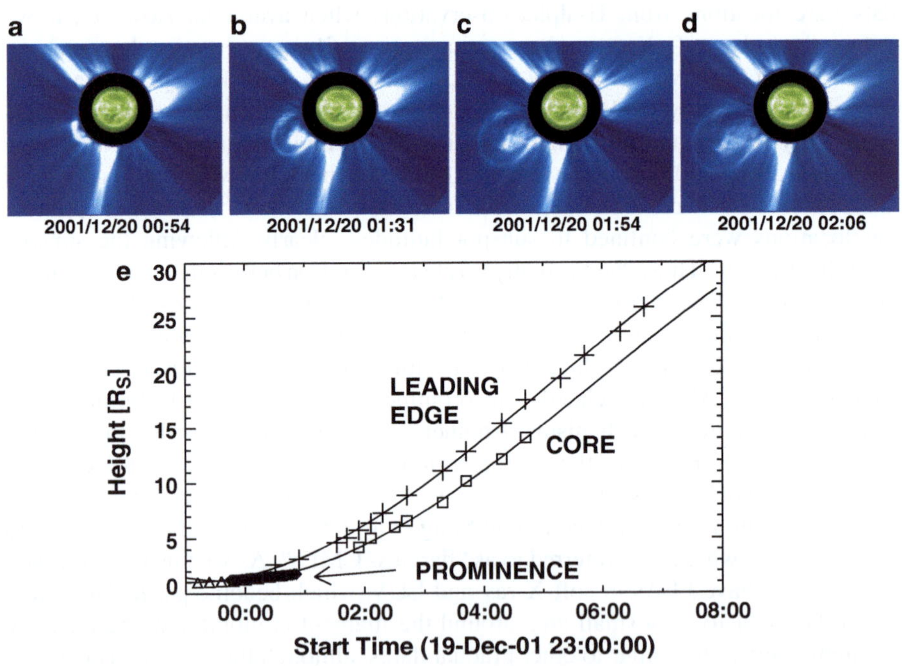

Fig. 15.3 *Top*: A CME with three-part structure observed on 2001 December 20 by SOHO/LASCO. *Bottom*: Height-time measurements of the CME, prominence in the NoRH FOV and the prominence core in the LASCO FOV (Data source: SOHO/LASCO)

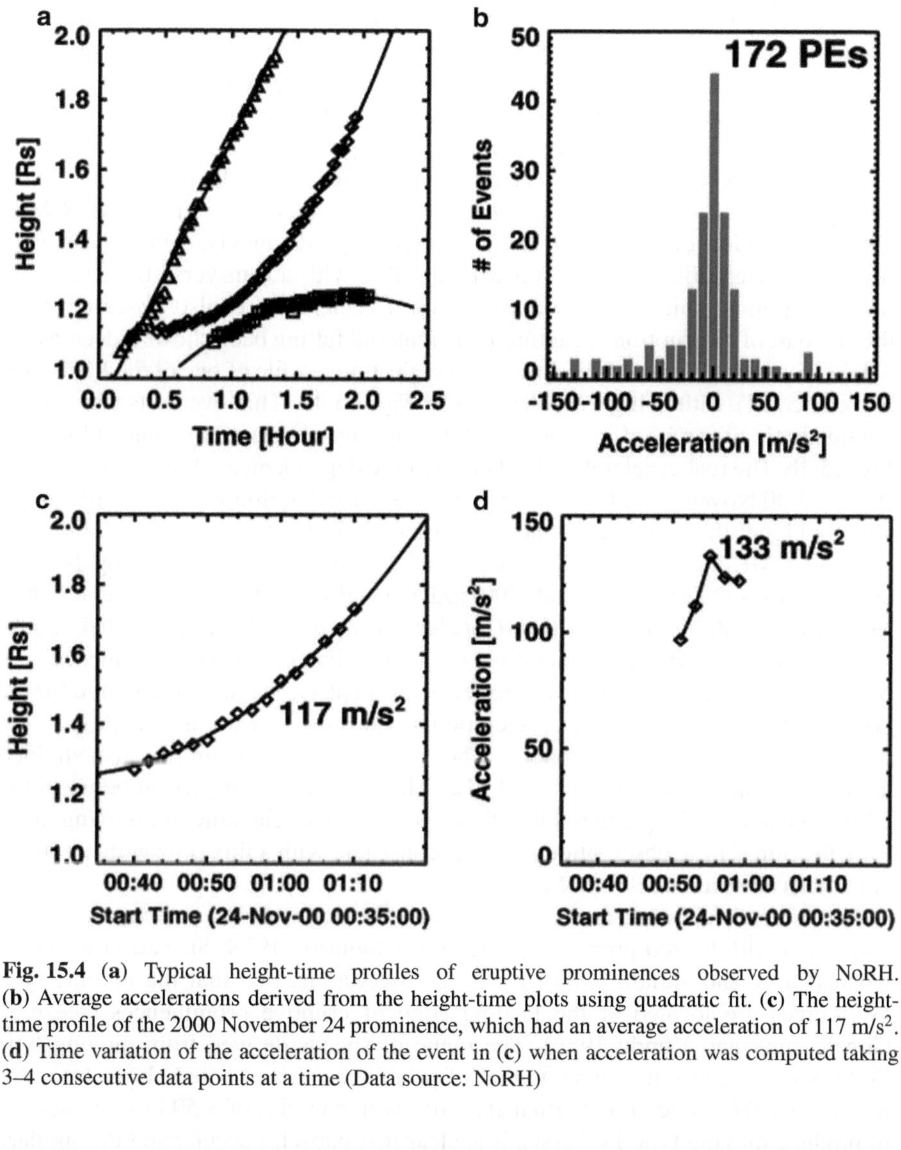

Fig. 15.4 (**a**) Typical height-time profiles of eruptive prominences observed by NoRH. (**b**) Average accelerations derived from the height-time plots using quadratic fit. (**c**) The height-time profile of the 2000 November 24 prominence, which had an average acceleration of 117 m/s^2. (**d**) Time variation of the acceleration of the event in (**c**) when acceleration was computed taking 3–4 consecutive data points at a time (Data source: NoRH)

(Gopalswamy et al. 2003a): (1) The CME core speed (average ~348 km/s) is always greater than the PE speed (average ~81 km/s) because of the continued acceleration. (2) Faster the PEs, the faster are the white-light cores. (3) The CME LE speeds are larger than the core speeds by ~43 % (Maričić et al. 2009 found ~30 % higher LE speed, but only for 18 events).

The acceleration of PEs shows a lot of variations, as can be seen from the typical height-time plots shown in Fig. 15.4. The first two profiles in Fig. 15.4a in which the height continues to increase correspond to PEs that generally leave the Sun and become part of CME cores. The profile with decreasing height at later times corresponds to transverse PEs that do not get very far from the Sun. Tandberg-Hanssen et al. (1980) made a similar comparison between the height-time history of flare sprays and eruptive prominences. They pointed out the height-time plots are similar except that the initial acceleration phase is too quick to be observed. Engvold (1980) showed a number of height-time profiles representing the full range of accelerations and decelerations. Within $\sim$2 Rs, the prominences have accelerations and decelerations as shown by the histogram in Fig. 15.4b. CMEs with acceleration close to zero attain constant speed quickly, while those with positive acceleration continue to accelerate. PEs with a transverse trajectory and failed eruptions typically show deceleration. Deceleration is also observed when the end part of an eruption is captured; the material falling back shows a decrease in height with time. Figure 15.4c shows the height-time profile of one of the PEs (2000 November 24) with a high acceleration in Fig. 15.4b. This event also illustrates the quadratic fitting used in order to get the average acceleration values plotted in Fig. 15.4b. The real acceleration is of course time dependent, as shown in Fig. 15.4d for the 2000 November 24 event. The maximum acceleration was $\sim$133 m/s^2, only slightly higher than the average acceleration ($\sim$117 m/s^2). These values fall in the range of CME leading edge (LE) accelerations (Wood et al. 1999; Gopalswamy and Thompson 2000; Zhang et al. 2001; Zhang and Dere 2006; Vršnak et al. 2007; Maričić et al. 2009; Bein et al. 2011; Gopalswamy et al. 2012a). Maričić et al. (2009) showed that the CME LE acceleration was higher than that of the prominence core by a factor of $\sim$2. The duration of the acceleration phase was about the same for the cores and CME LEs. The peak acceleration had an anti-correlation with the duration of acceleration for both components. The acceleration maximum was also attained around the same time for cores and LEs. The kinematic comparison between the CME core and the LE suggests that they evolve as a single structure moving away from the Sun. These observations are thus consistent with a flux rope with entrained cool material as a model for CMEs.

Nonthermal radio bursts that indicate plasma motion in the corona are closely associated with heated prominence material (Robinson 1978; Stewart et al. 1982; Gopalswamy and Kundu 1989). Imaging observations find that the moving type IV sources are located at the leading edge of eruptive prominences (see e.g., Gopalswamy and Kundu 1989). The range of speeds derived from moving type IV bursts is roughly the same as that of prominence cores noted above (see also Robinson 1978). Since nonthermal electrons with energies of $\sim$50 keV are needed to produce moving type IV bursts, it is clear that particles accelerated during flare reconnection have access to the prominence structure and the surrounding flux ropes. In another case, nonthermal microwave emission was observed from the core and CME in a very fast event on 2001 April 18 (Gopalswamy 2006a). Hard X-ray emission from the prominence core was also observed in this event (Hudson et al. 2001). The heated plasma from flare reconnection also enters flux ropes, observed in situ as the high-charge state plasma.

15.2.4 Prominences in the Heliosphere and Their Earth Impact

Even though significant amount of material drains along the legs of eruptive prominences, one can track prominence cores readily to the edge of the LASCO FOV. In many cases, the cores retain their initial shape throughout the LASCO FOV. Figure 15.5 shows an eruptive prominence that became the core of the 2013 September 29 CME. The core maintained its shape all the way to the edge of the LASCO FOV (~32 Rs) and probably beyond. The north end of the core left the LASCO FOV at ~5 UT, while the south end moved out of the FOV ~ 7 h later.

Early observations indicated that prominence material remained at low temperatures to large distances. From H-alpha observations of a prominence core, Schmahl and Hildner (1977) reported that the core was at a temperature of only $\sim 2 \times 10^4$ K at a distance of ~3 Rs. In some cases, the filament gets heated to coronal temperatures much sooner (Webb and Jackson 1981). When filaments erupt, the microwave brightness temperature typically increases to $\sim 10^4$ K from ~8,000 K, and remains roughly the same near the Sun. This should make the filament disappear because the quiet Sun at 17 GHz has a brightness temperature of $\sim 10^4$ K (decreased contrast). However, when the heated "invisible" filament moved over a nearby plage, it obscured the plage for the duration of the transit of the filament over the plage (Hanaoka and Shinkawa 1999). Similarly, Gopalswamy and Yashiro (2013) reported that a heated eruptive filament obscured the PEA of a nearby flare. These observations suggest that the core of the filament remains at ~8,000 K but the outer sheath is heated to transition region temperatures ($\sim 10^5$ K). Since the sheath plasma is optically thin, it contributes only a few times 1,000 K to the microwave brightness temperature, which explains the observed 10^4 K. These observations suggest that a slow evaporation of the prominence occurs at least during the early phase of the eruption.

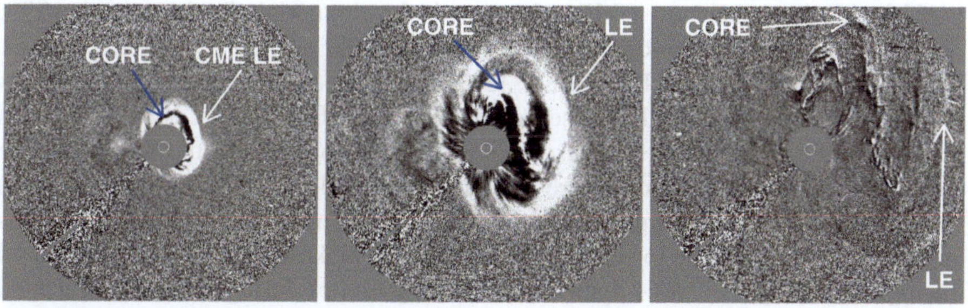

Fig. 15.5 Three snapshots of the CME-prominence core system for the 2013 September 29 CME. In the image (**c**), part of the CME leading edge (LE) has left the LASCO FOV. (**a**) 2013/09/29 23:18. (**b**) 2013/09/30 01:30. (**c**) 2013/09/30 04:18 (Data source: SOHO/LASCO)

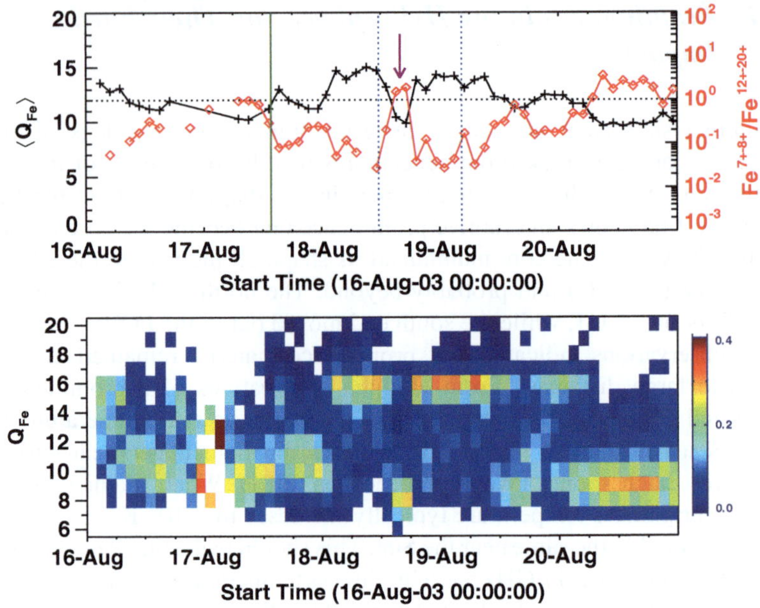

Fig. 15.6 *Top*: Average Fe charge state (<QFe>) and the ratio of low-to-high Fe charge states ($QFe^{7+-8+}/QFe^{12+-20+}$ ratio—*red*) during 17–19 August 2003 interplanetary CME (ICME) as observed by ACE/SWICS. The shock (*green*) and ICME (*blue*) times are marked. The narrow structure within the ICME (*arrow*) with low charge states is likely to be prominence material. *Bottom*: Individual ion charge state abundance (relative to the total abundance of Fe), from which the top curves were derived. Heavy ion charge states connect solar and in situ observations (Data source: ACE/SWICS)

CMEs are observed throughout the heliosphere as flux ropes (see e.g. Richardson et al. 2006). One would certainly expect prominence material to be found inside the interplanetary CMEs (ICMEs). Prominence material is often observed at 1 AU inside interplanetary CMEs along with flare material (Burlaga et al. 1998; Gopalswamy et al. 1998; Reinard 2008; Lepri and Zurbuchen 2010; Gilbert et al. 2012; Gruesbeck et al. 2012). Figure 15.6 shows the high and low charge states within a magnetic cloud (MC) observed by Wind on 2003 August 19. The interval of elevated Fe charge states corresponds to the flare plasma. In the middle of the enhanced charge state region, there is a small interval (12–18 UT on August 18) where the average Fe charge state drops to +10. In order to further explore this interval, we examined the lower Fe charge states. To make it definitive, we compared the high ($\geq$+12) and low (+7 and +8) Fe charge states during the interval around the MC. The low-to-high Fe charge state ratio ($QFe^{7+-8+}/QFe^{12+-20+}$) exceeds 1 in the narrow interval where the Fe charge state dropped to +10 (see the bottom panel of Fig. 15.6). The charge state observations confirm the basic CME morphology: frontal structure, coronal void (flux rope), and prominence core; a shock in addition if the CME is fast (as is the case in Fig. 15.6).

Sharma and Srivastava (2012) reported a similar depression in ion charge states and elevated He^{+}/He^{2+} ratio in intervals identified as filament material at the rear of two MCs. One of the MCs was from the rise phase of solar cycle 24 and the other from the declining phase of cycle 23, but they showed similar filament signatures. Identifying the filament material using elevated He^{+}/He^{2+} ratio, Kozyra et al. (2013) reported that the filament material in the 2005 January 21 CME reached the magnetosphere, allowing the formation of a cold dense plasma sheet from within the magnetosphere from that material (see also Sharma et al. 2013; Dmitriev et al. 2014).

15.2.5 High Latitude Prominences and Prominence Eruptions

The latitude distribution of filaments is intimately connected to solar activity and hence important in understanding the long-term behavior of solar magnetism. In particular the high-latitude filaments that form the polar crown are of interest because they occur only during the maximum phase of solar cycles (see e.g., Ananthakrishnan 1952). Disappearance of the bipolar regions of the polar crown is essential for the sign reversal at the poles. This can happen when the polar crown filaments (PCFs) erupt as part of polar CMEs (Gopalswamy et al. 2003a), providing a way to track the PCFs without observing all of them (see below for an update). The polar CMEs are also important in understanding the eruption mechanism for CMEs because their source regions are purely bipolar regions. The PCF situation is similar to that of quiescent filament regions at lower latitudes. Thus CME eruption mechanisms applicable to low-latitude quiescent filament regions (e.g. Moore and Sterling 2006) are likely to be valid for the polar CMEs also. The same mechanism may be applicable for active region CMEs as well because these regions also contain filaments overlying polarity inversion lines (see e.g., Vemareddy et al. 2012). Mechanisms that require multipolar configuration (e.g. Antiochos et al. 1999) may not apply to eruptions from bipolar regions.

15.2.5.1 Signatures of the Rush to the Pole Phenomenon

The rush to the poles (RTTP) phenomenon refers to filaments that appear in the 40–50° latitude just before sunspot minimum and then systematically move toward the poles in both hemispheres (Lockyer 1931). RTTP was graphically demonstrated by Ananthakrishnan (1952) for the period from 1905 to 1950 (for cycles 14–18). Waldmeier (1960) and Hyder (1965) demonstrated the synchronism between the high-latitude filaments and the sign reversal at solar poles (see also Howard and Labonte 1981; Fujimori 1984; Lorenc et al. 2003; McIntosh 2003). The PCF disappearance lagged the reversal by several months, while the redevelopment of polar coronal holes (PCH) lagged by a few additional months (see also Webb et al. 1984).

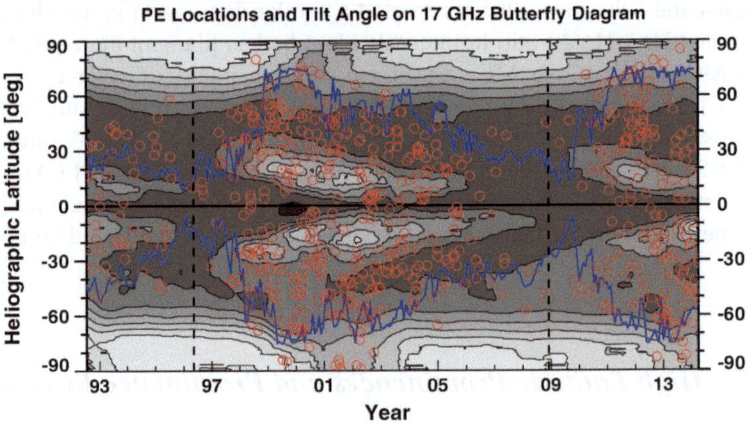

Fig. 15.7 Several indicators of solar cycle phases. (1) The 17 GHz brightness temperature (*contours*) averaged longitudinally for each Carrington rotation and stacked together to see the time variation (the Microwave butterfly diagram). (2) The locations of PEs (*red circles*) detected automatically from NoRH images. These are limb events, so the projection effects are minimal and hence the eruption latitudes are known. (3) The tilt angle of the heliospheric current sheet (*blue line*) obtained from the Wilcox Solar Observatory. The *vertical dashed lines* denote the start times of cycles 23 (May 1996) and 24 (December 2008) (Data source: NoRH)

Figure 15.7 illustrates the relationship among polar eruptions (PEs and CMEs), PCH and the tilt angle of the heliospheric current sheet. The distinct bright patches in microwaves at the poles (contour and gray-scale) correspond to PCH (Kosugi et al. 1986; Gopalswamy et al. 1999, 2012b; Shibasaki 2013). The polar microwave brightness enhancement is proportional to the polar magnetic field strength and corresponds to the chromosphere inside PCHs (Gopalswamy et al. 2012b). The disappearance of polar microwave emission corresponds to solar maximum phases. The low-latitude emission patches correspond to the active region emission (the microwave butterfly diagram). High tilt angles (>60°) correspond to the solar maximum phases. Note that the cycle 23/24 minimum is much extended and the onset of cycle 24 is delayed with respect to the time of peak polar brightness. In the north, the maxima of cycle 23 and 24 can be readily discerned. In the south, the cycle 23 maximum is bracketed by the end of cycle-22 maximum and the beginning of cycle-24 maximum. The locations of PEs (red circles) plotted on the chart are locations of prominence eruptions detected automatically from microwave images. PEs at latitudes >60° occur mainly during the maximum phase (indicated by high tilt angles), which is a representation of the RTTP phenomenon, except that we are tracking PEs rather than filaments or prominences. The cessation of PE activity at high latitudes marks the polarity reversal and the end of the maximum phase. There is clear north–south asymmetry in RTTP and polar sign reversal (see also Altrock 2014; Wang et al. 2002). In cycle 21 the PCF disappearance occurred first in the north, a trend that continued in cycles 22–24. Svalgaard and Kamide (2013) examined the hemispheric sunspot numbers since 1945 and concluded that the asymmetric polar sign reversal is a consequence of the hemispheric asymmetry in the sunspot activity: the hemisphere with dominant activity before the SSN maximum reverses first. Note that the sunspot asymmetry switched in cycle 20 (Svalgaard and Kamide 2013), while the reversal asymmetry occurred in cycle 21. It is not clear why the switch in the reversal asymmetry happens with a lag of one cycle and what implications it may have for dynamo models (Leighton 1969).

15.2.5.2 Polar CMEs vs. Regular CMEs

Helmet streamers, coronal cavities, and filament channels are all related entities that define the pre-eruption environment of a prominence (see Engvold 1989). Helmet streamers overlie cavities containing filaments as seen in eclipse pictures (e.g., Saito and Tandberg-Hanssen 1973). PCFs are no exception. When polar prominences appear at high latitudes, streamers can be found overlying them. Hansen et al. (1969) showed that the white-light brightness peak of the corona migrated from 50° to 80° during 1964–1967, consistent with the RTTP phenomenon (see Zhukov et al. 2008 for a polar streamer observed by SOHO/LASCO). In Kleczek's (1964) catalog of PEs, there were eruptions from various latitudes, including one from the polar zone that reached the largest height (>2 Rs).

Sheeley et al. (1980) were the first to report on a high-latitude CME observed by the Solwind coronagraph on board the P78-1 satellite on 1979 September 27. Sheeley et al. also speculated that there should be more such high-latitude CMEs citing the RTTP phenomenon. However, they were not able to find a solar source—neither a flare nor a prominence eruption—so they suggested that a change in the magnetic field configuration might have caused this CME. Sterling and Moore (2003) reported on a soft X-ray arcade from Yohkoh during the 1999 February 2 PCF eruption, although they did not study the CME association. Our examination revealed a relatively fast CME (average speed ~853 km/s). The acceleration was high (~60 m/s^2) in the LASCO FOV so the speed exceeded ~1,000 km/s before the CME left the coronagraph FOV. Details of this polar CME can be found in: http://cdaw.gsfc.nasa.gov/CME_list/UNIVERSAL/1999_02/htpng/19990209.013005.p049s.htp.html.

The solar source of the 2012 March 12 CME is shown in Fig. 15.11. The PEA as observed by SDO/AIA (193 Å) is in the southeast quadrant because the filament extended beyond the east limb and appeared as a long east–west eruption in STB/EUVI (195 Å) FOV. The variations of the intensity (I) and its derivative dI/dt show the familiar pattern of gradual flares, except that the intensity is very low. The peak acceleration of the CME and core agree with the first dI/dt peak. The acceleration profile of the CME core is similar to that of the LE, but the magnitude is slightly smaller. The CME observation ended before the second peak. The I and dI/dt variations of the PEA are in good agreement between SDO and STB images. The acceleration of the CME LE peaked at ~200 m/s^2. The peak acceleration occurred when the CME LE was at a heliocentric distance of ~2.3 Rs, which is similar to the statistical value obtained by Bein et al. (2011) for a set of ~100 CMEs from low latitudes.

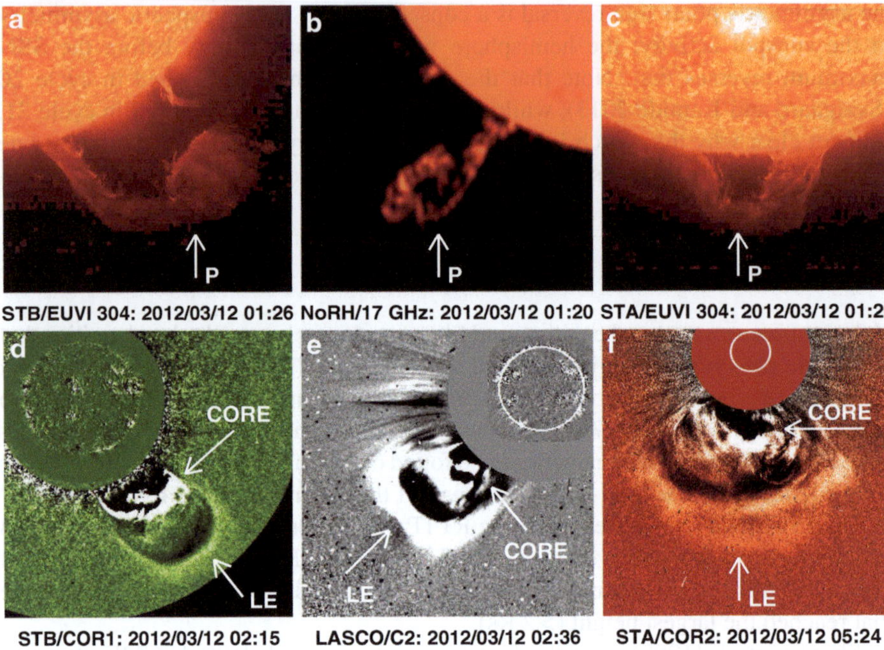

Fig. 15.8 PCF eruption on 2012 March 12 (**a–c**) and the associated CME (**d–f**) from STEREO, NoRH and SOHO observations. The STEREO/EUVI 304 Å images show that it is truly from the polar crown. The prominence (P) becomes CME core (below the CME leading edge LE) in the outer corona as observed by STEREO/COR1, LASCO/C2 and STEREO/COR2. Movies can be found in the CME catalog (Gopalswamy et al. 2009c) (http://cdaw.gsfc.nasa.gov/CME_list/daily_movies/2012/03/12/) (Data source: STEREO/EUVI, STEREO/COR1, STEREO/COR2, NoRH, and SOHO/LASCO)

Gopalswamy (2013) showed that polar CMEs do have near-surface signatures such as two-ribbon flare structure and PEA. High-latitude eruptions started occurring in late 2010 for cycle 24, so we have scores of polar CMEs that can be compared with low-latitude CMEs (see Fig. 15.7). In addition to SOHO, we now have the Solar TErrestrial RElations Observatory (STEREO) and Solar Dynamic Observatory (SDO) observations to study polar CMEs and identify their solar sources unambiguously. One such PCF eruptions (2012 March 12) from Gopalswamy (2013) is shown in Fig. 15.8. The eruption occurred in the south polar zone near the east limb in Earth view as seen in the NoRH 17 GHz image. The prominence was also observed by SDO/AIA at 304 Å (not shown). At the time of the eruption, the STEREO-Behind (STB) spacecraft was located at E117. Therefore, the eruption was observed as a disk event close to the south pole in STB view. It was a backside event in STEREO-Ahead (STA) and the filament can be seen moving straight south in STA/EUVI 304 Å images. The CME was observed in the STB's inner coronagraph (COR1) FOV with a clear 3-part structure. The PEA formed at the initial location of the PCF as is evident from the STB/EUVI image superposed on the COR1 image. The CME appeared in the LASCO FOV at 01:26 UT and was observed until it crossed the FOV about half a day later. The CME was accelerating ($\sim$9 m/s^2) in the LASCO FOV and had an average speed of $\sim$640 km/s. At the time of leaving the LASCO FOV, the CME had a speed of $\sim$715 km/s. In the outer coronagraph (COR2) images, the CME was viewed broadsided and hence showed the full extent. The CME appearance was similar in STB/COR1 and COR2 images.

There is another aspect of this eruption worth mentioning. The filament in the 2012 March 12 CME actually started rising towards the end of the previous day. SDO/AIA images taken before the filament rise showed the lower part of the prominence cavity. Figure 15.9 shows the evolution of the cavity and prominence in three 171 Å SDO/AIA images. The LASCO/C2 images show the pre-eruption streamer overlying the cavity and prominence and two snapshots of the CME. The

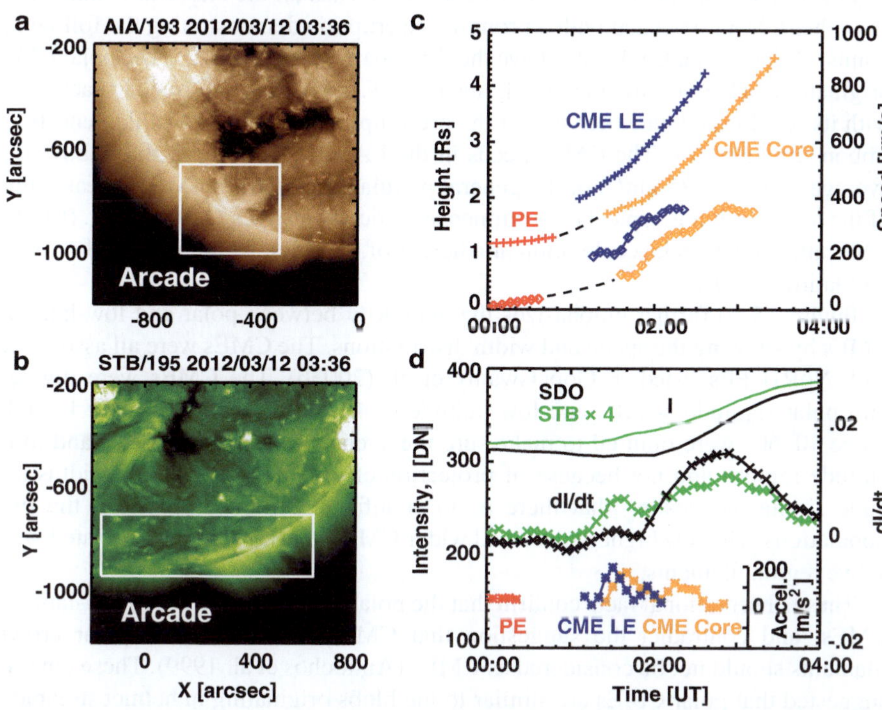

Fig. 15.9 *Top*: SDO/AIA images at 171 Å showing the pre-eruption prominence and cavity (March 11, 13:58 UT), the slowly-rising cavity and prominence (March 11 22:59 UT) and the prominence leg after the cavity has left the FOV (March 12 01:29 UT). *Bottom*: Three LASCO/C2 images showing the polar streamer (March 12 00:00 UT), the early phase of the CME when the prominence core is still below the occulting disk (02:00 UT) and the whole CME with all the substructures: Leading edge (LE), cavity, and prominence core (03:12 UT). The fine thread that crosses the CME in the latitudinal direction is likely to be a bundle of field lines indicating the flux rope structure. The prominence core is the lateral section of the long filament that extends into the plane of the figure, curving to the right because it was observed so in STB view (Data source: SDO/AIA, and STEREO/EUVI)

erupted cavity is seen in the LASCO images with the CME LE flattened. The fine thread-like feature near the outer edge of the cavity going over the prominence core of the CME is indicative of the flux rope structure. Such threads have been identified as field-line bundles that make up the flux rope (cavity) in white-light images (Chen et al. 1997). The flux rope was deformed somewhat between the 02:00 and 03:12 UT images. The cavity and prominence extended into the plane of the figure curving to the right as evidenced by the long east–west filament that erupted as observed by STB/EUVI images (see Fig. 15.9b). A similar prominence/cavity eruption was reported by Régnier et al. (2011), who concluded that the prominence is located at the bottom of the flux rope. It is now confirmed that cavities are ubiquitously found in quiescent filament regions and in the polar zone (Low and Hundhausen 1995; Gibson et al. 2010) and only occasionally in active regions. Detailed discussion on cavities and their relationship to prominences and CMEs can be found in Chap. 13.

In summary, the polar CME of 2012 March 12 has all the classical features like any other CME associated with a prominence eruption demonstrating the following points. (1) The polar CMEs also have the three-part morphology. (2) The polar CME originates in a helmet streamer overlying the PCF. (3) A PEA is formed in each case, with its feet located on either side of the pre-eruption location of the filament (two-ribbon structure). (4) The CME speeds in the LASCO FOV are slightly above the average value ($\sim$475 km/s) of the general population of CMEs. (5) The peak value of the acceleration is typical of prominence associated CMEs ($>$100 m/s^2). (6) The CME attains the peak acceleration at a height of $\sim$2.3 Rs, which is typical of most low-latitude CMEs.

Figure 15.10 further emphasizes the similarity between polar and low-latitude CMEs by showing the speed and width distributions. The CMEs were all associated with NoRH PEs listed in Gopalswamy et al. (2003b). The CMEs were divided into polar (latitude $> 60°$) and low-latitude ($\leq$40°) CMEs. CMEs in the latitude range 40–60° were omitted to make sure the events are truly from polar and low-latitude regions and not because of projection effects. The speeds and widths are quite similar suggesting that there is no significant difference between the two populations. The few higher-speed and wider CMEs are likely to be associated with active region filaments.

These observational facts confirm that the polar CMEs are similar to low-latitude CMEs, and contradict the suggestion that CMEs associated with polar crown filaments should not be considered as CMEs (Antiochos et al. 1999). These authors suggested that polar CMEs are similar to the blobs originating in helmet streamers at a heliocentric distance of about 3–4 Rs and accelerating slowly ($\sim$4 m/s^2) from $\sim$150 km/s at 5 Rs to 300 km/s at $\sim$25 Rs (Sheeley et al. 1997). Clearly, the peak acceleration of polar CMEs is larger by two orders of magnitude (see Fig. 15.11), unlike the Sheeley blobs, and similar to regular CMEs. Karpen et al. (2012) concluded that CMEs do occur even without flare reconnection, but the eruption will be slow, more like a streamer blowout (e.g., Sheeley et al. 1997) than a fast CME. They predicted a clear difference in the early acceleration profile between CMEs with and without impulsive flares. The observations presented here do indicate that the acceleration profile and magnitude are similar to that of regular

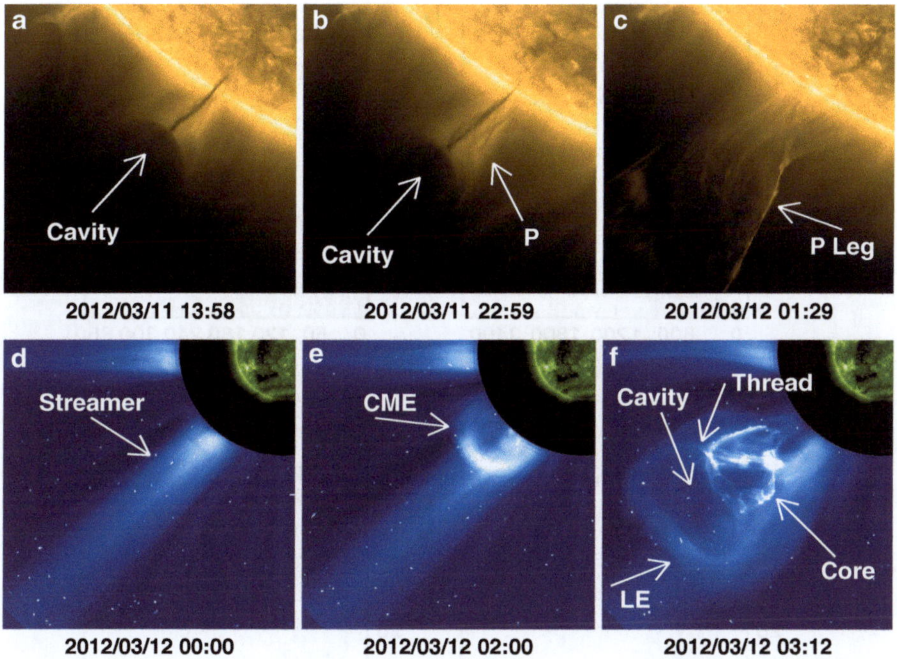

Fig. 15.10 (*top*) SDO/AIA images at 171 Å showing the pre-eruption prominence and cavity (March 11, 13:58 UT), the slowly-rising cavity and prominence (March 11 22:59 UT) and the prominence leg after the cavity has left the FOV (March 12 01:29 UT). (*bottom*) Three LASCO/C2 images showing the polar streamer (March 12 00:00 UT), the early phase of the CME when the prominence core is still below the occulting disk (02:00 UT) and the whole CME with all the substructures: Leading edge (LE), cavity, and prominence core (03:12 UT). The fine thread that crosses the CME in the latitudinal direction is likely to be a bundle of field lines indicating the flux rope structure. The prominence core is the lateral section of the long filament that extends into the plane of the figure, curving to the right because it was observed so in STB view (Data source: SDO/AIA and SOHO/LASCO)

CMEs (see also Joshi and Srivastava 2011). The polar CMEs are also associated with flare reconnection as evidenced by the PEAs. Thus polar CMEs carry mass, kinetic energy, and helicity from the source region into the interplanetary medium and remove the "PCF barrier" leading to the completion of the polarity reversal.

15.3 Non-radial Motion of Eruptive Prominences and CMEs

A systematic equatorward deflection of CMEs observed during 1973–1974 by $\sim 2°$ in the inner corona was reported by Hildner (1977), who concluded that (1) there must be a nonradial (equatorward) force acting on the CMEs and (2) this must result in an enhanced effect of CMEs on the near-ecliptic IP medium. MacQueen et al.

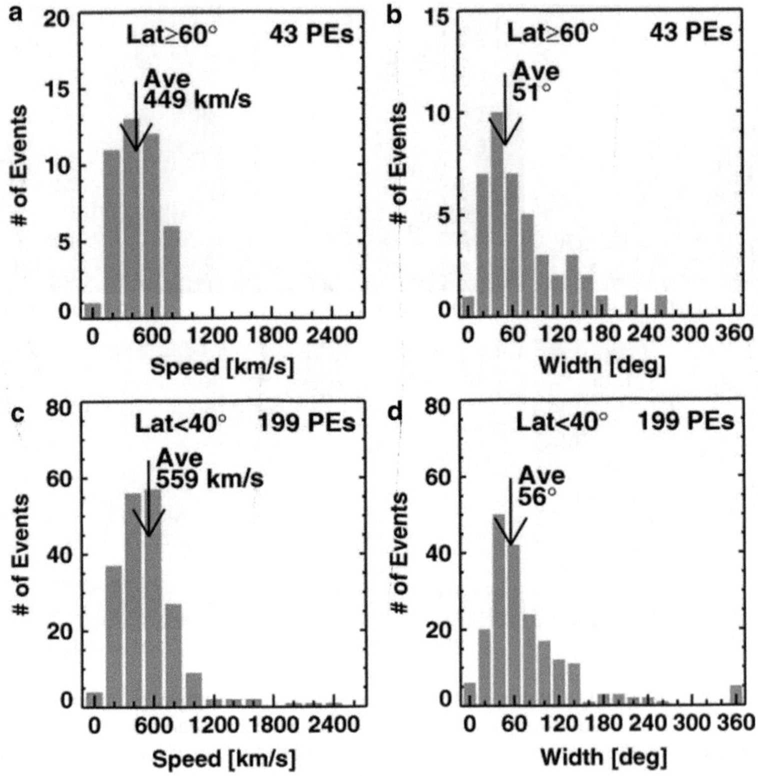

Fig. 15.11 Speed and width distributions of cycle-23 polar CMEs (*top*) and low-latitude CMEs (*bottom*). Data from Gopalswamy et al. (2003b)

(1986) confirmed the deflection (average ~2°, maximum ~10°), but dismissed the possibility that it may have enhanced CME impact on the near-ecliptic IP medium. Decades of CME observations have confirmed the importance of CME deflection in understanding the propagation and geo-impact of CMEs.

Prominences were also found to have nonradial motion (Gopalswamy et al. 2000; Gopalswamy and Thompson 2000; Simnett 2000). Gopalswamy and Thompson (2000) found that both the prominence and the CME showed the deflection, suggesting that the CME deflected as a whole. The initial position angle of the prominence eruption was 120° (or S30 in latitude). When the CME was observed in the LASCO/C3 FOV, both the CME and the prominence were at the position angle PA = 90° (at the equator). Thus the equatorward deflection was ~30°, which was the PA offset between the initial PE location and the CME nose.

Figure 15.12 illustrates a CME deflection event observed in full detail by the STEREO coronagraphs COR1 and COR2. The prominence erupted at a position angle of 256° at 01:50 UT. In the next 2 h, the CME nose and the eruptive prominence had a significant movement toward the equator to PA ~ 267°. Finally,

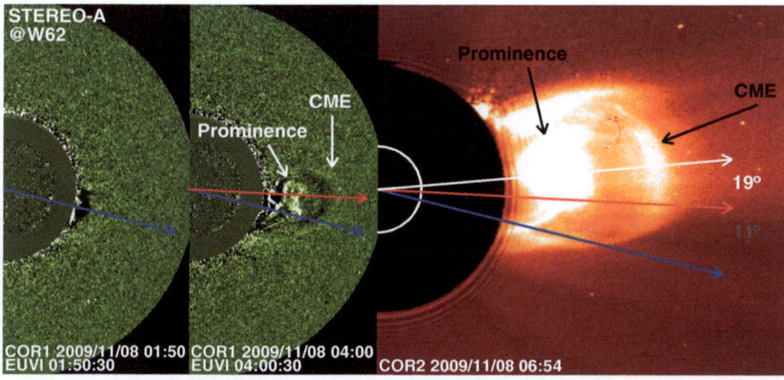

Fig. 15.12 Deflection of the 2009 November 08 CME and prominence from the southwest initial direction (*blue arrow*) to the west (*white arrow*). The *red arrow* (11° away from the *blue* one) represents the CME position angle before it left the COR1 FOV. The images are from STA/EUVI, COR1, and COR2. Total deflection was by ~19° over a period of ~5 h (Data source: STEREO EUVI, COR1, COR2)

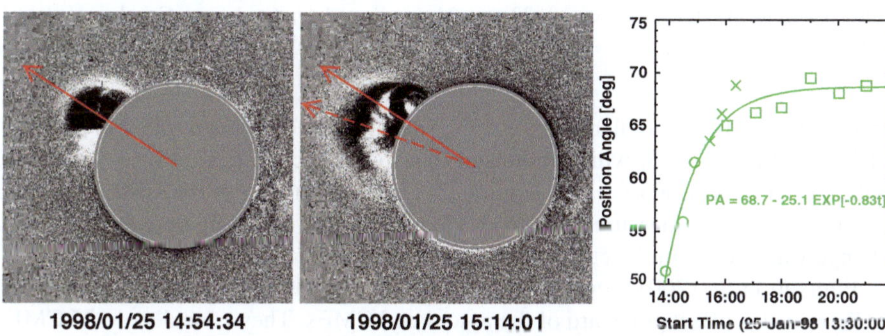

Fig. 15.13 Two SOHO/LASCO/C1 images (*left, middle*) of the corona showing the 1998 January 25 CME deflecting toward the equator from a position angle of ~55° (*solid arrow*) to ~61° (*dashed arrow*). *Right*: Variation of the CME central position angle as a function of time. *Circles*, *crosses*, and *squares* represent measurements, respectively from C1, C2, and C3 coronagraphs of SOHO/LASCO. The *solid line* is the fit to the data points, showing that the nonradial motion stopped within a time T ~ 2 h from the beginning of the eruption (Data source: SOHO/LASCO)

the CME was near the equatorial plane (PA ~ 275°) by 06:54 UT as observed by STEREO/COR2. The latitudinal movement of the CME can be quantified as 3°.75 per hour. This is typical of CMEs deflected toward equator in the rise phase of cycles 23 and 24. One of the consequences of this deflection is that relatively more magnetic clouds are observed near Earth during the rise phase of solar cycles resulting in intense geomagnetic storms (Gopalswamy et al. 2008).

Figure 15.13 shows the CME and prominence deflection in the 1998 January 25 event close to the Sun because SOHO's inner coronagraph LASCO/C1 was still operating. In the 14:54 UT frame, only the CME LE was observed. In the 15:14 UT frame, both the CME and the prominence core were visible. Between these two

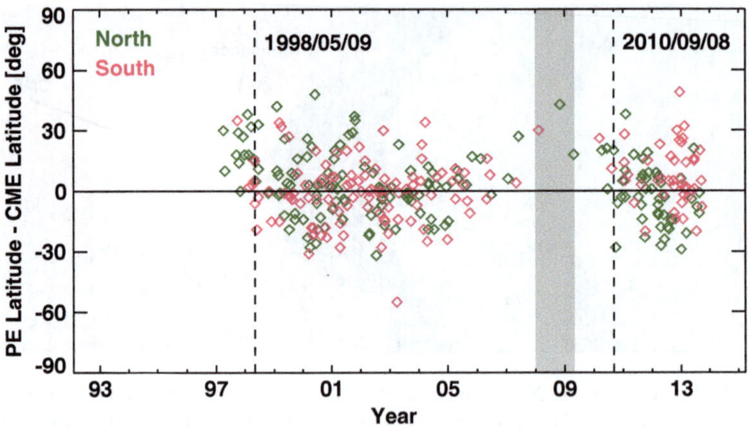

Fig. 15.14 The latitude offset between PE (NoRH) and CME (LASCO). PEs originating in the northern and southern hemispheres are distinguished. The *vertical dashed lines* mark the times when the systematic poleward offset of CMEs with respect to the PEs ended. The *shaded region* corresponds to the time when cycles 23 and 24 overlapped (updated from Gopalswamy et al. 2012b)

frames, the CME was deflected by $\sim$12°. The CME was further tracked in the FOV of LASCO/C2 and C3, which indicated that the deflection ceased by $\sim$16:00 UT, with a total deflection of $\sim$ 17°. Figure 15.13 shows the PA of the CME nose as a function of time (t, measured in h from 13:30 UT). The e-folding time for trajectory change was $\sim$1.2 h in the early phase.

Plunkett et al. (2001) reported that the initial location of eruptions observed in EUV was also offset poleward of the associated CMEs. They considered 135 CMEs during April–December 1997 and found a bimodal distribution for eruption latitudes in EUV, while the corresponding CME latitudes were unimodal. These observations also indicated an average offset of $\sim$30°. Gopalswamy et al. (2003b) investigated more than 200 PEs detected by NoRH, which revealed offsets as large as $\sim$40° (see their Fig. 15.12). The offset was systematically poleward of the associated CMEs in the rise phase. The offset was not systematic during the maximum phase as was noted by MacQueen et al. (1986). The systematic offset found for the rise phase of cycle 23 was again confirmed in the rise phase of cycle 24 as shown in Fig. 15.14 (see also Gopalswamy et al. 2012b).

Now we examine why Hildner (1977) and MacQueen et al. (1986) observed much smaller deflection. One possibility is that these authors measured the deflection in the narrow radial range of 2–3 Rs. In order to see if the deflection increases with radial distance, we have plotted the CME position angle as it moved out. Figure 15.15 shows the change in the PA of the CME nose as a function of the CME heliocentric distance (H) for two CMEs: the 1998 January 25 CME from cycle 23 (Fig. 15.14) and the 2009 May 5 CME from cycle 24. Clearly the PA changed until the CME reached a certain height and then became stable. The PA vs. H curves have

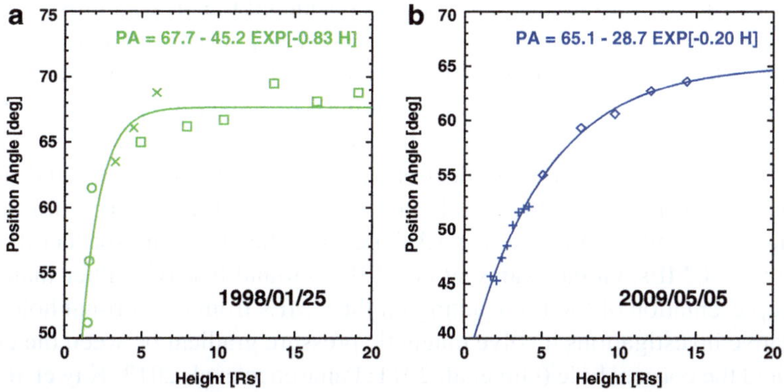

Fig. 15.15 CME Position angle (PA) at various heliocentric distances (H). (**a**) The 1998 January 25 event with data points from SOHO/LASCO telescopes C1 (*circles*), C2 (*crosses*) and C3 (*squares*). (**b**) The 2009 May 5 event with data points from STEREO COR1 (*crosses*) and COR2 (*diamonds*). The *solid curves* are fits to the data points in the form PA = A − B exp(−H/H*), where A, B, and H* are coefficients of the fit given in the plots (Data source: SOHO/LASCO and STEREO/COR1, COR2)

the form, PA = A − B exp(−H/H*) where A and B are constants and H* is the e-folding distance. In the 1998 January 25 case, the maximum PA was attained when H was ~5 Rs (H* ~1.2 Rs). On the other hand, the deflection occurred over a much larger height and gradually (H* ~5 Rs) in the 2009 May 5 event.

The large deflections noted above support the original suggestion by Hildner (1977) that the deflection may have implications for the plasma in the equatorial region. The deflection towards the equator has been suggested as the reason for the relative higher rate of detection of magnetic clouds during the rise phase of solar cycle 23 (Gopalswamy 2006b; Riley et al. 2006; Gopalswamy et al. 2008). All CMEs are likely to have a flux rope structure (i.e., magnetic cloud) and are observed so at 1 AU only when the observing spacecraft passes through the central part of the flux rope. The equatorward deflection during the rise phase thus allows CMEs to be detected as flux ropes at 1 AU.

Both Hildner (1977) and MacQueen et al. (1986) attributed the nonradial motion to the global pattern of magnetic field and flow in the corona, which are distinct during solar minima and maxima. Filippov et al. (2001) proposed a simple axisymmetric model of the global magnetic field configuration with embedded flux rope to explain the nonradial motion of a prominence reported in Gopalswamy et al. (2000). During the minimum phase, PCHs are prominent and contain open magnetic fields of high strength. Active regions emerge at higher latitudes (~40°) during this phase, so CMEs erupt in the vicinity of the PCHs. The current thinking is that the magnetic field in coronal holes is responsible for CME deflection toward the equator. Coronal holes also occur at low latitudes, so similar deflection should happen when a CME erupts near low-latitude coronal holes. Gopalswamy et al. (2004b, 2005) reported coronal-hole influence on CMEs, with deflections toward and away from the Sun–Earthline. Cremades et al. (2006) considered the area and distance to the eruption region of coronal holes (but not the magnetic field) to quantify the influence on CMEs. Gopalswamy et al. (2009a) introduced the photospheric magnetic field (B) inside coronal holes as another key parameter in determining the coronal hole influence on CMEs. They were able to show that many disk-center CMEs did not arrive at Earth because of deflection by large coronal holes located near the eruption region. However, the shocks associated with these CMEs did arrive at Earth and were called "driverless shocks" because the observing spacecraft did not intercept the CMEs. Gopalswamy et al. (2010b) found that B^2, rather than B, is a better representation of the force acting on the CMEs from the coronal holes. More quantitative investigations involve magnetic pressure gradient between the eruption region and the coronal hole (Gui et al. 2011; Panasenco et al. 2013; Kay et al. 2013), but the derived extent of deflection is similar to the previous work. Deflection from a high-field region can also occur in active regions, when the eruption occurs near large sunspots as described by Sterling et al. (2011b) for the 2006 December 13 CMEs. We also would like to point out that systematic nonradial motion can also be found in over-and-out CMEs that do not involve major PEAs (Moore and Sterling 2007).

The deflections to the extent of 20–30° away from the Sun–Earth line are adequate to make the CMEs behave like limb CMEs and miss Earth (Gopalswamy et al. 2010b). Such driverless shocks were mostly observed during the declining phase of solar cycle 23, consistent with the abundance of low-latitude coronal holes in this phase. When the deflection is less severe, disk-center CMEs arrive at Earth as non-cloud ICMEs because the observing spacecraft pass through the edges of the flux ropes due to deflection by coronal holes (Gopalswamy et al. 2013; Xie et al. 2013; Mäkelä et al. 2013). So far we discussed the deflection of prominences and CMEs, which correspond to the main body of CMEs. Coronal holes can also deflect the shock surrounding CMEs, which may be observed as reflected waves in EUV (Long et al. 2008; Gopalswamy et al. 2009b; Olmedo et al. 2012). Coronal hole deflection can also result in the lack of alignment between the ejecta and shock (Wood et al. 2012).

15.4 Implications for Models

The observations presented in this chapter indicate that eruptive prominences are an integral part of CMEs and typically occupy a small volume compared to the entire CME. Therefore, prominence eruptions need to be modeled along with the CME flux rope in which the prominence is embedded. There have been many observational signatures that confirm that almost all CMEs in the interplanetary medium have the flux rope structure. Therefore, the flux rope structure is fundamental to CMEs. When the flux rope is fast enough, it can drive a shock from certain distance in the corona. This has something to do with the interaction of the flux rope with the ambient medium. Observations of high and low charge states in flux ropes at 1 AU point to an intimate connection to the hot (flare) and cool (prominence) plasmas originating near the Sun. The high charge state material is a characteristic of the IP flux rope irrespective of the primary near-surface feature (flare or prominence eruption). The approximate equality between flare reconnection flux and the azimuthal magnetic flux of IP flux ropes (Qiu et al. 2007) point to formation of flux ropes as part of the eruptive process. On the other hand, the presence of prominence cavity also indicates the presence of a flux rope, probably formed in a non-eruptive manner. The example shown in Sect. 15.2.4 indicates that the cavity rises and leaves the Sun as a flux rope. This means an initial flux rope with some added flux seems to be the best possible scenario. Different proportions of existing and eruptive components of the flux are expected in different flux ropes in the IP medium.

The free energy that can be stored in closed magnetic regions on the Sun accounts for the observed range of CME speeds ($<$100 to $>$3,000 km/s) and accelerations (up to $\sim$10 km/s^2). The acceleration can vary by two orders of magnitude from (hundreds of m/s^2 for eruptions from quiescent filament regions to $>$1 km/s^2 for active region eruptions). Thus the lowest CME acceleration is of the order of surface gravity of the Sun, which is negligible for the most energetic eruptions. Given the observed CME mass of up to 10^{16} g, these accelerations give an idea of the magnitude of the force involved in the eruptions.

The fact that polar CMEs behave like any other CME is important in so far as it can help defining models. The polar CMEs originate from purely bipolar regions formed by the approaching of insurgent and incumbent fluxes in the polar zone. The bipolar nature of quiescent filament regions at low latitudes is no different. Many active regions also have bipolar structure and so produce CMEs. Models that require bipolar regions have the universal appeal, while those requiring multipolar configuration may work only in certain regions.

Finally, propagation models need to account for deflections by other large-scale structures in the corona and IP medium. This is especially important for Earth-directed CMEs because the deflection can channel a CME toward or away from the Sun–Earth line. What is important is the deflection close to the Sun, which may not be properly taken into account by models such as ENLIL (Odstrcil and Pizzo 2009) whose inner boundary is at $\sim$21 Rs. By the time CMEs reach this distance, the deflection effect might have disappeared.

Acknowledgments I thank the editors, Oddbjørn Engvold and Jean-Claude Vial, for inviting this contribution. I thank David Webb, Pertti Mäkelä, Sarah Gibson, Ronald Moore, and Alphonse Sterling for helpful comments. I also thank S. Yashiro, S. Akiyama, and P. Mäkelä for their help with the figures. Work supported by NASA's LWS TR&T program.

References

Altrock, R. C. (2014). Forecasting the maxima of solar cycle 24 with coronal Fe xiv emission. *Solar Physics, 289*, 623–629.

Ananthakrishnan, R. (1952). Prominence activity and the sunspot cycle. *Nature, 170*, 156–158.

Antiochos, S. K., Devore, C. R., & Klimchuk, J. A. (1999). A model for solar coronal mass ejections. *The Astrophysical Journal, 510*, 485–493.

Aulanier G (2014) The physical mechanisms that initiate and drive solar eruptions. In B. Schmieder, J.-M. Malherbe, & S. T. Wu (Eds.), *Nature of Prominences and their role in Space Weather. Proceedings of the International Astronomical Union, IAU Symposium* (Vol 300, pp. 184–196).

Bein, B. M., Berkebile-Stoiser, S., Veronig, A. M., et al. (2011). Impulsive acceleration of coronal mass ejections. I. Statistics and coronal mass ejection source region characteristics. *The Astrophysical Journal, 738*, 191–204.

Burlaga, L., Fitzenreiter, R., Lepping, R., Ogilvie, K., Szabo, A., Lazarus, A., et al. (1998). A magnetic cloud containing prominence material – January 1997. *Journal of Geophysical Research, 10*, 277.

Chen, P. F., & Shibata, K. (2000). An emerging flux trigger mechanism for coronal mass ejections. *The Astrophysical Journal, 545*, 524–531.

Chen, J., Howard, R. A., Brueckner, G. E., Santoro, R., Krall, J., Paswaters, S. E., Cyr, O. C., Schwenn, R., Lamy, P., & Simnett, G. M. (1997). Evidence of an erupting magnetic flux rope: LASCO coronal mass ejection of 1997 April 13. *The Astrophysical Journal, 490*, L191.

Chifor, C., Tripathi, D., Mason, H. E., & Dennis, B. R. (2007). X-ray precursors to flares and filament eruptions. *Astronomy and Astrophysics, 472*, 967–979.

Cremades, H., Bothmer, V., & Tripathi, D. (2006). Properties of structured coronal mass ejections in solar cycle 23. *Advances in Space Research, 38*, 461–465.

Cyr, O. C., & Webb, D. F. (1991). Activity associated with coronal mass ejections at solar minimum – SMM observations from 1984–1986. *Solar Physics, 136*, 379–394.

Dmitriev, A. V., Suvorova, A. V., Chao, J.-K., Wang, C. B., Rastaetter, L., Panasyuk, M. I., Lazutin, L. L., Kovtyukh, A. S., Veselovsky, I. S., & Myagkova, I. N. (2014). Anomalous dynamics of the extremely compressed magnetosphere during 21 January 2005 magnetic storm. *Journal of Geophysical Research, 119*, 877–896.

Engvold, O. (1980). Energy and mass injected by flares and eruptive prominences. *Solar and interplanetary dynamics* (pp. 173–187). Dordrecht: D. Reidel Publishing Co.

Engvold, O. (1989). Prominence environment. *Dynamics and structure of quiescent solar prominences* (pp. 47–76). Dordrecht: Kluwer Academic Publishers.

Fan, Y.-H. (2014). Magnetism and dynamics of prominences: MHD equilibria and triggers for eruption. In J.-C. Vial, & O. Engvold (Eds.), *Solar prominences, ASSL* (Vol. 415, pp. 295–320). Springer.

Feynman, J., & Martin, S. F. (1995). The initiation of coronal mass ejections by newly emerging magnetic flux. *Journal of Geophysical Research, 100*, 3355–3367.

Filippov, B. P., Gopalswamy, N., & Lozhechkin, A. V. (2001). Non-radial motion of eruptive filaments. *Solar Physics, 203*, 119–130.

Fujimori, K. (1984). Latitude distribution of solar prominences in the years 1975–1981. *Publications of the Astronomical Society of Japan, 36*, 189–190.

Gibson, S. E. (2014). Coronal cavities: Observations and implications for the magnetic environment of prominences. In J.-C. Vial, & O. Engvold (Eds.), *Solar prominences, ASSL* (Vol. 415, pp. 321–351). Springer.

Gibson, S. E., Kucera, T. A., Rastawicki, D., et al. (2010). Three-dimensional morphology of a coronal prominence cavity. *The Astrophysical Journal, 724*, 1133–1146.

Gilbert, H. R., Holzer, T. E., Burkepile, J. T., & Hundhausen, A. J. (2000). Active and eruptive prominences and their relationship to coronal mass ejections. *The Astrophysical Journal, 537*, 503–515.

Gilbert, J. A., Lepri, S. T., Landi, E., & Zurbuchen, T. H. (2012). First measurements of the complete heavy-ion charge state distributions of C, O, and Fe associated with interplanetary coronal mass ejections. *The Astrophysical Journal, 751*, 20–27.

Gopalswamy, N. (1999). X-ray and microwave signatures of coronal mass ejections. In T. Bastian, N. Gopalswamy, & K. Shibasaki (Eds.), *Solar physics with radio observations, NRO report no. 479* (pp. 141–152).

Gopalswamy, N. (2004). A global picture of CMEs in the inner heliosphere. In G. Poletto & S. T. Suess (Eds.), *The sun and the heliosphere as an integrated system: Astrophysics and space science library* (Vol. 317, p. 201). Dordrecht, The Netherlands: Kluwer Academic Publishers.

Gopalswamy, N. (2006a). *Radio observations of solar eruptions. Solar physics with the Nobeyama Radioheliograph* (pp. 81–94). Nobeyama: Nobeyama Solar Radio Observatory.

Gopalswamy, N. (2006b). Properties of interplanetary coronal mass ejections. *Space Science Reviews, 124*, 145–168.

Gopalswamy, N. (2013). Observations of CMEs and models of the eruptive corona. In *SOLAR WIND 13: Proceedings of the thirteenth international solar wind conference. AIP conference proceedings* (Vol. 1539, pp. 5–10).

Gopalswamy, N., & Kundu, M. R. (1989). A slowly moving plasmoid associated with a filament eruption. *Solar Physics, 122*, 91–110.

Gopalswamy, N., & Thompson, B. J. (2000). Early life of coronal mass ejections. *Journal of Atmospheric and Solar-Terrestrial Physics, 62*, 1457–1469.

Gopalswamy, N., & Yashiro, S. (2013). Obscuration of flare emission by an eruptive prominence. *Publications of the Astronomical Society of Japan, 65*, S11.

Gopalswamy, N., Hanaoka, Y., Kosugi, T., Lepping, R. P., Steinberg, J. T., Plunkett, S., Howard, R. A., Thompson, B. J., Gurman, J., Ho, G., Nitta, N., & Hudson, H. S. (1998). On the relationship between coronal mass ejections and magnetic clouds. *Geophysical Research Letters, 25*, 2485.

Gopalswamy, N., Shibasaki, K., Thompson, B. J., Gurman, J., & DeForest, C. (1999). Microwave enhancement and variability in the elephant's trunk coronal hole: Comparison with SOHO observations. *Journal of Geophysical Research, 104*, 9767–9780.

Gopalswamy, N., Hanaoka, Y., & Hudson, H. S. (2000). Structure and dynamics of the corona surrounding an eruptive prominence. *Advances in Space Research, 25*, 1851–1854.

Gopalswamy, N., Lara, A., Yashiro, S., & Howard, R. A. (2003a). Coronal mass ejections and solar polarity reversal. *The Astrophysical Journal, 598*, L63–L66.

Gopalswamy, N., Shimojo, M., Lu, W., Yashiro, S., Shibasaki, K., & Howard, R. A. (2003b). Prominence eruptions and coronal mass ejection: A statistical study using microwave observations. *The Astrophysical Journal, 586*, 562–578.

Gopalswamy, N., Nunes, S., Yashiro, S., & Howard, R. A. (2004a). Variability of solar eruptions during cycle 23. *Advances in Space Research, 34*, 391–396.

Gopalswamy, N., Yashiro, S., Krucker, S., Stenborg, G., & Howard, R. A. (2004b). Intensity variation of large solar energetic particle events associated with coronal mass ejections. *Journal of Geophysical Research, 109*, A12105.

Gopalswamy, N., Yashiro, S., Michalek, G., Xie, H., Lepping, R. P., & Howard, R. A. (2005). Solar source of the largest geomagnetic storm of cycle 23. *Geophysical Research Letters, 32*, L12S09.

Gopalswamy, N., Mikić, Z., Maia, D., Alexander, D., Cremades, H., Kaufmann, P., Tripathi, D., & Wang, Y.-M. (2006). The pre-CME sun. *Space Science Reviews, 123*, 303–339.

Gopalswamy, N., Akiyama, S., Yashiro, S., Michalek, G., & Lepping, R. P. (2008). Solar sources and geospace consequences of interplanetary magnetic clouds observed during solar cycle 23. *Journal of Atmospheric and Solar-Terrestrial Physics, 70*, 245–253.

Gopalswamy, N., Mäkelä, P., Xie, H., Akiyama, S., & Yashiro, S. (2009a). CME interactions with coronal holes and their interplanetary consequences. *Journal of Geophysical Research, 114*, A00A22.

Gopalswamy, N., Yashiro, S., Temmer, M., Davila, J., Thompson, W. T., Jones, S., McAteer, R. T. J., Wuelser, J.-P., Freeland, S., & Howard, R. A. (2009b). EUV wave reflection from a coronal hole. *The Astrophysical Journal, 691*, L123–L127.

Gopalswamy, N., Yashiro, S., Michalek, G., Stenborg, G., Vourlidas, A., Freeland, S., & Howard, R. (2009c). The SOHO/LASCO CME catalog. *Earth, Moon, and Planets, 104*, 295–313.

Gopalswamy, N., Akiyama, S., Yashiro, S., & Mäkelä, P. (2010a). Coronal mass ejections from sunspot and non-sunspot regions. In S. S. Hasan & R. J. Rutten (Eds.), *Magnetic coupling between the interior and atmosphere of the sun* (Astrophysics and Space Science Proceedings, pp. 289–307). Berlin: Springer.

Gopalswamy, N., Mäkelä, P., Xie, H., Akiyama, S., & Yashiro, S. (2010b). Solar sources of "driverless" interplanetary shocks. In *Twelfth International Solar Wind Conference. AIP Conference Proceedings* (Vol. 1216, pp. 452–458).

Gopalswamy, N., Xie, H., Yashiro, S., Akiyama, S., Mäkelä, P., & Usoskin, I. G. (2012a). Properties of ground level enhancement events and the associated solar eruptions during solar cycle 23. *Space Science Reviews, 171*, 23–60.

Gopalswamy, N., Yashiro, S., Mäkelä, P., Michalek, G., Shibasaki, K., & Hathaway, D. H. (2012b). Behavior of solar cycles 23 and 24 revealed by microwave observations. *The Astrophysical Journal, 750*, L42–L47.

Gopalswamy, N., Mäkelä, P., Akiyama, S., Xie, H., Yashiro, S., & Reinard, A. A. (2013). The solar connection of enhanced heavy ion charge states in the interplanetary medium: Implications for the flux-rope structure of CMEs. *Solar Physics, 284*, 17–46.

Gopalswamy, N., Akiyama, S., Yashiro, S., Xie, H., Mäkelä, P., & Michalek, G. (2014). Anomalous expansion of coronal mass ejections during solar cycle 24 and its space weather implications. *Geophysical Research Letters, 41*, 2673–2680.

Gosling, J. T., Hildner, E., MacQueen, R. M., Munro, R. H., Poland, A. I., & Ross, C. L. (1976). The speeds of coronal mass ejection events. *Solar Physics, 48*, 389–397.

Gruesbeck, J. R., Lepri, S. T., & Zurbuchen, T. H. (2012). Two-plasma model for low charge state interplanetary coronal mass ejection observations. *The Astrophysical Journal, 760*, 141.

Gui, B., Shen, C., Wang, Y., Ye, P., Liu, J., Wang, S., & Zhao, X. P. (2011). Quantitative analysis of CME deflections in the corona. *Solar Physics, 271*, 111–139.

Guo, Y., Ding, M. D., Schmieder, B., Li, H., Török, T., & Wiegelmann, T. (2010). Driving mechanism and onset condition of a confined eruption. *The Astrophysical Journal, 725*, L38–L42.

Hanaoka, Y., & Shinkawa, T. (1999). Heating of erupting prominences observed at 17 GHz. *The Astrophysical Journal, 510*, 466–473.

Hansen, R. T., Garcia, C. J., Hansen, S. F., & Loomis, H. G. (1969). Brightness variations of the white light corona during the years 1964 67. *Solar Physics, 7*, 417–433.

Hildner, E. (1977). Mass ejections from the corona into the interplanetary space. In M. A. Shea, D. F. Smart, & S. T. Wu (Eds.), *Study of traveling interplanetary phenomena* (pp. 3–21). Hingham, MA: Reidel.

Hildner, E., Gosling, J. T., Hansen, R. T., & Bohlin, J. D. (1975). The sources of material comprising a mass ejection coronal transient. *Solar Physics, 45*, 363–376.

Hori, K., & Culhane, J. L. (2002). Trajectories of microwave prominence eruptions. *Astronomy and Astrophysics, 382*, 666–677.

Howard, R. A., & Labonte, B. A. (1981). Surface magnetic fields during the solar activity cycle. *Solar Physics, 74*, 131–145.

Hudson, H. S., Kosugi, T., Nitta, N., & Shimojo, M. (2001). Hard X-radiation from a fast coronal ejection. *The Astrophysical Journal, 561*, L211–L214.

Hundhausen, A. J. (1993). Sizes and locations of coronal mass ejections – SMM observations from 1980 and 1984–1989. *Journal of Geophysical Research, 98*, 13177–13200.

Hyder, C. L. (1965). The "polar crown" of filaments and the sun's polar magnetic fields. *The Astrophysical Journal, 141*, 271–273.

Ji, H., Wang, H., Schmahl, E. J., Moon, Y.-J., & Jiang, Y. (2003). Observations of the failed eruption of a filament. *The Astrophysical Journal, 595*, L135–L138.

Joshi, A. D., & Srivastava, N. (2011). Kinematics of two eruptive prominences observed by EUVI/STEREO. *The Astrophysical Journal, 730*, 104–114.

Kahler, S. W., Cliver, E. W., Cane, H. V., McGuire, R. E., Stone, R. G., & Sheeley, N. R., Jr. (1986). Solar filament eruptions and energetic particle events. *The Astrophysical Journal, 302*, 504–510.

Karpen, J. T., Antiochos, S. K., & DeVore, C. R. (2012). The mechanisms for the onset and explosive eruption of coronal mass ejections and eruptive flares. *The Astrophysical Journal, 760*, 81–95.

Kay, C., Opher, M., & Evans, R. M. (2013). Forecasting a coronal mass ejection's altered trajectory: ForeCAT. *The Astrophysical Journal, 775*, 5–21.

Kleczek, J. (1964). Occurrence of eruptive prominences. *Bulletin of the Astronomical Institutes of Czechoslovakia, 15*, 41.

Kosugi, T., Ishiguro, M., & Shibasaki, K. (1986). Polar-cap and coronal-hole-associated brightenings of the sun at millimeter wavelengths. *Publications of the Astronomical Society of Japan, 38*, 1–11.

Kozyra, J. U., Manchester, W. B., Escoubet, C. P., Lepri, S. T., Liemohn, M. W., Gonzalez, W. D., Thomsen, M. W., & Tsurutani, B. T. (2013). Earth's collision with a solar filament on 21 January 2005: Overview. *Journal of Geophysical Research, 118*, 5967–5978.

Leighton, R. B. (1969). A magneto-kinematic model of the solar cycle. *The Astrophysical Journal, 156*, 1–26.

Lepri, S. T., & Zurbuchen, T. H. (2010). Direct observational evidence of filament material within interplanetary coronal mass ejections. *The Astrophysical Journal, 723*, L22–L27.

Liu, K., Wang, Y., Shen, C., & Wang, S. (2012). Critical height for the destabilization of solar prominences: Statistical results from STEREO observations. *The Astrophysical Journal, 744*, 168–177.

Lockyer, W. J. S. (1931). On the relationship between solar prominences and the forms of the corona. *MNRAS, 91*, 797–809.

Long, D. M., Gallagher, P. T., McAteer, R. T. J., & Bloomfield, D. S. (2008). The kinematics of a globally propagating disturbance in the solar corona. *The Astrophysical Journal, 680*, L81–L84.

Lorenc, M., Pastorek, L., & Rybanský, M. (2003). Magnetic field reversals on the sun and the N-S asymmetry. In A. Wilson (Ed.), *Solar variability as an input to the Earth's environment* (pp. 129–132). ESA SP-535. Noordwijk: ESA Publications Division.

Low, B. C., & Hundhausen, J. R. (1995). Magnetostatic structures of the solar corona. 2: The magnetic topology of quiescent prominences. *The Astrophysical Journal, 443*, 818–836.

Lugaz, N. (2014). Eruptive prominences and their impact on the earth: The impacts on our earth and our life. In J.-C. Vial, & O. Engvold (Eds.), *Solar prominences, ASSL* (Vol. 415, pp. 431–451). Springer.

Mackay, D. H., Gaizauskas, V., Rickard, G. J., & Priest, E. R. (1997). Force-free and potential models of a filament channel in which a filament forms. *The Astrophysical Journal, 486*, 534–549.

MacQueen, R. M., & Fisher, R. R. (1983). The kinematics of solar inner coronal transients. *Solar Physics, 89*, 89–102.

MacQueen, R. M., Hundhausen, A. J., & Conover, C. W. (1986). The propagation of coronal mass ejection transients. *Journal of Geophysical Research, 91*, 31–38.

Mäkelä, P., Gopalswamy, N., Xie, H., Mohamed, A. A., Akiyama, S., & Yashiro, S. (2013). Coronal hole influence on the observed structure of interplanetary CMEs. *Solar Physics, 284*, 59–75.

Maričić, D., Vršnak, B., & Roša, D. (2009). Relative kinematics of the leading edge and the prominence in coronal mass ejections. *Solar Physics, 260*, 177–189.

Marqué, C., Lantos, P., Klein, K. L., & Delouis, J. M. (2001). Coronal restructuring and electron acceleration following a filament eruption. *Astronomy and Astrophysics, 374*, 316–325.

Martin, S. F. (1973). The evolution of prominences and their relationship to active centers (a review). *Solar Physics, 31*, 3–21.

McAllister, A. H., Dryer, M., McIntosh, P., Singer, H., & Weiss, L. (1996). A large polar crown coronal mass ejection and a "problem" geomagnetic storm: April 14–23. *Journal of Geophysical Research, 101*, 13497–13516.

McIntosh, P. S. (2003). Patterns and dynamics of solar magnetic fields and HeI coronal holes in cycle 23. In A. Wilson (Ed.), *Solar variability as an input to the earth's environment* (pp. 807–818), ESA SP-535. Noordwijk: ESA Publications Division.

Moon, Y. J., Choe, G. S., Wang, H., Park, Y. D., Gopalswamy, N., Yang, G., & Yashiro, S. (2002). A statistical study of two classes of coronal mass ejections. *The Astrophysical Journal, 581*, 694–702.

Moore, R. L., & Sterling, A. C. (2006). Initiation of coronal mass ejections. In N. Gopalswamy, R. Mewaldt, & J. Torsti (Eds.), *Solar eruptions and energetic particles* (pp. 43–57), Geophysical monograph 165, AGU, Washington, DC.

Moore, R. L., & Sterling, A. C. (2007). The coronal-dimming footprint of a streamer-puff coronal mass ejection: Confirmation of the magnetic-arch-blowout scenario. *The Astrophysical Journal, 661*, 543–550.

Moore, R. L., Sterling, A. C., Hudson, H. S., & Lemen, J. R. (2001). Onset of the magnetic explosion in solar flares and coronal mass ejections. *The Astrophysical Journal, 552*, 833–848.

Munro, R. H., Gosling, J. T., Hildner, E., MacQueen, R. M., Poland, A. I., & Ross, C. L. (1979). The association of coronal mass ejection transients with other forms of solar activity. *Solar Physics, 61*, 201–215.

Odstrcil, D., & Pizzo, V. J. (2009). Numerical heliospheric simulations as assisting tool for interpretation of observations by STEREO heliospheric imagers. *Solar Physics, 259*, 297–309.

Olmedo, O., Vourlidas, A., Zhang, J., & Cheng, X. (2012). Secondary waves and/or the "Reflection" from and "Transmission" through a coronal hole of an extreme ultraviolet wave associated with the 2011 February 15 X2.2 flare observed with SDO/AIA and STEREO/EUVI. *The Astrophysical Journal, 756*, 143–155.

Panasenco, O., Martin, S. F., Velli, M., & Vourlidas, A. (2013). Origins of rolling, twisting, and non-radial propagation of eruptive solar events. *Solar Physics, 287*, 391–413.

Parenti, S. (2014). Solar prominences: Observations. *Living Reviews in Solar Physics, 11*, 1–88.

Plunkett, S. P., Thompson, B. J., Cyr, O. C., & Howard, R. A. (2001). Solar source regions of coronal mass ejections and their geomagnetic effects. *Journal of Atmospheric and Solar-Terrestrial Physics, 63*, 389–402.

Qiu, J., Hu, Q., Howard, T. A., & Yurchyshyn, V. B. (2007). On the magnetic flux budget in low-corona magnetic reconnection and interplanetary coronal mass ejections. *The Astrophysical Journal, 659*, 758–772.

Régnier, S., Walsh, R. W., & Alexander, C. E. (2011). A new look at a polar crown cavity as observed by SDO/AIA structure and dynamics. *Astronomy and Astrophysics, 533*, L1–L4.

Reinard, A. A. (2008). Analysis of interplanetary coronal mass ejection parameters as a function of energetics, source location, and magnetic structure. *The Astrophysical Journal, 682*, 1289.

Richardson, J. D., Liu, Y., Wang, C., & Burlaga, L. F. (2006). ICMES at very large distances. *Advances in Space Research, 38*, 528–534.

Riley, P., Schatzman, C., Cane, H. V., Richardson, I. G., & Gopalswamy, N. (2006). On the rates of coronal mass ejections: Remote solar and in situ observations. *The Astrophysical Journal, 647*, 648–653.

Robinson, R. D. (1978). Observations and interpretation of moving type IV solar radio bursts. *Solar Physics, 60*, 383–398.

Saito, K., & Tandberg-Hanssen, E. (1973). The arch systems, cavities, and prominences in the helmet streamer observed at the solar eclipse, November 12, 1966. *Solar Physics, 31*, 105–121.

Schmahl, E., & Hildner, E. (1977). Coronal mass-ejections-kinematics of the 19 December 1973 event. *Solar Physics, 55*, 473–490.

Schmieder, B., Démoulin, P., & Aulanier, G. (2013). Solar filament eruptions and their physical role in triggering coronal mass ejections. *Advances in Space Research, 51*, 1967–1980.

Schrijver, C. J., Elmore, C., Kliem, B., Török, T., & Title, A. M. (2008). Observations and modeling of the early acceleration phase of erupting filaments involved in coronal mass ejections. *The Astrophysical Journal, 674*, 586–595.

Sharma, R., & Srivastava, N. (2012). Presence of solar filament plasma detected in interplanetary coronal mass ejections by in situ spacecraft. *Journal of Space Weather and Space Climate, 2*, A10.

Sharma, R., Srivastava, N., Chakrabarty, D., Möstl, C., & Hu, Q. (2013). Interplanetary and geomagnetic consequences of 5 January 2005 CMEs associated with eruptive filaments. *Journal of Geophysical Research, 118*, 3954–3967.

Sheeley, N. R., Jr., Howard, R. A., Koomen, M. J., Michels, D. J., & Poland, A. I. (1980). The observation of a high-latitude coronal transient. *The Astrophysical Journal, 238*, L161–L164.

Sheeley, N. R., Jr., Wang, Y.-M., & Hawley, S. H. (1997). Measurements of flow speeds in the corona between 2 and 30 Ro. *The Astrophysical Journal, 484*, 472–478.

Sheeley, N. R., Walters, J. H., Wang, Y.-M., & Howard, R. A. (1999). Continuous tracking of coronal outflows: Two kinds of coronal mass ejections. *Journal of Geophysical Research, 104*, 24739–24768.

Shibasaki, K. (2013). Long-term global solar activity observed by the Nobeyama Radioheliograph. *Publications of the Astronomical Society of Japan, 65*, S17–S22.

Shimojo, M. (2013). Unusual migration of prominence activities in the southern hemisphere during cycles 23–24. *Publications of the Astronomical Society of Japan, 65*, S16.

Shimojo, M., Yokoyama, T., Asai, A., Nakajima, H., & Shibasaki, K. (2006). One solar-cycle observations of prominence activities using the Nobeyama Radioheliograph 1992–2004. *Publications of the Astronomical Society of Japan, 58*, 85–92.

Simnett, G. M. (2000). The relationship between prominence eruptions and coronal mass ejections. *Journal of Atmospheric and Solar-Terrestrial Physics, 62*, 1479–1487.

Song, H. Q., Chen, Y., Ye, D. D., Han, G. Q., Du, G. H., Li, G., Zhang, J., & Hu, Q. (2013). A study of fast flareless coronal mass ejections. *The Astrophysical Journal, 773*, 129–138.

Sterling, A. C., & Moore, R. L. (2003). Tether-cutting energetics of a solar quiet-region prominence eruption. *The Astrophysical Journal, 599*, 1418–1425.

Sterling, A. C., Moore, R. L., & Freeland, S. L. (2011a). Insights into filament eruption onset from solar dynamics observatory observations. *The Astrophysical Journal, 731*, L3.

Sterling, A. C., Moore, R. L., & Harra, L. K. (2011b). Lateral offset of the coronal mass ejections from the X-flare of 2006 December 13 and its two precursor eruptions. *The Astrophysical Journal, 743*, 63–73.

Stewart, R. T., Dulk, G. A., Sheridan, K. V., House, L. L., Wagner, W. J., Illing, R., & Sawyer, C. (1982). Visible light observations of a dense plasmoid associated with a moving type IV solar radio burst. *Astronomy and Astrophysics, 116*, 217–223.

Svalgaard, L., & Kamide, Y. (2013). Asymmetric solar polar field reversals. *The Astrophysical Journal, 763*, 23–28.

Tandberg-Hanssen, E., Martin, S. F., & Hansen, R. T. (1980). Dynamics of flare sprays. *Solar Physics, 65*, 357–368.

Vemareddy, P., Maurya, R. A., & Ambastha, A. (2012). Filament eruption in NOAA 11093 leading to a two-ribbon M1.0 class flare and CME. *Solar Physics, 277*, 337–354.

Vršnak, B., Maričić, D., Stanger, A. L., Veronig, A. M., Temmer, M., & Roša, D. (2007). *Solar Physics, 241*, 85.

Waldmeier, M. (1960). Zirkulation und Magnetfeld der solaren Polarzone. Mit 7 Textabbildungen. *Zeitschrift für Astrophysik, 49*, 176–185.

Wang, Y.-M., & Sheeley, N. R., Jr. (1999). Filament eruptions near emerging bipoles. *The Astrophysical Journal, 510*, L157–L160.

Wang, Y.-M., Sheeley, N. R., Jr., & Andrews, M. D. (2002). Polarity reversal of the solar magnetic field during cycle 23. *Journal of Geophysical Research, 107*, SH10–SH11.

Webb, D. F. (2014). Eruptive prominences and their association with coronal mass ejections. In J.-C. Vial, & O. Engvold (Eds.), *Solar prominences, ASSL* (Vol. 415, pp. 409–430). Springer.

Webb, D. F., & Howard, R. A. (1994). The solar cycle variation of coronal mass ejections and the solar wind mass flux. *Journal of Geophysical Research, 99*, 4201–4220.

Webb, D. F., & Hundhausen, A. J. (1987). Activity associated with the solar origin of coronal mass ejections. *Solar Physics, 108*, 383–401.

Webb, D. F., & Jackson, B. V. (1981). Kinematical analysis of flare spray ejecta observed in the corona. *Solar Physics, 73*, 341–361.

Webb, D. F., Davis, J. M., & McIntosh, P. S. (1984). Observations of the reappearance of polar coronal holes and the reversal of the polar magnetic field. *Solar Physics, 92*, 109–132.

Wood, B. E., Karovska, M., Chen, J., Brueckner, G. E., Cook, J. W., & Howard, R. A. (1999). Comparison of two coronal mass ejections observed by EIT and LASCO with a model of an erupting magnetic flux rope. *The Astrophysical Journal, 512*, 484–495.

Wood, B. E., Wu, C. C., Rouillard, A. P., Howard, R. A., & Socker, D. G. (2012). A coronal hole's effects on coronal mass ejection shock morphology in the inner heliosphere. *The Astrophysical Journal, 755*, 43–52.

Xie, H., Gopalswamy, N., & Cyr, O. C. (2013). Near-sun flux-rope structure of CMEs. *Solar Physics, 284*, 47–58.

Zhang, J., & Dere, K. P. (2006). A statistical study of main and residual accelerations of coronal mass ejections. *The Astrophysical Journal, 649*, 1100–1109.

Zhang, J., Dere, K. P., Howard, R. A., Kundu, M. R., & White, S. M. (2001). On the temporal relationship between coronal mass ejections and flares. *The Astrophysical Journal, 559*, 452–462.

Zhukov, A. N., Saez, F., Lamy, P., Llebaria, A., & Stenborg, G. (2008). The origin of polar streamers in the solar corona. *The Astrophysical Journal, 680*, 1532–1541.

Chapter 16
Eruptive Prominences and Their Association with Coronal Mass Ejections

David F. Webb

Abstract We discuss the origins and characteristics of solar eruptive phenomena focusing on coronal mass ejections (CMEs) and their associated phenomena, particularly erupting prominences (EPs). Statistically, CMEs are most frequently associated with EPs and X-ray long-duration events. In a few large events the masses of the EP and CME have been separately measured, with the EP mass comprising a large fraction of the total CME mass. EP and CME near-surface precursors include the development of sigmoids, the darkening and broadening of filaments, and their slow rise and Doppler shifts, and the cancellation of magnetic flux near filament channels. Prominences exist within coronal cavities which themselves are embedded in helmet streamers extending to high heights. This entire structure can erupt bodily to become a CME; indeed the most massive and energetic CMEs appear to be of this type. CMEs carry into the heliosphere large quantities of coronal magnetic fields and plasma which are detected by remote sensing and measured in-situ at spacecraft. The most important in-situ CME signature is a magnetic cloud, considered to be the flux rope embedded in most if not all CMEs. Although most CMEs are frequently associated with EPs near the Sun, it is still not known why the prominence material is only rarely identified in-situ. In the last decade, however, we have had heliospheric imaging observations that are helping to distinguish prominence material from the rest of a CME.

16.1 Introduction

Coronal mass ejections (CMEs) consist of large structures containing plasma and magnetic fields that are expelled from the Sun into the heliosphere. Most, if not all CMEs are now considered to contain and, indeed, be driven by magnetic flux ropes that are ejected from the Sun. Such helical fields are also associated with prominences, which often are embedded at the base of a flux rope as the CME ensemble moves outward from the Sun. In white light coronagraphs the erupting

D.F. Webb (✉)
Institute for Scientific Research, Boston College, Newton, MA, USA
e-mail: david.webb@bc.edu

J.-C. Vial, O. Engvold (eds.), *Solar Prominences*, Astrophysics and Space Science Library 415, DOI 10.1007/978-3-319-10416-4_16

prominence (EP) appears as the bright core following a bright front in perhaps half of all CMEs. This was first confirmed during the SMM and Solwind eras by the direct detection of Hα emission from the CME cores (e.g., Sheeley et al. 1980; Illing and Hundhausen 1985; Webb and Hundhausen 1987). EPs are also statistically associated with at least 70 % of all CMEs (Munro et al. 1979; Webb and Hundhausen 1987; Gopalswamy et al. 2003a). These results suggest that at least a major class of, and maybe a majority of CMEs have their origin in the destabilization and eruption of a magnetic flux rope, and its prominence and overlying coronal canopy.

Chapters 15–17 describe EPs and their impact on the Earth. Chapter 15 (Gopalswamy 2014) discusses the dynamical aspects of prominences including kinematics, long-term, solar-cycle patterns and spatial associations with other solar activity, and Chap. 17 (Lugaz 2014) emphasizes the observations and numerical simulations of the propagation of prominences in the heliosphere and their impact at Earth, i.e., effects on space weather. Since EPs are intimately related to CMEs, there is some overlap among all three chapters. However, this chapter focuses on the observational aspects of EPs that are clearly associated with CMEs, with emphasis on the role of EPs in the origins of CMEs and of observing prominence material in the heliosphere.

16.2 The General Properties of CMEs and Associated Prominences

CMEs are a key aspect of coronal and interplanetary dynamics in that they inject large amounts of mass and magnetic flux into the heliosphere, causing major transient disturbances (see Webb and Howard 2012, for a recent review). CMEs can drive interplanetary shocks, a key source of solar energetic particles, and are known to be the major contributor to severe space weather at the Earth. They are of interest for both scientific and technological reasons. Scientifically they are of interest because they are responsible for the removal of built-up magnetic energy and plasma from the solar corona, and technologically they are of interest because they are responsible for the most extreme space weather effects at Earth, as well as at other locations throughout the heliosphere. Most of the ejected material comes from the low corona, although prominences with considerable amounts of cooler, denser material probably of chromospheric origin are often imbedded in the CME. The CME plasma is entrained on an expanding magnetic field, which commonly has the form of helical field lines with changing pitch angles, i.e., a flux rope.

Historically, images of CMEs had been made near the Sun primarily by coronagraphs on board spacecraft. Coronagraphs view the outward flow of density structures emanating from the Sun by observing Thomson-scattered sunlight from the free electrons in coronal and heliospheric plasma. This emission has an angular dependence which must be accounted for in the measured brightness (e.g., Billings

1966; Vourlidas and Howard 2006; Howard and Tappin 2009). They are faint relative to the background corona, but much more transient, so some form of background subtraction is typically applied to identify them. CME-related phenomena include flares and prominence eruptions, Type II and IV radio bursts, coronal and interplanetary shocks and energetic particles. The first spacecraft coronagraph observations of CMEs were made by the OSO-7 coronagraph in the early 1970s, followed by increasingly better quality and longer periods of CME observations from Skylab (1973–1974, P78-1 [Solwind]) and SMM (1980; 1984–1989), and since 1996 SOHO with two of its three LASCO coronagraphs still operating, and the STEREO CORs operating since 2007. These observations have been complemented by white light data from the ground-based Mauna Loa Solar Observatory (MLSO) K-coronameters viewing from $\sim$1.2 to as high as 2.9R_S.

Early on, interplanetary transients were observed using interplanetary radio scintillation (IPS; 1964–present) and from the zodiacal light photometers on the twin Helios spacecraft (1975–1983). The Helios photometers observed regions in the inner heliosphere from 0.3 to 1.0 AU but with a very limited field of view. Within the last decade a new class of detector, the heliospheric imager (HI), was developed to directly image CMEs in the heliosphere. Two HIs have been flown, the first the Solar Mass Ejection Imager (SMEI—Eyles et al. 2003; Jackson et al. 2004) launched on board the Coriolis spacecraft early in 2003 and then the Heliospheric Imagers (HIs—Howard et al. 2008; Eyles et al. 2009) launched on the twin STEREO spacecraft in late 2006. These imagers have been used to track dense material in the heliosphere and, in rare cases, the prominence material has been isolated from its parent CME and their masses determined.

CMEs can exhibit a variety of forms, some having the classical "three-part" structure (Illing and Hundhausen 1985), usually interpreted as compressed plasma ahead of a flux rope followed by a cavity surrounding a bright prominence (Fig. 16.1 shows two examples). Other CMEs display a more complex geometry. Some CMEs appear as narrow jets, some arise from pre-existing coronal streamers (the so-called streamer blowouts), while others appear as wide almost global eruptions. CMEs spanning very large angular ranges have a large component along the Sun-observer line and appear large by perspective. These so-called halo CMEs can completely encircle the occulting disk (Howard et al. 1982). The LASCO CDAW CME catalog (Yashiro et al. 2004) defines a "partial halo" as a CME with an apparent position angle range $>120°$.

Many statistical association studies have indicated that erupting prominences (EPs) and X-ray events, especially of long duration, are the most common near-surface activity associated with CMEs. More recently, Gopalswamy et al. (2003b) showed that 73 % of microwave EPs, and nearly all those attaining high heights, were associated with CMEs, confirming results first found during Skylab (Munro et al. 1979; Webb and Hundhausen 1987). There is a strong correspondence between X-ray ejecta and CMEs. Nitta and Akiyama (1999) found that flares with X-ray ejecta were always associated with CMEs and X-ray ejecta could be associated with CME cores, likely dense, heated prominence material.

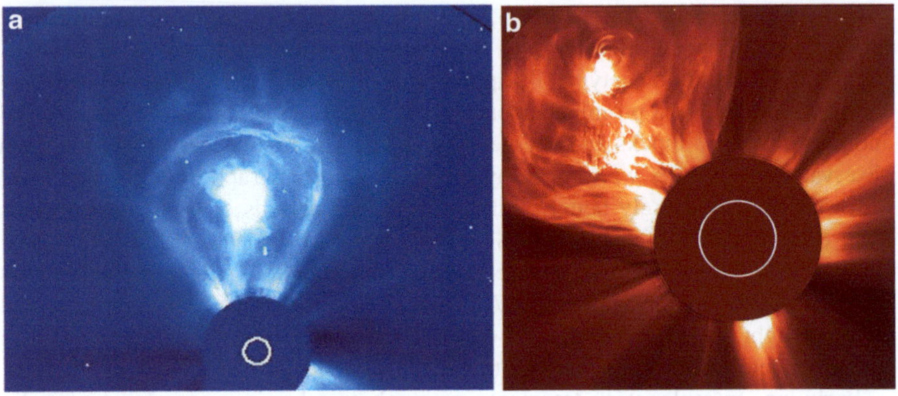

Fig. 16.1 (**a**) LASCO C3 images of a *lightbulb-shaped* CME on 27 February 2000. "Classic" three-part structure with outer shell, void and inner bright structure, in this case an erupting prominence. From SOHO online movie gallery: http://sohowww.nascom.nasa.gov/gallery/Movies/flares.html. (**b**) LASCO C2 image from 4 January 2002 image of a CME showing detail in the ejected material. Available from SOHO online image gallery: http://sohowww.nascom.nasa.gov/gallery/bestofsoho.html. In both images the solar limb is represented by the *white circle*. Also in Webb and Howard (2012)

There is no one-to-one relationship between CMEs and flares, and most optical flares occur independently of CMEs. When CMEs and flares occur together, the CME onsets can precede the flares, and the CMEs generally contain more total energy than that radiated by the flare itself. It is now generally accepted that CMEs and flares are part of a single magnetically-driven "event" and, therefore, it is more appropriate to consider a unified model that accounts for both. This "standard" flare model has been developed and refined over the last few decades and now includes Flux Cancellation and Catastrophe models (e.g., Svestka and Cliver 1992; Shibata et al. 1995; Lin and Forbes 2000; Lin 2004). In this model a stressed magnetic arcade that may contain a prominence and magnetic flux rope at its core begins to rise. A current sheet develops beneath it as external pressure causes oppositely directed magnetic field lines to converge and reconnect. Some of the energy liberated heats the CME plasma, while other energy is directed downward in the form of shock waves, energetic particles and/or rapidly moving plasma. This can heat the low-lying or reconnecting magnetic loops and travel down the loops to the chromosphere, producing the flare. In some cases, especially if a prominence lifts off slowly, there may be too little energy deposited in underlying structures to produce a detectable surface brightening, or flare. Typical flares are "confined" or "compact" and do not have sufficient energy or magnetic topology to open up the ambient field and produce an eruption or ejection. However, Shibata and colleagues have argued that impulsive, compact flares might also have narrow, plasma ejections yielding small CMEs. Models pertaining to EPs and cavities as they relate to CMEs are further discussed in, e.g., Fan (2014), Gibson (2014), and Karpen (2014).

16.3 Erupting Prominences as a Signature of CME Origin

16.3.1 White-Light Coronagraph Observations

Prominences can often be observed in coronagraph fields of view ($\geq 1.5 R_S$) as the bright, central core of the CME structure (the filament component of the classic three-part CME). They are also observed by instruments that image the solar disk, so EPs provide a direct comparison between coronagraphic and solar data. Prominences are believed to be caused by the formation of a flux rope low in the magnetic structure that eventually erupts to form the CME. Many CME onset models [e.g., Flux Cancellation, Mass Loading—see Karpen (2014); Fan (2014)] require the presence or formation of a prominence in order for the CME to erupt.

Coronal cavities are discussed in detail by Gibson (2014). Cavities can extend for long distances above photospheric neutral lines, some of which are filament channels associated with filaments [see Engvold (2014) and Martin (2014)]. At the limb cavities appear as dark ellipses and, like a tunnel, can be viewed for long time periods as the Sun rotates. Cavities along high latitude or polar crown filament channels often provide the best views because these channels are aligned east-west, are away from bright active regions, and can wrap nearly around the Sun. Cavities may or may not contain obvious prominence material, just as CMEs with interior voids may or may not have a bright core. An excellent example of what might be called an erupting cavity-CME is shown in Gibson (2014, Figure 12). Such eruptions are usually very slow or gradual, and may be similar to a bubble or balloon being released once its tether is broken or its equilibrium destroyed (e.g., Srivastava et al. 1999).

During the rise of cycle 23 the latitude distribution of LASCO CMEs peaked at the equator, but the distribution of EIT EUV activity including prominence eruptions associated with these CMEs was bimodal with peaks 30° north and south of the equator (Plunkett et al. 2002). This pattern indicates that many CMEs involve more complex, multiple-polarity systems (Webb et al. 1997) such as those modeled by Antiochos et al. (1999). Prominences themselves tend to be offset to one side of the CME axis and, occasionally, two prominences can erupt under the same CME canopy (Webb et al. 1997; Simnett 2000). Overall the latitude distribution of the central position angles of CMEs tends to cluster about the equator around solar minimum but broadens over all latitudes near solar maximum. Hundhausen (1993) first noted that this CME latitude variation more closely parallels that of streamers and prominences than of active regions or sunspots. This pattern also is closely linked to the variation of the global solar magnetic field, as exemplified by the tilt angle of the heliospheric current sheet (HCS) when the Sun makes its transition from solar minimum to maximum. Gopalswamy (2010) and Gopalswamy et al. (2010) confirmed the close match between CMEs, prominence eruptions and the HCS over cycle 23 (Fig. 16.2). On this figure also note the sharp decrease in the rate of CMEs and prominence eruptions in $\sim$2006 when the HCS became flatter below 30° solar latitude.

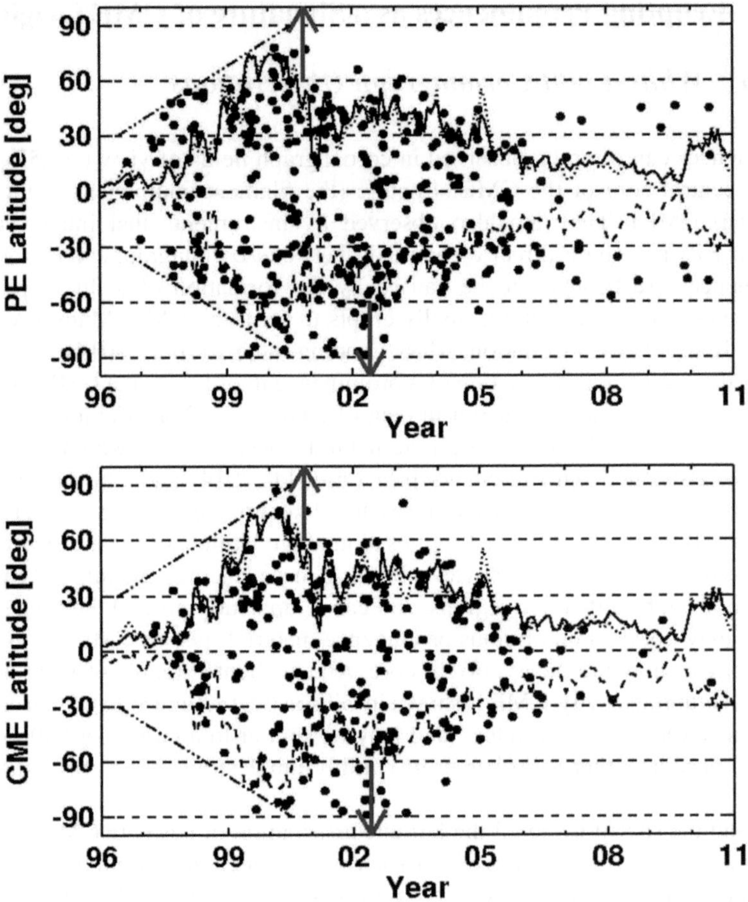

Fig. 16.2 Latitudes of LASCO CMEs (*filled circles*) with solar surface associations, in this case microwave prominence eruptions, plotted vs time, by Carrington Rotation number. The *dotted* and *dashed curves* represent the tilt angle of the heliospheric current sheet in the northern and southern hemispheres, respectively; the *solid curve* is the average of the two. The *up* and *down* arrows denote the times when the polarity in the north and south solar poles, resp., reversed. Note that the high latitude CMEs and EPs are confined to the solar maximum phase and their occurrence is asymmetric in the northern and southern hemispheres. EPs at latitudes <40° may arise from active regions or quiescent filament regions, but those at higher latitudes are always from the latter. Adapted from Gopalswamy et al. (2003b, 2010) and updated by S. Yashiro (priv. comm., 2011) for Webb and Howard (2012)

Using SOHO LASCO, EIT and MDI and ground-based Hα CME data, Cremades and Bothmer (2004) concluded that a simple scheme can be used to relate CME white light topology to the heliographic position and orientation of the underlying magnetic neutral line. When the neutral line is approximately parallel to the solar limb, the CME appears as a linear feature parallel to the limb having a broad, diffuse inner core. When the neutral line is approximately perpendicular to the solar limb, the CME is observed along its symmetry axis, and the core material lies along the line of sight. The frontside neutral line will typically lie perpendicular to the east limb and parallel to the west limb. The neutral line and CME orientations are reversed for the solar backside, so backside CMEs are viewed predominately orthogonally to frontside CMEs at each limb. These CME orientations are generally valid only for CMEs with source regions in the active region belts, <50° heliolatitude. The CME orientations will be different for polar crown filaments (McAllister et al. 2002; Gopalswamy et al. 2003b) or for CME source regions outside the active regions, where the neutral lines do not obey Joy's law. However, in an older, related study using SMM data, Webb (1988) found no clear pattern between the orientation of filaments, i.e., neutral lines, and the morphology or widths of associated CMEs.

With the accumulating 3-D observations, such as from STEREO, and modeling and simulations of CMEs, the consensus appears to be that most, if not all CMEs have magnetic flux ropes at their cores. From a detailed study of 16 years of LASCO observations, Vourlidas et al. (2013) concluded that the observations are consistent with all CMEs being flux ropes although only 40 % of LASCO CMEs exhibited clear rope structure. Other evidence based on CME–ICME studies were reported by Xie et al. (2013) and Gopalswamy et al. (2013). If most CMEs have flux ropes, then the bright fronts are likely preexisting plasma that piles up and is pushed outward by the strong, closed magnetic fields of the flux rope. This is supported by MHD and numerical simulations of CMEs [e.g., Lugaz (2014)].

There have been several recent studies of the kinematics and rotations of prominences using STEREO EUVI data. Joshi and Srivastava (2011) used a stereoscopic reconstruction technique to study the motions of two polar crown prominences. They found evidence of two different motions, a helical twist in the prominence spine and overall non-radial equatorward motion of the entire prominence structure (also see Gopalswamy and Thompson 2000, and Webb et al. 2014), and two phases of acceleration during the eruptions. Bemporad et al. (2011) used the tiepointing technique with COR1 and EUVI data to reconstruct the three-dimensional (3-D) shape and trajectory of an erupting prominence. They found evidence for a progressive clockwise rotation of the prominence by $\sim 90°$, and helical motion providing evidence for the conversion of twist into writhe. Finally, (Vourlidas et al. 2011a) used SECCHI and LASCO data with a forward-fitting model to determine the 3-D orientation of a three-part CME with embedded prominence. They found that the CME had a fast rotation rate, and suggested it was possibly due to disconnection of one of the CME footpoints.

16.3.2 *Prominence Activity as a CME Precursor*

A currently popular paradigm is that the activation of coronal magnetic fields leading to a CME begins well before the appearance of any associated surface activity such as flares or erupting prominences. Some of the energy released during a CME could drive precursor activity, and there is some evidence of such activity tens of minutes to hours before the onset of surface activity and even before CME onset (see Webb 1992, and Gopalswamy et al. 2006, for a recent review called "The Pre-CME Sun").

An S or reverse S-shaped structure called a sigmoid sometimes develops in association with filaments, and can be associated with a filament's activation (see Martin 2014). Like the filaments themselves, sigmoids are indicative of sheared coronal magnetic fields. Since many CME onset models require a magnetic shear to be established for the field to erupt, these sigmoids may be a precursor of a CME. Since erupting prominences are the most common type of surface activity associated with CMEs and appear as bright cores within many CMEs (Webb and Hundhausen 1987), pre-eruptive filament activity is a form of CME precursor. Tens of minutes before their eruption, large filaments can darken and get broader (e.g., Martin 1980). They begin to slowly rise and can exhibit Doppler shift patterns (Martin 2014). The cancellation of magnetic flux near filament channels can also build energy prior to an eruption, a process already referred to as Flux Cancellation (Martin and Livi 1992; Martin et al. 2012; Fan 2014). Kahler et al. (1988) found that the eruption of Hα filaments began before the onset of associated flare impulsive phases, suggesting that these erupting filaments, and by analogy the CMEs associated with them, were driven before and independently of the flare and its impulsive phase. On the other hand, using the SOHO LASCO C1 coronagraph and EIT telescopes and GOES soft X-ray observations, Zhang et al. (2001) found that CME initiation in the low corona coincided with the onset of X-ray emission before the flare impulsive phase.

Some of the most massive and energetic CMEs are the so-called streamer blowout events, first described in detail by Sheeley et al. (1982) and Illing and Hundhausen (1986). Preliminary statistics of streamer-blowout CMEs observed by LASCO were presented by Vourlidas et al. (2002b). In such events, a pre-existing streamer typically increases in brightness (density) for one to several days before erupting as a CME (Fig. 16.3). Following the CME, the pre-existing helmet streamer is often replaced by a thin ray and later a reforming streamer (Kahler and Hundhausen 1992; Webb et al. 2003). These events appear on white light synoptic charts as "bugles": portions of the streamer belt that brighten and widen with time until they disappear during a CME (Hundhausen 1993). Most streamer blowouts involve a pre-existing prominence sitting within a coronal void or cavity; this then erupts to form the classic "three-part" CME structure. Thus, the early prominence activations discussed above are probably related to streamer swellings and blowouts. The event illustrated in Fig. 16.3 is a particularly nice example of this class, with twisted bright prominence threads nestled within and at the base of the circular flux rope within the erupting cavity.

The physics of cavity activations is likely intertwined with that of prominence precursor activity and activation. As reviewed in Gibson (2014), there is a tendency for higher cavities to erupt in the EUV and in white light cavities above $1.6R_S$ nearly always erupt. This is related to the Munro et al. (1979) observation with Skylab data that nearly all prominences rising above $1.2R_S$ erupted as a transient. The slowly rising cavities also tend to narrow, becoming more teardrop-shaped with enhanced

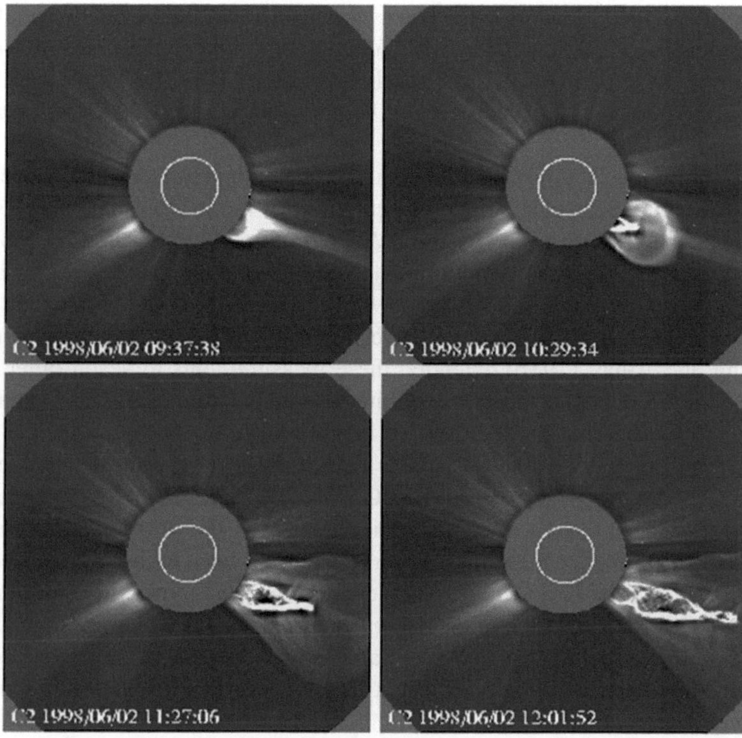

Fig. 16.3 Evolution of a three-part CME observed by the LASCO C2 coronagraph on 2 June 1998. Note the circular structures just above the prominence, suggesting a flux rope. Image reproduced with permission from Plunkett et al. (2000), copyright by Springer. Also in Webb and Howard (2012)

substructures. They often narrow at their base to a V or U-shape with a "stem", consistent with formation of a current sheet. Topology changes of the assumed flux rope within the cavity, likely driven by reconnections, lead to eruption. These changes are likely related to the associated streamer swelling and brightening in white light which precede a CME.

16.3.3 Mass and Energy Budgets

Mass and energy calculations of CMEs require difficult instrument calibrations and have large uncertainties. The average mass of CMEs derived from older coronagraph data was a few times 10^{15} g. LASCO results indicate a slightly lower average CME mass, 1.6×10^{15} g, likely because LASCO can measure smaller masses down to the order of 10^{13} g (Vourlidas et al. 2002a, 2010, 2011b; Kahler 2006). The older CME masses may have been underestimated because mass outflow may continue well after the CME's leading edge leaves the instrument field of view. Analysis of the mass density of LASCO CMEs as a function of height suggests that this density rises up to $\sim 7R_S$, then levels off (Vourlidas et al. 2010). The implication is that CMEs with larger masses reach greater heights and are more likely to escape the Sun.

In a fairly unique set of measurements, a group at the High Altitude Observatory in Boulder, CO, USA made detailed calculations of the masses and energies of the prominence material observed by the SMM C/P coronagraph. The C/P had an Hα filter and both the Hα and continuum data were used to study a few well-observed EP/CME events by Athay and Illing (1986). They made detailed calculations for the massive 18 August 1980 event described by Illing and Hundhausen (1986), and Illing and Athay (1986) extended these measurements to seven other events. Together they found that the EPs exhibited lower N_e, higher T_e, an expansion in volume and net heating as functions of height and time. The prominence masses ranged from 1.6 to 16×10^{15} g. Similar mass ranges have been determined from microwave and X-ray observations (e.g., Gopalswamy and Hanaoka 1998). It should be noted that these mass values do not include all of the original EP mass, since much of the prominence material is observed to drain back to the solar surface. The mass of the EP in the large August 1980 event was found to provide more than half of the total CME mass. Much of the energy expended in this event was considered to be increased potential energy estimated at $2\text{–}3 \times 10^{31}$ ergs (Athay and Illing 1986; Illing and Hundhausen 1986). See Gilbert (2014) for further discussion of the energy balance of prominences.

16.4 In-situ and Remotely Viewed Signatures of Prominences Within CMEs: Observing Prominence Material in the Heliosphere

As noted above most CMEs observed near the Sun are associated with EPs, but in interplanetary space rarely has the prominence material been clearly identified. Particularly since the launch of ACE in 1998 and, more recently, STEREO in 2006, the identification of prominence plasma has been aided by compositional signatures measured in-situ (Lepri and Zurbuchen 2010; Lee and Raymond 2012) and by the SMEI and STEREO HI heliospheric imagers viewing CMEs remotely (e.g., Sharma et al. 2013a, b). These signatures are reviewed by Lugaz (2014), but here we specifically discuss some recent observations in which prominence material may have been imaged within CMEs in the heliosphere and tracked from the Sun to 1 AU.

Sharma and Srivastava (2012) identified prominence plasma in two ICMEs using magnetic, plasma and compositional signatures. One CME was associated with an active region eruptive filament on 18 November 2003, and the other with one of several quiescent filament eruptions on 1 August 2010. These revealed low temperature, high density plasma in pressure-balanced regions in the trailing portions of magnetic clouds. Compositional signatures included low ion charge states, high ion and helium densities, and He^+ ions. As also shown statistically by Lepri and Zurbuchen (2010), these two events had a mixture of cold and hot plasma suggesting source or processing material from both flares and prominences. It is possible that filament material is present in most CMEs but not usually identified in-situ because it is no longer "cold" in an ionizational sense (Skoug et al. 1999; Sharma and Srivastava 2012). Also it is likely that the filament material is not encountered in many single-spacecraft observations, especially if it is narrowly confined at the back of the CME as is often observed near the Sun.

Trailing dense, likely prominence material has been associated with geostorm activity. A classic case was the well-studied January 1997 event (Fox et al. 1998; Burlaga et al. 1998) that had a very dense "plug" of material trailing the flux rope, resulting in the compression of the magnetopause to within geosynchronous orbit. Another CME observed by ACE on 2–4 May 1998 contained very high He abundance throughout and a prolonged period of He^+ through most of a magnetic cloud and behind it (Skoug et al. 1999). The event was associated with an EP and halo CME, suggesting the prominence material extended through much of the CME by the time it reached 1 AU.

Early observations by the Helios spacecraft permitted coarse imaging of CMEs in the heliosphere and a few mass estimates. During 1979–1980 several CMEs, some with EPs, were observed by both the Solwind coronagraph and the Helios zodiacal light photometers and their masses estimated (Jackson 1985; Jackson et al. 1985; Jackson and Leinert 1985; Jackson and Hick 1994). The May 24–25, 1979 EP/CME (Sheeley et al. 1980) was one of the most massive ever measured with a total mass of 5×10^{16} g, a large fraction of which was prominence material (see discussion in Sect. 16.3). In the last decade the SMEI and STEREO HI heliospheric imagers have been used to determine the heliospheric masses of more CMEs. For SMEI data the 3-D rendering of associated plasma structures is done by the use of a time-dependent reconstruction algorithm (e.g., Jackson et al. 2006) that allows a CME to be isolated from surrounding heliospheric material and its 3-D extent determined. Yao et al. (2010) recently linked a filament eruption/CME observed in Solwind on 8 May 1979 with Helios-2 in-situ measurements at 0.3 AU on 9 May of a CME and magnetic cloud having evidence of prominence material at its rear: high proton density, low temperature and He^+. A separate EP-CME was observed off the limb in Hα and later by Solwind leaving the Sun on 7 May 1979. This CME was also imaged remotely on 8–9 May by the Helios-2 photometers (Jackson 1985; Jackson and Leinert 1985; Jackson et al. 1988), but it was mostly north of the ecliptic so it is unclear which CME crossed the Helios-2 spacecraft.

In other recent events HI 3-D reconstructions have been helpful in separating apparent prominence material from the main CME in the inner heliosphere. Using LASCO and SMEI 3-D reconstructions of density and IPS velocity data, Jackson et al. (2006) were able to separately estimate the heliospheric masses of the EP and CME for the exceptional 28 October 2003 "Halloween" event. Using a 3-D volume-cube method to isolate the CME elements, Jackson et al. calculated a total excess-above-ambient mass of 7.1×10^{16} g for the prominence material, half of the total excess CME mass of 14×10^{16} g. The 3-D reconstructions also showed that the bulk of the CME passed to the northeast of the Earth and that the prominence passed mostly to the south of and missed Earth, while breaking up into separate pieces.

In the well-studied Sun-to-Earth event of 20–22 January 2005, a very fast X7 flare and three-part CME with a bright (filament) core was connected to an intense geostorm, despite the B_Z field in the CME and flux rope being mostly northward. Kozyra et al. (2013) concluded that this was due to very dense prominence material, including He^+, *near the CME front* that led to a cold overdense plasma sheet and strong stretching of the magnetotail. SMEI reconstructions of the density structures near 1 AU show two dense lobes, one at the longitude of the parent active region and the other towards Earth, matching the Wind and ACE shock/density profiles (Jackson et al. 2009). The dense material and shock observed in-situ at 1 AU may have been associated with two solar events, the first slower event being swept up and compressed by the fast X7 CME.

A new aspect of the well-studied series of events on 7–9 November 2004 was that a long transequatorial (TE) filament erupted with an X2 flare and CME and was tracked to and identified at 1 AU. The solar disk location and eruption of the event are shown in Fig. 16.4. IPS (Harra et al. 2007; Bisi et al. 2009), SMEI and Enlil (Bisi et al. 2008) 3-D reconstructions were made for this period. The SMEI reconstructions are shown in Fig. 16.5; they reveal three density patches that were associated with three in-situ density enhancements during this period at ACE (Fig. 16.6—top row) and Wind (Fig. 16.6—bottom row). The IPS data revealed a tongue of high-speed material just north of the ecliptic likely driven by the TE filament-CME (Harra et al. 2007). The first two early and large in-situ peaks on November 8 and 10 were also likely associated with the TE filament and its CME and maybe one or two others were likely merging. Each CME had a clear magnetic cloud and very low proton temperature and bidirectional electron streams, and drove intense storms with peak Dst values of −374 and −263 nT, respectively.

On 5 January 2005, LASCO observed two CMEs associated with eruptive filaments with different initial velocities and acceleration. The second CME accelerated much faster than the first and the resulting interaction appeared in-situ at L1 (ACE and Wind) as the presence of magnetic troughs at the border of the two distinct magnetic clouds associated with the CMEs. At this interface, these magnetic clouds had embedded filament plasma within complex magnetic flux ropes that were modeled by the Grad-Shafranov reconstruction technique. These flux ropes were associated with a moderate geomagnetic storm (Dst = −96 nT). During the main phase polarity reversals associated with the magnetic/filament structures may have triggered two substorms, as analyzed in detail by Sharma et al. (2013a). Along with the analysis of two other events, Sharma et al. (2013c) conclude that filament plasma in CMEs may contribute directly to the triggering of substorms.

Subsequently, Sharma et al. (2013b) compared the shape and extent of the filament plasma remnants in the magnetic clouds on 8 January 2005 as revealed by 3-D reconstruction analysis from the SMEI data. These results provide an overview of the two eruptive filaments on 5 January 2005 and their interplanetary propagation. Figure 16.7 shows the possible mapping of the filaments onto the SMEI reconstructed density map at 1 AU and two different viewing planes through

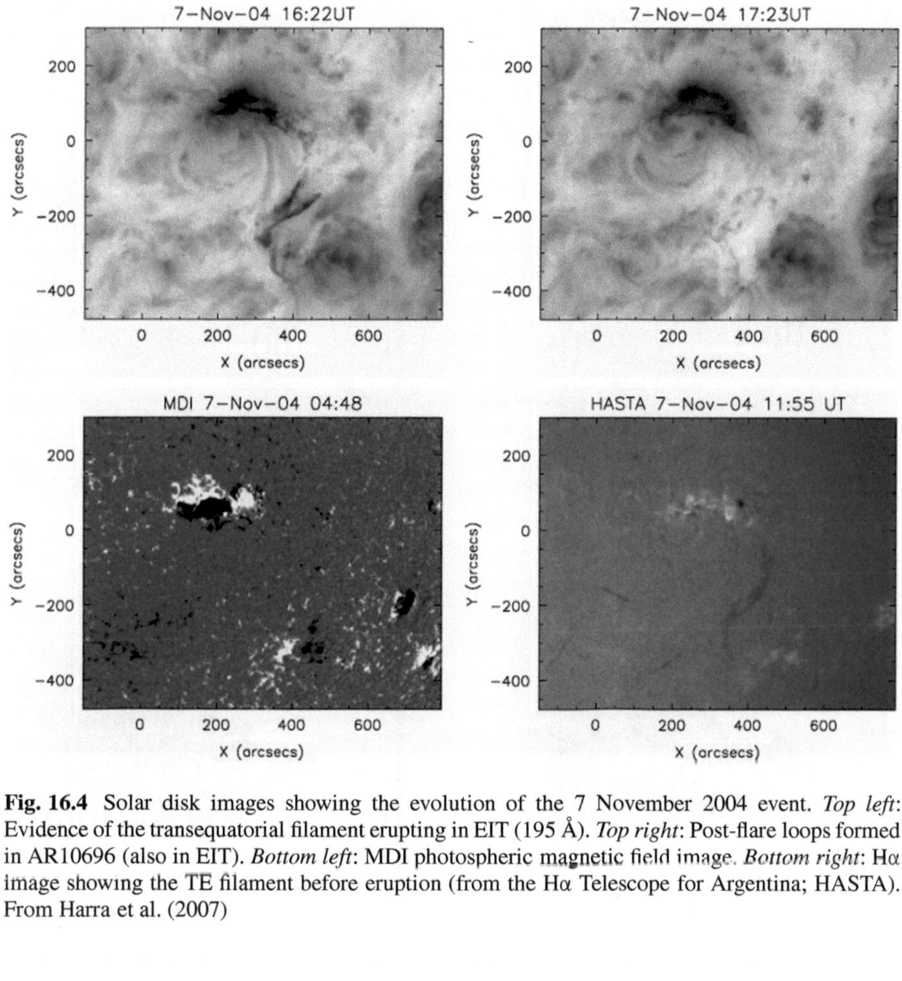

Fig. 16.4 Solar disk images showing the evolution of the 7 November 2004 event. *Top left*: Evidence of the transequatorial filament erupting in EIT (195 Å). *Top right*: Post-flare loops formed in AR10696 (also in EIT). *Bottom left*: MDI photospheric magnetic field image. *Bottom right*: Hα image showing the TE filament before eruption (from the Hα Telescope for Argentina; HASTA). From Harra et al. (2007)

the data. The reconstructions suggest that the filament structure(s) near 1 AU had expanded in latitudinal and longitudinal extent from the solar surface to nearly double that size and volume at 1 AU. Sharma et al. identified the CME associated with the first EP to the west and the in-situ remnant with the dense material near Earth. The second filament may have been the denser structure extending to the northwest and over half an AU from Earth. A qualitative comparison with in-situ data suggests that spacecraft (ACE/Wind) encountered a dense flank of the filament structures, associated with 5 January 2005 solar eruptions.

Recently a new processing pipeline involving the STEREO SECCHI data that reduces many sources of noise (e.g., starfields, F-corona) from the dataset has been developed at Southwest Research Institute (SwRI—DeForest et al. 2011). This allows the tracking and measurement of features that were previously inaccessible. Analyses of early results using this pipeline-processed data include observations and measurements of CME flux ropes (Howard and DeForest 2012) and disconnection

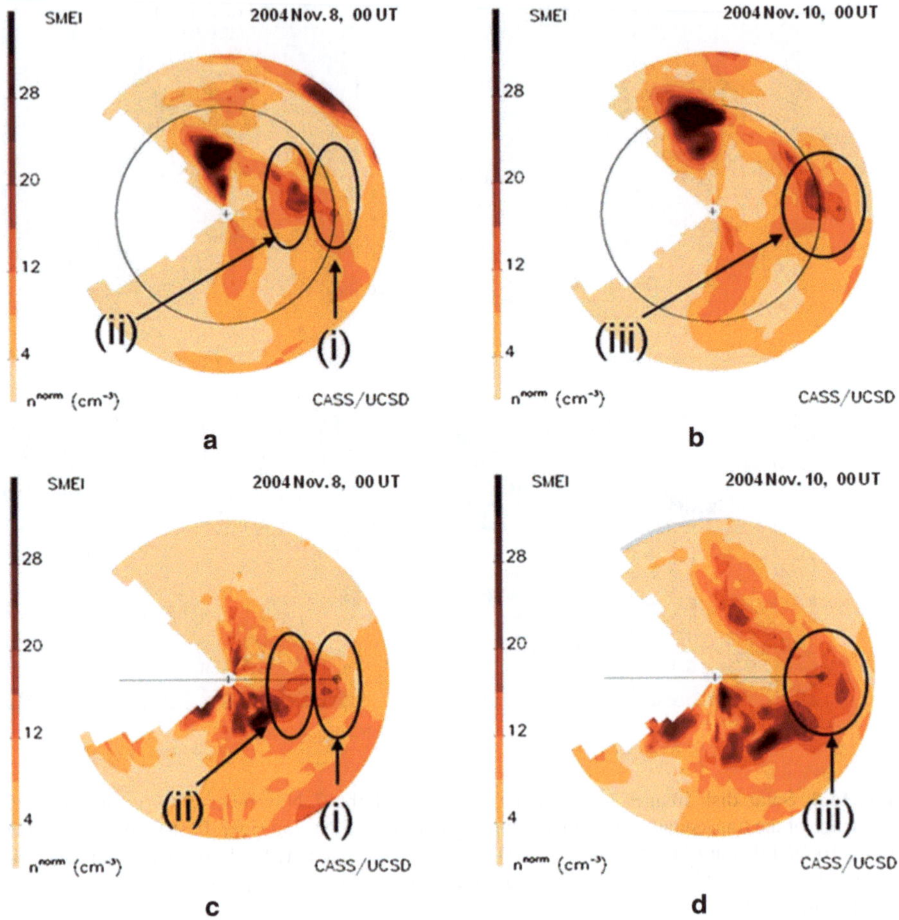

Fig. 16.5 Summary of ecliptic (**a** and **b**) and meridional (**c** and **d**) cuts through the SMEI 3-D density reconstructions on 8 and 10 November 2004 out to 1.5 AU at the times shown. Various features are *circled* in the images, which are also compared in Fig. 16.6 with the ACE and Wind in-situ density plots. Earth's orbit is shown as a *circle* or *line* with the Earth, *circled plus*, indicated on each plot. The expected r^{-2} density falloff scaling is used to normalize structures at different radii. Density contours to the left of each image are scaled to 1 AU with darker colors denoting higher density. Adapted from Bisi et al. (2008)

events (DeForest et al. 2012). The 12 December 2008 event was the first STEREO space weather event to be studied extensively with the SECCHI data. Davis et al. (2009) found that enhancements and voids in the HI data for both spacecraft matched well with in-situ data at Wind and ACE. Recently, the SwRI group have run the new pipeline for this period and tracked the CME front and void or cavity, likely the magnetic cloud, from the Sun to Earth (DeForest et al. 2013). They have identified the filament eruption and its disconnection within a flux rope from the Sun (DeForest et al. 2013; Howard and DeForest 2014), and found some

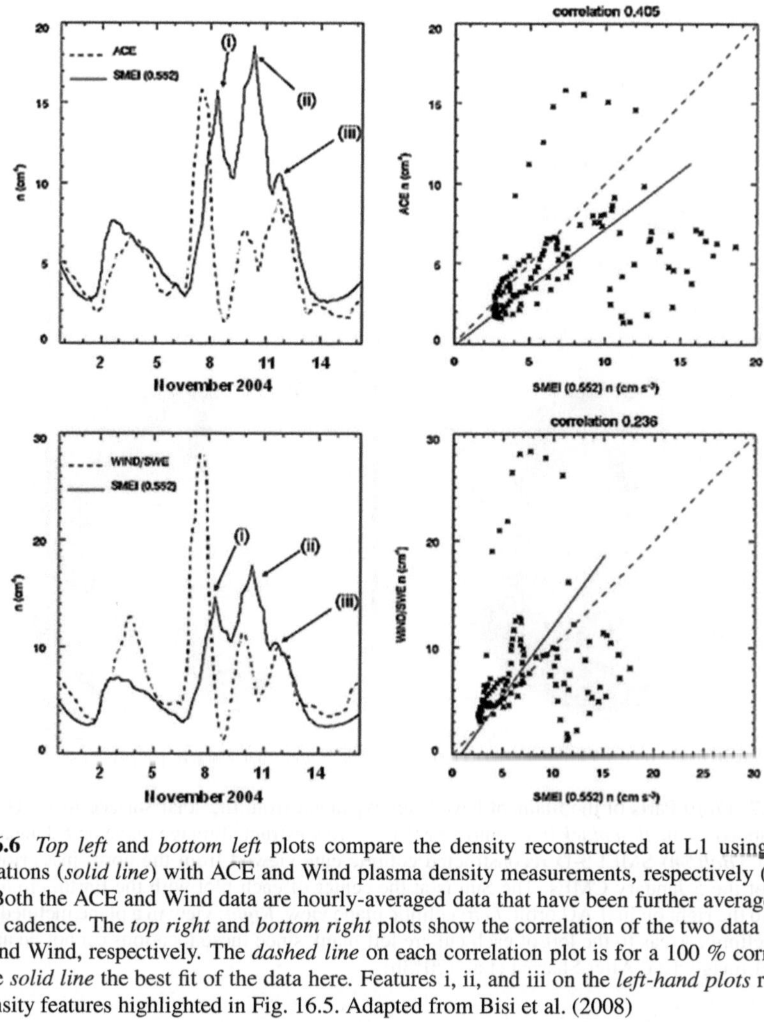

Fig. 16.6 *Top left* and *bottom left* plots compare the density reconstructed at L1 using SMEI observations (*solid line*) with ACE and Wind plasma density measurements, respectively (*dashed line*). Both the ACE and Wind data are hourly-averaged data that have been further averaged with a daily cadence. The *top right* and *bottom right* plots show the correlation of the two data sets for ACE and Wind, respectively. The *dashed line* on each correlation plot is for a 100 % correlation and the *solid line* the best fit of the data here. Features i, ii, and iii on the *left-hand plots* relate to the density features highlighted in Fig. 16.5. Adapted from Bisi et al. (2008)

evidence of bright, dense material trailing the flux rope. This matches in-situ density enhancements at Wind and STEREO-B behind the cloud void on 17 December (Figures 1 and 2 of Howard and DeForest 2012).

Conclusions

We have discussed the origins and characteristics of solar eruptive phenomena focusing on coronal mass ejections and their associated phenomena, particularly erupting prominences. The rate of occurrence of CMEs tracks solar

(continued)

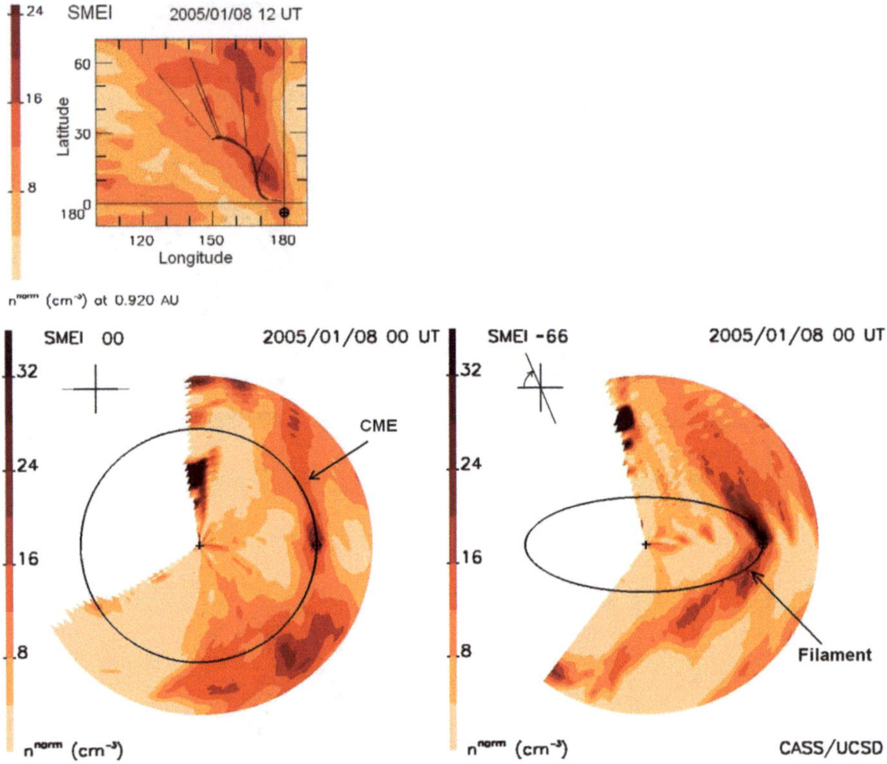

Fig. 16.7 (*Top*) Parts of the filament have been expanded from the solar surface to 1 AU and are here mapped as the *thin black lines* onto a SMEI 3-D reconstructed image plane on 8 January 2005 at 1 AU. (*Bottom*) SMEI 3-D reconstructed ecliptic cuts viewed from the north just prior to the arrival of the 5 January CMEs. The Sun is at the center of each plot with the Earth, *circled plus*, shown to the right on its 1 AU orbit. *Left*: ecliptic plane view. *Right*: view in a plane inclined by 66° to the ecliptic. Areas to the left of each cut are left blank since these locations cannot be accessed in SMEI image data. After Sharma et al. (2013b)

activity, but the size scales of CMEs are much larger and their latitude distributions different than those of near-surface smaller-scale activity like flares or active regions. Indeed, CMEs arise from large-scale closed coronal structures, especially helmet streamers, which overlie neutral lines and filament channels. Statistically, CMEs are most frequently associated with EPs and X-ray long-duration events. In a few large events the masses of the EP and CME have been separately measured, with the EP mass comprising a large fraction of the total CME mass. Large-scale coronal arcades frequently appear near the surface beneath the outgoing CME, and may result from reconnection of

(continued)

closed field systems opened by CMEs. Long current sheet-like structures are now observed connecting the arcade to the rising EP and CME. The size scales and field strengths of these associated systems are a function of latitude. Thus, CMEs involve the destabilization of large-scale coronal structures which result in reconfiguration of the larger-scale, weaker fields at higher latitudes and of the smaller-scale, stronger fields at low latitudes. The magnetic structures involved with the source regions of CMEs can be complex and multipolar. Against the disk the arcades are preceded by sigmoid structures associated with and aligned along the axis of filaments, which erupt forming cusp-shaped arcades.

EP and CME near-surface precursors include the development of sigmoids, the darkening and broadening of filaments, and their slow rise and Doppler shifts, and the cancellation of magnetic flux near filament channels. Prominences exist within coronal cavities which themselves are embedded in helmet streamers extending to high heights. This entire structure can erupt bodily to become a CME; indeed the most massive and energetic CMEs appear to be of this type. Precursors to such eruptions include the slow rise of the prominence and its cavity and an elongation of the cavity with enhanced substructures. During this process the white light streamer itself swells and becomes brighter (denser) until it erupts.

CMEs carry into the heliosphere large quantities of coronal magnetic fields and plasma which are detected by remote sensing and measured in-situ at spacecraft. The most important in-situ CME signature is a magnetic cloud, considered to be the flux rope embedded in most if not all CMEs. The low-density cavity, or void, surrounding the flux rope has now been imaged in the inner heliosphere. Although most CMEs are frequently associated with EPs near the Sun, it is still not known why the prominence material is only rarely identified in-situ. In the last decade, however, we have had heliospheric imaging observations that are helping to distinguish prominence material from the rest of a CME. In some of these cases the mass of the prominence itself could be determined.

Acknowledgments I am grateful to the editors, Oddbjørn Engvold and Jean-Claude Vial, for inviting me to contribute to this book. I thank Noe Lugaz, Nat Gopalswamy, Sarah Gibson, Mario Bisi, Tim Howard, Nandita Srivastava, and Bernard Jackson for helpful comments and references for my chapter.

References

Antiochos, S. K., DeVore, C. R., & Klimchuk, J. A. (1999). A model for solar coronal mass ejections. *Astrophysical Journal, 510*, 485–493.

Athay, R. G., & Illing, R. M. E. (1986). Analysis of the prominence associated with the coronal mass ejection of August 18, 1980. *Journal of Geophysical Research, 91*, 10,961–10,973.

Bemporad, A., Mierla, M., & Tripathi, D. (2011). Rotation of an erupting filament observed by the STEREO EUVI and COR1 instruments. *Astronomy & Astrophysics, 531*, A147.

Billings, D. E. (1966). *A guide to the solar corona*. New York: Academic.

Bisi, M. M., Jackson, B. V., Hick, P. P., Buffington, A., Odstrcil, D., & Clover, J. M. (2008). Three-dimensional reconstructions of the early November 2004 coordinated data analysis workshop geomagnetic storms: Analyses of STELab IPS speed and SMEI density data. *Journal Geophysical Research, 113*, A00A11.

Bisi, M. M., Jackson, B. V., Clover, J. M., Manoharan, P. K., Tokumaru, M., Hick, P. P., & Buffington, A. (2009). 3-D reconstructions of the early-November 2004 CDAW geomagnetic storms: Analysis of Ooty IPS speed and density data. *Annales Geophysicae, 27*, 4479–4489.

Burlaga, L., et al. (1998). A magnetic cloud containing prominence material: January 1997. *Journal of Geophysical Research, 103*, 277–285.

Cremades, H., & Bothmer, V. (2004). On the three-dimensional configuration of coronal mass ejections. *Astronomy and Astrophysics, 422*, 307–322.

Davis, C. J., Davies, J. A., Lockwood, M., Rouillard, A. P., Eyles, C. J., & Harrison, R. A. (2009). Stereoscopic imaging of an Earth-impacting solar coronal mass ejection: A major milestone for the STEREO mission. *Geophysical Research Letters, 36*, L08102.

DeForest, C. E., Howard, T. A., & Tappin, S. J. (2011). Observations of detailed structure in the solar wind at 1 AU with STEREO/HI-2. *Astrophysical Journal, 738*, 103–115.

DeForest, C. E., Howard, T. A., & McComas, D. J. (2012). Disconnecting open solar magnetic flux. *Astrophysical Journal, 745*, 36–44.

DeForest, C. E., Howard, T. A., & McComas, D. J. (2013). Tracking coronal features from the low corona to Earth: A quantitative analysis of the 2008 December 12 coronal mass ejection. *Astrophysical Journal, 769*, 43.

Engvold, O. (2014). Description and classification of prominences. In J.-C. Vial, & O. Engvold (Eds.), *Solar prominences, ASSL* (Vol. 415, pp. 31–60). Springer.

Eyles, C. J., Simnett, G. M., Cooke, M. P., Jackson, B. V., Buffington, A., Hick, P. P., Waltham, N. R., King, J. M., Anderson, P. A., & Holladay, P. E. (2003). The solar mass ejection imager (SMEI). *Solar Physics, 217*, 319–347.

Eyles, C. J., et al. (2009). The heliospheric imagers onboard the STEREO mission. *Solar Physics, 254*, 387–445.

Fan, Y. (2014). Magnetism and dynamics of prominences: MHD equilibria and triggers for eruption. In J.-C. Vial, & O. Engvold (Eds.), *Solar prominences, ASSL* (Vol. 415, pp. 295–320). Springer.

Fox, N. J., Peredo, M., & Thompson, B. J. (1998). Cradle to grave tracking of the January 6–11, 1997 Sun–Earth connection event. *Geophysical Research Letters, 25*(14), 2461–2760.

Gibson, S. (2014). Coronal cavities: Observations and implications for the magnetic environment of prominences. In J.-C. Vial, & O. Engvold (Eds.), *Solar prominences, ASSL* (Vol. 415, pp. 321–351). Springer.

Gilbert, H. (2014). Energy balance. In J.-C. Vial, & O. Engvold (Eds.), *Solar prominences, ASSL* (Vol. 415, pp. 155–176). Springer.

Gopalswamy, N. (2010). Coronal mass ejections: A summary of recent results. In I. Dorotovic (Ed.), *Proc. 20th Slovak Nat. Solar Phys. Workshop*. Slovak Central Observatory.

Gopalswamy, N. (2014). The dynamics of eruptive prominences. In J.-C. Vial, & O. Engvold (Eds.), *Solar prominences, ASSL* (Vol. 415, pp. 379–408). Springer.

Gopalswamy, N., & Hanaoka, Y. (1998). Coronal dimming associated with a giant prominence eruption. *Astrophysics Journal, 498*, L179–L182.

Gopalswamy, N., & Thompson, B. J. (2000). Early life of coronal mass ejections. *Journal of Atmospheric and Solar-Terrestrial Physics, 62*, 1457–1469.

Gopalswamy, N., Lara, A., Yashiro, S., Nunes, S., & Howard, R.A. (2003a). Coronal mass ejection activity during solar cycle 23. In *Proc. Solar Variability as an Input to the Earth's Environment* (p. 403). ESA-SP 535.

Gopalswamy, N., Shimojo, M., Lu, W., Yashiro, S., Shibasaki, K., & Howard, R. A. (2003b). Prominence eruptions and coronal mass ejection: A statistical study using microwave observations. *Astrophysical Journal, 586*, 562–578.

Gopalswamy, N., Mikić, Z., Maia, D., Alexander, D., Cremades, H., Kaufmann, P., Tripathi, D., & Wang, Y.-M. (2006). The pre-CME Sun. *Space Science Reviews, 123*, 303–339.

Gopalswamy, N., Akiyama, S., Yashiro, S., & Makela, P. (2010). Coronal mass ejections from sunspot and non-sunspot regions. In S. S. Hasan, & R. J. Rutten (Eds.), *Magnetic coupling between the interior and the atmosphere of the sun, astrophysics and space science proceedings* (pp. 289–307). Berlin: Springer.

Gopalswamy, N., Mäkelä, P., Akiyama, S., Xie, H., Yashiro, S., & Reinard, A. A. (2013). The solar connection of enhanced heavy ion charge states in the interplanetary medium: Implications for the flux-rope structure of CMEs. *Solar Physics, 284*, 17–46.

Harra, L. K., Crooker, N. U., Mandrini, C. H., van Driel-Gesztelyi, L., Dasso, S., Wang, J., Elliott, H., Attrill, G., Jackson, B. V., & Bisi, M. M. (2007). How does large flaring activity from the same active region produce oppositely directed magnetic clouds? *Solar Physics, 244*, 95–114.

Howard, T. A., & DeForest, C. E. (2012). Inner flux rope evolution via imaging of coronal mass ejections. *Astrophysics Journal, 746*, 64–75.

Howard, T. A., & DeForest, C.E. (2014). The formation and launch of a coronal mass ejection flux rope: A narrative based on observations. *Astrophysics Journal* (submitted).

Howard, T. A., & Tappin, S. J. (2009). Interplanetary coronal mass ejections observed in the heliosphere: 1. Review of theory. *Space Science Reviews, 147*, 31–54.

Howard, R. A., Michels, D. J., Sheeley, N. R., Jr., & Koomen, M. J. (1982). The observation of a coronal transient directed at Earth. *Astrophysics Journal, 263*, L101–L104.

Howard, R. A., et al. (2008). Sun Earth Connection Coronal and Heliospheric Investigation (SECCHI). *Space Science Reviews, 136*, 67–115.

Hundhausen, A. J. (1993). Sizes and locations of coronal mass ejections: SMM observations from 1980 and 1984–1989. *Journal of Geophysical Research, 98*, 13,177–13,200.

Illing, R. M. E., & Athay, G. (1986). Physical conditions in eruptive prominences at several solar radii. *Solar Physics, 105*, 173–190.

Illing, R. M. E., & Hundhausen, A. J. (1985). Observation of a coronal transient from 1.2 to 6 solar radii. *Journal of Geophysical Research, 90*, 275–282.

Illing, R. M. E., & Hundhausen, A. J. (1986). Disruption of a coronal streamer by an eruptive prominence and coronal mass ejection. *Journal of Geophysical Research, 91*, 10,951–10,960.

Jackson, B. V. (1985). Imaging of coronal mass ejections by the Helios spacecraft. *Solar Physics, 100*, 563–574.

Jackson, B. V., & Hick, P. (1994). Three-dimensional reconstruction of coronal mass ejections. In *Proc. of the third SOHO workshop* (p. 199). Estes, Park, CO, ESA-SP-373.

Jackson, B. V., & Leinert, C. (1985). Helios images of solar mass ejections. *Journal of Geophysical Research, 90*, 10759.

Jackson, B. V., Howard, R. A., Sheeley, N. R., Jr., Michaels, D. J., Koomen, M. J., & Illing, R. M. E. (1985). Helios spacecraft and Earth perspective observations of three looplike solar mass ejection transients. *Journal of Geophysical Research, 90*, 5075.

Jackson, B. V., Rompolt, B., & Svestka, Z. (1988). Solar and interplanetary observations of the mass ejection on 7 May 1979. *Solar Physics, 115*, 327–343.

Jackson, B. V., et al. (2004). The Solar Mass Ejection Imager (SMEI); The Mission. *Solar Physics, 225*, 177–207.

Jackson, B. V., Buffington, A., Hick, P. P., Wang, X., & Webb, D. (2006). Preliminary 3D analysis of the heliospheric response to the 28 October 2003 CME using SMEI white-light observations. *Journal of Geophysical Research, 111*, A04S91.

Jackson, B. V., Hick, P. P., Buffington, A., Bisi, M. M., Clover, J. M., Hamilton, M. S., Tokumaru, M., & Fujiki, K. (2009). 3D reconstruction of density enhancements behind interplanetary shocks from the solar mass ejection imager white-light observations. In M. Maksimovic, et al. (Eds.), *CP1216, Twelfth international solar wind conference* (pp. 659–662). Melville, NY: American Institute of Physics.

Joshi, A. D., & Srivastava, N. (2011). Kinematics of two eruptive prominences observed by EUVI/STEREO. *Astrophysics Journal, 730*, 104–114.

Kahler, S. W. (2006). Observational properties of coronal mass ejections. In N. Gopalswamy, R. Mewaldt, J. Torsti (Eds.), *Solar eruptions and energetic particles, Vol. 165 of geophysical monograph series* (pp. 21–32). Washington, DC: AGU.

Kahler, S. W., & Hundhausen, A. J. (1992). The magnetic topology of solar coronal structures following mass ejections. *Journal of Geophysical Research, 97*, 1619–1631.

Kahler, S. W., Moore, R. L., Kane, S. R., & Zirin, H. (1988). Filament eruptions and the impulsive phase of solar flares. *Astrophysics Journal, 328*, 824–829.

Karpen, J. (2014). Plasma structure and dynamics. In J.-C. Vial, & O. Engvold (Eds.), *Solar prominences, ASSL* (Vol. 415, pp. 235–255). Springer

Kozyra, J. U., Manchester, W. B., IV, Escoubet, C. P., Lepri, S. T., Liemohn, M. W., Gonzalez, W. D., Thomsen, M. W., & Tsurutani, B. T. (2013). Earth's collision with a solar filament on 21 January 2005: Overview. *Journal of Geophysical Research, 118*, 5967–5978. doi:10.1002/jgra.50567.

Lee, J.-Y., & Raymond, J. C. (2012). Low ionization state plasma in coronal mass ejections. *Astrophysics Journal, 758*, 116.

Lepri, S. T., & Zurbuchen, T. H. (2010). Direct observational evidence of filament material within interplanetary coronal mass ejections. *Astrophysics Journal, 723*, L22–L27.

Lin, J. (2004). CME-flare association deduced from catastrophic model of CMEs. *Solar Physics, 219*, 169–196.

Lin, J., & Forbes, T. G. (2000). Effects of reconnection on the coronal mass ejection process. *Journal of Geophysical Research, 105*, 2375–2392.

Lugaz, N. (2014). Eruptive prominences and their impact on the Earth: The impacts on our Earth and our life. In J.-C. Vial, & O. Engvold (Eds.), *Solar prominences, ASSL* (Vol. 415, pp. 431–451). Springer.

Martin, S. F. (1980). Preflare conditions, changes and events. *Solar Physics, 68*, 217–236.

Martin, S. F. (2014). The magnetic field structure of prominences from direct and indirect observations. In J.-C. Vial, & O. Engvold (Eds.), *Solar Prominences, ASSL* (Vol. 415, pp. 203–233). Springer.

Martin, S. F., & Livi, S. H. B. (1992). The role of cancelling magnetic fields in the buildup to erupting filaments and flares. In Z. Svestka, B. V. Jackson, & M. E. Machado (Eds.), *Eruptive solar flares* (pp. 33–45). Berlin: Springer.

Martin, S. F., Panasenco, O., Berger, M. A., Engvold, O., Lin, Y., Pevtsov, A. A., & Srivastava, N. (2012). The build-up to eruptive solar events viewed as the development of chiral systems. In Rimmele, T., et al. (Eds.), *The second ATST-EAST meeting: Magnetic fields from the photosphere to the corona. ASP conference proceedings* (Vol. 463, p. 157). San Francisco: Astronomical Society of the Pacific.

McAllister, A. H., Mackay, D. H., & Martin, S. F. (2002). The skew of high-latitude X-ray arcades in the declining phase of cycle 22. *Solar Physics, 211*, 155–163.

Munro, P. H., et al. (1979). The association of coronal mass ejection transients with other forms of solar activity. *Solar Physics, 61*, 201.

Nitta, N., & Akiyama, S. (1999). Relation between flare-associated X-ray ejections and coronal mass ejections. *Astrophysics Journal, 525*, L57.

Plunkett, S. P., Vourlidas, A., Šimberová, S., Karlický, M., Kotrč, P., Heinzel, P., Kupryakov, Y. A., Guo, W. P., & Wu, S. T. (2000). Simultaneous SOHO and ground-based observations of a large eruptive prominence and coronal mass ejection. *Solar Physics, 194*, 371–391.

Plunkett, S. P., Michels, D. J., Howard, R. A., Brueckner, G. E., St Cyr, O. C., Thompson, B. J., Simnett, G. M., Schwenn, R., & Lamy, P. (2002). New insights on the onsets of coronal mass ejections from SOHO. *Advances in Space Research, 29*, 1473–1488.

Sharma, R., & Srivastava, N. (2012). Presence of solar filament plasma detected in interplanetary coronal mass ejections by *in-situ* spacecraft. *Journal of Space Weather and Space Climate, 2*(A10), 10.1051.

Sharma, R., Srivastava, N., Chakrabarty, D., Moestl, C., & Hu, Q. (2013a). Interplanetary and geomagnetic consequences of January 5, 2005 CMEs associated with solar eruptive filaments. *Journal of Geophysical Research, 118*(7), 3954–3967.

Sharma, R., Srivastava, N., Jackson, B., Chakrabarty, D., Luckett, N., Yu, H.-S., Hu, Q., & Moestl, C. (2013b). Evolution of the 5 January 2005 CMEs associated with eruptive filaments in inner heliosphere. In B. Schmieder, J.-M. Malherbe, & S. T. Wu (Eds.), *Nature of prominences and their role in space weather, proceedings of IAU symposium No. 300* (p. 491). Cambridge, UK: Cambridge University Press.

Sharma, R., Srivastava, N., & Chakrabarty, D. (2013c). Role of filament plasma remnants in ICMEs leading to geomagnetic storms. In B. Schmieder, J.-M. Malherbe, & S. T. Wu (Eds.), *Nature of prominences and their role in space weather, proceedings of IAU symposium No. 300* (p. 493). Cambridge, UK: Cambridge University Press.

Sheeley, N. R., Jr., Michels, D. J., Howard, R. A., & Koomen, M. J. (1980). Initial observations of with the Solwind coronagraph. *Astrophysics Journal, 237*, L99–L101.

Sheeley, N. R., Jr., Howard, R. A., Koomen, M. J., Michels, D. J., Harvey, J. W., & Harvey, K. L. (1982). Observations of coronal structure during sunspot maximum. *Space Science Reviews, 33*, 219–231.

Shibata, K., Masuda, S., Shimojo, M., Hara, H., Yokoyama, T., Tsuneta, S., Kosugi, T., & Ogawara, Y. (1995). Hot-plasma ejections associated with compact-loop solar flares. *Astrophysics Journal, 451*, L83–L85.

Simnett, G. M. (2000). The relationship between prominence eruptions and coronal mass ejections. *Journal of Atmospheric and Terrestrial Physics, 62*, 1479–1487.

Skoug, R. M., et al. (1999). A prolonged He^+ enhancement within a coronal mass ejection in the solar wind. *Geophysical Research Letters, 26*, 161–164.

Srivastava, N., Schwenn, R., Inhester, B., Stenborg G., & Podlipnik, B. (1999). Measurements of flow speeds and acceleration in gradually evolving solar mass ejections as observed by LASCO. In *Solar wind-9, AIP conference proceedings* (Vol. 471, pp. 115–118).

Svestka, Z., & Cliver, E. W. (1992). History and basic characteristics of eruptive flare. In Z. Svestka, B. V. Jackson, & M. E. Machado (Eds.), *Proc. Int. Astron. Union*. New York: Springer.

Vourlidas, A., & Howard, R. A. (2006). The proper treatment of coronal mass ejection brightness: A new methodology and implications for observations. *Astrophysics Journal, 642*, 1216–1221.

Vourlidas, A., Buzasi, D., Howard, R.A., & Esfandiari, E. (2002a). Mass and energy properties of LASCO CMEs. In A. Wilson (Ed.), *Solar variability: from core to outer frontiers, vol. SP-506 of ESA* (pp. 91–94). Noordwijk: ESTEC.

Vourlidas, A., Howard, R.A., Morill, J.S., & Munz, S. (2002b). Analysis of LASCO streamer blowout events. In H. Wang, R.L. Xu (Eds.) *Solar-terrestrial magnetic activity and space environment, Proceedings of the COSPAR Colloquium held in the NAOC in Beijing, China, September 10–12, 2001, vol. 14 of COSPAR Colloquia Series* (pp. 201–208). Boston, Amsterdam: Pergamon.

Vourlidas, A., Howard, R. A., Esfandiari, E., Patsourakos, S., Yashiro, S., & Michalek, G. (2010). Comprehensive analysis of coronal mass ejection mass and energy properties over a full solar cycle. *Astrophysics Journal, 722*, 1522–1538.

Vourlidas, A., Colaninno, R., Nieves-Chinchilla, T., & Stenborg, G. (2011a). The first observation of a rapidly rotating coronal mass ejection in the middle corona. *Astrophysics Journal, 733*, L23.

Vourlidas, A., Howard, R. A., Esfandiari, E., Patsourakos, S., Yashiro, S., & Michalek, G. (2011b). Erratum: 'Comprehensive analysis of coronal mass ejection mass and energy properties over a full solar cycle'. *Astrophysics Journal, 730*.

Vourlidas, A., Lynch, B. J., Howard, R. A., & Li, Y. (2013). How many CMEs have flux ropes? Deciphering the signatures of shocks, flux ropes, and prominences in coronagraph observations of CMEs. *Solar Physics, 284*, 179.

Webb, D. F. (1988). Erupting prominences and the geometry of coronal mass ejections. *Journal of Geophysical Research, 93*, 1749–1758.

Webb, D. F. (1992). The solar sources of coronal mass ejections. In Z. Svestka, B. V. Jackson, & M. E. Machado (Eds.), *Eruptive solar flares* (pp. 234–247). Berlin: Springer.

Webb, D. F., & Howard, T. A. (2012). Coronal mass ejections: observations. *Living Reviews Solar Physics, 9*, 3. http://www.livingreviews.org/lrsp-2012-3.

Webb, D. F., & Hundhausen, A. J. (1987). Activity associated with the solar origin of coronal mass ejections. *Solar Physics, 108*, 383–401.

Webb, D., Kahler, S., McIntosh, P., & Klimchuk, J. (1997). Large-scale structures and multiple neutral lines associated with coronal mass ejections. *Journal of Geophysical Research, 102*, 24,161–24,174.

Webb, D. F., Burkepile, J., Forbes, T. G., & Riley, P. (2003). Observational evidence of new current sheets trailing coronal mass ejections. *Journal of Geophysical Research, 108*, 1440.

Webb, D. F., Bisi, M. M., de Koning, C. A., Farrugia, C. J., Jackson, B. V., Jian, L. K., Lugaz, N., Marubashi, K., Möstl, C., Romashets, E. P., Wood B. E., & Yu, H.-S. (2014). An ensemble study of a January 2010 CME: Connecting a non-obvious solar source with its ICME/magnetic cloud. *Solar Physics, 289*, 4173–4208. doi:10.1007/s11207-014-0571-1.

Xie, H., Gopalswamy, N., & St. Cyr, O. C. (2013). Near-sun flux-rope structure of CMEs. *Solar Physics, 284*, 47–58.

Yao, S., Marsch, E., Tu, C.-Y., & Schwenn, R. (2010). Identification of prominence ejecta by the proton distribution function and magnetic file structure in interplanetary coronal mass ejections in the inner heliosphere. *Journal of Geophysical Research, 115*, A05103.

Yashiro, S., Gopalswamy, N., Michalek, G., St Cyr, O. C., Plunkett, S. P., Rich, N. B., & Howard, R. A. (2004). A catalog of white light coronal mass ejections observed by the SOHO spacecraft. *Journal of Geophysical Research, 109*, A07105.

Zhang, J., Dere, K. P., Howard, R. A., Kundu, M. R., & White, S. M. (2001). On the temporal relationship between coronal mass ejections and flares. *Astrophysics Journal, 559*, 452–462.

Chapter 17
Eruptive Prominences and Their Impact on the Earth and Our Life

Noé Lugaz

Abstract Following prominence eruptions (see Chap. 16: Webb, *Solar prominences*. New York: Springer, 2014), the associated coronal mass ejections (CMEs) propagate into the solar wind and interplanetary medium. While the complex interactions with the magnetic field and plasma in the corona (see Chaps. 15 and 16: Gopalswamy, *Solar prominences*. New York: Springer, 2014; Webb, *Solar prominences*. New York: Springer, 2014) have been observed for decades, it has only been in the last decade that the interaction of CMEs with the interplanetary medium can be directly imaged. As CMEs and prominences impact Earth, Earth's magnetosphere may be disrupted through reconnection and/or compression, resulting in geomagnetic storms. In the most extreme cases, CMEs and prominences may have a global effect on man-made technologies and human beings, especially if they are in space. The conditions in the near-Earth environment directly affected by the Sun and the solar activity are known as space weather and will be discussed here.

In this chapter, we review different types of measurements and observations of prominences and CMEs as they propagate between the Sun and the Earth, as well as recent advances in numerical modeling and theoretical ideas related to CME propagation. We also discuss the potential effects of CMEs and prominences on Earth's magnetosphere and atmosphere and the very direct impact it may exert on our lives.

17.1 The Propagation of Prominences and CMEs in the Inner Heliosphere

As prominences erupt, they first interact with the coronal magnetic field and the solar coronal plasma. Understanding these complex interactions involves studying the reconnection between the prominence and neighboring magnetic structures, the interplay between the Lorentz force, the gravitational force and the prominence

N. Lugaz (✉)
Space Science Center and Department of Physics, University of New Hampshire, 8 College Rd, Durham, NH 03824, USA
e-mail: noe.lugaz@unh.edu

J.-C. Vial, O. Engvold (eds.), *Solar Prominences*, Astrophysics and Space Science Library 415, DOI 10.1007/978-3-319-10416-4_17

momentum in a rapidly varying medium and background (see for examples works by Filippov et al. 2001, Kilpua et al. 2009, Zuccarello et al. 2012, Panasenco et al. 2013, Fan 2014, Gopalswamy 2014, Webb 2014). When the difference in speed between a CME and the solar wind is greater than the local fast magnetosonic speed, CMEs drive a fast magnetosonic shock wave. Typically, for fast CMEs (with speed of 800 km s^{-1} or higher), the CME-driven shock forms within 2–6 $R_\odot$ (Raymond et al. 2000; Vourlidas et al. 2003; Roussev et al. 2004; Bemporad and Mancuso 2010) and may affect the propagation of the CMEs (or its observations). In the low corona, the prominences and CMEs may be observed in extreme ultraviolet (EUV) with a cadence of a few tens of seconds with the Solar Dynamics Observatory (SDO, Lemen et al. 2012) and the density structures associated with these eruptions have been routinely imaged with coronagraphs since the 1980s up to distances of 10–30 $R_\odot$.

The propagation is somewhat simpler in the inner heliosphere (at distances of 20 $R_\odot$ and farther), since the gravitational force can be ignored and the solar wind speed is more or less constant. However, the observations themselves are sparser and are often more complicated to analyze, as discussed below. Numerical simulations may be used to fill the void of knowledge between the upper corona and the near-Earth environment and to complement existing observations. In addition, density structures associated with CMEs have been routinely imaged in the heliosphere by instruments since the mid-2000s.

17.1.1 Observing the Propagation of Density Structures in the Inner Heliosphere

Remote-sensing observations of coronal mass ejections, prominences and associated density structures (CME sheath, for example) were pioneered by the zodiacal light experiment onboard Helios (Jackson and Leinert 1985). It is not until the 2000s that two follow-up instruments were launched: the Solar Mass Ejection Imager onboard *Coriolis* (SMEI: Eyles et al. 2003; Jackson et al. 2004) and the Heliospheric Imagers (HIs: Eyles et al. 2009) part of SECCHI (Howard et al. 2008) onboard the twin STEREO spacecraft. Tappin et al. (2004) and Harrison et al. (2008) reported the first observations of a CME propagating in the heliosphere by SMEI and HIs, respectively. Reviews of SMEI and HI observations have been published in Howard et al. (2013a) and Harrison et al. (2009).

These instruments work in a way similar to coronagraphs. The main differences are: (1) the fact that the signal-to-noise ratio is very low due to the star field and the F-corona (created by light scattered by dust particles), and, (2) the locus of the maximum signal intensity (e.g., see DeForest et al. 2011). Contrary to coronagraphs, for which observations can be assumed to originate from the "plane of sky", Thomson scattering geometry must be taken into account for wide-angle heliospheric imagers. While the locus of maximum Thomson scattering signal is on the so-called Thomson sphere (Vourlidas and Howard 2006), the fact that CMEs are wide density structures combined with the fact that the scattering efficiency is more or less constant for $\pm 30°$ around the Thomson sphere mean that CMEs may be well observed even when far from the Thomson sphere (Howard and DeForest 2012b). In fact, CMEs have been imaged into HI-2 field-of-view while propagating more than 130° away from the Sun-observer line (Lugaz et al. 2012b; Howard and DeForest 2012b) at distances corresponding to 0.5 AU or more. Figure 17.1 shows two examples of Earth-directed CMEs propagating during the early and later years of the STEREO missions, as the separation between the observing spacecraft and Earth increases from 45° to 130°. These images show the Thomson scattered light, a measure of the electron density, and a long-term average image is subtracted to remove the contribution from the F-corona.

Howard and DeForest (2012a) and DeForest et al. (2013) have published the most detailed analyses of HI observations of a CME and associated density structures to date for the 2008 December 12 CME event. This same event was also studied by Davis et al. (2009), Liu et al. (2010), Byrne et al. (2010) and Lugaz et al. (2010), among others. In wide-angle white-light imagers, it is possible to track the bright region corresponding to the high-density structure of the CME sheath but also the dark cavity corresponding to the low-β, low-density magnetic cloud itself. Through its interaction with the solar wind, the CME gets sometimes distorted and this effect can be directly observed (see also Savani et al. 2010). The high-density region associated with filament material over-expands inside the cavity as it propagates in the corona (Bothmer and Schwenn 1998) and it is unusual to image it into the heliosphere. Note that white-light measurements depend on the electron density and material cold enough not to be ionized, would not contribute to the signal. Careful

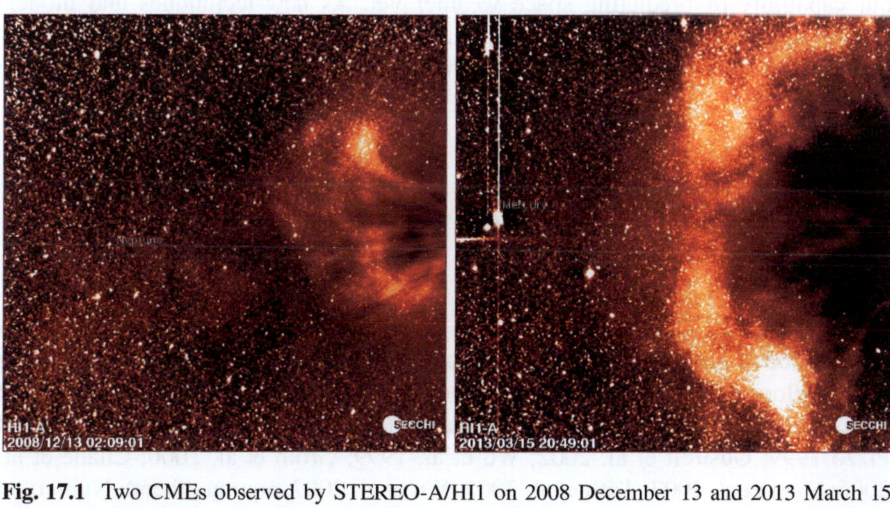

Fig. 17.1 Two CMEs observed by STEREO-A/HI1 on 2008 December 13 and 2013 March 15. Although STEREO-A was more than 130° away from the source region of the CME in 2013, it was able to observe it until it impacted Earth

analyses have revealed the presence of filament-like dense material at the back of the cavity for the December 12, 2008 CME (Byrne et al. 2010; DeForest et al. 2013).

In the low corona, flux ropes and CMEs often deflect and rotate due to magnetic forces, ideal instabilities such as the kink instability, as well as reconnection (Török and Kliem 2003; Cremades and Bothmer 2004; Gopalswamy et al. 2009; Yurchyshyn et al. 2009; Isavnin et al. 2013). However, it is expected that magnetic forces become small as the CME reaches the upper corona (see, e.g. Kay et al. 2013) and that the CME trajectory becomes close to radial with distortion and acceleration/deceleration associated with hydrodynamical-like interactions with the solar wind (Riley et al. 1997; Odstrčil and Pizzo 1999; Vršnak 2001; Cargill 2004; Shen et al. 2011). Remote-sensing imaging of CMEs have revealed that, in some cases, deflection and rotation may continue into the heliosphere (Nieves-Chinchilla et al. 2012; Liu et al. 2013). It is still unclear at this time how common these phenomena are and whether it is due to interaction with solar wind streams, to magnetic forces or to the momentum associated with deflection and rotation in the corona.

In addition to these cases where only one CME is involved, recent HI observations have revealed that the interaction of successive CMEs may affect their trajectory (Lugaz et al. 2012a; Shen et al. 2012) and their kinematics (Temmer et al. 2012; Shen et al. 2012). In absolute terms, these effects are relatively small: deflections of $\sim 10°$ and decelerations of a few m s^{-2}, while acceleration, if present, should be only a few tenths of m s^{-2} for the leading CME or may be reflected in a CME deceleration occurring more slowly than expected. Decelerations of this magnitude are close to impossible to measure with current analysis techniques. This is because these changes are typically smaller than the uncertainty in predicting the CME direction and speed from remote-sensing observations and do not affect our capability of predicting space weather yet. As new techniques and models are developed to improve prediction, changes in the CME properties during their propagation may need to be taken into account in order to perform accurate forecasts.

17.1.2 Numerical Simulations of CMEs and Prominence Propagation

In order to fill the gap between the upper corona and the near-Earth environment in the absence of direct remote-sensing measurements, advanced numerical modeling was developed in the 1990s. Simulations of CME propagation have been performed by a number of groups in the United States, in Europe and in Asia (Odstrčil and Pizzo 1999; Odstrčil et al. 2002; Wu et al. 1999; Groth et al. 2000; Chané et al. 2006; Shen et al. 2007; Tóth et al. 2005; Wu et al. 2007, among others). Numerical simulations have been central to the study of the interaction between a CME and the solar wind, which results in deceleration for fast CMEs and acceleration for slow CMEs. This interaction can be somewhat approximated using the paradigm of aerodynamical drag but with a coefficient that also reflects the magnetic interaction between CME and solar wind (Cargill 2004; Vršnak et al. 2010). Using numerical simulations, Riley et al. (1997) and Manchester et al. (2004b) predicted that CMEs get distorted as they propagate; such simulations have also been used to investigate the over-expansion of the filamentary mass inside the magnetic ejecta (Manchester et al. 2004b; Lugaz et al. 2005). Both phenomena have been confirmed by recent observations, although the extreme CME distortion as seen in numerical simulations is only rarely observed remotely (Savani et al. 2010). In a numerical study, Lugaz et al. (2005) have shown that the filament mass remains approximately constant as the CME propagates, but the total CME mass increases by a factor of 2–5 from the corona to 1 AU (see left panel of Fig. 17.2). This mass increase is due to the development of the CME sheath and has been recently observed by STEREO/HIs (right panel of Fig. 17.2). Figure 17.2 shows that the total CME mass increases by a factor of $\sim$2 from the upper corona (20 $R_\odot \sim 0.1$ AU) to 0.8 AU based on a MHD simulation and HI measurements. As no mass is added to the prominence at such distances from the Sun, all of the mass increase is due to solar wind material swept up by the CME or shocked by the CME-driven shock.

These simulations now complement observations of CMEs by LASCO, SMEI or the STEREO/HIs (e.g., see Manchester et al. 2008; Webb et al. 2009; Odstrcil and Pizzo 2009; Lugaz et al. 2009b; Webb et al. 2013; Shen et al. 2013). One feature of many of the most recent simulations is the ability to produce synthetic remote-sensing observations and *in situ* measurements. Simulated spacecraft measurements have helped to show that a flux rope initiated at the Sun develops into the classical magnetic cloud at 1 AU (Manchester et al. 2004b), but has also revealed that a

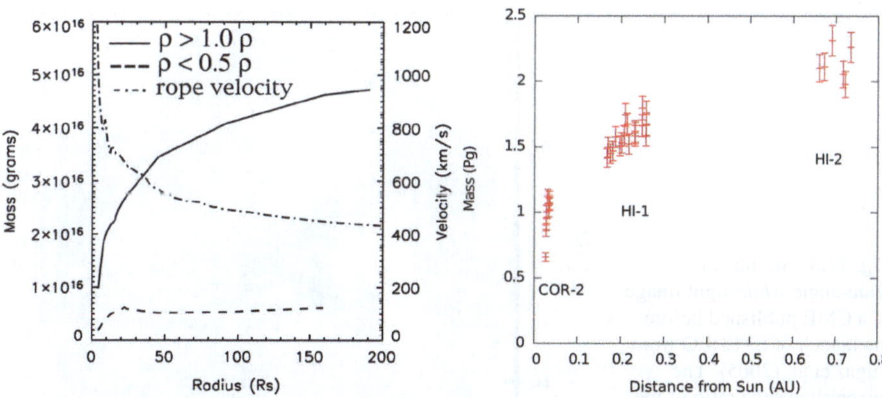

Fig. 17.2 CME mass evolution from a numerical simulation (*left*, from Lugaz et al. 2005) and from HI measurements (*right*, from DeForest et al. 2013) for the 2008 December 12 CME. The *left plot* shows the CME mass with *solid line* and the mass in the cavity with a *dash line*. The prominence material is part of the cavity once the CME reaches 20 $R_\odot$. The x axis in *both panels* shows the radial distance from the Sun (1 AU $\sim$ 215 R_s). Reproduced by permission of the AAS

flux rope may or may not be detected as a magnetic cloud, depending on where it crosses a spacecraft (Riley et al. 2004). Synthetic coronagraphic observations have been created with MHD codes in the past 10 years (Wu et al. 1997; Manchester et al. 2004a; Riley et al. 2007). They have been useful in associating CME morphology as observed by coronagraphs with the actual three-dimensional structure of CMEs. For example, CMEs are often observed as having a 3-part structure: a bright sheath followed by a dark cavity and a dense core. With the help of numerical simulations, it is relatively straight-forward to confirm that this morphology corresponds to the dense sheath of the CME surrounding the flux rope into which the filament is embedded (Gibson and Low 1998; Manchester et al. 2004a; Lugaz et al. 2005).

A normal continuation of these works is the synthesis of wide-angle white-light images such as those from SMEI and STEREO/HIs (Lugaz et al. 2005; Odstrcil and Pizzo 2009). Synthetic images as originally published before the launch of STEREO are shown in Fig. 17.3. This figure shows the typical structure of a CME as observed remotely in the heliosphere: a bright sheath followed by a dark cavity. Comparisons with 3-D simulations have shown that the bright sheath corresponds to the dense sheath formed by swept-up solar wind and coronal material as well as material compressed by the CME-driven shock. The cavity corresponds to the magnetic cloud and the filament material is usually not visible due to over-expansion. As the filament expands with respect to the background solar wind, its density decreases and it becomes less visible in white-light imaging. Dense material sometimes observed at the back of the CME corresponds to compressed material associated with reconnection with the solar wind and the expansion of the magnetic ejecta. It is often observed behind real observations of CMEs (Harrison et al. 2008; Savani et al. 2010). In addition to SMEI-like and HI-like images, it is possible to

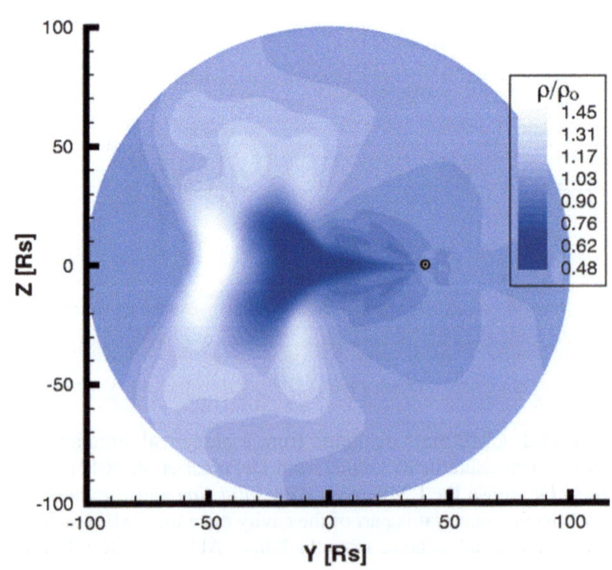

Fig. 17.3 Simulated wide-angle *white*-light image of a CME published before the launch of STEREO from Lugaz et al. (2005). The image shows the ratio of the Thomson scattered signal 24 h after the launch of the CME to a pre-event image. The Sun is marked by the *black* disk. Reproduced by permission of the AAS

perform post-processing on simulated images in a way similar to what is done for real observations, for example by creating time-elongation maps [also referred to as J-maps (Rouillard et al. 2008; Davies et al. 2009)]. With the help of such simulated J-maps, it is possible to test new analysis techniques for observed CMEs (Lugaz et al. 2009a, 2011; Rollett et al. 2013), to prepare for future instrumentation (Howard et al. 2013b; Xiong et al. 2013) and to better understand complex events (Lugaz et al. 2009b; Webb et al. 2009). These combined simulations and remote-sensing analysis studies can be further completed by *in situ* measurements at Earth, as discussed in the following sections.

17.2 Measurements and Impact on Our Earth

Whereas an average of three coronal mass ejections are observed by coronagraphs per day (up to five, on average, per day during solar maximum), coronal mass ejections are only measured *in situ* about once or twice per month at Earth. These measurements are currently made by the Advanced Composition Explorer (ACE: Stone et al. 1998) and Wind (Lepping et al. 1995; Lin et al. 1995) spacecraft at the L1 Lagrangian point (since 2004 for Wind) but also by the STEREO spacecraft on orbits around the Sun between 0.95 and 1.08 AU, as well as planetary missions such as MESSENGER, MAVEN and Venus Express during their cruise to Mercury, Mars, or Venus, or during the part of their orbits in the solar wind.

17.2.1 Overview of in situ Measurements of CMEs and Geo-Effectiveness

In situ measurements of CMEs have been made since the 1960s and 1970s with missions such as Pioneer, Helios, Voyager, or ISEE. The nature and cause of solar transients measured *in situ* was a matter of intense debate during these early years of the space age (see, for example Gosling 1993). The idea of a coronal mass ejection as a flux tube consisting of closed magnetic field lines connected to the Sun was initially proposed very early in the space age (e.g., see Gold 1962) and actually predates the first detailed measurements of the interplanetary magnetic field (Ness et al. 1964). Properties of magnetic clouds, a subset of CMEs measured *in situ* with well-defined characteristics, were discussed in Burlaga et al. (1981) and Klein and Burlaga (1982): they are (1) low proton density, temperature and low proton β, (2) enhanced magnetic field strength, and, (3) smooth and large rotation of the magnetic field vector. It was further shown that, to a good approximation, magnetic clouds can be fitted with linear force-free twisted magnetic flux ropes (Burlaga 1988) and, as a consequence, they are often thought of as large magnetic flux ropes. A CME containing a typical magnetic cloud as observed by ACE in 29 May 2001 is shown

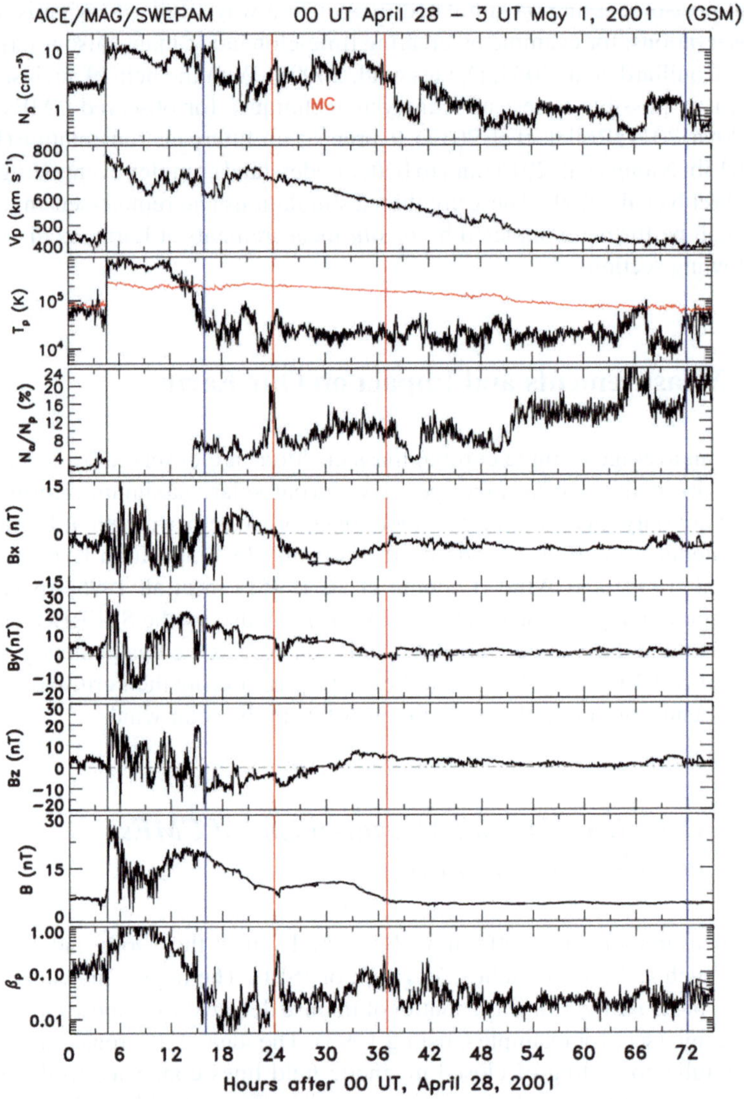

Fig. 17.4 A CME measured by ACE at 1 AU. The panels from *top to bottom* show the proton density, velocity and temperature (in *red*, the expected temperature, following Lopez 1987), the alpha-to-proton ratio, the magnetic field component in GSM and the magnetic field strength and the proton β. The dense and hot sheath stops at the first *vertical blue line* and the magnetic cloud interval is delimited by the two *red vertical lines*. The CME finishes at the second *vertical blue line* following (Richardson and Cane 2010)

in Fig. 17.4. As is commonly observed for fast CMEs, the transient consists of a fast-mode shock, a dense sheath and the magnetic cloud. The fast forward shock is marked with the black line, the CME interval is delimited by the two blue lines and the portion between the two red lines satisfy all the properties of a magnetic cloud, following Burlaga et al. (1981). On average, about one third of CMEs observed at 1 AU contain a magnetic cloud (Gosling 1990), although the proportion varies with the phase of the solar cycle (Richardson and Cane 2004b, 2010).

When the interplanetary magnetic field (IMF) makes a large angle with the Earth magnetic field, reconnection can occur on the day side of Earth's magnetosphere. Reconnection is particularly efficient when the IMF has a strong component anti-parallel to Earth's magnetic field, i.e. when it contains southward B_z in the Geocentric Solar Magnetospheric (GSM) coordinate system. Reconnection at the dayside of the magnetosphere allows solar wind particles to enter Earth's magnetosphere and create a current system which closes onto Earth's ionosphere and upper atmosphere (Axford 1969; Kan and Lee 1979; Akasofu 1981). Particle precipitation onto newly open field lines result in auroral emissions, which are typically seen in high latitude regions but may move down to lower-latitude regions during periods of strong solar wind driving typically associated with a strong southward B_z component. In general, conditions in the near-Earth space environment which lead to impact on human technology are referred to as space weather and they will be further discussed in Sect. 17.2.3. During geomagnetic storms, a strong depression of the Earth's magnetic field results from an enhancement of the ring current due to reconnection. The ring current is an electric current due to charged particles inside Earth's magnetosphere. One of the most common indices used to quantify the intensity of geomagnetic storms is the disturbance storm time (Dst) index, which measures the strength of the ring current around Earth based on mid-latitude ground-based magnetometers (Sugiura and Chapman 1960; Gonzalez et al. 1994). The lower the Dst, the stronger the storm, with the largest measured value during the space age of -640 nanoTesla (nT $= 10^{-9}$ T) in March 1989. Other geomagnetic indices, such as the Kp and AE indices complement the Dst index in characterizing Earth's magnetosphere, aurora and ring current.

On the most basic level, the effect of a solar wind stream on Earth's magnetic field depends on the dawn-to-dusk electric field, $V\,B_s$ (Kan and Lee 1979; Akasofu 1981), where V and B_s are the solar wind velocity and southward component of the magnetic field, respectively. As such, fast CMEs with a large southward magnetic field component have a strong geo-effectiveness. Indeed, CMEs account for a large majority ($\sim$85 %) of intense geomagnetic storms as defined by a peak Dst of less than -100 nT (Zhang et al. 2007) and a plurality ($\sim$40 %) of moderate storms of Dst between -100 and -50 nT (Echer et al. 2013). In the past few decades, more complex coupling functions, representing the energy deposition in the magnetosphere by the interplanetary medium, have been introduced, which take into account other parameters such as the IMF clock angle (i.e. the polar angle in the GSM YZ plane), the plasma β, the solar wind Mach number and density (Newell et al. 2007; Borovsky 2008). Among these additional parameters, it has been recognized that the solar wind density plays the most important role: high-density plasma may load the sheet-like region of the magnetosphere in the anti-sunward direction referred to as the plasmasheet. The high-density plasmasheet provides more seed particles for the ring current after they are accelerated earthward in the Earth's electric field (Borovsky and Steinberg 2006; Farrugia et al. 2006). As such, prominence material, while rarely measured *in situ* may play an important role for some extreme events, as described in the next section.

17.2.2 In-Situ Measurements of Prominence Materials Inside CMEs

As described in previous chapters, prominence eruptions are one of the main sources of CMEs (Bothmer and Schwenn 1994; Gopalswamy et al. 2003). However, at 1 AU, they are rarely measured *in situ*. Prominence material is characterized by its cool temperature (in the 10^{4-5} K range). Because the ion charge states become "frozen-in" in the corona (e.g., see von Steiger et al. 2000), filament material may be detected in the inner heliosphere as ions with low charge states such as single-charged and double-charged carbon and oxygen ions, and single-charged helium ions. In fact, CMEs at 1 AU typically consist of elevated ion charge states, such as O^{8+}, O^{7+}, C^{6+} (e.g. Lepri et al. 2001; Richardson and Cane 2004a; Zurbuchen and Richardson 2006). This is due to reconnection during the eruption, which may heat the erupting plasma to temperature in excess of 10^6 K. Observations near the Sun in the ultraviolet (UV) have revealed that CMEs may contain ions of low as well as high charge states (Ciaravella et al. 2000; Raymond et al. 2003). Ionic charge states have been also recently investigated by ways of numerical simulations (Lynch et al. 2011; Reinard et al. 2012). Such models typically predict that low charge states should be observed towards the back of CMEs measured *in situ* at 1 AU.

Schwenn et al. (1980) reported of possible filament material in the form of enhanced He^+ measured by Helios at 0.95 AU. Burlaga et al. (1998) and Gloeckler et al. (1999) reported one event observed by ACE on 1998 May 1–2 with an enhancement of He^+, and Lepri and Zurbuchen (2010) published the first systematic search for cold ions inside CME measurements at 1 AU. They found a total of 11 events with cold ions in a 11-year period, 8 of which were further analyzed in Gilbert et al. (2012). This corresponds to about 4 % of the CMEs measured *in situ* at 1 AU. Figure 17.5 shows an example of low charge states at the back of a CME. In fact, these studies have found no preference regarding the location of the prominence material within the CME. Another event was recently added to this list by detecting an enhancement of single-charged helium ions during the CME of 2005 January 21 (Kozyra et al. 2013; Sharma et al. 2013), where the filament material was at the front of the magnetic ejecta. This was explained by the lower rate of deceleration of the prominence as compared to that of rest of the CME (Kozyra et al. 2013).

Yao et al. (2010) reported the detection of prominence material inside 3 CMEs measured by Helios in the 1970s at distances less than 1 AU. Results for the three events are consistent with a much lower expansion of the filament material as compared to the expansion of the magnetic cloud inside which it is embedded. Such results may explain why filament material is rarely measured *in situ* at 1 AU

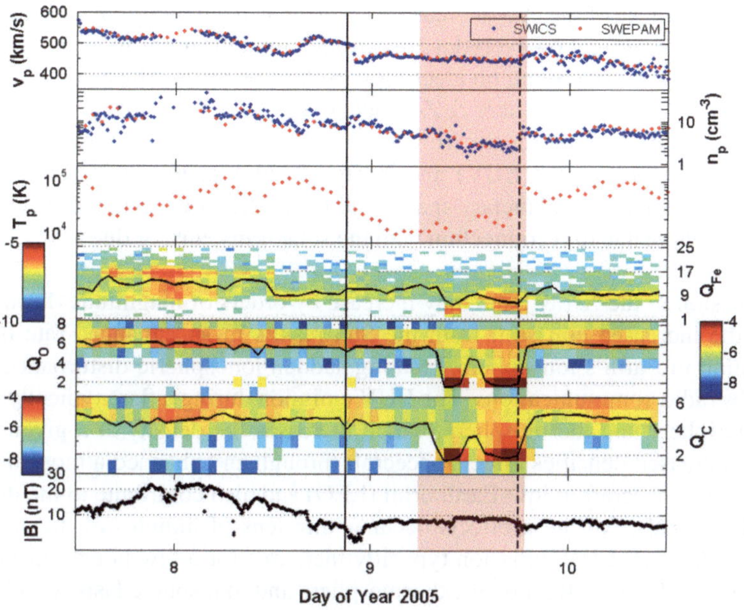

Fig. 17.5 *In situ* measurements of a CME in 2005 January 9 from Gilbert et al. (2012). *Panels* show proton velocity, density and temperature; iron, oxygen and carbon charge states; and magnetic field strength from *top to bottom*. A period with very low charge states in all ions is highlighted in *pink* at the back of the CME. Reproduced by permission of the AAS

when the typical magnetic cloud size is 0.21 AU ($\sim$45 $R_\odot$). The influence of the presence of cold and dense plasma within a CME on its geo-effectiveness was further discussed in detail for the May 1998 CME/filament in Farrugia et al. (2002).

17.2.3 Space Weather and Impact on Life

During solar cycle 23 (1995–2006) there were 93 intense geo-magnetic storms, more than 80 of which were due to CMEs (Zhang et al. 2007). The connection between solar transients and geomagnetic storms was made first for the Carrington event in 1859 when a extreme disruption of Earth's magnetic field was preceded 18 h earlier by a white-light flare on the Sun (Carrington 1859). The connection was made during the space age, first between the interplanetary magnetic field and geomagnetic disturbances as manifested by aurorae as early as the 1960s (Dungey 1961), and in the 1980s between the magnetic ejecta of a CME and the disruption in Earth's magnetic field (e.g., see Gonzalez and Tsurutani 1987; Wilson 1987; Tsurutani et al. 1988). Solar eruptions primarily affect man-made technology without any direct influence on human health (except astronauts).

However, because of the ever increasing reliance on technology, space weather may have very direct effect on everyday life, from communications to airplane travel, or possibly any system relying on electricity (Solar and Space Physics: A Science for a Technological Society, Baker & Zurbuchen (eds) 2012).

Space weather effects results from three main sources: (1) solar radiation, mostly from extreme-ultraviolet (EUV) and X-ray, (2) high-energy particles accelerated in the corona and interplanetary space, and (3) particles accelerated inside Earth's magnetosphere due to reconnection with the solar transients, or due to wave-particle interactions.

Emission in the X-ray band may increase by orders of magnitude during a solar flare; the increased emission from the Sun results in an enhanced rate of photo-ionization in Earth's ionosphere, creating sudden ionospheric disturbances, which disturbs radio communications. The EUV emission during a flare typically heats up the ionosphere and thermosphere, which increases the density at a given altitude. These increased densities affect spacecraft through an enhanced atmospheric drag, especially spacecraft in low-Earth orbit (LEO) and also contribute to the disruption of radio signals. All these effects occur within tens of minutes of the start of the event at the Sun. EUV emission typically increases for a few hours during a large X-class flare but the effect on the thermosphere and ionosphere lasts for a few days.

High-energy particles are primarily (1) solar energetic particles (SEPs) accelerated in the corona by the CME-driven shock or reconnection and transported onto magnetic field lines to Earth; and (2) solar proton events (SPEs), which are particles locally accelerated by the CME-driven shock when it passes over Earth. The most energetic SEPs may reach Earth only 20 min after the start of an event, making any prediction nearly impossible at this stage. On the other hand, SPEs arrive at Earth 1–3 days after the start of an event at the Sun; this delay is the typical propagation time between the Sun and the Earth of a fast CME. Often, these locally accelerated particles lead to a peak in measured flux as the CME passes Earth. While potentially the most dangerous, these can be forecasted well in advance. Both these types of energetic particles may affect communications with satellites in orbit around Earth, such as the Global Positioning System (GPS), which are used by a large variety of industries and technologies, from agriculture to air traffic. Energetic particles may cause spacecraft anomalies as well as, over time, degrade the solar panels of satellites. SEPs may directly affect astronauts in the space station, or on cruise during a planetary mission (to the Moon, and potentially to Mars or other objects in our solar system). One of the largest solar energetic particle events occurred in August 1972, about half-way between Apollo 16 and Apollo 17 lunar missions (e.g., see Lockwood and Hapgood 2007). Had an astronaut been in cruise between the Earth and the Moon at the time, the radiation dose that he or she would have received would have been life-threatening.

The last source of energetic particles is associated with Earth's radiation belts. As a CME impacts Earth, the reconnection between the CME's magnetic field and Earth's magnetic field creates high-energy particles in the inner magnetosphere, radiation belts and ring current. These particles precipitate onto open field lines into Earth's upper atmosphere (Hardy et al. 1985). The currents associated with these energetic particles can result in ground-induced currents (GICs), which may overload the grid system and corrode pipelines. Large electric transformers, because they are grounded to Earth are particularly vulnerable to GICs (Kappenman 2003). A Hydro-Québec transformer failure in March 1989 associated with a large geomagnetic storm is the most often cited space weather effect in the space age, yielding a global blackout. These particles also contribute to the energy input onto the top of Earth's atmosphere. Solar radiation is typically the main source of energy deposited into Earth's atmosphere (Roble et al. 1987) but energy from the magnetosphere may become the dominant source during geomagnetic storms (Fuller-Rowell et al. 1994). Aurora account for a small share of this deposited energy, which is mostly associated with Joule heating (e.g., see Foster et al. 1983; McHarg et al. 2005). The increase in Joule heating in the high latitude creates gravity waves, which globally affect Earth's atmosphere.

17.3 Open Questions and Perspectives

The central question related to prominence in the near-Earth environment is the reasons behind the relative lack of direct measurements of filament material *in situ*. Since prominence eruption is one of the main sources of CMEs as observed in the corona, the lack of direct measurements near Earth must reflect the effect of the interplanetary propagation. With remote-sensing observations of CMEs in the heliosphere now routine, it is possible to follow density structures as they propagate between the Sun and the Earth, and these observations will help us determine the fate of prominences. Prominences have mass of the order of few 10^{14} g, or about one order of magnitude lower than the total CME mass in the upper corona. In addition, the CME mass may increase by a factor of 2–3 between the upper corona and Earth due to swept-up material and the expansion of the dense sheath ahead of the magnetic ejecta. One possible reason for the lack of *in situ* measurements of prominences may be associated with the differences in expansion and mass increase during the propagation, which would "dilute" the filamentary material inside a magnetic ejecta. Another reason might be that filament material is present but its characteristics have been modified during the propagation, for example, cold material may have been heated due to reconnection with the solar wind. In that case, it would not be that filament material is not present, but rather than we do not know yet how to identify it. Combined with remote-sensing observations in the EUV, analyses of measurements of ionic charge states at 1 AU may give us a partial view of the heating history of a CME (e.g., see Rakowski et al. 2011; Gruesbeck et al. 2011).

For these cases when filament material has been identified near Earth, the main open question is how it influences the interaction between the CME and Earth's magnetosphere. Because of the paucity of cases, there has not been yet a dedicated study of the effect of filament material on Earth. A few works (Farrugia et al. 2002; Sharma et al. 2013; Kozyra et al. 2013) have considered the effect of filament material on the geo-effectiveness of a CME on a case-by-case basis. Due to their large density, prominences are expected to enhance the effect of the magnetic field inside a CME. Recent works on coupling between the IMF and the magnetosphere have, for example, illustrated the importance of the density in addition to the dawn-to-dusk electric field, $V B_s$ to understand that coupling. Reconnection between CMEs and the magnetosphere is one of the three main sources of space weather effects, the others being EUV and X-ray radiation, and energetic particles accelerated during an eruption. The effects associated with the bulk plasma motion of CMEs are, in some sense, the easiest to forecast because CMEs take between one and 3 days to propagate between the Sun and the Earth and are directly imaged as they propagate (Davis et al. 2011). It is therefore possible to predict their arrival time with typical uncertainties of ± 6 h. This is somewhat mitigated by the relative lack of progress in forecasting the internal structure of CMEs, and especially the strength of the southward magnetic field component. Overall, geomagnetic storms are the major cause of space weather affecting the ground, from ground-induced currents, to aurora and effects on Earth's upper atmosphere. The effect of CMEs may become more important when discussing close-in extrasolar planets or planets around young stars with intense dynamo, as will be discussed in the next chapter.

Acknowledgements N. L. would like to thank Charles Farrugia for fruitful conversations and suggestions about the chapter as well as the LOC and SOC of IAU 300 for the organization of that symposium on prominences and their role on space weather.

References

Akasofu, S. I. (1981). Energy coupling between the solar wind and the magnetosphere. *Space Science Reviews, 28*, 121–190. doi:10.1007/BF00218810.

Axford, W. I. (1969). Magnetospheric convection. *Reviews of Geophysics and Space Physics, 7*, 421–459. doi:10.1029/RG007i001p00421.

Bemporad, A., & Mancuso, S. (2010). First complete determination of plasma physical parameters across a coronal mass ejection-driven shock. *The Astrophysical Journal, 720*, 130–143. doi:10.1088/0004-637X/720/1/130.

Borovsky, J. E. (2008). The rudiments of a theory of solar wind/magnetosphere coupling derived from first principles. *Journal of Geophysical Research, 113*, A08228. doi:10.1029/2007JA012646.

Borovsky, J. E., & Steinberg, J. T. (2006). The "calm before the storm" in CIR/magnetosphere interactions: Occurrence statistics, solar wind statistics, and magnetospheric preconditioning. *Journal of Geophysical Research, 111*, A07S10. doi:10.1029/2005JA011397

Bothmer, V., & Schwenn, R. (1994). Eruptive prominences as sources of magnetic clouds in the solar wind. *Space Science Reviews, 70*, 215–220. doi:10.1007/BF00777872.

Bothmer, V., & Schwenn, R. (1998). The structure and origin of magnetic clouds in the solar wind. *Annales Geophysicae, 16*, 1–24. doi:10.1007/s00585-997-0001-x.

Burlaga L, Sittler E, Mariani F, & Schwenn R. (1981). Magnetic loop behind an interplanetary shock—Voyager, Helios, and IMP 8 observations. *Journal of Geophysical Research, 86*, 6673–6684.

Burlaga L., Fitzenreiter R., & Lepping R., et al. (1998). A magnetic cloud containing prominence material—January 1997. *Journal of Geophysical Research, 103*, 277. doi:10.1029/97JA02768.

Burlaga, L. F. (1988). Magnetic clouds and force-free fields with constant alpha. *Journal of Geophysical Research, 93*, 7217–7224. doi:10.1029/JA093iA07p07217.

Byrne, J. P., Maloney, S. A., McAteer, R. T. J., Refojo, J. M., & Gallagher, P. T. (2010). Propagation of an earth-directed coronal mass ejection in three dimensions. *Nature Communications, 1*, 74. doi:10.1038/ncomms1077.

Cargill, P. J. (2004). On the aerodynamic drag force acting on interplanetary coronal mass ejections. *Solar Physics, 221*, 135–149. doi:10.1023/B:SOLA.0000033366.10725.a2.

Carrington, R. C. (1859). Description of a singular appearance seen in the sun on September 1, 1859. *Monthly Notice Royal Astronomical Society, 20*, 13–15.

Chané, E., van der Holst, B., Jacobs, C., Poedts, S., & Kimpe, D. (2006). Inverse and normal coronal mass ejections: Evolution up to 1 AU. *Astronomy and Astrophysics, 447*, 727–733. doi:10.1051/0004-6361:20053802.

Ciaravella, A., Raymond, J. C., & Thompson, B. J., et al. (2000). Solar and heliospheric observatory observations of a helical coronal mass ejection. *The Astrophysical Journal, 529*, 575–591. doi:10.1086/308260.

Cremades, H., & Bothmer, V. (2004). On the three-dimensional configuration of coronal mass ejections. *Astronomy Astrophysics, 422*, 307–322. doi:10.1051/0004-6361:20035776.

Davies, J. A., Harrison, R. A., & Rouillard, A. P., et al. (2009). A synoptic view of solar transient evolution in the inner heliosphere using the Heliospheric Imagers on STEREO. *Geophysical Research Letters, 36*, L02,102. doi:10.1029/2008GL036182.

Davis, C. J., Davies, J. A., Lockwood, M., Rouillard, A. P., Eyles, C. J., & Harrison, R. A. (2009). Stereoscopic imaging of an earth-impacting solar coronal mass ejection: A major milestone for the STEREO mission. *Geophysical Research Letters, 36*, L08,102. doi:10.1029/2009GL038021.

Davis, C. J., de Koning, C. A., & Davies, J. A., et al. (2011). A comparison of space weather analysis techniques used to predict the arrival of the earth-directed CME and its shockwave launched on 8 April 2010. *Space Weather, 9*, S01,005. doi:10.1029/2010SW000620.

DeForest, C. E., Howard, T. A., & Tappin, S. J. (2011). Observations of detailed structure in the solar wind at 1 AU with STEREO/HI-2. *The Astrophysical Journal, 738*, 103. doi:10.1088/0004-637X/738/1/103.

DeForest, C. E., Howard, T. A., & McComas, D. J. (2013). Tracking coronal features from the low corona to earth: A quantitative analysis of the 2008 December 12 coronal mass ejection. *The Astrophysical Journal, 769*, 43. doi:10.1088/0004-637X/769/1/43.

Dungey J. W. (1961). Interplanetary magnetic field and the auroral zones. *Physical Review Letters, 6*, 47–48. doi:10.1103/PhysRevLett.6.47

Echer, E., Tsurutani, B. T., & Gonzalez, W. D. (2013). Interplanetary origins of moderate ($-100\,nT > Dst > -50\,nT$) geomagnetic storms during solar cycle 23 (1996–2008). *Journal of Geophysical Research, 118*, 385–392. doi:10.1029/2012JA018086.

Eyles, C. J., Simnett, G. M., & Cooke, M. P., et al. (2003). The solar mass ejection imager (Smei). *Solar Physics, 217*, 319–347. doi:10.1023/B:SOLA.0000006903.75671.49.

Eyles, C. J., Harrison, R. A., & Davis, C. J., et al. (2009). The heliospheric imagers onboard the STEREO mission. *Solar Physics, 254*, 387–445. doi:10.1007/s11207-008-9299-0.

Fan, Y. (2014). Chapter 12: Magnetism and dynamics of prominences: MHD equilibria and triggers for eruption. In J.-C. Vial & O. Engvold (Eds.), *Solar prominences, ASSL* (Vol. 415, pp. 295–320). New York: Springer.

Farrugia, C. J., Popecki, M., & MöBius, E., et al. (2002). Wind and ACE observations during the great flow of 1–4 May 1998: Relation to solar activity and implications for the magnetosphere. *Journal of Geophysical Research, 107*, 1240. doi:10.1029/2001JA000188.

Farrugia, C. J., Jordanova, V. K., Thomsen, M. F., Lu, G., Cowley, S. W. H., & Ogilvie, K. W. (2006). A two-ejecta event associated with a two-step geomagnetic storm. *Journal of Geophysical Research, 111*. doi:10.1029/2006JA011893.

Filippov, B. P., Gopalswamy, N., & Lozhechkin, A. V. (2001). Non-radial motion of eruptive filaments. *Solar Physics, 203*, 119–130.

Foster, J. C., St-Maurice, J. P., & Abreu, V. J. (1983). Joule heating at high latitudes. *Journal of Geophysical Research, 88*, 4885–4897. doi:10.1029/JA088iA06p04885.

Fuller-Rowell, T. J., Codrescu, M. V., Moffett, R. J., & Quegan, S. (1994). Response of the thermosphere and ionosphere to geomagnetic storms. *Journal of Geophysical Research, 99*, 3893–3914. doi:10.1029/93JA02015.

Gibson, S. E., & Low, B. C. (1998). A time-dependent three-dimensional magnetohydrodynamic model of the coronal mass ejection. *The Astrophysical Journal, 493*, 460.

Gilbert, J. A., Lepri S. T., Landi, E., & Zurbuchen, T. H. (2012). First measurements of the complete heavy-ion charge state distributions of C, O, and Fe associated with interplanetary coronal mass ejections. *The Astrophysical Journal, 751*, 20. doi:10.1088/0004-637X/751/1/20.

Gloeckler, G., Fisk, L. A., Hefti, S., Schwadron, N. A., Zurbuchen, T. H., & Ipavich, F. M., et al. (1999). Unusual composition of the solar wind in the 2–3 May 1998 CME observed with SWICS on ACE. *Geophysical Research Letters, 26*, 157–160. doi:10.1029/1998GL900166.

Gold, T. (1962). Magnetic storms. *Space Science Reviews, 1*, 100–114. doi:10.1007/BF00174637.

Gonzalez, W. D., & Tsurutani, B. T. (1987). Criteria of interplanetary parameters causing intense magnetic storms (Dst of less than −100 nT). *Planetary and Space Science, 35*, 1101–1109. doi:10.1016/0032-0633(87)90015-8.

Gonzalez, W. D., Joselyn, J. A., Kamide, Y., Kroehl, H. W., Rostoker, G., & Tsurutani, B. T., et al. (1994). What is a geomagnetic storm? *Journal of Geophysical Research, 99*, 5771–5792. doi:10.1029/93JA02867.

Gopalswamy, N., Yashiro, S., Kaiser, M. L., & Howard, R. A. (2003). Coronal mass ejection interaction and particle acceleration during the 2001 April 14–15 events. *Advances in Space Research, 32*, 2613–2618. doi:10.1016/j.asr.2003.09.038.

Gopalswamy, N., Mäkelä, P., Xie, H., Akiyama, S., & Yashiro, S. (2009). CME interactions with coronal holes and their interplanetary consequences. *Journal of Geophysical Research, 114*. doi:10.1029/2008JA013686.

Gopalswamy N. (2014). Chapter 15: The dynamic of eruptive prominences. In J.-C. Vial & O. Engvold (Eds.), *Solar prominences, ASSL* (Vol. 415, pp. 379–408). New York: Springer.

Gosling JT. (1990). Coronal mass ejections and magnetic flux ropes in interplanetary space. *AGU Geophysical Monograph Series, 58*, 343–364.

Gosling JT. (1993). The solar flare myth. *Journal of Geophysical Research, 98*, 18,937–18,950. doi:10.1029/93JA01896.

Groth, C. P. T., De Zeeuw, D. L., Gombosi, T. I., & Powell, K. G. (2000). Global three-dimensional MHD simulation of a space weather event: CME formation, interplanetary propagation, and interaction with the magnetosphere. *Journal of Geophysical Research, 105*, 25,053–25,078.

Gruesbeck, J. R., Lepri, S. T., Zurbuchen, T. H., & Antiochos, S. K. (2011). Constraints on coronal mass ejection evolution from in situ observations of ionic charge states. *The Astrophysical Journal, 730*, 103. doi:10.1088/0004-637X/730/2/103.

Hardy, D. A., Gussenhoven, M. S., & Holeman, E. (1985). A statistical model of auroral electron precipitation. *Journal of Geophysical Research, 90*, 4229–4248. doi:10.1029/JA090iA05p04229.

Harrison, R. A., Davis, C. J., & Eyles, C. J., et al. (2008). First imaging of coronal mass ejections in the heliosphere viewed from outside the sun earth line. *Solar Physics, 247*, 171–193. doi:10.1007/s11207-007-9083-6.

Harrison, R. A., Davies, J. A., & Rouillard, A. P., et al. (2009). Two years of the STEREO heliospheric imagers. Invited review. *Solar Physics, 256*, 219–237. doi:10.1007/s11207-009-9352-7.

Howard, R. A., Moses, J. D., & Vourlidas, A., et al. (2008). Sun earth connection coronal and heliospheric investigation (SECCHI). *Space Science Reviews, 136*, 67–115. doi:10.1007/s11214-008-9341-4.

Howard, T. A., & DeForest, C. E. (2012a). Inner heliospheric flux rope evolution via imaging of coronal mass ejections. *The Astrophysical Journal, 746*, 64. doi:10.1088/0004-637X/746/1/64.

Howard, T. A, & DeForest, C. E. (2012b). The Thomson surface. I. Reality and myth. *The Astrophysical Journal, 752*, 130. doi:10.1088/0004-637X/752/2/130.

Howard, T. A., Bisi, M. M., & Buffington, A., et al. (2013a). The solar mass ejection imager and its heliospheric imaging legacy. *Space Science Reviews, 180*, 1–38. doi:10.1007/s11214-013-9992-7.

Howard, T. A., Tappin, S. J., Odstrcil, D., & DeForest, C. E. (2013b). The Thomson surface. III. Tracking features in 3D. *The Astrophysical Journal, 765*, 45. doi:10.1088/0004-637X/765/1/45.

Isavnin, A., Vourlidas, A., & Kilpua, E. K. J. (2013). Three-dimensional evolution of erupted flux ropes from the sun (2 - 20 $R_\odot$) to 1 AU. *Solar Physics, 284*, 203–215. doi:10.1007/s11207-012-0214-3.

Jackson, B. V, & Leinert, C. (1985). HELIOS images of solar mass ejections. *Journal of Geophysical Research, 90*, 10,759. doi:10.1029/JA090iA11p10759.

Jackson, B. V., Buffington, A., & Hick, P. P., et al. (2004). The solar mass-ejection imager (SMEI) mission. *Solar Physics, 225*, 177–207. doi:10.1007/s11207-004-2766-3

Kan, J. R., & Lee, L. C. (1979). Energy coupling function and solar wind-magnetosphere dynamo. *Geophysical Research Letters, 6*, 577–580. doi:10.1029/GL006i007p00577.

Kappenman JG. (2003). Storm sudden commencement events and the associated geomagnetically induced current risks to ground-based systems at low-latitude and midlatitude locations. *Space Weather, 1*, 1016. doi:10.1029/2003SW000009.

Kay, C., Opher, M., & Evans, R. M. (2013). Forecasting a coronal mass ejection's altered trajectory: ForeCAT. *The Astrophysical Journal, 775*, 5. doi:10.1088/0004-637X/775/1/5.

Kilpua, E. K. J., Pomoell, J., Vourlidas, A., Vainio, R., Luhmann, J., & Li, Y., et al. (2009). STEREO observations of interplanetary coronal mass ejections and prominence deflection during solar minimum period. *Annales Geophysicae, 27*, 4491–4503. doi:10.5194/angeo-27-4491-2009.

Klein, L. W., & Burlaga, L. F. (1982). Interplanetary magnetic clouds at 1 AU. *Journal of Geophysical Research, 87*, 613–624. doi:10.1029/JA087iA02p00613.

Kozyra, J. U., Manchester, W. B., Escoubet, C. P., Lepri, S. T., Liemohn, M. W., & Gonzalez, W. D. (2013). Earth's collision with a solar filament on 21 January 2005. Overview. *Journal of Geophysical Research, 118*, 5967–5978. doi:10.1002/jgra.50567.

Lemen, J. R., Title, A. M., & Akin, D. J., et al. (2012). The atmospheric imaging assembly (AIA) on the solar dynamics observatory. (SDO). *Solar Physics, 275*, 17–40. doi:10.1007/s11207-011-9776-8.

Lepping, R. P., Acuna, M. H., & Burlaga, L. F. (1995). The wind magnetic field investigation. *Space Science Reviews, 71*, 207–229.

Lepri, S. T., & Zurbuchen, T. H. (2010). Direct observational evidence of filament material within interplanetary coronal mass ejections. *The Astrophysical Journal Letters, 723*, L22–L27. doi:10.1088/2041-8205/723/1/L22.

Lepri, S. T., Zurbuchen, T. H., Fisk, L. A., Richardson, I. G., Cane, H. V., & Gloeckler, G. (2001). Iron charge distribution as an identifier of interplanetary coronal mass ejections. *Journal of Geophysical Research, 106*, 29,231–29,238. doi:10.1029/2001JA000014.

Lin, R. P., Anderson, K. A., & Ashford, S., et al. (1995). A three-dimensional plasma and energetic particle investigation for the wind spacecraft. *Space Science Reviews, 71*, 125–153. doi:10.1007/BF00751328.

Liu, Y., Davies, J. A., Luhmann, J. G., Vourlidas, A., Bale, S. D., & Lin, R. P. (2010). Geometric triangulation of imaging observations to track coronal mass ejections continuously out to 1 AU. *The Astrophysical Journal Letters, 710*, L82–L87. doi:10.1088/2041-8205/710/1/L82.

Liu, Y. D., Luhmann, J. G., Lugaz, N., Möstl, C., Davies, J. A., & Bale, S. D. (2013). On sun-to-earth propagation of coronal mass ejections. *The Astrophysical Journal, 769*, 45. doi:10.1088/0004-637X/769/1/45.

Lockwood, M., & Hapgood, M. (2007). The rough guide to the moon and mars. *Astronomy and Geophysics, 48*, 6.11–6.17. doi:10.1111/j.1468-4004.2007.48611.x.

Lopez RE. (1987). Solar cycle invariance in solar wind proton temperature relationships. *Journal of Geophysical Research, 92*, 11,189–11,194. doi:10.1029/JA092iA10p11189.

Lugaz, N., Manchester, W. B., & Gombosi, T. I. (2005). The evolution of coronal mass ejection density structures. *The Astrophysical Journal, 627*, 1019–1030.

Lugaz, N., Vourlidas, A., & Roussev, I. I. (2009a). Deriving the radial distances of wide coronal mass ejections from elongation measurements in the heliosphere—application to CME–CME interaction. *Annales Geophysicae, 27*, 3479–3488.

Lugaz, N., Vourlidas, A., Roussev, I. I., & Morgan, H. (2009b). Solar-terrestrial simulation in the STEREO era: The January 24-25, 2007 Eruptions. *Solar Physics, 256*, 269–284. doi:10.1007/s11207-009-9339-4.

Lugaz, N., Hernandez-Charpak, J. N., Roussev, I. I., Davis, C. J., Vourlidas, A., & Davies, J. A. (2010). Determining the azimuthal properties of coronal mass ejections from multi-spacecraft remote-sensing observations with STEREO SECCHI. *The Astrophysical Journal, 715*, 493–499. doi:10.1088/0004-637X/715/1/493.

Lugaz, N., Roussev, I. I., & Gombosi, T. I. (2011). Determining CME parameters by fitting heliospheric observations: Numerical investigation of the accuracy of the methods. *Advances in Space Research, 48*, 292–299. doi:10.1016/j.asr.2011.03.015.

Lugaz, N., Farrugia, C. J., Davies, J. A., Möstl, C., Davis, C. J., & Roussev, I. I. (2012a). The deflection of the two interacting coronal mass ejections of 2010 May 23–24 as revealed by combined in situ measurements and heliospheric imaging. *The Astrophysical Journal, 759*, 68. doi:10.1088/0004-637X/759/1/68.

Lugaz, N., Kintner, P., Möstl, C., Jian, L. K., Davis, C. J., & Farrugia, C. J. (2012b). Heliospheric observations of STEREO-directed coronal mass ejections in 2008–2010: Lessons for future observations of earth-directed CMEs. *Solar Physics, 279*, 497–515. doi:10.1007/s11207-012-0007-8.

Lynch, B. J., Reinard, A. A., Mulligan, T., Reeves, K. K., Rakowski, C. E., & Allred, J. C. (2011). Ionic composition structure of coronal mass ejections in axisymmetric magnetohydrodynamic models. *The Astrophysical Journal, 740*, 112. doi:10.1088/0004-637X/740/2/112.

Manchester, W. B., Gombosi, T. I., Roussev, I., De Zeeuw, D. L., Sokolov, I. V., & Powell, K. G. (2004a). Three-dimensional MHD simulation of a flux rope driven CME. *Journal of Geophysical Research, 109*, 1102.

Manchester, W. B., Gombosi, T. I., Roussev, I., Ridley, A., De Zeeuw, D. L., & Sokolov, I. V. (2004b). Modeling a space weather event from the sun to the earth: CME generation and interplanetary propagation. *Journal of Geophysical Research, 109*, 2107.

Manchester, W. B., Vourlidas, A., Tóth, G., Lugaz, N., Roussev, I. I., & Sokolov, I. V. (2008). Three-dimensional MHD Simulation of the 2003 October 28 Coronal Mass Ejection: Comparison with LASCO Coronagraph Observations. *The Astrophysical Journal, 684*, 1448–1460. doi:10.1086/590231.

McHarg, M., Chun, F., Knipp, D., Lu, G., Emery, B., & Ridley, A. (2005). High-latitude Joule heating response to IMF inputs. *Journal of Geophysical Research, 110*, A08309. doi:10.1029/2004JA010949.

Ness, N. F., Scearce, C. S., & Seek, J. B. (1964). Initial Results of the Imp 1 Magnetic Field Experiment. *Journal of Geophysical Research, 69*, 3531–3569. doi:10.1029/JZ069i017p03531.

Newell, P. T., Sotirelis, T., Liou, K., Meng, C. I., & Rich, F. J. (2007). A nearly universal solar wind-magnetosphere coupling function inferred from 10 magnetospheric state variables. *Journal of Geophysical Research, 112*, A01206. doi:10.1029/2006JA012015.

Nieves-Chinchilla, T., Colaninno, R., Vourlidas, A., Szabo, A., Lepping, R. P., & Boardsen, S. A. (2012). Remote and in situ observations of an unusual earth-directed coronal mass ejection from multiple viewpoints. *Journal of Geophysical Research, 117*, A06106. doi:10.1029/2011JA017243.

Odstrcil, D., & Pizzo, V. J. (2009). Numerical heliospheric simulations as assisting tool for interpretation of observations by STEREO heliospheric imagers. *Solar Physics, 259*, 297–309. doi:10.1007/s11207-009-9449-z.

Odstrčil, D., & Pizzo, V. J. (1999). Three-dimensional propagation of CMEs in a structured solar wind flow: 1. CME launched within the streamer belt. *Journal of Geophysical Research, 104*(13), 483–492.

Odstrčil, D., Linker, J. A., Lionello, R., Mikic, Z., Riley, P., & Pizzo, V. J., et al. (2002). Merging of coronal and heliospheric numerical two-dimensional MHD models. *Journal of Geophysical Research, 107*, 14–1.

Panasenco, O., Martin, S. F., Velli, M., & Vourlidas, A. (2013). Origins of rolling, twisting, and non-radial propagation of eruptive solar events. *Solar Physics, 287*, 391–413. doi:10.1007/s11207-012-0194-3.

Rakowski, C. E., Laming, J. M., & Lyutikov, M. (2011). In situ heating of the 2007 May 19 CME ejecta detected by stereo/PLASTIC and ACE. *The Astrophysical Journal, 730* 30. doi:10.1088/0004-637X/730/1/30.

Raymond, J. C., Thompson, B. J., St Cyr, O. C., Gopalswamy, N., Kahler, S., & Kaiser, M. (2000). SOHO and radio observations of a CME shock wave. *Geophysical Research Letters, 27*, 1439–1442.

Raymond, J. C., Ciaravella, A., Dobrzycka, D., Strachan, L., Ko, Y. K., & Uzzo, M. (2003). Far-ultraviolet spectra of fast coronal mass ejections associated with X-class flares. *The Astrophysical Journal, 597*, 1106–1117. doi:10.1086/378663.

Reinard, A. A., Lynch, B. J., & Mulligan, T. (2012). Composition structure of interplanetary coronal mass ejections from multispacecraft observations, modeling, and comparison with numerical simulations. *The Astrophysical Journal, 761*, 175. doi:10.1088/0004-637X/761/2/175.

Richardson, I. G., & Cane, H. V. (2004a). Identification of interplanetary coronal mass ejections at 1 AU using multiple solar wind plasma composition anomalies. *Journal of Geophysical Research, 109*, A09104. doi:10.1029/2004JA010598.

Richardson, I. G., & Cane, H. V. (2004b). The fraction of interplanetary coronal mass ejections that are magnetic clouds: Evidence for a solar cycle variation. *Geophysical Research Letters, 31*, L18804. doi:10.1029/2004GL020958.

Richardson, I. G., & Cane, H. V. (2010). Near-earth interplanetary coronal mass ejections during solar cycle 23. (1996–2009): Catalog and summary of properties. *Solar Physics, 264*, 189–237. doi:10.1007/s11207-010-9568-6.

Riley, P., Gosling, J. T., & Pizzo, V. J. (1997). A two-dimensional simulation of the radial and latitudinal evolution of a solar wind disturbance driven by a fast, high-pressure coronal mass ejection. *Journal of Geophysical Research, 102*, 14,677–14,686. doi:10.1029/97JA01131.

Riley, P., Linker, J. A., & Lionello, R., et al. (2004). Fitting flux ropes to a global MHD solution: a comparison of techniques. *Journal of Atmospheric and Solar-Terrestrial Physics, 66*,1321–1331. doi:10.1016/j.jastp.2004.03.019.

Riley, P., Lionello, R., Mikić, Z., Linker, J., Clark, E., & Lin, J. (2007). "Bursty" reconnection following solar eruptions: MHD simulations and comparison with observations. *The Astrophysical Journal, 655*, 591–597. doi:10.1086/509913.

Roble, R. G., Ridley, E. C., & Dickinson, R. E. (1987). On the global mean structure of the thermosphere. *Journal of Geophysical Research, 92*, 8745–8758. doi:10.1029/JA092iA08p08745.

Rollett, T., Temmer, M., Möstl, C., Lugaz, N., Veronig, A. M., & Möstl, U. V. (2013). Assessing the constrained harmonic mean method for deriving the kinematics of ICMEs with a numerical simulation. *Solar Physics, 283*, 541–556. doi:10.1007/s11207-013-0246-3.

Rouillard, A. P., Davies, J. A., & Forsyth, R. J., et al. (2008). First imaging of corotating interaction regions using the STEREO spacecraft. *Geophysical Research Letters, 35*. doi:10.1029/2008GL033767.

Roussev, I. I., Sokolov, I. V., Forbes, T. G., Gombosi, T. I., Lee, M. A., & Sakai, J. I. (2004). A numerical model of a coronal mass ejection: Shock development with implications for the acceleration of GeV protons. *The Astrophysical Journal Letters, 605*, L73–L76.

Savani, N. P., Owens, M. J., Rouillard, A. P., Forsyth, R. J., & Davies, J. A. (2010). Observational evidence of a coronal mass ejection distortion directly attributable to a structured solar wind. *The Astrophysical Journal Letters, 714*, L128–L132. doi:10.1088/2041-8205/714/1/L128.

Schwenn, R., Rosenbauer, H., & Muehlhaeuser, K. H. (1980). Singly-ionized helium in the driver gas of an interplanetary shock wave. *Geophysical Research Letters, 7*, 201–204. doi:10.1029/GL007i003p00201.

Sharma, R., Srivastava, N., Chakrabarty, D., Möstl, C., & Hu, Q. (2013). Interplanetary and geomagnetic consequences of 5 January 2005 CMEs associated with eruptive filaments. *Journal of Geophysical Research, 118*, 3954–3967. doi:10.1002/jgra.50362.

Shen, C., Wang, Y., Wang, S., Liu, Y., & Liu, R. (2012). Super-elastic collision of large-scale magnetized plasmoids in the heliosphere. *Nature Physics, 8*, 923–928. doi:10.1038/nphys2440.

Shen, F., Feng, X., Wu, S. T., & Xiang, C. (2007). Three-dimensional MHD simulation of CMEs in three-dimensional background solar wind with the self-consistent structure on the source surface as input: Numerical simulation of the January 1997 Sun-Earth connection event. *Journal of Geophysical Research, 112*, 6109. doi:10.1029/2006JA012164.

Shen, F., Feng, X. S., Wang, Y., Wu, S. T., Song, W. B., Guo, J. P., & Zhou, Y. F. (2011). Three-dimensional MHD simulation of two coronal mass ejections' propagation and interaction using a successive magnetized plasma blobs model. *Journal of Geophysical Research, 116*, A09103. doi:10.1029/2011JA016584.

Shen, F., Shen, C,, Wang, Y., Feng, X., & Xiang, C. (2013). Could the collision of CMEs in the heliosphere be super-elastic? Validation through three-dimensional simulations. *Geophysical Research Letters, 40*, 1457–1461. doi:10.1002/grl.50336.

Solar and Space Physics: A Science for a Technological Society. (2012). Committee on a decadal strategy for solar and space physics (Heliophysics). In D. Baker & T. Zurbuchen (Eds.), The National Academies Press.

Stone, E. C., Frandsen, A. M., Mewaldt, R. A., Christian, E. R., Margolies, D., & Ormes, J. F., et al. (1998). The advanced composition explorer. *Space Science Reviews, 86*, 1–22. doi:10.1023/A:1005082526237.

Sugiura, M., & Chapman, S. (1960). The average morphology of geomagnetic storms with sudden commencement. Abandl Akad Wissen Goettingen Math Phys Kl.

Tappin, S. J., Buffington, A., & Cooke, M. P., et al. (2004). Tracking a major interplanetary disturbance with SMEI. *Geophysical Research Letters, 31*, L02802. doi:10.1029/2003GL018766.

Temmer, M., Vršnak, B., & Rollett, T., et al. (2012). Characteristics of kinematics of a coronal mass ejection during the 2010 August 1 CME-CME Interaction Event. *The Astrophysical Journal, 749*, 57. doi:10.1088/0004-637X/749/1/57.

Török, T., & Kliem, B. (2003). The evolution of twisting coronal magnetic flux tubes. *Astronomy Astrophysics, 406*, 1043–1059. doi:10.1051/0004-6361:20030692.

Tóth, G,, Sokolov, I. V., & Gombosi, T. I., et al. (2005). Space weather modeling framework: A new tool for the space science community. *Journal of Geophysical Research, 110*. doi:10.1029/2005JA011126.

Tsurutani, B. T., Smith, E. J., Gonzalez, W. D., Tang, F., & Akasofu, S. I. (1988). Origin of interplanetary southward magnetic fields responsible for major magnetic storms near solar maximum (1978–1979). *Journal of Geophysical Research, 93*, 8519–8531. doi:10.1029/JA093iA08p08519.

von Steiger, R., Schwadron, N. A., Fisk, L. A., Geiss, J., Gloeckler, G., & Hefti, S., et al. (2000). Composition of quasi-stationary solar wind flows from Ulysses/solar wind ion composition spectrometer. *Journal of Geophysical Research, 105*, 27,217–27,238. doi:10.1029/1999JA000358.

Vourlidas, A., & Howard, R. A. (2006). The proper treatment of coronal mass ejection brightness: A new methodology and implications for observations. *The Astrophysical Journal, 642*, 1216–1221. doi:10.1086/501122.

Vourlidas, A., Wu, S. T., Wang, A. H., Subramanian, P., & Howard, R. A. (2003). Direct detection of a coronal mass ejection-associated shock in large angle and spectrometric coronagraph experiment white-light images. *The Astrophysical Journal, 598*, 1392–1402.

Vršnak, B. (2001). Deceleration of coronal mass ejections. *Solar Physics, 202*, 173–189.

Vršnak, B., Žic, T., Falkenberg, T. V., Moëstl, C., Vennerstrom, S., & Vrbanec, D. (2010). The role of aerodynamic drag in propagation of interplanetary coronal mass ejections. *Astronomy Astrophysics, 512*, A43. doi:10.1051/0004-6361/200913482.

Webb, D. F., Howard, T. A., & Fry, C. D., et al. (2009). Study of CME propagation in the inner heliosphere: SOHO LASCO, SMEI and STEREO HI observations of the January 2007 Events. *Solar Physics, 256*, 239–267. doi:10.1007/s11207-009-9351-8.

Webb, D. F., Möstl, C., & Jackson, B. V., et al. (2013). Heliospheric imaging of 3D density structures during the multiple coronal mass ejections of late July to Early August 2010. *Solar Physics, 285*, 317–348. doi:10.1007/s11207-013-0260-5.

Webb, D. F. (2014). Chapter 16: Eruptive prominences and their association with coronal mass ejections. In J.-C. Vial & O. Engvold (Eds.), *Solar prominences, ASSL* (Vol. 415, pp. 155–186). New York: Springer.

Wilson, R. M. (1987). Geomagnetic response to magnetic clouds. *Planetary and Space Science, 35*, 329–335. doi:10.1016/0032-0633(87)90159-0.

Wu, C. C., Fry, C. D., Wu, S. T., Dryer, M., & Liou, K. (2007). Three-dimensional global simulation of interplanetary coronal mass ejection propagation from the Sun to the heliosphere: Solar event of 12 May 1997. *Journal of Geophysical Research, 112*. doi:10.1029/2006JA012211.

Wu, S. T., Guo, W. P., & Andrews, M. D., et al. (1997). MHD interpretation of LASCO observations of a coronal mass ejection as a disconnected magnetic structure. *Solar Physics, 175*, 719–735. doi:10.1023/A:1004923016322.

Wu, S. T., Guo, W. P., Michels, D. J., & Burlaga, L. F. (1999). MHD description of the dynamical relationships between a flux rope, streamer, coronal mass ejection, and magnetic cloud: An analysis of the January 1997 sun-earth connection event. *Journal of Geophysical Research, 104*, 14,789–14,802.

Xiong, M., Davies, J. A., Feng, X., Owens, M. J., Harrison, R. A., & Davis, C. J., et al. (2013). Using coordinated observations in polarized white light and Faraday rotation to probe the spatial position and magnetic field of an interplanetary sheath. *The Astrophysical Journal, 777*, 32. doi:10.1088/0004-637X/777/1/32.

Yao, S., Marsch, E., Tu, C. Y., & Schwenn, R. (2010). Identification of prominence ejecta by the proton distribution function and magnetic fine structure in interplanetary coronal mass ejections in the inner heliosphere. *Journal of Geophysical Research, 115*, A05103. doi:10.1029/2009JA014914

Yurchyshyn, V., Abramenko. V., & Tripathi, D. (2009). Rotation of white-light coronal mass ejection structures as inferred from LASCO coronagraph. *The Astrophysical Journal, 705*, 426–435. doi:10.1088/0004-637X/705/1/426.

Zhang, J., Richardson, I. G., & Webb, D. F., et al. (2007). Solar and interplanetary sources of major geomagnetic storms (Dst < -100 nT) during 1996–2005. *Journal of Geophysical Research, 112*. doi:10.1029/2007JA012321.

Zuccarello, F. P., Bemporad, A., Jacobs, C., Mierla, M., Poedts, S., & Zuccarello, F. (2012). The role of streamers in the deflection of coronal mass ejections: Comparison between STEREO three-dimensional reconstructions and numerical simulations. *The Astrophysical Journal, 744*, 66. doi:10.1088/0004-637X/744/1/66.

Zurbuchen, T. H., & Richardson, I. G. (2006). In-situ solar wind and magnetic field signatures of interplanetary coronal mass ejections. *Space Science Reviews, 123*, 31–43. doi:10.1007/s11214-006-9010-4.

Chapter 18
Stellar Activity and CMEs: Important Factors of Planetary Evolution

Maxim L. Khodachenko
In collaboration with the "Pathway to Habitabillity" Team

Abstract CME activity of the Sun is known to be an important impacting factor for the magnetospheres, atmospheres, and surfaces of solar system planets. Following an idea of a solar-stellar analogy, CME phenomena are expected on other stars as well. The main planetary impact factors of the stellar CMEs include the associated interplanetary shocks, plasma density and velocity disturbances, energetic particles accelerated in the shock regions, as well as distortions of the magnetic field direction and modulus. All these factors should be properly taken into account during the study of evolutionary processes on exoplanets and their atmospheric and plasma environments. The planetary impact of the stellar CME activity may vary depending on stellar age, stellar spectral type and the orbital distance of a planet. Because of the relatively short range of propagation of the majority of CMEs, they affect most strongly the magnetospheres and atmospheres of close-orbit (<0.1 AU) exoplanets. In this chapter we discuss an issue of the stellar CME activity in the context of several actual problems of modern exoplanetology, including planetary atmosphere mass loss, planet survival at close orbits, and definition of a criterion for habitability.

18.1 Introduction

Due to its vicinity, the Sun allows us the detailed investigation of its activity phenomena. The knowledge on solar activity is often used for extrapolation to general stellar activity parameters. Broad observational material, together with the growing body of information about the Sun, shaped up the currently accepted

M.L. Khodachenko (✉)
Space Research Institute, Austrian Academy of Sciences, 8042 Graz, Austria

Skobeltsyn Institute of Nuclear Physics, Moscow State University, 119992 Moscow, Russia

"Pathways to Habitability" Team of the FWF-NFN Project S116:
Helmut Lammer, Manuel Güdel, Igor I. Alexeev, Elena S. Belenkaya, Ildar F. Shaikhislamov, Kristina G. Kislyakova, Petra Odert, Martin Leitzinger, Oleksiy V. Arkhypov, Yury Sasunov, Zoltan Vörös
e-mail: maxim.khodachenko@oeaw.ac.at

J.-C. Vial, O. Engvold (eds.), *Solar Prominences*, Astrophysics and Space Science Library 415, DOI 10.1007/978-3-319-10416-4_18

paradigm regarding the stellar magnetic dynamo and its connection with the stellar activity in cool stars (Parker 1955, 1993). According to this paradigm, rotation of a star and gas motions in its convective layer generate the stellar magnetic fields that give rise to the observed activity. Based on the assumption that the same underlying mechanisms as in the Sun are responsible for the observed stellar phenomena, the solar-stellar analogy hypothesis is widely used for the study and interpretation of the stellar activity processes. At the same time, stellar activity in the form of flares has been observed since the 1940s on UV Ceti type stars. In the 1970s stellar flares were detected also in X-rays, and today we know, thanks to sensitive instruments, that every star that holds a corona is also variable in X-rays. A stellar flare is the result of a short-term (minutes to hours) explosive release of non-potential energy accumulated in the coronal magnetic fields, ultimately derived from convective energy at the stellar surface. During flares, plasma in the stellar atmosphere is rapidly heated, and radiation flux increases across the electromagnetic spectrum, sometimes exceeding the entire quiescent stellar output (Osten et al. 2010). Flare radiation amplitudes are particularly high in the short-wavelength (XUV, gamma-ray) range, and also in the radio. It has been found by various investigators that stellar flares occur at rates depending on the flare amplitude or released total (radiative) energy E in the form of a power law: $dN/dE \propto E^{-\alpha}$, where $\alpha \approx 2 \pm 0.4$, i.e., small flares occur at a much higher rate than the rare big flares (Güdel 2004, and references therein).

Another highly energetic stellar activity phenomenon are the Coronal Mass Ejections (CMEs), first discovered for the Sun in coronagraph observations on the two spacecrafts OSO-7 (Tousey 1973) and Skylab (Gosling et al. 1974). CMEs are large-scale magnetized plasma expulsions carrying billions of tons of material that erupt from a star and propagate in the stellar heliosphere, interacting in multiple ways with the stellar wind. On the Sun, the likelihood of CME events increases with the size of the related flare-event (Kahler 1992). Therefore, it is assumed that the powerful flares on magnetically active, late-type stars are accompanied by strong CME events. Stellar activity in the form of plasma ejections is only indirectly detectable, which is more difficult through direct observations. Indications of stellar mass ejections were found several times in active M dwarfs as Balmer line asymmetries (Houdebine et al. 1990; Fuhrmeister and Schmitt 2004; Leitzinger et al. 2010), showing projected Doppler velocities of $\sim$ 100, $\sim$ 5,800 and $\sim$ 90 $\mathrm{km s^{-1}}$, respectively. The signs of expansion of a dark, solar-like filament at the beginning stages of a stellar flare event were also found on AD Leo (Houdebine et al. 1993a,b). The main planetary impact factors of the stellar CMEs include the associated interplanetary shocks, plasma density and velocity disturbances, energetic particles accelerated in the shock regions, as well as distortions of the magnetic field direction and modulus.

An additional special interest to the stellar flaring and CME activity is caused by their strong planetary effectiveness. The discovery of hundreds of exoplanets with a constantly growing number of low-mass Earth-like rocky planets strengthens the major question, which has to be answered in the future,—if the appropriate conditions for appearance of life may evolve on an Earth-like exoplanet outside the Solar System. A circumstellar area, where a planet could have the necessary conditions for development and maintaining any kind of life is called a Habitable Zone (HZ). The classical concept of HZ around a star primarily relies on the stellar radiation flux allowing liquid water to exist on the surface of an Earth-like planet with a suitable atmosphere (Kasting et al. 1993, Kasting 1997). However, numerous further stellar and planetary properties influence and constrain this radiation-based habitability criterion. In particular, apart from "geophysical" processes depending on the internal structure and composition of a planet, a complex array of stellar factors has to be additionally taken into consideration (Khodachenko et al. 2007a,b; Cowen 2014). Among these are variable stellar UV, EUV, and X-ray (so-called XUV) radiations, stellar and interplanetary magnetic fields, ionized winds and CMEs, energetic particles which control the constitution of upper planetary atmospheres and their physical and chemical evolution. All these stellar factors are closely connected with the stellar activity phenomena, making the last to be a very important aspect of a more general and complete definition of habitability.

For better understanding of the principles governing the long-time habitable planetary environments similar to that on Earth, a global consideration of the whole complex of stellar-planetary relations with the variety of its internal mutual connections and interactions is needed. The planetary atmosphere evolution processes and their climatologic, as well as the habitability consequences should always be considered in context with the fact that the radiation and plasma environment produced by the host star may vary significantly over the evolutionary timescales. Therefore, the shape and location of a HZ around a star are very likely to change throughout the star's lifetime as the stellar luminosity and activity evolve. This evolution is different for different star types, and depends also on their age. That makes the type and age of a host star to be the important factors, which affect the ways of possible development of some of their planets into the potentially habitable worlds.

Below we consider the issue of the stellar CME activity in the context of several actual problems of modern exoplanetology, including planetary atmosphere mass loss, planet survival at close orbits, and definition of a criterion for habitability.

18.2 Exoplanets

Exoplanetology is one of the fastest growing fields in present day astrophysics and space science. Almost two decades after the discovery of 51 Peg b, the first Jupiter-type gas giant outside our solar system, more than 1,000 exoplanets have been discovered mainly by ground-based radial velocity (RV) measurements and space- and ground-based photometric transit surveys (http://exoplanetarchive.ipac.caltech.edu/). In addition, about 3,900 "*planet candidates*" have been found by the Kepler mission (Batalha et al. 2013). The constantly growing number of discovered exoplanets and accumulation of data regarding their physical and orbital characteristics provide an empirical platform for a more detailed study of general principles and major trends of the formation and evolution of planets and planetary systems (including the planetary potential habitability aspect).

18.2.1 Major Features: Mass-Orbit Distribution; Planet Types

Although most of the exoplanets discovered so far are thought to be gas or ice giants, like Jupiter or Neptune, some potentially rocky planets have been identified around M stars (Bonfils et al. 2013). The discovery of these planets became possible due to several international ground-based transit search projects, as well as the *COROT* and *Kepler* space observatories. More than a half of known exoplanets have orbits around their host stars shorter than 0.6 AU (Fig. 18.1). By this, an evident maximum in the orbital distribution of exoplanets takes place in the vicinity of 0.05 AU, with two well pronounced major sub-populations there corresponding to the giant type planets ($0.2M_J < m_p < 8M_J$), so called *Hot Jupiters*, and less massive ($0.008M_J < m_p < 0.08M_J$), *Neptune-* and *Super-Earth* type planets. Here M_J stays for the mass of Jupiter. Altogether the Hot Jupiters comprise about 30 % of the total number of known exoplanets.

The detection of exoplanets at orbital distances $\leq$ 0.05 AU rises questions regarding their upper atmosphere structure, the planet interaction with the extreme stellar plasma environment, the role of possible magnetospheres for atmospheric protection, destructive tidal forces between the host star and the planet, the formation of plasma torii, comet-like escaping planetary plasma tails, as well as the

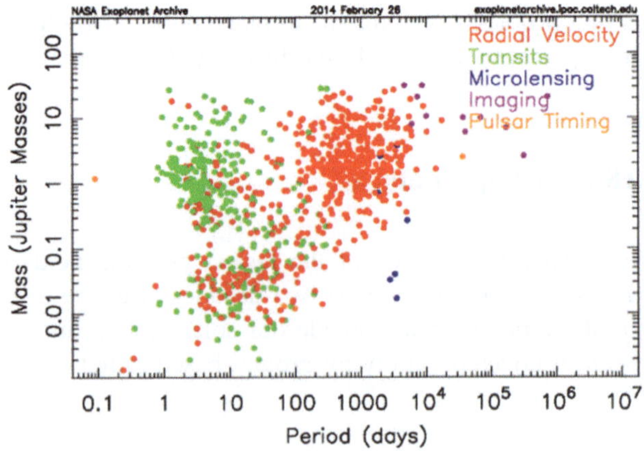

Fig. 18.1 Mass-Orbit distribution of confirmed exoplanets (http://exoplanetarchive.ipac.caltech.edu/exoplanetplots/)

stability of atmospheres against different erosion and mass loss processes. The study of close-orbit exoplanets under extreme stellar radiation and plasma conditions also helps to understand how terrestrial planets and their atmospheres, including early Venus, Earth and Mars evolved during the active early evolution phase of their host stars.

18.2.2 Habitability and Habitable Zone

One of the major questions of the recent and future exoplanet search missions (COROT, Kepler, CHEOPS, PLATO, etc.) is, which of the main sequence star-types (M, K, G, F) may be good or at least preferred candidates for hosting habitable terrestrial type planets? Obviously, the search for the Earth-like exoplanets should not be limited only to the Sun-like G-type stars. It needs to be extended also to the lower mass M- and K-type stars, as well as to slightly more massive F-type stars. Since all these stars have masses lower than $2M_{Sun}$, where M_{Sun} stays for the solar mass, they have therefore lifetimes longer than 1 Gyr. The last is considered as a crucial condition for the possible origin of life. By this, 95 % of the stars in the mass range between $(0.1-2)M_{Sun}$, are M-type dwarf stars, which are also the most numerous stars in the galaxy. Moreover, technical limitations of the present day exoplanet finding methods restrict selection of the target stars to nearby objects in the distance range up to 10–25 pc, where the majority of stars as well fall in the lower mass domain. Altogether, this makes the study of the main sequence low mass stars and their potential impacts on the planetary environments to be an actual task in view of the characterization of exoplanets and their possible habitability.

With the discovery of low mass exoplanets, the question as to whether life could evolve on a planet outside our Solar System has taken on a new urgency. Answering this question requires a complex study of a variety of internal and external factors which may influence the conditions on a planet in order that life could evolve there. The only criterion, used so far, for definition of boundaries of a HZ in the vicinity of a star is based on the possibility for a planet with an atmosphere to have climate and geophysical conditions which allow the existence on its surface of liquid water over geological time periods. The boundaries of a HZ were estimated by Kasting et al. (1993) based on 1-D climate modeling calculations for Earth-like planets. These calculations were done with a cloud-free model and under the assumption of a fully saturated troposphere. A strict limit of the inner edge of the HZ is the so-called "runaway greenhouse" limit. At this limit the entire water ocean of an Earth-analog planet would reside within the atmosphere due to the self-enhancing water vapor feedback cycle. The outer edge of the HZ is defined by the "maximum greenhouse" limit, beyond which a $CO_2 - H_2O$ greenhouse is no longer capable of maintaining a warm surface. This limit is based on the assumption that an Earth-like planet will have volcanoes that emit CO_2 and that CO_2 will accumulate in the planet's atmosphere as its surface becomes cold. In such calculations, the "maximum greenhouse" limit occurs at $\sim$ 1.67 AU for the Sun full analog star. Recently, a more distant outer edge has been suggested for the HZ by Seager (2013), based on a calculation by Pierrehumbert and Gaidos (2011) showing that super-Earths with dense captured H_2 atmospheres could remain habitable out to as far as 10 AU in a system of the Sun "twin" (i.e. full analogue with the same type, size, mass, and age) star. Such planets will remain speculative, however, until they are observed.

At the same time, the fact that a planet lies within the HZ of its parent star does not guarantee that it will be habitable. The width and circumstellar distance range of the HZ, defined as described above, depend mainly on the stellar luminosity (Fig. 18.2) which evolves during the life time of a star and influences the planetary surface temperature. Moreover, in the most populated family of the low-mass main sequence stars (M- and K- type dwarfs) the HZs appear to be very close to the stars (at radial distances $d < 0.2$ AU). This results in the importance of the additional "short range" (in astrophysical scales) planetary impacting factors of the stellar activity such as interplanetary shocks, magnetic clouds (MCs) and CMEs propagating in the dense stellar winds (Khodachenko et al. 2007a; Lammer et al. 2007, 2009a,b), as well as more intensive near the star, stellar cosmic rays (Grießmeier et al. 2004, 2005), as well as X-ray and extreme ultraviolet radiations. These factors directly influence the magnetospheres, atmospheres, and possible biospheres of exposed exoplanets. Based on the experiences with our Sun one may also conclude that the activity of a G-type star remains a crucial factor for the planetary environment evolution within its HZ.

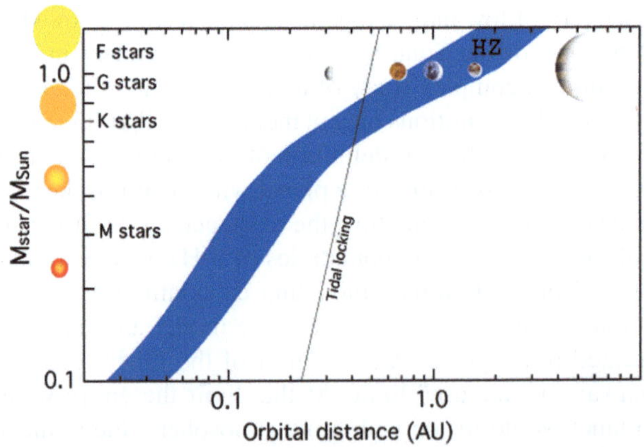

Fig. 18.2 A circumstellar distance range of the Habitable Zone (HZ) of the main sequence stars (adopted from Lammer et al. 2009a)

18.3 Impact of Stellar Radiation and Plasma Flows on Planetary Atmospheres

Close location of the majority of known exoplanets to their host stars results in intensive heating, ionization, and chemical modification of their upper atmospheres by the stellar X-ray/EUV (XUV) radiation with the subsequent expansion of the ionized atmospheric material and its loss due to interaction with the stellar wind (Lammer et al. 2009b; Khodachenko et al. 2007a,b). A number of actual questions regarding the evolutionary paths of planetary systems and the key factors, influencing them, are nowadays under continuous tackling. Among these questions a prominent position belongs to the problem of stellar—planetary interactions, including consideration of influences of stellar radiation and plasma flows (e.g., stellar wind and CMEs) on planetary environments and related erosion of upper atmospheres of exoplanets and their mass loss.

The action of intensive stellar radiation and stellar winds on planetary environments consists of the following effects.

(1) XUV radiation of a host star affects the energy budget of the planetary thermosphere, resulting in the heating and expansion of the upper atmosphere, which under certain conditions could be so large that the majority of light atmospheric constituents overcome the gravitational binding and escape from the planet in the form of a hydrodynamic wind (Yelle 2004; Tian et al. 2005; Penz et al. 2008; Koskinen et al. 2010, 2013; Erkaev et al. 2013; Lammer et al. 2013). This effect contributes to the so-called thermal mass loss of the atmospheric material. High upper atmospheric temperatures and the resulting hydrodynamic escape have crucial impact on the atmospheric stability of terrestrial-type planets (Kulikov et al. 2007) and the evolution of the planet's water inventory. Simultaneously with the direct radiative heating of the upper atmosphere, the processes of ionization and recombination as well as production of energetic neutral atoms by sputtering and various photo-chemical and charge exchange reactions take place (Lammer et al. 2013; Shematovich 2012; Guo 2011, 2013). Such processes result in the formation of extended (in some cases) coronas around planets, filled with hot neutral atoms.

(2) The expanding, XUV heated and photo-chemically energized, upper planetary atmospheres and hot neutral coronae may reach and even exceed the boundaries of the planetary magnetospheres. In this case they will be directly exposed to the plasma flows of the stellar wind and CMEs with the consequent loss due to ion pick-up mechanism. That contributes to the non-thermal mass-loss process of atmosphere (Lichtenegger et al. 2009; Khodachenko et al. 2007a). As a crucial parameter here appears the size of the planetary magnetosphere, characterized by the magnetopause stand-off distance R_S (Khodachenko et al. 2007a,b; Kislyakova et al. 2013, 2014). Altogether, this makes the planetary magnetic field and the structure of magnetosphere, as well as the parameters of the stellar wind (e.g., density n_{sw} and speed v_{sw}) to be very important for the processes of atmospheric erosion and mass-loss of a planet.

Interaction of short-periodic exoplanets with the stellar wind plasma and high XUV flux at close-orbit distances plays a crucial role regarding the ionization and ion loss processes of atmospheric species. Grießmeier et al. (2004) modelled the stellar wind interaction with short-period gas giants for the present and for early evolutionary stages of solar-like stars. They found that it may be possible that short periodic gas giants can have an ionosphere—stellar wind interaction similar to that one on Venus, because the internal magnetic moments of such exoplanets orbiting close to their host stars are expected to be weaker compared to Jupiter due to their tidal locking. Erkaev et al. (2005) showed that short-periodic gas giants, similar to HD209458b, at orbital distances < 0.1–0.2 AU around solar-like G-type stars with ages comparable to our Sun, may have stellar wind induced H^+ ion loss rates in the orders of about 10^8–10^9 g s^{-1}, which are lower than the observation-based and modelled neutral hydrogen loss rates of about 10^{10}–10^{12} g s^{-1} (Lammer et al. 2003; Yelle 2004; Tian et al. 2005). In that respect, the interaction of these exoplanets with stellar CMEs appears to be an important process, which is central to a better understanding of the non-thermal atmosphere mass loss.

18.4 Stellar Radiation and Plasma Environment

18.4.1 Stellar Activity/Radiation

Activity in late-type stars (i.e., spectral types G, K, M) has been the subject of intense studies for many years. The relevant physical phenomena of stellar activity and their observational manifestations include modulations of the stellar photospheric light due to stellar spots, intermittent and energetic flares, CMEs, stellar cosmic rays, enhanced XUV emissions (see Scalo et al. 2007 and references therein). Evaluation of flaring rates and intensities usually require long-duration monitoring. So, proxies for the flaring activity are used, such as optical Ca H and K emission cores, H-alpha and Mg II emission, soft X-ray continuous emission, and a large number of UV-to soft X-ray emission lines (Ayres 1997; Gershberg 2005).

Observations of stars in clusters have revealed that single late-type stars spin down monotonically with their age because of angular momentum loss (Skumanich 1972). For a given age of star t, the stellar rotation period can be estimated as (Newkirk 1980): $P_{\rm rot} \propto \left(1 + \frac{t}{\tau}\right)^{0.7}$, where $\tau = 2.56 \times 10^7$ yrs is a time constant calculated by Newkirk (1980). At the same time, early studies already pointed out a strong correlation between the rotation rate of a star and its activity level (Wilson 1966). This correlation means that there must be dependence between stellar activity and age. For solar-type stars this has been studied within the "Sun in Time" project (Ribas et al. 2005). Based on the analysis of a large amount of X-ray, EUV and UV observations of a homogeneous sample of single nearby G0-5 stars with known rotation periods, luminosity and ages, it has been concluded that during the fist 100 Myr after the Sun arrived at the Zero-Age Main-Sequence (ZAMS), the integrated XUV flux was up to 100 times higher than today. After this very active stage, XUV flux of a solar-type star decreases with the time: $\propto t[\text{Gyr}]^{-1.72}$ (Ribas et al. 2005).

According to the currently accepted paradigm, the wide range of activity levels and related phenomena observed in different stars is directly connected with operation of the stellar magnetic dynamo. By this, two basic parameters: (i) stellar rotation rate and (ii) depth of the convective zone, are believed to control the stellar dynamo efficiency, which increases with increasing of any, or both of these quantities. Since the stellar convective envelope becomes thicker with decreasing stellar mass, it is straightforward to infer that, at a given rotation period (i.e. age), the low-mass M- and K-stars should be more active than a solar-type G-star. This fact has many observational confirmations. For example, a relatively old (~ 5.5 Gyr) dwarf M-star, Proxima Centauri, experiences measurable flares at a rate of about one flare per hour (Walker et al. 1981).

A stellar flare is the result of short-term (minutes to hours) explosive energy release tapped from non-potential energy in the coronal magnetic fields, ultimately derived from convective energy at the stellar surface. During flares, plasma in the stellar atmosphere is rapidly heated, and therefore fluxes increase across the electromagnetic spectrum, in very rare cases exceeding the entire quiescent stellar output (Osten et al. 2010). Flare radiation amplitudes are particularly high in the short-wavelength (XUV, gamma-ray) range, and also in the radio regime. Audard et al. (2000) found that the energy of flares correlates with the stellar activity, characterized by $L_{\text{X}}/L_{\text{bol}}$, where L_{X} and L_{bol} are X-ray and bolometric luminosities of a star, respectively. The evolution of $log(L_{\text{X}}/L_{\text{bol}})$ with time for stars of various masses is shown in Fig. 18.3, provided by Scalo et al. (2007). According to this activity-age diagram, the solar-type G- stars stay at saturated emission levels only until ages of $\sim$ 100 Myr, and then their XUV luminosities rapidly decrease with age: $\propto (t[\text{Gyr}])^{-1.72}$. On the other hand, M- stars have saturated emission periods up to 0.5-1 Gyr, and then their luminosity decreases in a way similar to the solar-type stars.

According to Audard et al. (2000), the rate of high-energy ($E > 10^{32}$ erg) flares per day as $logN|_{E>10^{32}\,\text{erg}} = -26.7 + 0.95 logL_X$, which in the case of M-stars with a saturated activity level $L_X = 7 \times 10^{28}$ erg/s implies $\sim$ 6 strong flares per day. Therefore, the powerful flares occur more often in X-ray bright stars. Altogether it has been found (Audard et al. 2000; Ribas et al. 2005; Scalo et al. 2007) that early K-stars and early M-stars may have XUV emissions level, and therefore flaring rates, of $\sim$(3–4) and $\sim$(10–100), respectively, times higher than solar-type G-stars of the same age. Moreover, the rate of large EUV flares exceeding some given radiative energy threshold is proportional to the average, long-term X-ray luminosity of the star (Audard et al. 2000), suggesting a very high rate of super-flares in young, active stars. This fact, in particular, indicates that super-flares are not occurring at a higher rate in M dwarfs than in solar analogs. Therefore, it is the closer-in HZs around M dwarfs combined with the slower decay of M-dwarf activity that subjects potentially habitable planets around M-dwarfs to much higher

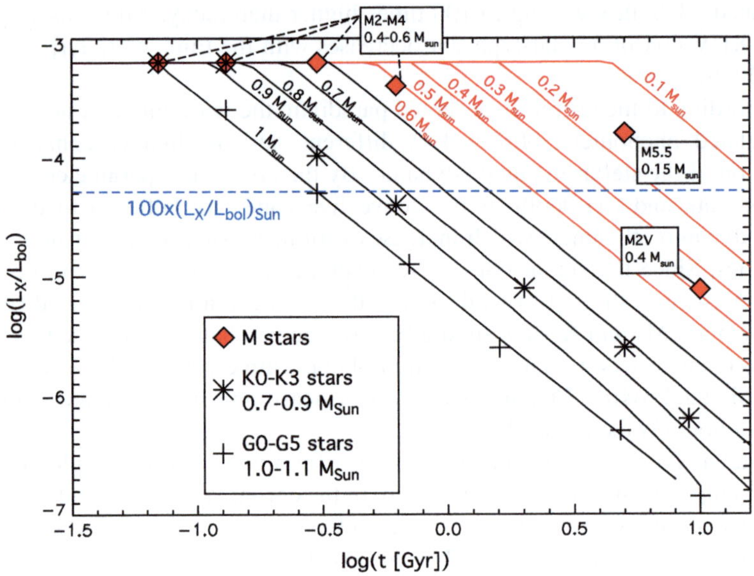

Fig. 18.3 $L_X/L_{\rm bol}$ as a function of age for stars with masses $< M_{\rm Sun}$. Symbols represent stars from the "Sun in Time" program (adopted from Scalo et al. 2007)

flare-radiation doses for a longer time. Very energetic flares or super-flares may contribute to the ionization of upper planetary atmospheres (ionospheres) and upper-atmospheric chemistry, thus potentially changing planetary habitable conditions.

Schrijver et al. (2012) investigated the presence of super-flares on the Sun in historical times, using natural archives (nitrate in polar ice cores and various cosmogenic radionuclids) together with flare statistics and the historical record of sunspots. They conclude that flares in the past four centuries are unlikely to have exceeded the largest observed solar flare with total energy of $\approx 10^{33}$ erg. The solar magnetic field (and similarly, stellar magnetic fields) provides an upper bound to the maximum flare energy release simply by the amount of free magnetic energy that could coherently be converted. This limit is indeed about $\approx 10^{33}$. However, Schaefer et al. (2000) reported, from observations in various wavebands, nine super-flares with total radiative energies of $10^{33} - 10^{38}$ erg on solar (F-G type) analog stars, none of which is exceptionally young or extremely active, or a member of a close binary system. Some of the flaring objects were even older than the Sun. They concluded that the average recurrence time for super-flares on such stars may be decades to centuries although this would obviously be an overestimate for the Sun. With the new *Kepler* satellite data, Maehara et al. (2012) found rather many super-flares with bolometric energies of $10^{33} - 10^{36}$ erg, with an estimated recurrence time of ≈ 800 years for 10^{34} erg flares on old, solar-like stars. None of these stars is known to host close-in exoplanets that could induce such energy release.

The impact of stellar flares on the atmospheres of terrestrial extrasolar planets has been studied by, e.g. Segura et al. (2010) and Grenfell et al. (2012). The importance of such single events would be their potential to melt ice surfaces on habitable planets, the possible temporary breakup of the ionosphere and ozone depletion after creation of nitrous oxides in the irradiated atmosphere; an event of 10^{36} erg (ionizing energy) was estimated to result in a loss of 80 % of the total ozone content for more than one year, with the consequent increase in UV irradiation (Schaefer et al. 2000).

18.4.2 Stellar Winds and CMEs

In addition to being exposed to electromagnetic radiation from their host stars, exoplanets are also exposed to high-speed outflows of particles from the stellar atmosphere. For cool main sequence stars like the Sun, stellar winds arise in the hot coronas that represent the outermost atmospheres of the stars. Although the mechanisms of coronal heating and coronal wind acceleration remain hot topics of research, Parker (1958) demonstrated long ago that if once you have a hot corona, a wind much like that of the Sun arises naturally through thermal expansion. Thus, any star known to have a hot corona can be expected to possess a coronal wind. Observations from X-ray observatories such as *Einstein*, *ROSAT*, *Chandra*, and *XMM-Newton* have demonstrated that X-ray emitting coronas are ubiquitous among cool main sequence stars, so coronal winds can be expected to be a common feature as well. Unfortunately, detecting and studying these winds is much harder than detecting and studying the coronas in which they arise. Current observational capabilities do not yet allow us to directly detect solar-like coronal winds emanating from other stars.

Recently, there have been important developments towards indirect detections of stellar winds through their interactions with the surrounding interstellar medium. In particular, the stellar mass loss rates and related stellar wind parameters have been estimated by observing astrospheric absorption features of several nearby G- and K-stars. Comparison of the measured absorption to that calculated by hydrodynamic codes made it possible to perform empirical estimation of the evolution of the stellar mass loss rate as a function of stellar age (Wood et al. 2002, 2005) and to conclude about the dependence of n_{sw} and v_{sw} on the age of the stellar system. In particular, the younger solar-type G-stars appeared to have much denser and faster stellar winds as compared to the present Sun. Combining the stellar mass loss measurements of Wood et al. (2005) with the results of Newkirk (1980) for the age-dependence of stellar wind velocity, Grießmeier et al. (2007a) proposed a method for calculation of stellar wind density n_{sw} and velocity v_{sw} at a given orbital location of an exoplanet d for a given mass M_{*}, radius R_{*} and age t_{*} of a star. As an example, the values of the stellar wind plasma parameters for a solar-analog G-type star ($M_{*} = M_{Sun}$, age: 4 Gyr) at orbital distances of 0.045, 0.1, and 0.3 AU are given in Table 18.1.

Table 18.1 Stellar wind and CME parameters for a solar-analog G-type star ($M_* = M_{Sun}$, age: 4 Gyr) at different orbital distances

Orbital distance	n_{SW} [cm^{-3}]	$\tilde{v}_{SW}$ [km/s]	n_{CME}^{min} / n_{CME}^{max} [cm^{-3}]	$\tilde{v}_{CME}^{av}$ / $\tilde{v}_{CME}^{fast}$ [km/s]
0.045 AU	$9.1e3$	210	$9.1e3/7.8e4$	$520/810$
0.1 AU	$1.2e3$	260	$1.2e3/7.1e3$	$510/810$
0.3 AU	92	340	$92/2.6e2$	$500/800$

The values of $\tilde{v}_{CME}^{fast}$ and $\tilde{v}_{CME}^{av}$ are obtained using (18.2) with $v_0 = 800\,\text{km/s}$ and $v_0 = 500\,\text{km/s}$, respectively
All the velocities include a contribution of the Keplerian planetary orbital velocity V_K

Furthermore, it is known from observations of our Sun that flaring activity of a star is accompanied by eruptions of coronal mass (e.g. CMEs), occurring sporadically and propagating in the stellar wind as large-scale plasma-magnetic structures. Traveling outward from the star at high speeds (up to thousands km/s), CMEs create major disturbances in the interplanetary medium and produce strong impacts on the planetary environments and magnetospheres. Since CMEs can be directly observed on the Sun, the current knowledge on them comes from the study of the Sun and the heliosphere. On the Sun, CMEs are associated with flares and prominence eruptions and their sources are usually located in active regions and prominence sites. The likelihood of CME-events increases with the size and power of the related flare event. Kahler (1992) summarized various studies which found that larger and longer soft X-ray flares are well associated with CMEs, whereas smaller and shorter events are not. Generally, it is expected that the frequent and powerful flares on magnetically active flaring stars should be accompanied by an increased rate of CME production.

The direct signature of plasma moving towards an observer is Doppler shifted emission. The Balmer spectral lines have been proven to show Doppler shifted emission signatures related to mass ejected from stars. This fact forms the basis for the present day spectral diagnostics of stellar CMEs. Additionally to that, various observational signatures such as line asymmetries, specific radiative features, UV(EUV) dimmings, and long flare decay time are attributed to stellar CMEs.

CMEs cause density fluctuations (Kathiravan and Ramesh 2005) because they propagate through the stellar atmosphere. Spectral line ratios sensitive to the electron density, such as the $C_{III}(\lambda 1176\text{Å})/C_{III}(\lambda 977\text{Å})$ line ratio, provide information on the variation of electron density in the region from where these lines originate. Dupree et al. (1976) found that the ratio is enhanced for active solar regions. On stars, this ratio was found to be enhanced during flaring (Bloomfield et al. 2002; Christian et al. 2006). The $C_{III}(\lambda 977\text{Å})$ transition region line shows a significant increase in flux during solar CMEs, as shown by Ciaravella et al. (1997) from observations with the Ultra Violet Coronagraph Spectrometer onboard the SOlar Heliospheric Observatory (UVCS/SOHO). Whether this line ratio is affected by solar CMEs and if it could therefore be used as a diagnostic for stellar CMEs remains unclear.

In the past decades there have been several attempts to detect stellar analogs of the solar type II bursts which are known to be the signatures of the propagating shock waves associated with CMEs (Jackson et al. 1990; Abdul-Aziz et al. 1995; Abranin et al. 1998; Leitzinger et al. 2009). However, in the case of stellar type II bursts one has to consider that the atmospheres of G, K and M stars are different in their extents and density. As this kind of radio burst is related to the local plasma frequency (i.e. plasma density), the analog of the type II burst occurring in the decameter wavelength range on the Sun may not be detected in the same range on an M star. For such stars the occurrence of type II bursts might shift to shorter wavelength ranges such as the meter or centimeter ones. Due to these uncertainties, so far no clear detection of a stellar type II burst was reported. At the same time, the CMEs correlate in the spectra with the regions of decreased emission known as dimmings seen in X-ray and EUV wavelengths representing the early stage of a CME. Dimmings are known to last for hours; they correspond to regions of either plasma evacuation or changing temperature (Gopalswamy and Thompson 2000). On the Sun more than 80 % of CMEs can be tracked back to dimming regions (Bewsher et al. 2008). As shown by Aschwanden et al. (2009), from dimmings it is possible to deduce the parameters of CMEs. In principle it is technically possible to detect dimmings on stars. As dimmings last for hours, they should be seen in stellar X-ray and EUV light curves even if the exposure times need several hundreds of seconds.

The best observational indications of stellar CMEs come from V471 Tau, an eclipsing binary consisting of a K-dwarf and a white dwarf. Mullan et al. (1989) found discrete absorption features in the UV spectra of the white dwarf with velocities of 700–800 km s^{-1} relative to the K-dwarf. The observed material had temperatures below a few 10^4 K, which is comparable to the dense prominence part of solar CMEs. Since the corresponding velocities were also comparable to the solar CME speeds, Mullan et al. (1989) interpreted the observed features as signatures of CMEs from the K-star. Later, Bond et al. (2001) also found absorption features in Hubble Space Telescope (HST) spectra of V471 Tau with properties comparable to those of solar CMEs. They estimated a frequency of 100–500 CMEs per day, which would result in the star's mass-loss rate of $(5 - 25) \times 10^{-14} M_{Sun}$ per year by CMEs alone. Guenther and Emerson (1997) monitored a sample of classical and weak-line T Tauri stars (CTTS and WTTS, respectively) in the Taurus Auriga region and detected an extra emission in the blue wing of H_α on a WTTS which was identified as a mass ejected from the star. The authors deduced a projected velocity of 600 km s^{-1} and estimated a mass in the range of 10^{15}–10^{16} kg, which is a factor 10–100 higher than the most massive solar CMEs.

For M stars, indications of CMEs have been observed only on a few objects. The signature of a mass ejection was found by Ambruster et al. (1986), who interpreted an observed drop in the UV flux lasting for 1.5 h shortly after a violent flare on EV Lac as an episode of mass expulsion. Houdebine et al. (1990) analyzed a set of optical spectra of the young (0.2 Gyr) and active M 3.5 Ve star AD Leonis. The authors detected a distinct blue wing enhancement of the H_γ (434.1 nm) Balmer line, decreasing with time until it vanished completely. They deduced a Doppler velocity of 5,800 km s^{-1}, which is about twice the value of the fastest solar CME. Other estimated parameters for this event are 7.7×10^{14} kg mass of the material ejection and 5×10^{27} J for its kinetic energy. In consequent papers, Houdebine et al. (1993a,b) estimated the properties (minimum values) of a dark filament associated with the same flare. They derived a mass of 6.4×10^{13} kg, a kinetic energy of 2.7×10^{25} J and a loop radius of 50 Mm.

Extreme Ultraviolet Explorer (EUVE) observations of the young M-dwarf star AU Microscopii revealed a powerful flare with a decay time of a day. Cully et al. (1994) successfully reconstructed the strongest spectral lines by including the assumption of an ejected magnetically confined plasma in their model. The deduced mass and kinetic energy of this event were 10^{17} kg and 10^{29} J, respectively, which is about four orders of magnitude higher than analogous parameters of the solar CMEs. The initial ejection speed was found to be 1,200 km s^{-1}. Fuhrmeister and Schmitt (2004) detected a huge flare on the old dM9 star DENIS 104814.7–395606.1. They further found a blue wing asymmetry in the Balmer H_α and H_β lines, which they favour to be the signature of plasma that has been ejected into the observers direction.

Leitzinger et al. (2010, 2011) performed a search for highly energetic activity phenomena in a small sample of late-type main-sequence stars in the far ultraviolet (FUV) using data from the Far Ultraviolet Spectroscopic Explorer (FUSE). Because FUSE allows a simultaneous photometric and spectroscopic analysis, it was possible to analyze variations in the stellar flaring light curves and possible CME-related signatures (line asymmetries, enhancements, and shifts) in the spectra. The authors detected an extra emission in the blue wing of the first component of *OVI* doublet at 1,032 Å; they estimated the projected velocity of 84 km s^{-1}, which may be too low to be clearly assigned to a CME, although, just before the detected extra emission there was flare.

By considering the Sun as a typical representative of G- stars, it seems reasonable to assume a similarity of the basic parameters of the stellar winds of G- stars and those known for the Sun. Such a solar-stellar analogy principle is widely considered for the investigation of basic processes of the stellar wind—planet interaction. Based on the estimations of solar CME plasma density n_{CME}, using the in-situ spacecraft measurements (at distances > 0.4 AU) and the analysis of white-light coronagraph images (at distances $\leq 30 R_{\mathrm{Sun}} \approx 0.14$ AU), Khodachenko et al. (2007a) provided general power-law interpolations of n_{CME} dependence on the distance to a star:

$$n_{\mathrm{CME}}^{min}(d) = 4.88(d[\mathrm{AU}])^{-2.3}, \quad n_{\mathrm{CME}}^{max}(d) = 7.10(d[\mathrm{AU}])^{-3.0}, \tag{18.1}$$

Eq. (18.1) identify a typical maximum-minimum range of n_{CME}. The dependence of stellar CME speed v_{CME} on the orbital distance d can be approximated by the formula:

$$v_{\mathrm{CME}} = v_0 \left(1 - e^{\frac{2.8 R_{\mathrm{Sun}} - d}{8.1 R_{\mathrm{Sun}}}}\right)^{1/2}, \tag{18.2}$$

proposed in Sheeley et al. (1997) on the basis of the tracking of several solar wind density enhancements at close distances ($d < 0.1$ AU). For the approximation of average- and high-speed CMEs one may take in (18.2) $v_0 = 500\,\mathrm{km/s}$ and $v_0 = 800\,\mathrm{km/s}$, respectively. The average mass of CMEs is estimated as 10^{15} g, whereas their average duration at distances ~ 0.05 AU is close to 8 h. Table 18.1 provides an example of stellar CME plasma parameters for a solar-analog G-type star at orbital distances of 0.045 AU, 0.1 AU, and 0.3 AU. Because of the relatively short range of the propagation of majority of CMEs, they should strongly impact first of all the planets at close orbits (≤ 0.3 AU). Khodachenko et al. (2007a) have found that for a critical CME production rate $f^{cr}_{\mathrm{CME}} \approx 36$ CMEs per day (and higher) a close-orbit exoplanet appears under continuous action of the stellar CMEs plasma, so that each next CME collides with the planet during the time when the previous CME is still passing over it. This means in general harder conditions for the planetary environments than those in the case of a regular stellar wind. Therefore, the investigation of evolutionary paths of close-orbit exoplanets in potentially habitable zones around young active stars, besides of the higher XUV radiation, should take also into account the effects of "short range" (in astrophysical scales) planetary impacting factors of stellar activity such as relatively dense stellar winds and frequent magnetic clouds (MCs) and CMEs.

18.5 The Problem of Magnetospheric Protection of Close-Orbit Exoplanets

Planetary magnetic fields and related magnetospheres are widely recognized to play an important role in the context of the non-thermal mass loss of close-orbit exoplanets (Lammer et al. 2003; Lammer et al. 2007; Khodachenko et al. 2007a,b). Magnetosphere acts as an obstacle (magnetospheric obstacle), which interacts with the stellar wind, deflecting it and protecting planetary ionosphere and upper atmosphere against the direct impact of stellar wind plasmas and energetic particles (e.g., cosmic rays). The background magnetic field of planetary magnetosphere not only forms a barrier for the upcoming plasma flow of stellar wind, but it influences also the outflow of the escaping planetary plasma wind formed in the course of atmosphere heating and ionization by stellar XUV. For example, Adams (2011) considered outflows from close-in gas giants in the regime where the flow is most likely controlled by magnetic fields. In that respect it is important to note, that the processes of material escape and planetary magnetosphere formation have to be considered jointly in a self-consistent way in their mutual relation and influence. The expanding partially ionized plasma of a Hot Jupiter atmosphere interacts with the planetary intrinsic magnetic field and appears a strong driver in formation and shaping of the planetary magnetospheres (Adams 2011; Trammell et al. 2011; Khodachenko et al. 2012), which in turn influences the overall mass loss of a planet.

18.5.1 Planetary Magnetism

The intrinsic magnetic field of a planet, which influences the character of the magnetospheric obstacle, is generated by a magnetic dynamo. The existence and efficiency of the dynamo are closely related to the type of planet and its interior structure. Not all planets have intrinsic magnetic fields, or in other words, efficiently operating dynamos. Planetary magnetic dynamo requires the presence of an electrically conducting region (i.e. a liquid outer core for terrestrial planets, or a layer of electrically conducting liquid hydrogen for gas giants) with non-uniform flows organized in a certain manner, which create a self-sustaining magnetic field. According to dynamo theory, this flow should be convective in nature (Stevenson 1983). Therefore, convection can be regarded as a necessary requirement for a planetary magnetic field (Stevenson 2003).

Limitations of the existing observational techniques make direct measurements of the magnetic fields of exoplanets impossible. At the same time, a rough estimation of an intrinsic planetary magnetic dipole moment $\mathcal{M}$ can be obtained by simple scaling laws found by the comparison of different contributions in the governing equations of planetary magnetic dynamo theory (Farrell et al. 1999; Sánchez-Lavega 2004; Grießmeier et al. 2004; Christensen 2010). Most of these scaling laws reveal a connection between the intrinsic magnetic field and the rotation of a planet. Grießmeier et al. (2004) estimated the intrinsic planetary magnetic dipole moments of exoplanets and corresponding sizes of their magnetospheres using the following scaling laws for $\mathcal{M}$:

$$\begin{aligned}
&\mathcal{M} \propto \rho_c^{1/2}\omega_{\rm p} r_c^4 && \text{Busse 1976,}\\
&\mathcal{M} \propto \rho_c^{1/2}\omega_{\rm p}^{1/2} r_c^3 \sigma^{-1/2} && \text{Stevenson 1983,}\\
&\mathcal{M} \propto \rho_c^{1/2}\omega_{\rm p}^{3/4} r_c^{7/2} \sigma^{-1/4} && \text{Mizutani et al. 1992,}\\
&\mathcal{M} \propto \rho_c^{1/2}\omega_{\rm p} r_c^{7/2} && \text{Sano 1993.}
\end{aligned} \tag{18.3}$$

Here r_c is the radius of the dynamo region (also called the core radius), and $\omega_{\rm p}$ is the angular velocity of a planet rotation around its axis. The internal properties of a planet, such as the mass density and the conductivity of the dynamo region are denoted by ρ_c and σ, respectively (for details of the model parameters estimation see Grießmeier et al. 2004, 2007b). More recently, Reiners and Christensen (2010), based on scaling properties of convection-driven dynamos (Christensen and Aubert 2006), calculated the evolution of average magnetic fields of Hot Jupiters and found that (a) extrasolar gas giants may start their evolution with rather high intrinsic magnetic fields, which then decrease during the planet life time, and (b) the planetary magnetic moment may be independent of planetary rotation.

Equation (18.3) provide a range $\mathcal{M}_{\rm min} - \mathcal{M}_{\rm max}$ of reasonable planetary magnetic moment values. In spite of being different in details, all these models yield an increase of $\mathcal{M}$ with an increasing planetary angular velocity $\omega_{\rm p}$. In that respect it is necessary to take into account the fact, that close-orbit exoplanets such as Hot Jupiters very likely are tidally locked to their host stars. The angular rotation of

Table 18.2 The ranges of $\mathcal{M}$ given by (18.3) and $R_s = R_s^{(dip)}$ provided by (18.4) for a Jupiter-type exoplanet orbiting a solar-analogue G-type star at different distances d under the action of the stellar wind plasma flow described in Table 18.1

d [AU]	$\mathcal{M}$ [$\mathcal{M}_J$]	$R_s^{(dip)}(\mathcal{M}_{min})$ / $R_s^{(dip)}(\mathcal{M}_{max})$ [r_p]
0.045[a]	0.12 ... 0.3	4.3 ... 6.2
0.1[b]	0.04 ... 1.0	3.8 ... 12
0.3[c]	1.0 ... 1.0	15 ... 15

[a] Tidally locked
[b] Possible tidally locked
[c] Not tidally locked

a tidally locked planet is synchronized with its orbital motion so, that ω_p is equal to the orbital angular velocity Ω determined by Kepler's law. The time scale for tidal locking τ_{sync} depends on the planetary structure, orbital distance to the host star, and the stellar mass (Showman and Guillot 2002). By this, the planets for which $\tau_{sync} \leq 0.1$ Gyr, can be assumed to be tidally locked, since the age of a planet is at least an order of magnitude longer. On the other hand, the planets with $\tau_{sync} \geq 10$ Gyr are almost certainly tidally unlocked. The influence of tidal locking on the value of an expected planetary magnetic dipole was studied for different planets (giants and terrestrial-type) in Grießmeier et al. (2004, 2007b). It was shown, that the magnetic moments of slowly rotating tidally locked exoplanets usually are much smaller than those for similar, but freely rotating tidally unlocked planets. In Table 18.2 possible ranges of planetary magnetic dipole moments defined by (18.3) for a Jupiter-type ($M_p = M_J$; $r_p = R_J$) exoplanet orbiting around a solar-analogue G-type star ($M_* = M_{Sun}$, $R_* = R_{Sun}$) are provided. The values of $\mathcal{M}$ are scaled in units of the present time Jupiter magnetic moment $\mathcal{M}_J = 1.56 \times 10^{27}$ A m^2.

18.5.2 Magnetic Shielding

For an efficient magnetic shielding (i.e. magnetospheric protection) of a planet, the size of its magnetosphere characterized by the magnetopause stand-off distance R_s should be much larger than the height of the exobase. By this, the value of R_s is determined from the balance between the stellar wind ram pressure and the planetary magnetic field pressure at the sub-stellar point (Grießmeier et al. 2004; Khodachenko et al. 2007a). In most of studies so far, the investigation of an exoplanetary magnetospheric protection is performed within the highly simplifying assumption of a planetary *dipole-dominated* magnetosphere. This means that only the intrinsic magnetic dipole moment of an exoplanet $\mathcal{M}$ and the corresponding magnetopause electric currents (i.e., "screened magnetic dipole" case) are considered as the major magnetosphere forming factors. In this case, i.e. assuming $B(r) \propto \mathcal{M}/r^3$, the value of R_s has been defined by the following expression:

$$R_s \equiv R_s^{(dip)} = \left[\frac{\mu_0 f_0^2 \mathscr{M}^2}{8\pi^2 \rho_{sw} \tilde{v}_{sw}^2} \right]^{1/6}, \tag{18.4}$$

where μ_0 is the diamagnetic permeability of free space, $f_0 \approx 1.22$ is a form-factor of the magnetosphere caused by the account of the magnetopause electric currents, $\rho_{sw} = n_{sw}m$ is the mass density of the stellar wind, and $\tilde{v}_{sw}$ is the relative velocity of the stellar wind plasma which includes also the planetary orbital rotation velocity. For the tidally locked close-orbit exoplanets with weak magnetic moments exposed to a dense and/or fast stellar wind plasma flows, (18.4) yields rather small values for sizes of dipole-dominated magnetospheres, $R_s = R_s^{(dip)}$, compressed by the stellar wind plasma flow (see Table 18.2), which in the most extreme cases of colliding with CMEs may even shrink down to the planetary radius r_p. Therefore, the approach to estimation of the magnetosphere size based on (18.4) resulted in the commonly accepted conclusion, that in order to have an efficient magnetic shield, a planet needs a strong intrinsic magnetic dipole $\mathscr{M}$.

Khodachenko et al. (2007b) studied the mass loss of the Hot Jupiter HD 209458b due to the ion pick-up mechanism caused by stellar CMEs, colliding with the planet. In spite of the sporadic character of the CME-planetary collisions, in the case of the moderately active host star of HD 209458b, it has been shown that the integral action of the stellar CME impacts over the exoplanet's lifetime can produce significant effect on the planetary mass loss. The estimates of the non-thermal mass loss of the weakly magnetically protected Hot Jupiter, HD 209458b, due the stellar wind ion pick-up, lead to significant and sometimes unrealistic values—up to several tens of planetary masses M_p lost during a planet life time (Khodachenko et al. 2007b). In view of the fact that multiple close-in giant exoplanets, comparable in mass and size with the Solar System Jupiter exist, and that it is unlikely that all of them began their life as ten times, or even more massive objects, one may conclude that additional factors and processes have to be taken into consideration in order to explain the protection of close-in exoplanets against of destructive non-thermal mass loss. In view of that problem, Khodachenko et al. (2012) proposed a more complete model of the magnetosphere of a giant gas exoplanet, which due to its consequent account of the specifics of close-orbit exoplanets provides under similar conditions larger sizes for the planetary magnetospheric obstacles, than those given by the simple screened magnetic dipole model, traditionally considered so far in the literature.

18.6 Magnetodisk-Dominated Magnetosphere of a Hot Jupiter

The investigation of exoplanetary magnetospheres and their role in evolution of planetary systems forms a new and fast developing branch. Magnetosphere of a close-orbit exoplanet is a complex object, whose formation depends on different external and internal factors. These factors may be subdivided on two basic groups: (a) *stellar factors*, e.g., stellar radiation, stellar wind plasma flow, stellar magnetic field and (b) *planetary factors*, e.g., type of planet, orbital characteristics, escaping material flow, and planetary magnetic field. The structure of an exoplanetary magnetosphere depends also on the speed regime of the stellar wind plasma relative the planet (Erkaev et al. 2005; Ip et al. 2004). In particular, for an exoplanet at sufficiently large orbital distance when the stellar wind is super-sonic and super-Alfvénic, i.e. when the ram pressure of the stellar wind dominates the magnetic pressure, an Earth/Jupiter-type magnetosphere with a bow shock, magnetopause, and magnetotail, is formed. At the same time, in the case of an extremely close-orbit location of an exoplanet (e.g., $d < 0.03$ AU for the Sun analogue star), where the stellar wind is still under acceleration and remains sub-magnetosonic and sub-Alfvénic (Ip et al. 2004; Preusse et al. 2005), an Alfvénic wing-type magnetosphere without a shock in the upstream region is formed (Woodward and McKenzie 1999). The character of the stellar wind impact on the planetary nearby plasma environment and inner atmosphere is different for the super- and sub-Alfvénic types of the magnetosphere and in each particular planet case it has to be properly taken into account. In the present paper, further on, we do not consider the Alfvénic wing-type magnetospheres, aiming at moderately short orbit giant planets near solar-type stars, under the conditions of a super-Alfvénic stellar wind flow, i.e., with the magnetospheres having in a general case a bow shock, a magnetopause, a magnetotail, similar to the case of the solar system Jupiter.

18.6.1 Magnetodisk: A Key Element of Hot Jupiter Magnetosphere

To explain an obvious survival and sufficient magnetospheric protection of close-orbit Hot Jupiters under the extreme conditions of their host stars Khodachenko et al. (2012) proposed a more generic view of an exoplanetary magnetosphere. A key element in the proposed approach consists in taking into account the upper atmosphere of a planet as an expanding dynamical gas layer heated and ionized by the stellar XUV radiation (Johansson et al. 2009; Koskinen et al. 2010, 2013). Interaction of the outflowing plasma with the rotating planetary magnetic dipole field leads to the development of a current-carrying magnetodisk surrounding the exoplanet (Khodachenko et al. 2012). The inner edge of the magnetodisk is located at the so called Alfvénic surface ($r = R_A$) where the kinetic energy density of the moving plasma becomes equal to the energy density of the planetary magnetic field. This condition is equivalent also to the equality of the plasma ram pressure and magnetic pressure, or the Alfvén Mach number $M_A^2 = 1$. Beyond the Alfvénic surface the expanding plasma is not guided any more by the dipole magnetic field. It deforms the field lines leading to creation of a current-carrying magnetodisk which in turn entirely changes the topology of planetary middle and outer magnetosphere.

According to Khodachenko et al. (2012), a Hot Jupiter's magnetodisk can be formed by different mechanisms, acting simultaneously (see Fig. 18.4): (1) the thermal expansion of the escaping planetary plasma wind, heated by the stellar radiation, and (2) the centrifugal acceleration of plasma by rotating planetary magnetic field in the co-rotation region, with subsequent release of material in the vicinity of the Alfvénic surface (so called "sling" mechanism). A self-consistent description of both mechanisms represents an important and complex physical problem. So far, besides a qualitative treatment of origin and interconnection of the inner (dipole dominated) and outer (magnetodisk-dominated) parts of the magnetosphere of a Hot Jupiter (Khodachenko et al. 2012), also laboratory experiments on simulation of the magnetodisk formation under the conditions of outflowing plasma in the background magnetic dipole field has been performed (Antonov et al. 2013). Two major regions with the different topology of magnetic field can be distinguished in the magnetosphere of a Hot Jupiter driven by the escaping plasma flow (Mestel 1968). The first region corresponds to the inner magnetosphere, or so-called "dead zone", filled with closed dipole-type magnetic field lines. The magnetic field in the "dead zone" is strong enough to keep plasma locked with the planet. In the second region, so-called "wind zone", the expanding plasma drags and opens the magnetic field lines. These two regions are separated by Alfvénic surface $r = R_A$ (see Fig. 18.4). The plasma escaping along field lines beyond the Alfvénic surface not only deforms and stretches the original planetary dipole field, but also creates a thin disk-type current sheet in the equatorial region (Antonov et al. 2013). Altogether, this leads to the development of a new type of magnetodisk-dominated magnetosphere of a Hot Jupiter, which has no analogues among the solar system planets (Khodachenko et al. 2012).

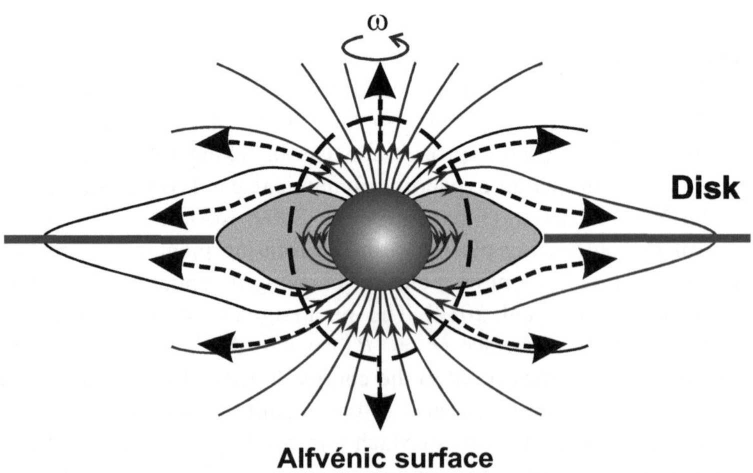

Fig. 18.4 Schematic view of magnetodisk formation (adopted from Khodachenko et al. (2012))

18.6.2 Scaling of a Magnetosphere with Magnetodisk

The more complete view of the Hot Jupiter magnetosphere structure proposed by Khodachenko et al. (2012), is based on the Paraboloid Magnetospheric Model (PMM). PMM is a semi-analytical approach to the modeling of planetary magnetosphere structure (Alexeev et al. 2003; Alexeev and Belenkaya 2005; Alexeev et al. 2006; Khodachenko et al. 2012). The name of the model is derived from its key simplifying assumption that the magnetopause of a planet may be represented by a paraboloid surface co-axial with the direction of the ambient stellar wind plasma. The PMM calculates the magnetic field generated by a variety of current systems located on the boundaries and within the boundaries of a planetary magnetosphere. Besides the intrinsic planetary magnetic dipole and magnetopause currents, the PMM has, among the main sources of magnetic field, also the electric current system of the magnetotail, and the induced ring currents of the magnetodisk. The model works without any restrictions imposed on the values of interplanetary medium parameters, enabling therefore the description of the whole variety of possible magnetosphere configurations caused by different intrinsic magnetic fields of exoplanets and various stellar wind conditions. As applied to the Hot Jupiters, PMM reveals that the electric currents induced in the plasma disk produce an essential effect on the overall magnetic field structure around the planet, resulting in the formation of a *magnetodisk-dominated* magnetosphere of a Hot Jupiter. Due to the extension of the plasma disks around close-in exoplanets, the sizes of their magnetodisk-dominated magnetospheres are usually larger than those, followed from the traditional estimates with the Eq. (18.4), based on the account of only the screened planetary magnetic dipoles (Grießmeier et al. 2004; Khodachenko et al. 2007a). In general, the role of magnetodisk may be attributed to an expansion of a part of the dipole magnetic flux from the inner magnetosphere regions outwards and a resulting increase of the magnetosphere size. The magnetic field produced by magnetodisk ring currents, dominates above the contribution of the intrinsic magnetic dipole of a Hot Jupiter and finally determines the size and shape of the whole magnetosphere. Khodachenko et al. (2012) provided an approximate formula for estimation of the magnetopause stand-off distance taking into account the contribution of the magnetodisk:

$$\frac{R_{\mathrm{s}}^{(dip+MD)}}{r_{\mathrm{p}}} \sim \frac{B_{d0J}^{1/2}(1+\kappa^2)^{1/4}}{(2\mu_0 p_{\mathrm{sw}})^{1/4}} \left(\frac{R_{\mathrm{AJ}}}{r_{\mathrm{p}}}\right)^{-1/2} \times \left(\frac{\omega_{\mathrm{p}}}{\omega_{\mathrm{J}}}\right)^{\frac{3k+1}{10}} \left(\frac{\mathrm{d}M_{\mathrm{p}}^{(th)}/\mathrm{d}t}{\mathrm{d}M_{\mathrm{J}}/\mathrm{d}t}\right)^{\frac{1}{10}}. \quad (18.5)$$

where R_{AJ}, $\frac{\mathrm{d}M_{\mathrm{J}}}{\mathrm{d}t}$, and B_{d0J} are the known values corresponding to the Alfvénic radius, mass load to the disk, and surface magnetic field for the solar system Jupiter. The coefficient $\kappa \approx 2.44$ is an amplifying factor of the inner magnetospheric field at the magnetopause (Alexeev et al. 2003), which is required to take into account the contribution of the magnetopause electric currents (i.e. Chapman-Ferraro field) at the substellar point. It is connected with the form-factor f_0 from (18.4) as $\kappa = 2f_0$.

Therefore, according to (18.5), for a given kinetic pressure of stellar wind, p_{sw}, the size of magnetosphere increases with the increasing planetary angular velocity ω_p and/or thermal mass loss rate $dM_p^{(th)}/dt$.

A slower, than the dipole-type decrease of magnetic field with distance comprises the essential specifics of the magnetodisk-dominated magnetospheres of Hot Jupiters. This results in their 40–70 % larger scales, as compared to those traditionally estimated by taking into account of only the planetary dipole. Such larger magnetospheres, extending well beyond the planetary exosphere height, provide better protection of close-in planets against of the erosive action of extreme stellar winds (Khodachenko et al. 2007a). Table 18.3 summarizes the values for a Hot Jupiter magnetopause stand-off distance at different orbits around a Sun full analogue star and gives for comparison the stand-off distance values, obtained with Eq. (18.4), i.e. in the case when the contribution of magnetodisk is ignored (e.g., a pure dipole case). A typical example of the magnetic field structure in the magnetosphere of a Hot Jupiter, obtained with PMM, is shown in Fig. 18.5.

Table 18.3 Hot Jupiter Alfvénic radius, R_A, and magnetopause stand-off distance for only a dipole controlled magnetosphere, $R_s^{(dip)}$, and a magnetosphere with magnetodisk, $R_s^{(dip+MD)}$, given by PMM

d [AU]	$R_s^{(dip+MD)}$ [r_p]	$R_s^{(dip)}$ [r_p]	R_A [r_p]
0.045[a]	8.0	5.76	3.30
0.1[a]	8.27	6.16	4.66
0.3[b]	24.2	15.0	7.30
5.2[c]	71.9	41.8	19.8

Full analog of the solar system Jupiter orbiting the Sun analog star at different orbits is considered
[a]: Tidally locked
[b]: Not tidally locked
[c]: Jupiter

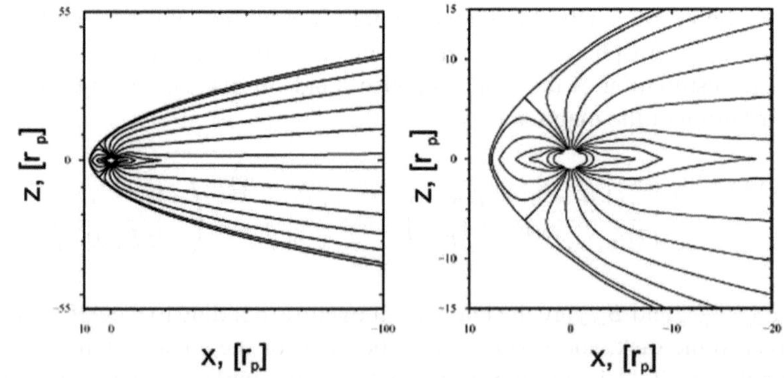

Fig. 18.5 Typical view of a magnetodisk-dominated magnetosphere (adopted from Khodachenko et al. 2013)

Conclusions

To summarize this paper we would like to emphasize that stellar XUV radiation and stellar wind plasma flow strongly impact the environments of close-orbit exoplanets, such as terrestrial-type exoplanets in the HZs of low mass stars and Hot Jupiters. We took into consideration the number density and velocity of stellar CMEs as a function of orbital distance and used these to model the effect of the stellar CME plasma flows on the size of possible magnetospheres of close-orbit exoplanets. In our investigation of the magnetospheric effects expected from CMEs, we considered only the density and velocity disturbances associated with CMEs, which change plasma flux parameters around the planetary obstacle. The extrapolation of existing solar observational data to the stellar CME case, with taking into account of the knowledge regarding activity of low mass G, K, and M stars, shows that exoplanets within close-in HZs should experience a continuous CME exposure over the entire active period of stellar history.

Given the fact that the complete or partial tidal locking of such close-orbit exoplanets may lead to relatively weak intrinsic planetary magnetic moments, the encountering stellar wind and CMEs will push the planetary magnetospheres down to the heights at which direct exposure of the planetary atmosphere to the stellar CMEs plasma flow takes place. This will result in the ionization and pick-up of the planetary atmospheric material by the stellar plasma flow leading to the intensive mass loss of a close-orbit exoplanet. Altogether that makes the stellar activity and planetary magnetospheric protection to play a crucial role for the whole complex of planetary evolution processes, including atmosphere erosion and mass loss, and finally—for the definition of the HZ criterion at close orbits around a star. Those may significantly limit the actual HZ range, as compared to that followed from the traditional HZ definition based on the pure climatological paradigm.

Large enough extended magnetospheres are needed to protect the upper atmospheric environments against stellar XUV and stellar wind/CMEs impacts. In that respect, an important role belongs to the expanding and escaping upper atmospheric gas heated and ionized by the stellar radiation. It contributes to the build-up of the magnetodisk around the planet, which constitutes the major specifics of a Hot Jupiter magnetosphere considered in this work. The magnetic field produced by magnetodisk ring currents, dominates above the contribution of intrinsic magnetic dipole of a Hot Jupiter and finally determines the size and shape of the whole magnetosphere. A more realistic structure of the magnetodisk-dominated magnetosphere of a Hot Jupiter predicted by the Paraboloid Magnetospheric Model and its significantly larger size, as compared to a dipole-type magnetosphere, have important consequences for the study of magnetospheric protection of close-orbit exoplanets.

Acknowledgements This work was supported by the NFN project S116 "Pathways to Habitability" of the Austrian Science Foundation (FWF) and its related subprojects S11606-N16, S11607-N16. The authors acknowledge the support of the FWF project P25587-N27 and RFBR grant 14-29-06036 and thank the ISSI team Characterizing stellar and exoplanetary environments as well as EU FP7 project IMPEx for providing collaborative environment for research and communication.

References

Abdul-Aziz, H., et al. (1995). Coordinated observations of the red dwarf flare star EV Lacertae in 1992. *Astronomy and Astrophysics Supplement, 114*, 509.

Abranin, E. P., et al. (1998). Coordinated Observations of the Red Dwarf Flare Star EV LAC in 1994 and 1995. *Astronomical and Astrophysical Transactions, 17*, 221.

Adams, F. C. (2011). Magnetically controlled outflows from hot Jupiters. *Astrophysical Journal, 730*, article id. 27.

Alexeev, I. I., Belenkaya, E. S., Bobrovnikov, S. Yu., et al. (2003). Modelling of the electromagnetic field in the interplanetary space and in the Earth's magnetosphere. *Space Science Review, 107*, 7.

Alexeev, I. I., & Belenkaya, E. S. (2005). Modeling of the Jovian Magnetosphere. *Annals of Geophysics, 23*, 809.

Alexeev, I. I., Kalegaev, V. V., Belenkaya, E. S., et al. (2006). A global magnetic model of Saturn's magnetosphere and a comparison with Cassini SOI data. *Geophysical Research Letters, 33*, L08101.

Ambruster, C. W., Pettersen, B. R., Hawley, S., et al. (1986). An episode of mass expulsions from the M-dwarf flare star EV Lacertae? In E. J. Rolfe (Ed.), *New Insights in Astrophysics. Eight Years of UV Astronomy with IUE* (p. 137). ESA SP-263

Antonov, V. M., Boyarinsev, E. L., Boyko, A. A., et al. (2013). Inflation of a dipole field in laboratory experiments: Toward an Understanding of magnetodisk formation in the magnetosphere of a hot Jupiter. *Astrophysical Journal, 769*, 28.

Aschwanden, M. J., Ntta, N. V., Wuelser, J. -P., et al. (2009). First measurements of the mass of coronal mass ejections from the euv dimming observed with STEREO EUVI A + B spacecraft. *Astrophysical Journal, 706*, 376.

Audard, M., Güdel, M., Drake, J. J., et al. (2000). Extreme-ultraviolet flare activity in late-type stars. *Astrophysical Journal, 541*, 396.

Ayres, T. R. (1997). Evolution of the solar ionizing flux. *Journal of Geophysical Research, 102*, 1641.

Batalha N. M., Rowe J. F., Bryson S. T., et al. (2013). Planetary Candidates observed by Kepler. III. Analysis of the first 16 months of data. *Astrophysical Journal Supplement Series, 204*, id.24.

Bewsher, D., Harrison, R. A., & Brown, D. S. (2008). The relationship between EUV dimming and coronal mass ejections. I. Statistical study and probability model. *Astronomy and Astrophysics*, 478, 897

Bloomfield, D. S., Mathioudakis, M., Christian, D. J., et al. (2002). Opacity in the upper atmosphere of AU Mic. *Astronomy and Astrophysics, 390*, 219.

Bond, H. E., Mullan, D. J., O'Brien, M. S., et al. (2001). Detection of coronal mass ejections in v471 tauri with the hubble space telescope. *Astrophysical Journal, 560*, 919.

Bonfils X., Delfosse X., Udry S., et al. (2013). The HARPS search for southern extra-solar planets. XXXI. The M-dwarf sample. *Astronomy and Astrophysics, 549*, A109.

Busse, F. H. (1976). Generation planetary magnetism by convection. *Physics of the Earth and Planetary Interiors, 12*(4), 350.

Christensen, U. R., & Aubert, J. (2006). Scaling properties of convection-driven dynamos in rotating spherical shells and application to planetary magnetic fields. *Geophysical Journal International*, *166*, 97.

Christensen, U. R. (2010). Dynamo scaling laws and applications to the planets. *Space Science Review*, *152*, 565.

Christian, D. J., Mathioudakis, M., Bloomfield, D. S., et al. (2006). Opacity in the upper atmospheres of active stars. II. AD Leonis. *Astronomy and Astrophysics*, *454*, 889.

Ciaravella, A., Raymond, J. C., Fineschi, S., et al. (1997). Ultraviolet coronagraph spectrometer observation of the 1996 December 23 coronal mass ejection. *Astrophysical Journal*, *491*, L59.

Cowen, R. (2014). Wind may deflate search for habitable planets. *Nature*. doi:10.1038/nature.2014.15335.

Cully, S. L., Fisher, G. H., Abbott, M. J. et al. (1994). A coronal mass ejection model for the 1992 July 15 flare on AU Microscopii observed by the extreme ultraviolet explorer. *Astrophysical Journal*, *435*, 435.

Dupree, A. K., Foukal, P. V., & Jordan, C. (1976). Plasma diagnostic techniques in the ultraviolet - The C III density-sensitive lines in the sun. *Astrophysical Journal*, *209*, 621.

Erkaev, N. V., Penz, T., Lammer, H., et al. (2005). Plasma and magnetic field parameters in the vicinity of short-periodic giant exoplanets. *Astrophysical Journal Supplement*, *157*, 396.

Erkaev N. V., Lammer H., Odert P., et al. (2013). XUV-exposed, non-hydrostatic hydrogen-rich upper atmospheres of terrestrial planets. Part I: Atmospheric expansion and thermal escape. *Astrobiology*, *13*, 1011.

Farrell, W. M., Desch, M. D., & Zarka, P. (1999). On the possibility of coherent cyclotron emission from extrasolar planets. *Journal of Geophysical Research*, *104*(E6), 14025.

Fuhrmeister, B., & Schmitt, J. H. M. M. (2004). Detection and high-resolution spectroscopy of a huge flare on the old M 9 dwarf DENIS 104814.7-395606.1. *Astronomy and Astrophysics*, *420*, 1079.

Gershberg, R. E. (2005). *A solar-type activity in main-sequence stars*. Berlin: Springer.

Gosling, J. T., Hildner, E., MacQueen, R. M., Munro, R. H., Poland, A. I., & Ross, C. L. (1974). Mass ejections from the sun A view from SKYLAB. *Journal of Geophysical Research*, *79*, 4581–4587.

Gopalswamy, N., & Thompson, B. J. (2000). Early life of coronal mass ejections. *Journal of Atmospheric and Solar-Terrestrial Physics*, *62*, 1457.

Grenfell J. L., Grießmeier J.-M., von Paris P., et al. (2012). Response of atmospheric biomarkers to NOx-induced photochemistry generated by stellar cosmic rays for earth-like planets in the habitable zone of M dwarf stars. *Astrobiology*, *12*, 1109.

Grießmeier, J. -M., Motschmann, U., Stadelmann, A., et al. (2004). The effect of tidal locking on the magnetospheric and atmospheric evolution of "Hot Jupiters". *Astronomy and Astrophysics*, *425*, 753.

Grießmeier, J. -M., Stadelmann, A., Motschmann, U., et al. (2005). Cosmic Ray Impact on Extrasolar Earth-Like Planets in Close-in Habitable Zones. *Astrobiology*, *5*(5), 587.

Grießmeier, J. -M., Zarka, P., & Spreeuw, H. (2007a). Predicting low-frequency radio fluxes of known extrasolar planets. *Astronomy and Astrophysics*, *475*, 359.

Grießmeier, J. -M., Preusse, S., Khodachenko, M. L., et al. (2007b). Exoplanetary radio emission under different stellar wind conditions. *Planetary and Space Science*, *55*, 618.

Guenther, E. W., & Emerson, J. P. (1997). Spectrophotometry of flares and short time scale variations in weak line, and classical T Tauri stars in Chamaeleon. *Astronomy and Astrophysics*, *321*, 803.

Guo, J. H. (2011). Escaping particle fluxes in the atmospheres of close-in exoplanets. I. Model of hydrogen. *Astrophysical Journal*, *733*, article id. 98.

Guo, J. H. (2013). Escaping particle fluxes in the atmospheres of close-in exoplanets. II. Reduced mass-loss rates and anisotropic winds. *Astrophysical Journal*, *766*, article id. 102.

Güdel, M. (2004). X-ray astronomy of stellar coronae. *Astronomy and Astrophysics Review*, *12*, 71.

Houdebine, E. R., Foing, B. H., & Rodono, M. (1990). Dynamics of flares on late-type dMe stars. I - Flare mass ejections and stellar evolution. *Astronomy and Astrophysics*, *238*, 249.

Houdebine, E. R., Foing, B. H., Doyle, J. G., et al. (1993a). Dynamics of flares on late type Dme-stars - Part Two - Mass motions and prominence oscillations during a flare on Ad-Leonis. *Astronomy and Astrophysics*, *274*, 245.

Houdebine, E. R., Foing, B. H., Doyle, J. G., et al. (1993b). Dynamics of flares on late-type dMe stars. 3: Kinetic energy and mass momentum budget of a flare on AD Leonis. *Astronomy and Astrophysics*, *278*, 109.

Ip, W. -H., Kopp, A., & Hu, J. -H. (2004). On the star-magnetosphere interaction of close-in exoplanets. *Astrophysical Journal*, *602*, L53.

Jackson, P. D., Kundu, M. R., & Kassim, N. (1990). Meter-decameter observations of dMe flare stars with the Clark lake radio telescope. *Solar Physics*, *130*, 391.

Johansson, E. P. G., Bagdonat, T., & Motschmann, U. (2009). Consequences of expanding exoplanetary atmospheres for magnetospheres. *Astronomy and Astrophysics*, *496*, 869.

Kahler, S. W. (1992). Solar flares and coronal mass ejections. *Annual Review Astronomy and Astrophysics*, *30*, 113.

Kasting J. F., Whitmire D. P., & Reynolds R. T. (1993). Habitable zones around main sequence stars. *Icarus*, *101*, 108.

Kasting, J. F. (1997). Habitable Zones around low mass stars and the search for extraterrestrial life. *Origins of Life and Evolution of Biospheres*, *27*(1/3), 291.

Kathiravan, C., & Ramesh, R. (2005). Identification of the source region of a "Halo" coronal mass ejection using meter-wavelength radio data. *Astrophysical Journal*, *627*, L77.

Khodachenko, M. L., Ribas, I., Lammer, H., et al. (2007a). Coronal Mass Ejection (CME) activity of low mass M stars as an important factor for the habitability of terrestrial exoplanets, Part I: CME impact on expected magnetospheres of Earth-like exoplanets in close-in habitable zones. *Astrobiology*, *7*, 167.

Khodachenko, M. L., Lammer, H., Lichtenegger, H. I. M., et al. (2007b). Mass loss of "Hot Jupiters" - Implications for CoRoT discoveries. Part I: The importance of magnetospheric protection of a planet against ion loss caused by coronal mass ejections. *Planetary and Space Science*, *55*, 631.

Khodachenko, M. L., Alexeev, I. I., Belenkaya, E. et al. (2012). Magnetospheres of 'Hot Jupiters': The importance of magnetodisks for shaping of magnetospheric obstacle. *Astrophysical Journal*, *744*, article id. 70.

Khodachenko, M. L., Sasunov, Yu., Arkhypov, O., et al. (2013). Stellar CME activity and its possible influence on exoplanets' environments: Importance of magnetospheric protection. In B. Schmieder & J.-M. Malherbe (Eds.), *Nature of prominences and their role in space weather*. IAUS 300, 335.

Kislyakova, K. G., Lammer, H., Holmström, M., et al. (2013). XUV-exposed, non-hydrostatic hydrogen-rich upper atmospheres of terrestrial planets. Part II: hydrogen coronae and ion escape. *Astrobiology*, *13*, 1030.

Kislyakova, K. G., Johnstone, C. P., Odert, P., et al. (2014). Stellar wind interaction and pick-up ion escape of the Kepler-11 "super-Earths". *Astronomy and Astrophysics*, *562*, id.A116.

Koskinen, T. T., Yelle, R. V., Lavvas, P., et al. (2010). Characterizing the thermosphere of HD209458b with UV transit observations. *Astrophysical Journal*, *723*, 116.

Koskinen T. T., Harris M. J., Yelle R. V., et al. (2013). The escape of heavy atoms from the ionosphere of HD209458b. I. A photochemical-dynamical model of the thermosphere. *Icarus*, *226*, 1678.

Kulikov Yu. N., Lammer H., Lichtenegger H. I. M., et al. (2007). A comparative study of the influence of the active young sun on the early atmospheres of earth, venus, and mars. *Space Science Review*, *129*, 207.

Lammer, H., Selsis, F., Ribas, I., Guinan, E. F., Bauer, S. J., Weiss, W. W. (2003). Atmospheric loss of exoplanets resulting from stellar X-ray and extreme-ultraviolet heating. *Astrophysical Journal*, *598*, L121–L124.

Lammer, H., Lichtenegger, H., Kulikov, Yu., et al. (2007). CME activity of low mass M stars as an important factor for the habitability of terrestrial exoplanets. Part II: CME induced ion pick up of earth-like exoplanets in close-in habitable zones. *Astrobiology*, *7*, 185.

Lammer, H., Bredehöft, J. H., Coustenis, A., et al. (2009a). What makes a planet habitable? *Astronomy and Astrophysics Review*, *17*, 181.

Lammer, H., Odert, P., Leitzinger, M., et al. (2009b). Determining the mass loss limit for close-in exoplanets: What can we learn from transit observations? *Astronomy and Astrophysics*, *506*, 399.

Lammer, H., Erkaev, N. V., Odert, P., et al. (2013). Probing the blow-off criteria of hydrogen-rich 'super-Earths'. *MNRAS*, *430*, 1247.

Leitzinger, M., Odert, P., Hansmeier, A., et al. (2009). Decametric observations of active M-dwarfs. In E. Stempels (Ed.), *American Institute of Physics Conference Series* (Vol. 1094, p. 680).

Leitzinger, M., Odert, P., Hanslmeier, A., et al. (2010). Spectral line enhancements as signatures for stellar activity: AD Leonis - an example. *International Journal of Astrobiology*, *9*(4), 235.

Leitzinger, M., Odert, P., Ribas, I., et al. (2011). Search for indications of stellar mass ejections using FUV spectra. *Astronomy and Astrophysics*, *536*, A62.

Lichtenegger, H. I. M., Gröller, H., Lammer, H. et al. (2009). On the elusive hot oxygen corona of Venus. *Geophysical Research Letters*, *36*, CiteID L10204.

Maehara H., Shibayama T., Notsu S., et al. (2012). Superflares on solar-type stars. *Nature*, *485*, 478.

Mestel, L. (1968) Magnetic braking by a stellar wind-I. *MNRAS*, *138*, 359.

Mizutani, H., Yamamoto, T., & Fujimura, A. (1992). A new scaling law of the planetary magnetic fields. *Advanced Space Research*, *12*(8), 265.

Mullan, D. J., Sion, E. M., Bruhweiler, F. C., et al. (1989). Evidence for a cool wind from the K2 dwarf in the detached binary V471 Tauri. *Astrophysical Journal*, *339*, L33.

Newkirk, G., Jr. (1980). Solar variability on time scales of 10 to the 5th years to 10 to the 9.6th years. *Geochimica et Cosmochimica Acta Supplement*, *13*, 293.

Osten R. A., Godet O., Drake S., et al. (2010). The mouse that roared: A superflare from the dMe flare star EV lac detected by swift and konus-wind. *Astrophysical Journal*, *721*, 785.

Parker, E. N. (1955). Hydromagnetic dynamo models. *Astrophysical Journal*, *122*, 293.

Parker, E. N. (1958). Dynamics of the interplanetary gas and magnetic fields. *Astrophysical Journal*, *128*, 664.

Parker, E. N. (1993). A solar dynamo surface wave at the interface between convection and nonuniform rotation. *Astrophysical Journal*, *408*, 707.

Penz, T., Erkaev, N. V., Kulikov, Yu. N. et al. (2008). Mass loss from "Hot Jupiters" Implications for CoRoT discoveries, Part II: Long time thermal atmospheric evaporation modeling. *Planetary and Space Science*, *56*, 1260.

Pierrehumbert, R., & Gaidos, E. (2011). Hydrogen Greenhouse Planets Beyond the Habitable Zone. *Astrophysical Journal*, *734*, id L13.

Preusse, S., Kopp, A., Büchner, J., et al. (2005). Stellar wind regimes of close-in extrasolar planets. *Astronomy and Astrophysics*, *434*, 1191.

Reiners, A., & Christensen, U. R. (2010). A magnetic field evolution scenario for brown dwarfs and giant planets. *Astronomy and Astrophysics*, *522*, A13.

Ribas, I., Guinan, E. F., Güdel, M., et al. (2005). Evolution of the Solar Activity over Time and Effects on Planetary Atmospheres. I. High-Energy Irradiances (1–1700 Å). *Astrophysical Journal*, *622*, 680.

Sánchez-Lavega, A. (2004). The magnetic field in giant extrasolar planets. *Astrophysical Journal*, *609*, L87.

Sano, Y. (1993). The magnetic fields of the planets: A new scaling law of the dipole moments of the planetary magnetism. *J. Geomag. Geoelectr.*, *45*, 65.

Scalo, J., Kaltenegger, L., Segura, A. G., et al. (2007). M stars as targets for terrestrial exoplanet searches and biosignature detection. *Astrobiology*, **7**, 85.

Schaefer, B. E., King, J. R., & Deliyannis, C. P. (2000). Superflares on ordinary solar-type stars. *Astrophysical Journal*, *529*, 1026.

Schrijver, C. J., Beer, J., Baltensperger, U., et al. (2012). Estimating the frequency of extremely energetic solar events, based on solar, stellar, lunar, and terrestrial records. *Journal of Geophysical Research, 117,* CiteID A08103.

Seager, S. (2013). Exoplanet habitability. *Science 340,* 577.

Segura, A., Walkowicz, L. M., Meadows, V., et al. (2010). The Effect of a strong stellar flare on the atmospheric chemistry of an earth-like planet orbiting an M Dwarf. *Astrobiology, 10,* 751.

Sheeley, N. R., Jr., Wang, Y.-M., Hawley, S. H., et al. (1997). Measurements of flow speeds in the corona between 2 and 30 R_{Sun}. *Astrophysical Journal, 484,* 472.

Shematovich, V. I. (2012). Formation of complex chemical species in astrochemistry (a review). *Solar System Research, 46,* 391.

Showman, A. P., & Guillot, T. (2002). Atmospheric circulation and tides of "51 Pegasus b-like" planets. *Astronomy and Astrophysics, 385,* 166.

Skumanich, A. (1972). Time scales for ca ii emission decay, rotational braking, and lithium depletion. *Astrophysical Journal, 171,* 565.

Stevenson, D. J. (1983). Planetary magnetic fields. *Reports on Progress in Physics, 46,* 555.

Stevenson, D. J. (2003). Planetary magnetic fields. Earth and Planetary Science Letters, *208,* 1.

Trammell, G. B., Arras, P., & Li, Z.-Y. (2011). Hot Jupiter magnetospheres. *Astrophysical Journal, 728,* 152.

Tian, F., Toon, O. B., Pavlov, A. A., et al. (2005). Transonic hydrodynamic escape of hydrogen from extrasolar planetary atmospheres. *Astrophysical Journal, 621,* 1049.

Tousey, R. (1973). The solar corona. In M.J. Rycroft & S. K. Runcorn (Eds.), *Space research XIII* (p. 713). Berlin: Akademie.

Walker, J. C. G., Hays, P. B., & Kasting, J. F. (1981). A negative feedback mechanism for the long-term stabilization of the earth's surface temperature. *Journal of Geophysical Research, 86,* 9776.

Wilson, O. C. (1966). Stellar convection zones, chromospheres, and rotation. *Astrophysical Journal, 144,* 695.

Wood, B. E., Müller, H.-R., Zank, G. P., et al. (2002). Measured mass-loss rates of solar-like stars as a function of age and activity. *Astrophysical Journal, 574,* 412.

Wood, B. E., Müller, H. -R., Zank, G. P., et al. (2005). New Mass-loss measurements from astrospheric Lya absorption. *Astrophysical Journal, 628,* L143.

Woodward, T. I., & McKenzie, J. F. (1999). Stationary incompressible MHD perturbations generated by a current source in a moving plasma. *Planetary and Space Science, 47,* 545.

Yelle, R. V. (2004). Aeronomy of extra-solar giant planets at small orbital distances. *Icarus, 170,* 167.

Index

J.-C. Vial, O. Engvold (eds.), *Solar Prominences*, Astrophysics and Space Science Library 415, DOI 10.1007/978-3-319-10416-4

MIX
Papier aus verantwortungsvollen Quellen
Paper from responsible sources
FSC® C105338

If you have any concerns about our products,
you can contact us on
ProductSafety@springernature.com

In case Publisher is established outside the EU,
the EU authorized representative is:
Springer Nature Customer Service Center GmbH
Europaplatz 3, 69115 Heidelberg, Germany

Printed by Libri Plureos GmbH
in Hamburg, Germany